Solid-Liquid Separation

Solid-Liquid Separation
Third Edition

Editor

Ladislav Svarovsky, Dipl Ing, PhD, CEng, FIChem E

Senior Lecturer in Chemical Engineering, University of Bradford

Butterworths

London Boston Durban Sydney Toronto Wellington

PART OF REED INTERNATIONAL P.L.C.

First published 1977
Reprinted 1979
Second edition 1981
Third edition 1990

© **Butterworth & Co. (Publishers) Ltd, 1990**

British Library Cataloguing in Publication Data

Solid–liquid separation. — 3rd ed
  1. Solid materials. Separation from liquid materials
  I. Svarovsky, Ladislav
  660.2′842

ISBN 0-408-03765-2

Library of Congress Cataloging in Publication Data

Solid–liquid separation/Ladislav Svarovsky, editor, — 3rd ed.
  p.23.4cm. — (Butterworths monographs in chemistry and chemical engineering)
  Includes bibliographies and index.
  1. Separation (Technology)
  I. Svarovsky, Ladislav. II. Series.
  TP156.S45S64 1990
  660′.2842—dc20   89-35567

  ISBN 0-408-03765-2

Typeset by Poole Typesetting (Wessex) Ltd., Bournemouth, Dorset
Printed in Great Britain at the University Press, Cambridge

Preface to the Third Edition

This book is now almost like a living thing: it has a character of its own, it is growing and maturing. The book is getting longer, yet the prefaces are getting shorter. Only a few chapters remain the same since the second edition—most have been appreciably revised in the light of recent developments and/or the changing requirements of the short course around which the book is centred. Some chapters have been replaced completely and new ones added. I would like to welcome Jim Watson of the University of Southampton and Hugh Stitt of the University of Bradford, who have contributed one new chapter each, on 'Magnetic Separation' and 'Membrane Separation', respectively. Changing times in solid–liquid separation have also been reflected in the other new (or replacement) chapters, on the 'Problems of Fine Particle Recycling', on 'Countercurrent Washing' and on 'Continuous Pressure Filters'. Another new chapter, on 'Particle–Fluid Interaction', has been added partly for the benefit of the short course, but it also makes the book more complete in its scope.

All I have to add is my sincere thanks to all contributors who made the painful process of revision very smooth by meeting the deadlines we set ourselves. I am also grateful to the sponsors of the short course out of which this book has grown, the Institution of Chemical Engineers, for their continuing support.

I am very happy with and proud of this edition.

L.S.

Preface to the First Edition

It is with great pleasure and excitement that I introduce this work consisting of contributions from myself and several leading specialists. In my opinion it is nearly impossible for a single author to cover in depth most of the field of solid–liquid separation; not only because the field is very wide—it extends over very different branches of technology—but also because the range of equipment and principles involved is enormous. That is why I regard myself as very fortunate to have had the benefit of the co-operation of a group of contributors with whom I have worked over the past two years on post-experience courses for industry.

The contributors include consultants, academics and industrialists who, despite their inevitably varied approaches, have produced a book of surprisingly compact and coherent structure. All those who have ever ventured into editing a multi-author publication of this kind will know what a headache it can be to ensure that everyone meets the given deadlines; I have been fortunate in this respect too, as the high level of discipline and responsibility of the contributors has made my task relatively easy.

The book comprises chapters on basic fundamentals, on principles and on equipment, as well as on various important aspects of solid–liquid separation such as filter aids, washing, flocculation, etc. The emphasis is on the use of equipment rather than on its design, although the latter is not ignored; consequently, the book will probably be most useful to chemical engineers and process engineers, particularly those in plant operation, plant design or equipment testing and commissioning. I hope that we have managed to strike a good balance between practical and academic considerations as both are equally important and cannot be separated. The book can be used as a textbook for both undergraduate and postgraduate teaching and for the post-experience courses from which it originated.

My long list of acknowledgements and thanks must start with the contributors because without them this book would never have been possible. I am therefore indebted to Dr M. A. Hughes, a colleague at Bradford University; Professor K. J. Ives of University College, London; Dr D. G. Osborne of Anglo-American Corporation Ltd., Coal Division, Republic of South Africa; Mr A. L. Masters of Johns–Manville (Great Britain) Ltd.; Mr H. G. W. Pierson of Pierson & Co. (Manchester) Ltd.; Dr. R. J. Wakeman of University of Exeter; and Dr K. Zeitsch of Alfa–Laval Separationstechnik GmbH, Germany.

My own contributions naturally reflect my professional career and experience, both of which have been greatly influenced by a number of people.

The late Professor J. Pulkrábek gave me the initial encouragement and opportunity in the field now called environmental engineering; Dr J. Smolik, also at the Technical University of Prague, first introduced me to particle–fluid separation and the work with him gave me a firm basis of knowledge in particle technology on which I have been building ever since. In Bradford, I have had the benefit of support and encouragement from Dr J. C. Williams together with a close association with Dr T. Allen, whose expertise in particle size measurement (and also in technical writing) has greatly benefited my own progress. I am also grateful to Professor C. Hanson and Professor W. L. Wilkinson both of whom have backed and encouraged my work in solid–liquid separation.

I must also acknowledge the useful work of a long string of undergraduate, postgraduate and extra-mural students who over the past few years have worked with me on various projects related to particle separation. Thanks are also due to many industrialists who have shared their experience with me and contributed indirectly to this publication. I am grateful to Professor Mullin, the series editor, for his guidelines and comments with regard to the manuscript.

Last but not least I am indebted to my own family who patiently suffered the inevitable reduction in my attention during the preparation of this book. My wife's engineering and computing expertise has enhanced my work and this is greatly appreciated.

L.S.

Contents

1

Introduction to Solid–Liquid Separation

L. Svarovsky

Department of Chemical Engineering, University of Bradford

As the title suggests. solid–liquid separation involves the separation of two phases, solid and liquid, from a suspension. It is used in many processes with the aim of

1. recovering the valuable solids (the liquid being discarded);
2. recovering the liquid (the solids being discarded);
3. recovering both the solids and the liquid;
4. recovering neither (but for example to prevent water pollution).

A perfect solid–liquid separation would result in a stream of liquid going one way and dry solids going another. Unfortunately, none of the separation devices works perfectly, all are imperfect in some way or other. Typically (see *Figure 1.1*) there may be some fine solids leaving in the liquid stream,

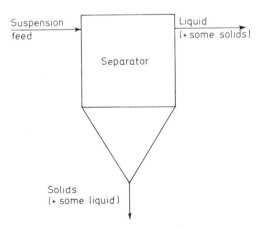

Figure 1.1. A schematic diagram of a separator

and some of the liquid may leave with the bulk of the solids, the latter being a somewhat more common problem. This imperfection of separation can be characterized in two ways. The mass fraction of the solids recovered is often called the separation efficiency, and is expressed as a percentage (in filtration this is also known as 'retention') whilst the dryness of the solids recovered may be characterized by the moisture content (% by weight). The concepts of efficiency and dewatering are further studied in chapters 3 and 14. Sometimes, in order to compensate for the fact that the solids stream entrains some liquid, washing is used in order to replace the mother liquor with a wash liquid.

In addition to solid–liquid separation it is often desirable to remove either the coarse or the fine particles from the product (de-gritting and de-sliming respectively). This process is referred to as classification or solid–solid separation and can be achieved in many types of solid–liquid separation equipment because of the particle-size-dependent nature of the principles employed in such equipment. Classification may also be made before separation in order that the material in each different size range may be treated by the type of equipment best suited to it.

1.1 SOLID–LIQUID SEPARATION PROCESSES

Solid–liquid separation processes may be classified according to the principles involved (see *Figure 1.2*). If the liquid is constrained and particles can move freely within it (due to fields of acceleration) we have sedimentation and flotation. For sedimentation, a density difference between the solids and the liquid is necessary. If particles are constrained by a medium and the liquid can flow through we have filtration and screening, for which a density difference is not necessary.

Further sub-division of these two main groups can be seen in *Figure 1.2*. Most of these processes are dealt with in some depth in this book; only a brief description is included here.

1.1.1 Flotation

This process is based on the generation of air (or other gas) bubbles within the suspension and attachment of the solids on to the bubbles. The solids are then transported to the surface through buoyancy and scraped off. Depending on the way the gas bubbles are generated, flotation is divided into dispersed air, dissolved air and electrolytic flotation.

Flotation has been used in the mineral separation field for some time and has become recognized as an effective means of solid–liquid separation in other applications such as paper-making, refineries or sewage treatment.

1.1.2 Gravity sedimentation

Gravity sedimentation is a process of solid–liquid separation that separates, under the effect of gravity, a feed slurry into an underflow slurry of higher

3

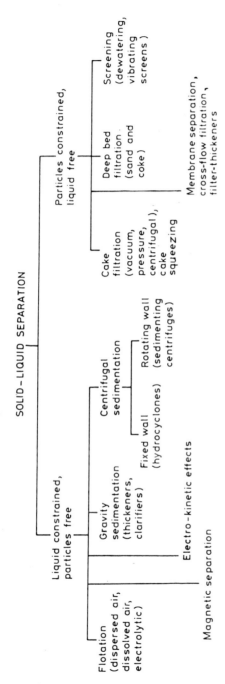

Figure 1.2. Classification of solid–liquid separation processes

solids concentration and an overflow of substantially clear liquid. Difference in density between the solids and the suspending liquid is a necessary prerequisite.

Nearly all commercial equipment is built for continuous sedimentation in relatively simple settling tanks. Distinction is commonly made depending on the purpose of the separation. If the clarity of the overflow is of primary importance, the process is called clarification and the feed slurry is usually dilute. If a thick underflow is the primary aim, the process is called thickening and the feed slurry is usually more concentrated. Flocculating agents are often used to enhance settling.

1.1.3 Centrifugal sedimentation

Centrifugation increases the mass forces on particles and thus extends sedimentation to finer particle sizes and to emulsions that are normally stable in the gravity field. Equipment available for centrifugation is divided into fixed-wall devices (hydrocyclones) and rotating wall devices (sedimenting centrifuges).

1.1.3.1 Hydrocyclones

These have no rotating parts; a vortex is produced by feeding the suspension through a tangential inlet. High velocity gradients exist in a hydrocyclone thus causing shear; this may cause breakage of possible agglomerates of flocs which is not desirable in separation but very suitable for classification. However, hydrocyclones are used extensively for both separation and classification because of their reliability and low cost; in separation their primary use is for thickening.

1.1.3.2 Sedimenting centrifuges

These have a bowl through which the suspension flows whilst rotating with the bowl. Lack of shear in the flow makes them most suitable for separation but they are often also used for classification. Of the five types available, the disc centrifuges (nozzle type) and the decanters (scroll type) are fully continuous in operation; the imperforate basket type are semi-continuous but their operation is usually fully automated. The moisture content of the separated solids can be relatively low, particularly with the basket type and also with the decanter type, and high separation efficiencies are usually achieved.

1.1.4 Cake filtration

In cake filtration particles are deposited on the face of a relatively thin permeable medium, due mostly to the screening principle. As soon as a layer of cake appears on the filter face, deposition shifts to the cake itself, and the medium acts only as a support.

In conventional cake filtration devices the cake is undisturbed and the particles, as well as the suspending medium, approach the filter medium at a right angle. There are several methods available where the cake growth is artificially limited by mechanical or hydraulic means in order to maintain high flow rates. Dewatering of compressible cakes can be further improved by mechanical squeezing of the cake and this is achieved in the so-called variable chamber filters.

Equipment for cake filtration is commonly divided according to the driving force used for making the fluid flow through the porous medium: vacuum, pressure and centrifugal filters.

1.1.5 Deep bed filtration

Very low concentrations of solids are separated in deep bed filters which employ a deep bed and the collection takes place within the bed rather than on the face of it. Particles recovered in a depth filter are generally smaller than the pores. Deep bed filtration is most often operated as a batch process in gravity-fed down-flow or up-flow arrangements. Some filters are pressure-fed and continuous filters are becoming available.

1.1.6 Screening

Similar to surface filters, screens rely on passing the liquid flow through the medium, which in this case is much more open, and the driving force is gravity. Although screening is usually used for classification of solids, screens are also useful for dewatering of coarse or highly flocculated suspensions. Vibration or some other type of motion is often employed to make sure that sooner or later all the material is introduced to the screen apertures and that the apertures do not block.

Also in this category are the strainers of various design, which are placed in the suspension flow to collect the occasional large particle that is undesirable for the process downstream.

1.1.7 Cartridge filters

These are filters that use an easily replaceable cartridge made of paper, cloth or various membranes with pore size down to 0.2 μm. The suspension is simply pumped through the filter and the cartridge is replaced when it is loaded with particles and the pressure drop becomes excessive.

It is difficult to say which is the prevailing filtration mechanism here, cake filtration or deep bed filtration, but in most cartridges it is most likely the latter.

In order to keep down the frequency of cartridge replacement, cartridge filtration is almost always limited to polishing of liquids with solids contents less than 0.01% by weight.

The most important characteristics of a cartridge filter are its rating (the largest spherical particle that will pass through it), the relationship between the pressure-drop and the dirt capacity, and the maximum allowable pressure-drop beyond which the cartridge will fail structurally.

1.1.8 Magnetic separation

Conventional magnetic separation is a long-established technique in mineral processing for removing tramp iron and for concentrating magnetic ores. It is generally restricted to the separation of strongly magnetic materials. The most widely used type of machine is the drum-type separator, in which magnetic and non-magnetic particles are sorted out.

The application of conventional magnetic separators can be extended to separation of small ferromagnetic particles by placing permanent magnet 'filters' into the flow of contaminated fluid. Such filters are very simple in design and mostly used for cleaning fluids in hydraulic circuits. Ceramic magnets, metal-alloy magnets or magnetized steel balls are generally used. They are cleaned manually by withdrawing the entire assembly, removing the magnets, and washing their collecting surfaces. At low flow rates, ferromagnetic particles down to 1 μm or less can be removed. Some non-ferromagnetic contaminants are also recovered because they agglomerate with the magnetic particles.

Enjoying considerable attention recently is high-gradient magnetic separation (HGMS), which maximizes magnetic forces by using electromagnets. Separation of very small and weakly paramagnetic particles can thus be achieved on a large scale.

High-gradient magnetic separators usually feature a filamentous ferromagnetic medium, such as stainless-steel wool, loosely contained at about 5% packing density in a uniform magnetic field. Weakly paramagnetic particles will be retained from a fluid flowing through the wire wool matrix.

When the magnetic field is removed, the separated particles are easily washed from the matrix because the magnetic forces are small at zero field. The high-intensity magnetic separators are therefore used in a batch mode of operation similar to that of conventional gravity deep bed filters.

Continuous feed can be treated with carousel-type separators. The importance and the potential of HGMS has been boosted recently by the advances in superconductivity.

1.1.9 Membrane separation, cross-flow filtration, filter thickeners

The most important disadvantage of conventional cake filtration is the declining rate due to the increasing pressure drop caused by the growth of the cake on the filter medium. The flow rate of the liquid through the medium can be kept high if no or little cake is allowed to form on the medium. This leads to the thickening of the slurry on the upstream part of the medium and filters based on this principle are sometimes called filter thickeners. The cake can be prevented from forming by hydraulic or mechanical means and such operation can be carried out over either part or the whole of the filtration cycle.

There is a number of different ways of keeping the cake away from the filter medium but the most important category is cross-flow filtration, as a natural extension of membrane separation (reverse osmosis, ultra-filtration), where the stationary membrane surface is swept by the fluid flowing across it. The second most popular way of limiting cake growth is in the so-called dynamic filters where the medium surface is rotated within an essentially stationary suspension, and much the same effect is achieved.

1.1.10 Electrokinetic effects

The application of direct current (d.c.) potential in filtration or sedimentation is known to have a beneficial effect on the separation. Although this has been known and studied since the beginning of the 19th century, its practical application and development have only recently accelerated and commercial application is likely in the near future.

There are several effects due to the existence of the double layer on the surface of most particles suspended in liquids, and they can all be used to measure the so-called zeta potential. A simplified summary of the effects is given in *Table 1.1* and further explanation and definitions are given in section 4.3.

Table 1.1. ELECTROKINETIC EFFECTS

Fluid	Particles	Electric field	Effect
Still	Moving	Applied	Electrophoresis
Still	Moving	Measured	Sedimentation (or migration) potential
Moving	Still	Applied	Electro-osmosis
Moving	Still	Measured	Streaming potential
Still	Still	Applied	Electro-osmotic pressure

The best known effect is electrophoresis. This term is given to the migration of small particles suspended in a polar liquid, in an electric field, towards an electrode. If a sample of the suspension is placed in a suitably designed cell (with a d.c. potential applied across the cell) and the particles are observed through a microscope, they can all be seen to move in one direction, towards one of the two electrodes. All the particles, regardless of their size, appear to move at the same velocity (as both the electrostatic force and resistance to particle motion depend on particle surface) and this velocity can be easily measured.

The second effect, which often accompanies electrophoresis, is electro-osmosis. This is the transport of the liquid past a surface or through a porous solid, which is electrically charged but immovable, towards the electrode with the same sign of charge as that of the surface. It can be said that electrophoresis reverts to electro-osmotic flow when the charged particles are made immovable, and, if the electro-osmotic flow is forcibly prevented, pressure builds up and is called electro-osmotic pressure.

1.1.10.1 Electrophoretic settling and electro-osmotic dewatering

This is a straightforward use of electrophoresis when settling velocities are enhanced by placing one electrode (usually negative and perforated) floating on top of the slurry and another at the bottom of the tank. Furthermore, electro-osmosis thickens the settled sludge at the same time. The settling consists in fact of three successive processes. In the first, suspended particles

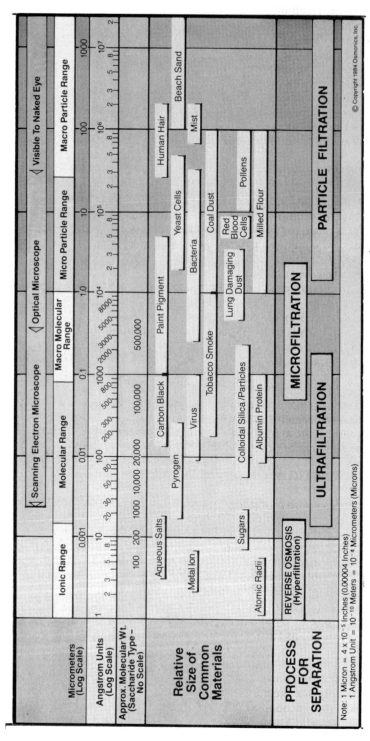

Figure 1.3. The spectrum of particle size. Note: $1 \mu m = 4 \times 10^{-5}$ in; $1 \overset{\circ}{A} = 10^{-10} m = 10^{-4} \mu m$. (Courtesy of Osmonics, Inc.)

settle with the resultant velocity of the gravitational and the electrophoretic components. In the second process, the sludge is thickened by compression and electro-osmosis. In the third process, when compression due to gravity is completed, the sludge is thickened only by electro-osmosis.

Enhancement of gravity settling by electrophoresis is also used with liquid–liquid dispersions in solvent extraction and for the separation of oil–water emulsions. Electro-osmosis on its own is a well-tested method for the gathering of ground water to facilitate its removal from foundations and other sites in civil engineering. It can also be applied to the dewatering of filter cakes as described in the following.

1.1.10.2 Electrokinetic filtration

Electrophoresis and electro-osmosis can also be used to enhance conventional cake filtration. Electrodes of suitable polarity are placed on either side of the filter medium (as most particles carry negative charge, the electrode upstream of the medium is usually positive) so that the incoming particles move towards the upstream electrode, away from the medium. The electric field can cause the suspended particles to form a more open cake or, in the extreme, to prevent cake formation altogether by keeping all particles away from the medium.

There is an additional pressure drop across the cake developed by electro-osmosis which leads to increased flow rates through the cake (and further dewatering at the end of the filtration cycle). The filtration theory proposed for electrofiltration assumes the simple superposition of electro-osmotic pressure on the hydraulic pressure drop.

Whether the recent growth in the theoretical and experimental study of the electrokinetic effects applied to solid–liquid separation will lead to greater commercial use remains to be seen; to date, there are only two commercial electrofilters on the market known to the author. However, electrofiltration is certainly an area of potential development.

1.2 THE SPECTRUM OF PARTICLE SIZE

This book deals with processes that can, between them, deal with particle size spanning several orders of magnitude. *Figure 1.3* shows the spectrum of particle size and where various common materials lie on the scale. It also gives the broad ranges of application of the membrane separation, filtration and screening processes (the right-hand side group in *Figure 1.2*), the separation efficiency of which is, within their respective range of application, more-or-less independent of particle size. The efficiency of the group on the left in *Figure 1.2* is highly size dependent (most of the separators in that group are often referred to as 'dynamic separators') and it is characterized by the 'grade efficiency curve', as further explained in chapter 3. *Figure 1.4* shows such curves for some types of equipment used at the fine end of the particle size spectrum.

This book aims to cover the whole of the size spectrum shown in *Figure 1.3* but, inevitably, the greater emphasis is given to a somewhat narrower range, most frequently encountered in the chemical and process industries, approximately between 0.05 and 500 μm.

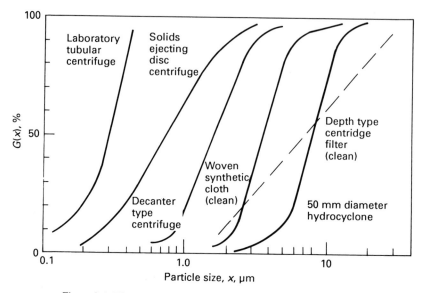

Figure 1.4. The grade efficiency curves of various types of equipment

2

Characterization of Particles Suspended in Liquids

L. Svarovsky

Department of Chemical Engineering, University of Bradford

2.1 INTRODUCTION, THE REASONS FOR PARTICLE CHARACTERIZATION

Particle characterization, i.e. the description of the primary properties of particles in a particulate system, underlies all work in particle technology. Primary particle properties such as the particle size distribution, particle shape, density, surface properties and others, together with the primary properties of the liquid (viscosity and density) and also with the concentration and the state of dispersion, govern the other. secondary properties such as the settling velocities of the particles, the permeability of a bed or the specific resistance of a filter cake. Knowledge of these properties is vital in the design and operation of equipment for solid–liquid separation.

One could of course argue that it may be not only simpler but often more reliable to measure the secondary properties directly without reference to the primary properties; this is of course done in practice but the ultimate aim is to be able to predict the secondary properties from the primary ones. After all, in fluid dynamics for example, we do not test the pipe resistance to flow every time we need to design a piece of pipework, we measure the primary properties of the liquid (viscosity and density) and of the pipeline (roughness) and determine the resistance from known relationships. As the relationships in solid–liquid separation are rather complex and in many cases not yet available. primary particle properties are mainly used for only a qualitative assessment of the behaviour of suspensions, for example as a selection guide. Taking particle size for instance, the finer the particle size the more difficult is the separation but the concentration of solids also plays an important role. Lloyd and Ward[1] have given a very informative diagram, presented in a slightly modified form in *Figure 2.1*, which shows schematically the range of solid–liquid separation equipment for different particle sizes.

The particle size greatly affects the permeability, or specific resistance, of packed beds (as can be clearly seen from equation 9.36, the specific cake resistance (α) is given by

$$\alpha \sim 1/x^2$$

where x is the particle size, because the specific surface of the particles making up the bed is inversely proportional to particle size), hence particle size can be used for a qualitative assessment of the permeability. In sedimentation, Stokes' diameter (as defined later) plays an important role in 'free settling' applications and in the Sigma theory used for sedimenting centrifuges (see chapter 7).

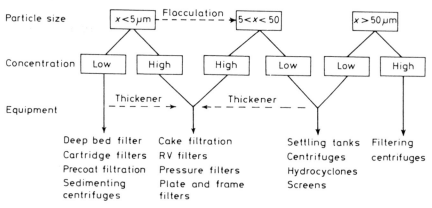

Figure 2.1. Particle size as a guide in the selection of solid–liquid separation equipment[1]

There are many properties of particulate systems other than those relevant to solid–liquid separation, which depend strongly on particle size: the activity of drugs, the setting time of cement, and the hiding power of pigments, to name just a few.

The characterization of solid particles, most of which are, in practice, irregular in shape, is usually made by analysing the particle size (the measure of size most relevant to the particle property which is under investigation) and its distribution. Other characteristic properties of the solid material may be included in the measure of size determined, for example Stokes' diameter combines size, density and shape all in one parameter; they can be characterized separately if necessary. British Standard BS2955 attempts to define shape qualitatively; a quantitative measure of particle shape can be obtained indirectly by analysing two or more measures of particle size and looking at different 'shape coefficients' that relate to those sizes.

Before a technique can be selected for particle size analysis, two important decisions have to be made about the variables we measure: i.e. the measure of particle size x and the type of size distribution Φ required.

2.2 DEFINITIONS OF PARTICLE SIZE

An irregular particle can be described by a number of sizes depending on what dimension or property is measured. There are basically three groups of sizes: 'equivalent sphere diameters', 'equivalent circle diameters' and 'statistical diameters'.

The first group of sizes are the diameters of a sphere which would have the same property as the particle itself (e.g. the same volume, the same projected area, the same settling velocity etc.)—see *Table 2.1*.

Table 2.1. A LIST OF DEFINITIONS OF 'EQUIVALENT SPHERE DIAMETERS'

Symbol	Name	Equivalent property of a sphere
x_v	Volume diameter	Volume
x_s	Surface diameter	Surface
x_{sv}	Surface volume diameter	Surface to volume ratio
x_d	Drag diameter	Resistance to motion in the same fluid at the same velocity
x_f	Free-falling diameter	Free-falling speed in the same liquid, same particle density
x_{St}	Stokes' diameter	Free-falling speed, if Stokes' Law is used ($Re < 0.2$)
x_A	Sieve diameter	Passing through the same square aperture

The second group of sizes are the diameters of a circle that would have the same property as the projected outline of the particles—see *Table 2.2*.

Table 2.2. A LIST OF DEFINITIONS OF 'EQUIVALENT CIRCLE DIAMETERS'

Symbol	Name	Equivalent property of a circle
x_a	Projected area diameter	Projected area if the particle is resting in a stable position
x_p	Projected area diameter	Projected area if the particle is randomly orientated
x_c	Perimeter diameter	Perimeter of the outline

The third group of sizes, the 'statistical diameters' are obtained when a linear dimension is measured (by microscopy) parallel to a fixed direction— see *Table 2.3*.

Table 2.3. A LIST OF DEFINITIONS OF 'STATISTICAL DIAMETERS'

Symbol	Name	Dimension measured
x_F	Feret's diameter	Distance between two tangents on opposite sides of the particle
x_M	Martin's diameter	Length of the line which bisects the image of the particle
x_{SH}	Shear diameter	Particle width obtained with an image shearing eyepiece
x_{CH}	Maximum chord diameter	Maximum length of a line limited by the contour of the particle

Different methods of particle size measurement determine different measures of size (see the list in *Table 2.7*) and great care must be taken when making a selection as to what size is most relevant to the property or process which is to be controlled. For example, in those methods of solid–liquid separation in which particle motion relative to the fluid is the governing mechanism (gravity or centrifugal sedimentation, hydrocyclones) it is of course most relevant to use a method which measures the free-falling diameter or, more often, the Stokes' diameter (sedimentation or fluid classification methods). In filtration on the other hand, it is the surface volume diameter (which is measured, for example, by permeametry) which is most relevant to the mechanism of separation.

2.3 TYPES OF PARTICLE SIZE DISTRIBUTION

For a given particulate matter, four different types of particle size distribution are defined (see *Figure 2.2a*). These are:

1. particle size distribution by number $f_N(x)$;
2. particle size distribution by length (not used in practice) $f_L(x)$;
3. particle size distribution by surface $f_S(x)$;
4. particle size distribution by mass (or volume) $f_M(x)$.

These distributions are related but conversion from one to another is possible only when the shape factor is constant, i.e. the particle shape is independent of size.

The following relationships show the basis of such conversions:

$$f_L(x) = k_1 x f_N(x) \tag{2.1}$$

$$f_S(x) = k_2 x^2 f_N(x) \tag{2.2}$$

$$f_M(x) = k_3 x^3 f_N(x) \tag{2.3}$$

The constants k_1, k_2 and k_3 contain a shape factor which may often be particle-size-dependent; this makes an accurate conversion impossible without a full quantitative knowledge of the shape factor's dependence on particle size. If the shape of the particles does not vary with size, the constants k_1, k_2 and k_3 can easily be found because, by definition of the distribution frequency

$$\int_0^\infty f(x)\,\mathrm{d}x = 1 \tag{2.4}$$

the areas under the curves of frequency against particle size should be equal to one. The actual procedure is best shown in the example in *Figure 2.2b* where the particle size distribution frequency by number, $f_N(x)$, is converted to a distribution by mass, $f_M(x)$, by multiplication of the values of f_N, corresponding to different sizes x, by x^3 (following equation 2.3). The resulting curve of $x^3 f_N(x)$ is then simply scaled down by a factor

$$k_3 = \frac{1}{\int_0^\infty x^3 f_N(x)\,\mathrm{d}x} = \frac{1}{A} \tag{2.5}$$

in order to give $f_M(x)$.

If the data are in a cumulative form, as cumulative percentage undersize or oversize (see section 2.5), conversions between the different particle size

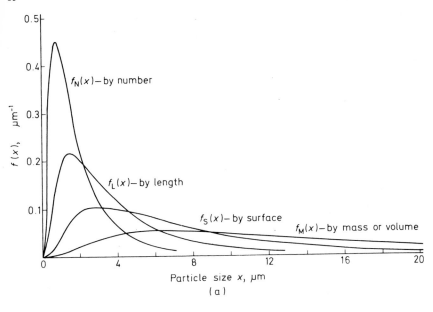

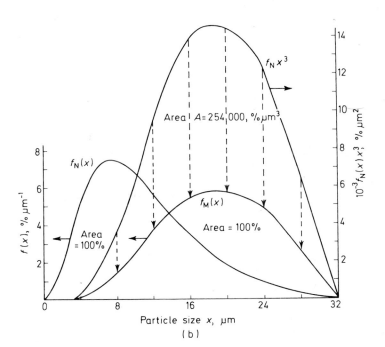

Figure 2.2. (a) Four particle size distributions of a given particle population; (b) Example of conversion of f_N to f_M (particle size distribution by number to particle size distribution by mass)

distributions can be made without having to differentiate the curves to obtain the size frequencies, as follows.

If, for example, the cumulative percentage undersize by surface in *Figure 2.3* is to be converted into a distribution by mass (a procedure often needed

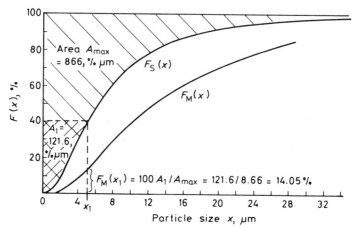

Figure 2.3. Example of conversion of F_S to F_M (surface to mass)

when using a photosedimentometer—see section 2.7 on methods), equations 2.2 and 2.3 combined give:

$$f_M(x) = kxf_S(x) \tag{2.6}$$

which if integrated for all particle sizes up to a given size x_1

$$\int_0^{x_1} f_M(x)\,dx = k\int_0^{x_1} xf_S(x)\,dx \tag{2.7}$$

gives

$$F_M(x_1) = k\int_0^{F_S(x_1)} x\,dF_S(x) \tag{2.8}$$

where F denotes the cumulative percentage such that

$$f_S(x) = \frac{dF_S(x)}{dx} \tag{2.9}$$

Equation 2.8 forms the basis for the conversion because

$$\int_0^{F_S(x_1)} x\,dF_S(x)$$

is simply the area A_1 under the curve up to the point corresponding to x_1 but integrated with respect to the vertical axis $F_S(x)$ as shown in *Figure 2.3*. Hence,

$$F_M(x_1) = kA_1 \tag{2.10}$$

and k may be found from the condition that

$$kA_{max} = 1 \qquad (2.11)$$

where

$$A_{max} = \int_0^1 x \, dF_S(x)$$

This conversion is carried out step by step as indicated in *Figure 2.3* and the resulting $F_M(x)$ can be plotted on the same graph.

To conclude this section, it must be emphasized that the conversions described above are to be avoided whenever possible (because of inherent errors in such procedures), by using a method which gives the desired type of distribution directly. Different methods give different particle size distributions (see *Table 2.7*, section 2.7) and the selection of a method is made on the basis of *both* the particle size and the type of distribution required. In most applications in solid–fluid separation, it is the particle size distribution by mass that is of interest because, usually, we are interested in gravimetric efficiencies. There are, however, cases such as liquid clarification where the turbidity of the overflow is of importance (clarification of beer etc.) and particle size distribution by surface or even by number is more relevant.

2.4 MEASURES OF CENTRAL TENDENCY

There are a great number of different average or mean sizes which can be defined for a given particle size distribution. The purpose of such measures of central tendency is to represent a population of particles by a single figure; this of course gives no indication of the width of the distribution but it may sometimes provide a useful guide for process control. The following relationships and definitions are based on general mathematical concepts, applied here specifically to particulate systems.

There are three most important measures of central tendency for a given size distribution (by a given quantity Φ)—see *Figure 2.4*: the mode, the median and the mean.

The mode is the most commonly occurring size, i.e. the size corresponding to the peak on the size distribution frequency curve. Some distributions may have more than one peak and are commonly referred to as multimodal distributions. The median or the 50% size is the size such that half the particles (by the given quantity Φ) are larger and half are smaller, i.e. the size which divides the area under the distribution frequency curve into two halves. The median is most easily determined from the cumulative percentage curves (see section 2.5) where it corresponds to 50%.

There are many mean diameters that can be defined for a given particle size distribution; their definition is, in a general form

$$g(\bar{x}) = \int_0^\infty g(x) f(x) \, dx \qquad (2.12)$$

where $f(x)$ is the particle size distribution frequency either by number,

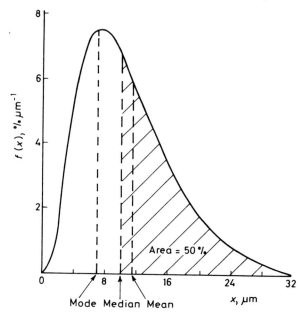

Figure 2.4. The mode, median and mean of a size distribution

length, area or mass, whichever may be of interest, and $g(x)$ is a certain function of particle size x. Depending on the form of this function we have several types of mean diameters $\bar{x}$ as shown in *Table 2.4*.

Table 2.4. TYPES OF MEAN DIAMETER $\bar{x}$

Form of g(x)	Name of mean diameter $\bar{x}$
$g(x) = x$	arithmetic mean, $\bar{x}_a$
$g(x) = x^2$	quadratic mean, $\bar{x}_q$
$g(x) = x^3$	cubic mean, $\bar{x}_c$
$g(x) = \log x$	geometric mean, $\bar{x}_g$
$g(x) = 1/x$	harmonic mean, $\bar{x}_h$

Comparing equation 2.12 for the arithmetic mean $(g(x) = x)$, quadratic mean $(g(x) = x^2)$ and cubic mean $(g(x) = x^3)$ with equations 2.1, 2.2 and 2.3 respectively, it may be shown that, for example, the number arithmetic mean relates the total length (if all particles are placed next to each other in a line) to the total number in the population and is therefore known as the number length mean x_{NL}; *Table 2.5* summarizes these relationships.

Evaluation of the various means required for a given size distribution is based on equation 2.12 which may also be written as

$$g(\bar{x}) = \int_0^1 g(x)\, dF \tag{2.13}$$

Table 2.5

$\bar{x}_{\text{a number}}$	$= x_{NL}$	number length mean
$\bar{x}_{\text{q number}}$	$= x_{NS}$	number surface mean
$\bar{x}_{\text{c number}}$	$= x_{NV}$	number volume (or mass) mean
$\bar{x}_{\text{a length}}$	$= x_{LS}$	length surface mean
$\bar{x}_{\text{q length}}$	$= x_{LV}$	length volume (or mass) mean
$\bar{x}_{\text{a surface}}$	$= x_{SV}$	surface volume (or mass) mean
$\bar{x}_{\text{a volume}}$	$= x_{VM}$	volume (or mass) moment mean

because (similarly to equation 2.9)

$$f(x) = dF/dx \tag{2.14}$$

If either $f(x)$ or $F(x)$ are available as analytical functions (see section 2.5), the desired mean diameters are evaluated by integration following equation 2.12. For example, if the arithmetic mean of a log–normal distribution (see equation 2.24, section 2.5) is required, equation 2.12 becomes

$$\bar{x}_a = A \int_0^\infty x \exp\left(-b \ln^2 \frac{x}{x_m}\right) dx$$

where

$$A = \left(\frac{b}{\pi}\right)^{1/2} \frac{\exp(-1/4b)}{x_m}$$

b is a constant (known as the steepness constant, see equation 2.25)

and

x_m is the mode (see section 2.5).

After integration this reduces to

$$\bar{x}_a = x_m \exp(3/4b)$$

or

$$\bar{x}_a = x_m c^{-3/2} \tag{2.15}$$

where

$$c = \exp(-1/2b) \tag{2.16}$$

If, however, no analytical function is fitted and the particle size distribution is in the form of a graph or a table, evaluation of mean diameters can best be done graphically.

As discussed in section 2.5, most particle size measurement techniques yield the cumulative percentage distribution $F(x)$, and these can be used in equation 2.13 directly as follows; if $F(x)$ is plotted against $g(x)$ for a number of corresponding sizes, $g(\bar{x})$ is then represented by the area under the curve with respect to the $F(x)$ axis as illustrated in *Figure 2.5*. The mean is evaluated from this area using the corresponding equation for $g(x)$—see *Table 2.4*.

It is immaterial whether $F(x)$ is plotted as percentage oversize or undersize.

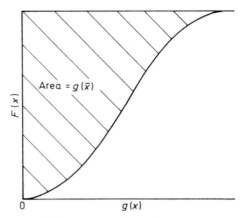

Figure 2.5. Evaluation of a mean $\bar{x}$ from the cumulative percentage $F(x)$

Table 2.6 and *Figure 2.6* give an example of such a graphical evaluation of $\bar{x}_a$, $\bar{x}_q$ and $\bar{x}_g$. Experimental values of the cumulative percentage $F(x)$ are given in *Table 2.6*. For evaluation of the arithmetic mean $\bar{x}_a$, $F(x)$ is plotted against x and the area measured, giving $g(\bar{x}_a) = \bar{x}_a = 7.64$ µm (see *Figure 2.6*). The quadratic mean $\bar{x}_q$ is determined from a plot of $F(x)$ against x^2; the area is $g(\bar{x}_q) = \bar{x}_q^2 = 74.89$ µm² thus $\bar{x}_q = 8.65$ µm. Similarly, the geometric mean $\bar{x}_g$ may be calculated from the corresponding area on the plot of $F(x)$ against $\log x$, giving $g(\bar{x}_g) = \log \bar{x}_g = 0.8293$ and $\bar{x}_g = 6.75$ µm.

There are a great number of different mean sizes and a question arises which of those is to be chosen to represent the population. The selection is of

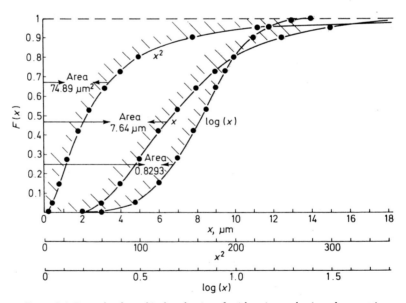

Figure 2.6. Example of graphical evaluation of arithmetic, quadratic and geometric means

course based on the application, namely what property is of importance and should be represented. In liquid filtration for example, it is the surface volume mean x_{sv} (surface arithmetic mean $\bar{x}_{a\,surface}$) because the resistance to flow through packed beds depends on the specific surface of the particles that make up the bed (see equation 9.36). It can be shown that x_{sv} is equal to the mass harmonic mean $\bar{x}_h$ (see Appendix 2.2). For distributions that follow closely the log–normal equation (see section 2.5) the geometric mean $\bar{x}_g$ is equal to the median.

Table 2.6

Cumulative percentage undersize, %	Particle size x, μm	x^2, μm²	log x
0.7	2	4	0.30
5.0	3	9	0.48
15.0	4	16	0.60
27.5	5	25	0.70
42.0	6	36	0.78
53.0	7	49	0.85
64.0	8	64	0.90
72.5	9	81	0.95
80.0	10	100	1.00
90.0	12.5	156	1.10
95.0	15	225	1.18
98.5	20	400	1.30
99.6	25	625	1.40

In conclusion it has to be emphasized that it is always best to represent a population of particles by the actual size distribution curve and only in cases when this is not possible or feasible should one resort to using a single number, a measure of central tendency, for characterizing a particulate system; in such cases care must be taken to select the type of mean size most relevant to the given application; some guidance is given in Appendix 2.1.

2.5 PRESENTATION OF DATA

Particle size distributions are presented either in an analytical form (as a function) or as a set of data in a table or a diagram. The distributions are given as either frequencies $f(x)$ or cumulative frequencies (expressed as fractions or percentages) $F(x)$ which mutually correspond because the frequency curve can be obtained by differentiation of the cumulative curve

$$f(x) = dF(x)/dx \qquad (2.17)$$

or, vice versa, the cumulative curve $F(x)$ can be obtained by integration of the frequency $f(x)$, i.e.

$$F(x) = \int f(x)\,dx \qquad (2.18)$$

The area under the frequency curve is by definition equal to 1 (see *Figure 2.7*) so that $F(x)$ goes from 0 to 1 or 100%. The cumulative percentages are given as either oversize or undersize, which mutually correspond because (see *Figure 2.7*)

$$F(x)_{\text{oversize}} = 1 - F(x)_{\text{undersize}} \tag{2.19}$$

Most particle size analysis methods do not produce continuous frequency curves but give percentages in a few ranges of particle size (e.g. with sieving).

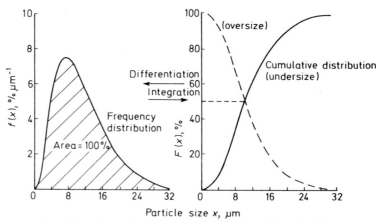

Figure 2.7. *Relationship between frequency and cumulative distributions*

Many textbooks recommend that such information may be plotted as a histogram, which is a step-like frequency distribution curve. Such a plot is a very crude approximation of the actual, continuous frequency curve and a smooth curve cannot be accurately drawn through histograms. The preferred method is to plot the data cumulatively as percentages undersize or oversize and, as these points do lie on the actual distributions, a smooth curve may be drawn through them. If a distribution frequency curve is required, it should be obtained by differentiation of the cumulative plots, rather than directly from histograms.

Differentiation is often unnecessary because the required data processing (such as grade efficiency evaluation—see section 3.2.2—or evaluation of mean diameters) may be made using the cumulative plots directly. The main advantage of the distribution frequency curve is, however, that it gives a good illustration of the size spectrum particularly if it happens to be multi-modal, i.e. if it has more than one local maximum.

Owing to the gradual increase in the availability and capacity of computers, it is becoming more and more convenient and feasible to fit an analytical function to the experimental particle size distribution data and then handle this function mathematically in further treatment. It is for example very much easier to evaluate mean sizes from analytical functions than from experimental data. Apart from the many curve-fitting techniques available, use can also be made of special graph papers which exist for many of the common analytical functions used. A brief review of the functions available is given in the following. Most of them are two-parameter functions but some three-parameter equations are also shown, the latter being of course more general and more likely to fit closely the practical empirical data. All of the functions should be treated as empirical equations as they very rarely have any theoretical relation to the process in which the particles were produced.

2.5.1 Normal distributions

The normal distribution is a symmetrical bell-shaped curve referred to in statistics as a Gaussian curve. It is a two-parameter function, one parameter is the mean, $\bar{x}_a$ which due to the symmetry of the curve coincides with the mode and median, and the other is the standard deviation σ, which is a measure of the width of the distribution. The normal distribution of particle size is given by

$$\frac{dF}{dx} = \frac{1}{\sigma\sqrt{(2\pi)}} \exp\left(\frac{(x - \bar{x}_a)^2}{2\sigma^2}\right) \tag{2.20}$$

Graph papers are available on which the integral function to equation 2.20 ($F(x)$) is plotted against particle size x; normal distributions give straight lines on such a grid and the two parameters for equation 2.20 can be easily determined; the mean (which is also the median) corresponds to 50%, the standard deviation is the 84% size minus the 50% size.

Real powders and suspensions rarely fit normal distributions closely because in practice most particle size distributions are skewed. The main theoretical criticism of the normal distribution is that it extends into the region of negative size.

2.5.2 Log–normal distribution

The log–normal distribution is probably the most widely used type of function; it is again a two-parameter function, it is skewed to the right and it gives equal probability to ratios of sizes rather than to size differences, as in the normal distribution. The log–normal distribution equation is obtained from the normal distribution in equation 2.20 by substitution of $\ln x$ for x, $\ln x_g$ for $\bar{x}_a$ and $\ln \sigma_g$ for σ, i.e.

$$\frac{dF}{d(\ln x)} = \frac{x \, dF}{dx} = \frac{1}{\ln \sigma_g \sqrt{(2\pi)}} \exp\left(-\frac{(\ln x - \ln x_g)^2}{2 \ln^2 \sigma_g}\right) \tag{2.21}$$

As the plot of $dF/d(\ln x)$ against $\ln x$ is a normal distribution and, therefore, symmetrical, it follows from the definitions of the geometric mean (equation 2.12 and *Table 2.4*) that x_g represents the geometric mean of the distribution dF/dx. It can be shown mathematically that x_g is in this case equal to the median x_{50}.

As equation 2.21 does not explicitly express dF/dx, it is often given in a different form where x from the left-hand side of the equation is absorbed into the exponential term in the right-hand side and the substitution

$$\ln x_m = \ln x_g - \ln^2 \sigma_g \tag{2.22}$$

is made so that equation 2.21 becomes

$$\frac{dF}{dx} = \frac{1}{x_m \ln \sigma_g \sqrt{(2\pi)}} \exp\left(-\ln^2 \sigma_g/2\right) \exp\left(-\frac{(\ln x - \ln x_m)^2}{2 \ln^2 \sigma_g}\right) \quad (2.23)$$

This form of equation 2.21 is mathematically more convenient because the variable x only appears once in it; x_m represents the mode because it is the size at which dF/dx has its maximum.

Equation 2.23 may be re-written as[2]

$$dF(x) = A \exp\left(-b \ln^2 \frac{x}{x_m}\right) dx \quad (2.24)$$

where

$$A = \left(\frac{b}{\pi}\right)^{1/2} \frac{\exp\left(-1/4b\right)}{x_m} \quad (2.25)$$

and b is a new parameter called the steepness constant, and replaces σ_g:

$$b = \frac{1}{2 \ln^2 \sigma_g} \quad (2.26)$$

The relationship between the mode and the median in equation 2.22 assumes the form

$$x_m = c x_g \quad (2.27)$$

where

$$c = \exp\left(-\frac{1}{2b}\right)$$

(see equation 2.16).

Graph papers are available on which the integral function $F(x)$ is plotted against $\ln x$ and distributions that follow the log–normal law give straight lines. The evaluation of the two necessary parameters from the plots can be done in two somewhat different ways.

The conventional method, quoted in many textbooks[3], is based on equation 2.21. The median size $x_{50} = x_g$ is read off the chart, σ_g is the ratio of the 84% size to the 50% size. Equation 2.22 may be used for evaluation of the mode and the mean (arithmetic) may be calculated from its definition (equation 2.12 with equation 2.22)

$$\ln \bar{x}_a = \ln x_g + 0.5 \ln^2 \sigma_g \quad (2.28)$$

Similar equations exist for other types of means (x_q, x_c, x_h).

The second method is based on equation 2.24 and the parameters x_m and b may be determined either from x_g and σ_g (found as described above) using equations 2.22 and 2.26 or directly from a graph paper which is furnished with two additional scales of b and c (see *Figure 2.8*); the mode x_m is then determined from equation 2.27. This method offers the advantage of a mathematically simpler equation (equation 2.24) together with simpler equations for the various mean diameters (derived from the definitions in equation 2.12

with equation 2.24)

$$x_m = cx_g \qquad (2.27)*$$
$$\bar{x}_a = c^{-1/2}x_g \qquad (2.29)$$
$$\bar{x}_q = c^{-1}x_g \qquad (2.30)$$
$$\bar{x}_c = c^{-3/2}x_g \qquad (2.31)$$
$$\bar{x}_h = c^{1/2}x_g \qquad (2.32)$$

Example 2.1

The data given in *Table 2.6*, when plotted in *Figure 2.8*, give a straight line. The parameters for the conventional form of the log–normal equation (equation 2.21) are $x_g = 6.75$ and

$$\sigma_g = x_{84\%}/x_{50\%} = 11.15/6.75 = 1.65$$

The mode is, according to equation 2.22

$$x_m = \text{antilog}\,(\ln 6.75 - \ln^2 1.65) = 5.27\ \mu m$$

and the mean, according to equation 2.28

$$\bar{x}_a = \text{antilog}\,(\ln 6.75 + 0.5 \ln^2 1.65) = 7.65\ \mu m$$

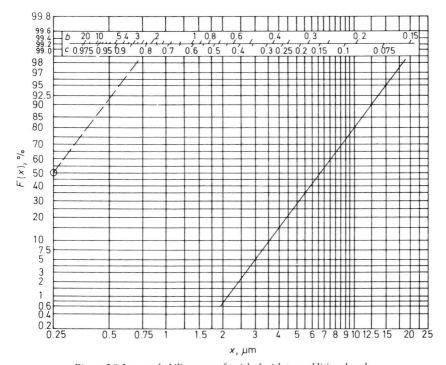

Figure 2.8. Log–probability paper furnished with two additional scales

If the additional scales in *Figure 2.8* are used for parameter determination, then $b = 2.0$ and $c = 0.78$, giving, according to equation 2.27

$$x_m = 0.78 \times 6.75 = 5.27 \, \mu\text{m}$$

and the complete size distribution function is then

$$dF(x) = 0.134 \exp\left[-2\ln^2\left(\frac{x}{5.27}\right)\right]dx$$

with $A = 0.134$ determined from b and equation 2.25.

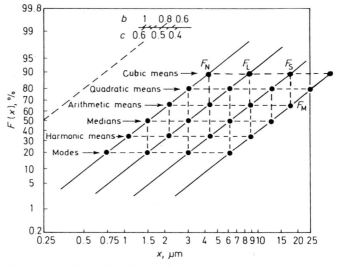

Figure 2.9. The four particle size distributions shown in Figure 2.2(a) plotted as cumulative percentages on log–probability paper; note the relationships between various means

Equations 2.29 to 2.32 give

$$\bar{x}_a = 6.75/\sqrt{(0.78)} = 7.64 \, \mu\text{m}$$

$$\bar{x}_q = 6.75/0.78 = 8.65 \, \mu\text{m}$$

$$\bar{x}_c = 6.75/\sqrt{(0.78)^3} = 9.80 \, \mu\text{m}$$

$$\bar{x}_h = 6.75\sqrt{(0.78)} = 5.96 \, \mu\text{m}$$

Another important advantage of log–normal distributions is the easy conversion from one type of size distribution to another. It can be shown mathematically that all four particle size distributions, by number, length, surface and volume (mass), when plotted on log-probability paper are parallel lines with equal linear spacing, as can be seen in *Figure 2.9*, because

$$x_{\text{g length}} = x_{\text{g number}} \times \frac{1}{c} \tag{2.33}$$

and also

$$x_{g\,\text{surface}} = x_{g\,\text{length}} \times \frac{1}{c} \qquad (2.34)$$

$$x_{g\,\text{mass}} = x_{g\,\text{surface}} \times \frac{1}{c} \qquad (2.35)$$

Equations 2.33, 2.34 and 2.35 are derived from equations 2.1, 2.2, 2.3 and 2.24. As the lines in *Figure 2.9* are parallel, equations 2.33, 2.34 and 2.35 must also apply to all mean sizes and to the mode; the steepness constant b (or geometric standard deviation σ_g) remain the same and only x_m (or x_g) varies in equation 2.24 (or 2.21) so that the expressions for the four size distributions can be written easily from one set of experimental data (bearing in mind the assumptions involved in the conversions as discussed in section 2.3). *Figure 2.9* gives an example and also shows graphically the relationships between the different mean values.

Although the log–normal distribution is primarily a two-parameter function, one or even two extra parameters can be added to define the lower and/or the upper bound to the range of values of the variable[4] (this being the case in many practical size distributions).

2.5.3 The Rosin–Rammler distribution

The Rosin–Rammler distribution function is another equation widely used in particle size measurement. It is a two-parameter function, usually given as cumulative percentage oversize:

$$F(x) = \exp\left[-\left(\frac{x}{x_R} \right)^n \right] \qquad (2.36)$$

where x_R is a constant giving a measure of the particle size range present and n is another constant characteristic of the material under analysis and gives a measure of the steepness of the cumulative curve. The frequency distribution is obtained by differentiation of equation 2.36.

Equation 2.36 can be reduced to

$$\log\left(\ln \frac{1}{F(x)} \right) = n \log x - n \log x_R \qquad (2.37)$$

which gives a straight line if $\log (\ln 1/F(x))$ is plotted against $\ln x$. This is the basis for the Rosin–Rammler graph paper; x_R can be easily found from the plot in the Rosin–Rammler graph paper because it is the size corresponding to $100/e = 36.8\%$ and n is the slope of the line.

Rosin–Rammler charts from German sources contain edge scales, which with the aid of a parallel rule allow direct estimation of n. For accurate work, the use of large size ($32.5 \times 25.4\,\text{cm}$) charts designed by Harris[5] is recommended.

The various mean sizes available may be easily calculated from the Rosin–Rammler equation; they all involve the Gamma functions[3] which can be readily evaluated from mathematical tables.

2.5.4 Harris's three-parameter equation

Harris[6] introduced a three-parameter equation which is very versatile and fits most empirical size distributions that I have come across[7]. He showed that most of the widely used two-parameter equations are in fact special cases of the equation

$$F(x) = \left[1 - \left(\frac{x}{x_0} \right)^s \right]^r \tag{2.38}$$

where $F(x)$ is cumulative percentage oversize, x_0 is the maximum size in the sample, s is a parameter concerned with the slope of the log–ln plot in the fine size region and r is a parameter concerned with the shape of the log–ln plot in the coarse region. Using the same transformation as for the Rosin–Rammler equation (equation 2.37) we get from equation 2.38

$$\log \left(\ln \frac{1}{F(x)} \right) = \log r + \log \left(\ln \frac{1}{1 - (x/x_0)^s} \right) \tag{2.39}$$

Harris developed a procedure for parameter determination based on equation 2.39 and it is similar in the region of extremely fine particles ($x/x_0 \ll 1$) to the Rosin–Rammler plot because for $x/x_0 \ll 1$:

$$\log \ln \frac{1}{1 - (x/x_0)^s} \simeq s \log x - s \log x_0 \tag{2.40}$$

(compare with equation 2.37).

The procedure makes use of the standard Rosin–Rammler plot (which is usually curved in the coarse region) which is overlaid with a transparent tracing of several standard curves of $F(x) = (x/x_0)^s$ (Gates–Gaudin–Schuhman equation as a special case of equation 2.38) furnished with an extra scale of r values; s is estimated from the slope of the fine end of the distribution and the values of x_0 and r are obtained from the relative shifts of the axes in the x and y directions respectively.

Harris's three-parameter equation is very flexible and can be closely fitted to most uni-modal distributions. As its mathematical form lends itself easily to further treatment it has been used in a relatively involved evaluation of results from scanning centrifugal sedimentation instruments for particle size analysis[7].

2.5.5 Other equations available

There is a wide range of other equations available for modelling particle size distributions, e.g. the Weibull function, the Gaudin–Meloy equation, Pearson's equations and many others. Harris presented a very comprehensive review of these[8] and after a very detailed study found that most of the well known distribution equations are in fact special cases of

$$\frac{dF}{dx} = c' x^\alpha \left[1 - \left(\frac{x}{x_0} \right)^\gamma \right]^{x_0^\gamma \beta'} \tag{2.41}$$

where c', α, β', γ and x_0 are parameters.

For the normal and the log–normal distributions, for example, $\alpha = 0$, $\gamma = 2$ and $x_0 \to \infty$: for the Rosin–Rammler distribution, $\alpha = n - 1$, $\gamma = n$ and $x_0 \to \infty$, and for Harris's three-parameter equation $\alpha = s - 1$, $x_0^{\gamma}\beta' = r - 1$ and $\gamma = s$.

2.6 SAMPLING

It is often said that particle size analysis can be only as good as the sampling technique used for collecting the sample. As all laboratory techniques (and also some on-line techniques) use a small sample taken from a production stream in which particle stratification or segregation often takes place, the correct sampling technique is critical for the accuracy of the eventual particle size analysis data.

There are two basic 'golden' rules of sampling:

1. sampling should be made preferably from a moving stream (this applies to both powders *and* suspensions);
2. a sample of the whole of the stream should be taken for many short periods rather than part of the stream for the whole of the time.

It is very likely that a sample collected from the main production stream is going to be too large for most particle size analysis techniques; the problem is then to ensure a good subdivision of the sample into several fractions, all representing closely the original sample. In liquids this is usually done by sampling a stirred suspension with a syringe. This is satisfactory for fine suspensions but with coarser suspensions concentration gradients and the stratification of particles resulting from the centrifugal motion of the liquid caused by the stirring action may lead to appreciable deviations in both the size distribution and concentration of solids in the sample. Burt *et al.*[9] developed a suspension sampler for such subdivisions of coarse suspensions, which divides a sample of from 10 to 100 ml into ten fractions; Burt found it to give significantly better results than the syringe withdrawal.

If the second rule cannot be satisfied for some reason and a small sample has to be withdrawn continuously from a moving suspension, then some thought has to be given to a sampling condition referred to as 'isokinetic sampling'. This is based on the aspiration of a sample of the suspension from the main stream through a sampling probe (with the sampling nozzle pointing towards the direction of flow) so that the velocity of flow into the nozzle is identical with the undisturbed local stream velocity. Isokinetic sampling is designed to ensure an accurate representation of the coarse particles in the sample because at higher or lower sampling rates than iso-kinetic, the proportion of coarse particles in the sample would, because of inertial effects, be lower or higher respectively.

Fine particles have a low inertia compared with the drag force, and follow the flow closely whatever the sampling rate.

Whilst isokinetic sampling is usually critical in dust-laden gas[10] sampling it does not have to be observed in most applications of liquid suspension sampling because of the high liquid viscosities. An inertia parameter (Stokes' number) is usually used to assess the necessity of isokinetic sampling; Parker[11] suggested the range of

$$0.05 < \Psi < 50 \qquad (2.42)$$

outside of which isokinetic sampling is not critical.

The inertia parameter used is the ratio between the particle stopping distance and the diameter of the sampling nozzle D so that

$$\Psi = \frac{ux^2\rho_s}{18\mu D} \qquad (2.43)$$

where u is velocity of flow, ρ_s is the density of solids and μ is the viscosity of the suspending liquid.

Equation 2.43, using the lower limit of $\Psi = 0.05$, gives $x_{critical} \simeq 60\ \mu m$ for $u = 1\ m\ s^{-1}$, $D = 1\ cm$, $\mu = 0.001$ (water) and $\rho_s = 2600\ kg\ m^{-3}$ (silica). It can therefore be concluded that isokinetic sampling of liquid suspensions only becomes important if the suspensions contain very coarse particles (in our example, coarser than 60 µm); equations 2.42 and 2.43 can be used to assess the necessity of isokinetic sampling for any given conditions.

For sampling of powders in a dry form, Allen[12] reviewed and tested most methods available and found the spinning riffler to be the best.

2.7 LABORATORY MEASUREMENT OF PARTICLE SIZE

There is an abundance of methods available for the measurement of particle size distributions and many excellent textbooks exist[3] which review the field in great depth. Because of the limited scope of this chapter only a short review of the methods particularly relevant to solid–liquid systems is given.

Table 2.7 gives a schematic review of the methods available, the size ranges covered and the types of particle size and size distribution measured. This information is given in order to assist an engineer faced with the problem of selection of the best method for a given application—only a preliminary selection can be attempted using Table 2.7, because it was impossible to list all the important factors influencing the choice, these are:

1. type of equivalent diameter required;
2. quantity to be measured (number, surface, mass);
3. size range;
4. quantity of sample available;
5. number of points on the distribution (or perhaps just some measure of central tendency) required;
6. number and frequency of analyses required;
7. operator's involvement and degree of experience necessary;
8. cost of accessories such as sample preparation, evaluation of data etc.;
9. degree of automation required.

It should be noted that many of these factors are inter-related and their relative importance varies in different applications.

As was stated in the introduction, Stokes' diameter x_{St} is usually used to characterize particle size in those applications where it is the behaviour of

Table 2.7. CLASSIFICATION OF LABORATORY METHODS OF PARTICLE SIZE MEASUREMENT*

Method	Approximate size range, μm	Type of particle size— see Tables 2.1, 2.2, 2.3	Type of size distribution
Sieving (wet or dry)			
Woven wire	37–4000	x_A	by mass
Electro-formed	5–120		
Microscopy			
Optical	0.8–150	x_a, x_F, x_M	by number
Electron	0.001–5	x_{SH}, x_{CH}	
Gravity sedimentation			
Incremental	2–100	x_{St}, x_f	by mass (by surface)
(except photosedimentation)			
Cumulative	2–100	x_{St}, x_f	by mass
Centrifugal sedimentation			
Two layer—incremental	0.01–10	x_{St}, x_f	by mass
—cumulative			
Homogeneous—incremental			
Flow classification			
Gravity elutriation (dry)	5–100	x_{St}, x_f	by mass
Centrifugal elutriation (dry)	2–50	x_{St}, x_f	by mass
Impactors (dry)	0.3–50	x_{St}, x_f	by mass or number
Cyclonic (wet or dry)	5–50	x_{St}, x_f	by mass
Coulter principle (wet)	0.8–200	x_v	by number
Particle counters (wet or dry)	0.3–100 or 2–9000	x_p, x_s	by number
Field flow fractionation	1 nm–100 μm	x_d	depends on detector
Hydrodynamic chromatography	0.01–50	x_d	depends on detector
Fraunhofer diffraction (laser)	1–2000	equivalent laser diameter	by volume
Mie theory light scattering (laser)	0.1–40	equivalent laser diameter	by volume
Photon correlation spectroscopy	0.003–3	equivalent laser diameter	by number
Scanning infrared laser	3–100	chord length	by number
Aerodynamic sizing in nozzle flow	0.5–30	x_d	by number
Mesh obscuration method	5–25	x_A	by number
Laser Doppler phase shift	1–10 000	equivalent laser diameter	mean only
Time-of-transition	150–1200	equivalent laser diameter	by number
Surface area determination		x_{sv}	mean only (surface volume)
Permeametry			
Hindered settling			
Gas diffusion			
Gas adsorption			
Adsorption from solution			
Flow microcalorimetry			

*For details of the newer methods in the latter part of this table, consult Lloyd[15].

particles in liquids that determines the separation efficiency and other operational characteristics of the separators (e.g. in sedimentation, centrifugation and hydrocyclones). Methods that measure Stokes' diameter, such as sedimentation or fluid classification, have therefore been used extensively in this field. Although preference is naturally given to wet methods, air classification is also widely used.

Most of gravity sedimentation techniques use an initially homogeneous suspension in which particles are allowed to settle under the influence of gravity. Two basic modes of operation can be found in practice: the fraction of particles which have a given settling velocity is determined from either the concentration measurements at a certain depth below the surface in a sedimentation cell (incremental techniques) or from measurements of the total mass of solids accumulated at the bottom (cumulative techniques). Sedimentation balances and β back-scattering techniques have been used for the cumulative measurements while sampling (Andreasen pipette method) or the absorption of radiation (photosedimentation or use of X-rays) are most frequently used for the incremental measurements. It seems that the incremental techniques, because of their advantages of easy operation and evaluation, and relatively simple instrumentation, have somewhat wider application.

Stokes' law is used for the evaluation of particle sizes corresponding to the settling velocities of the appropriate mass fractions measured; this law gives the well known equation

$$x_{St} = \sqrt{\left(\frac{18\mu v}{\Delta\rho g}\right)} \tag{2.44}$$

which requires knowledge of the liquid viscosity, μ; the difference between the density of solids and of the liquid, $\Delta\rho$; and the acceleration due to gravity g. There are many applications in which μ and $\Delta\rho$ are difficult to determine with a sufficient degree of certainty, say when the density of the solids varies with particle size, and in such cases the conversion from the settling velocity, v, to the particle size x_{St} in equation 2.44 is simply not carried out and the results are left in the form of the distribution of settling velocities (usually by mass). Equation 2.44 requires SI units or units in any other consistent system; many experienced particle size analysts use and remember its more convenient but dimensionally inconsistent form

$$x_{St} \simeq 175 \sqrt{\frac{\mu v}{\Delta\rho}} \tag{2.45}$$

where the units are x_{St}, μm; μ, P; v, cm min^{-1} and $\Delta\rho$, g cm^3. (The constant in equation 2.45 is, more accurately, 174.90.)

All of the gravity sedimentation techniques, with the exception of photosedimentometers which detect the projected area of the particles (which in turn is proportional to particle surface) measure size distribution by mass. Photosedimentometers can only give useful results for particles coarser than around 5 μm because for finer particles the wavelength of light becomes comparable to the particle size, the laws of geometric optics break down and the so-called light extinction coefficient becomes highly size dependent.

In centrifugal sedimentation the days of bottle or tube-type centrifuges seem to be over and the disc centrifuge almost entirely dominates the field. In addition to equipment which uses the homogeneous technique, some disc centrifuges are also designed to use a 'line start' or 'two layer' technique where a thin layer of suspension is introduced on top of a clear suspending fluid. This mode of operation leads to very simple evaluation of results but suffers from an effect called streaming, which results from particles breaking through the initial interface in 'streams'. This can be overcome by various means but observation by an operator is always required. A more serious disadvantage of the line start technique is that it is inherently unstable because at all times after the start of the analysis density inversions occur (higher suspension densities at small radii than 'lower down' at greater radii). Both the line start and the homogeneous mode of operation may be used either cumulatively or incrementally and the latter is again more frequent in use.

Instruments working with initially homogeneous suspensions are very simple in operation but evaluation of the results is mathematically complicated. Unlike the gravity settling, particles here settle along radial, i.e. diverging paths and this continuous dilution effect obscures the fall in concentrations resulting from particles falling out of the suspension. Several evaluation methods have been suggested including a semi-automated analogue evaluation[13] designed for an X-ray centrifugal sedimentometer. An equivalent to the Andreasen pipette method in gravity sedimentation is the Simcar centrifuge or the recently introduced Ladal Pipette Centrifuge[14] where samples are withdrawn at different times from a given depth in a rotating suspension. The latter offers increased versatility (0.01 to 5 μm) and requires only a small sample of 1 to 2 g.

In centrifugal sedimentation the basic equation for the conversion of settling velocity to particle size is analogous to equation 2.44:

$$x_{St} = \sqrt{\left(\frac{18\mu \ln R/R_0}{\Delta\rho\omega^2 t}\right)} \qquad (2.46)$$

where R is the radius of the measurement zone, R_0 is the radius of the liquid surface, ω is angular speed and t is time from the start of the analysis.

Equation 2.46 becomes:

$$x_{St} = 5477 \sqrt{\left(\frac{\mu \ln (R/R_0)}{\Delta\rho\omega^2 t}\right)} \qquad (2.47)$$

where the units are μ, P; $\Delta\rho$, g cm^{-3}; ω, s^{-1} and t, min.

Finally it should be pointed out that equations 2.44 and 2.46 have a limited application to small sizes (the minimum is 1 μm in a gravity field and much smaller in a centrifugal field depending on the speed of rotation) when Brownian motion effectively slows down settling rates and to large sizes when, for Reynolds numbers greater than about 0.2, increasing deviations from Stokes' law occur.

Elutriation and fluid classification methods are also highly relevant to solid–liquid separation problems—they use the same or similar mechanisms for analysis as many separators. Use is made here of the size-dependent nature of dynamic separation processes and most of these methods are based on the analytical cut size defined in chapter 3, 'Efficiency of Separation'.

In cake filtration, where the surface volume diameter is of interest, methods for surface area determination are relevant (these measure only mean sizes) particularly, permeametry, gas diffusion and hindered settling methods.

Other methods of particle size measurement are also widely used in the characterization of suspensions, e.g. particle counters or the Coulter principle in filter rating, microscopy for general particle investigations and screening for coarse solids (above 75 μm).

Table 2.7, particularly the lower part, also contains some more recent additions to the available laboratory measurement methods. Many of these methods involve the use of a laser and the interaction between its light output and an assembly of particles. Most of these methods give a fast response and are thus suitable for automation and for application of the statistical measurement control techniques (see section 2.9). It cannot be said, however, that any of the newer methods are particularly suitable in solid–liquid separation; their increasing popularity in industry owes more to their speed and ease of use than to their fundamental suitability for solid–liquid applications. Only a very brief account of the newer methods is given below, in the same order as they appear in *Table 2.7*, and for further reading and details the reader is referred to Lloyd.[15]

In field flow fractionation, a sample is injected into a narrow channel where stratification according to particle size takes place under the effects of gravity or centrifugal fields. The coarse particles are eluted from the channel later than the fine ones due to the flow being slower lower down, near the wall, where the coarse fraction prevails. The concentration at the outlet of the channel is monitored over time using a variety of detectors, mostly based on light extinction.

Hydrodynamic chromatography relies on different particle velocities in laminar flow through capillaries or packed columns. Larger particles move faster with the flow than do fine ones because they are, on average, further away from the capillary wall. The operation and the equipment are the same as in liquid chromatography: colloidal particles are injected into a column packed with beads and a suitable detector (ultraviolet light detector or a spectrophotometer) monitors the flow from the column. Both field flow fractionation and hydrodynamic chromatography are most suitable for nearly mono-sized particle systems.

A number of expensive but very powerful analysers available on the market are based on Fraunhofer diffraction. An assembly of particles in a liquid or in a gas are illuminated by an expanded laser beam and the particle size distribution is derived from the measurement of the spatial distribution of the diffraction patterns on a flat detector behind the sample.

A similar principle but different hardware and software are used in the Mie theory light-scattering-based laser analysers, extending the range into the submicron region.

Photon correlation spectroscopy measures the interference between coherent light and fine particles diffusing in a liquid. The interference is modulated by the random motion of the particles which depends on their size; a correlation computer evaluates the size distribution from the autocorrelation function of the signal measured in one or more defined scattering directions.

The scanning infrared laser method sends an oscillating laser beam into a suspension and detects back-scattered infrared light. The random chords within particles are thus measured and this information is related to the particle size distribution of the suspension. This method can be used to measure more concentrated suspensions than can most other light-based methods.

The equivalent aerodynamic diameter of dry powder or aerosol particles can be measured using aerodynamic sizing in nozzle flow. This measures the transit times between two focussed laser beams in accelerated air flow seeded with particles.

The mesh obscuration method is a quick way of, for example, measuring the size of the contaminant in lubricating liquids. The contaminated fluid is passed through a fully characterized filter mesh and the development of a pressure drop is monitored. The number of particles larger than the mesh size is deduced from the pressure drop; three filter mesh units are used in series.

A mean particle size can be measured using laser Doppler phase shift. Two laser beams cross in a measurement volume and scattered light is measured; in addition to the usual velocity measurement obtained from the frequency shift, the phase shift can provide the mean particle size.

Finally, the recently proposed time-of-transition method scans a focussed, narrow laser beam in a measurement zone, in a circular fashion. The interaction pulses are detected by a photodiode behind the beam. The pulse width is a measure of particle size, and particle shape analysis is also possible.

2.8 ON-LINE MEASUREMENT TECHNIQUES

Automation of process control has created a need for continuous monitoring of the particle size of particulate matter in process streams. Some 'on-line' particle size analysis instrumentation has been developed recently to meet this need; it can initiate regulatory or shut-down signals in control systems. The basic requirements for such instrumentation are that it must operate automatically and continuously under preset instructions, and the response time from observation to readout must be so short as to be nearly instantaneous.

This relatively new range of techniques follows the general pattern found in the whole subject of particle size measurement: a variety of techniques differing in

1. principle;
2. suitability for different systems, materials and particle sizes;
3. the types of particle diameters they measure;
4. the number of points on the distribution they are capable of determining;
5. degree of truly 'on-stream' operation.

Some methods only give a measure of central tendency (a mean diameter), others give one or more points on the size distribution. They may be truly on-line in operating on the whole process stream, or they may need a partial sample stream taken off the main stream or they may merely be automated rapid-response batch techniques.

On-line measurement is a buoyant area which is undergoing fast growth. Reviews in the literature become quickly out of date; there are two excellent review papers published by Hinde and Lloyd[16] and by Stanley-Wood[17], the latter also listing some methods that are merely fast-response laboratory techniques.

Equipment may be broadly divided into two categories: 'stream scanning' and 'field scanning'. Stream scanning is generally applied to dilute systems; the particles are sent in single file past a detecting device, i.e. a particle counter based on the Coulter method, a light interruption or scattering method, or the Langer acoustic method. All of these require a small sample withdrawn continuously from the main stream, with all the associated sampling problems.

Field scanning usually applies to concentrated systems in which some size-dependent behaviour of the bulk material is monitored and the particle size is deduced from theoretical or calibrated relationships. Ultrasonic attenuation, echo measurements, laser attenuation, on-line viscometry, electronic noise correlation techniques, X-ray attenuation and X-ray fluorescence are examples of such field scanning methods. Most relevant to solid–liquid separation problems are those field scanning instruments which deduce particle size from the separation efficiency of some separational equipment, using the concept of the analytical cut size (see chapter 3 'Efficiency of Separation of Particles from Fluids'). Firstly, there are automatic wet sieving machines, one developed by Hinde and Lloyd[16] and another described by Schönert et al.[18] Secondly, the separation performance of hydrocyclones which may already form parts of some industrial plants, has been suggested as a means of giving an indication of changes in particle size in the feed suspension[19]; particle size measurement is thus reduced to solids concentration measurement.

The use of cyclones or other separators, in the sampling mode, for on-line particle size measurement is now well established in both dry[20-23] and wet applications[24, 25]. This involves taking a continuous sample stream through a small cyclone or hydroclone under a given set of operating conditions, and monitoring the recovery of the cyclone. The task of particle size measurement is thus reduced to measurement of solids concentrations in two material streams.

A similar idea forms the basis for the operation of the Mintex/RSM slurry analyser[26,35] based on the work of Holland-Batt[27]; the slurry is pumped through a single-turn helical tube of rectangular cross-section and β-ray attenuation measurements before the helix and at different positions just after the helix (where particle stratification occurs due to centrifugal forces) yield a point on size distribution data. The instrument requires calibration, which can be made at any selected reference size in the range of 10 to 105 µm.

2.9 STATISTICAL MEASUREMENT CONTROL

The recent proliferation of fast and convenient instruments for particle size analysis has brought with it, for the first time in this subject, a practical opportunity to apply the techniques of statistical measurement control. Repeat analyses with such instruments are easy and statistical techniques can thus be used to evaluate precision and accuracy, and keep the measurement under statistical control. We can no longer justify making comparisons or conclusions on the basis of one or two measurements, as was so often the case in the past.

A measurement process is said to be under statistical control when all the critical parameters and conditions are under sufficient control such that any variation in data from repeated measurements does not change over an extended period of time. The variation (scatter) must also be demonstrated to occur in a random fashion. The best way of checking this is to use control charts[28-34] in which the mean values and standard deviations of replicate measurements are plotted against time and the data are checked to be within statistically derived limits.

The replicate measurements (at least two each time) are made over a prolonged period of time (at least 40 times), on a 'check-standard' sample which must remain constant and homogeneous during the measurement cycle. Software packages are available to assist in the evaluation of results and in plotting the control charts, and any measurements outside the control limits are automatically highlighted. Visual inspection of the charts may reveal systematic trends on drifts. The variability is then analysed, its causes identified and brought under control and reduced.

Apart from data on precision (repeatability, reproducibility), control charts may be used to check or monitor accuracy, i.e. to what extent the measured value of a quantity (such as the mean particle size, for example) agrees with the accepted value for that quantity, bearing in mind that the 'true value' is never known. In testing accuracy, the standard test powders available on the market[33] may be used.

It is beyond the scope of this book to give more than a quick review of this important subject. The interest in and application of the statistical-measurement control techniques in particle size measurement, however, are certain to continue to grow in the future.

APPENDIX 2.1 THE CHOICE OF A MEAN PARTICLE SIZE

As shown in section 2.4, there is a bewildering variety of different definitions of 'mean' particle size. Most textbooks quote the definitions either in the form of summations or as integrals, the latter being more appropriate if the particle size distributions are generated as continuous curves. Few textbooks, however, give any hints as to which of the definitions are to be used and when. This is a serious omission because the correct choice of the most appropriate mean is vital in most applications.

As can be seen from *Figure 2.A1.1*, two different size distributions may have one mean the same (arithmetic mean in the example) but all the other means may be different. Thus, if the comparison is made on the basis of that one mean (arithmetic in this case), the distributions would be judged to be the same. If any of the other means were chosen, the conclusion would be very different: according to the higher order means such as the quadratic or cubic means, the top distribution in *Figure 2.A1.1* is the coarser of the two, whilst according to the median or harmonic mean the top distribution is finer. The wrong choice of the mean may, therefore, lead to incorrect or misleading conclusions and examples of this may be found in many branches of particle technology.

The mean particle size is rarely quoted in isolation: it is usually related to some application and used as a single number to represent the full size distribution. It represents the distribution by some property which is vital to the application or process under study; if two size distributions have the same mean, the two materials are likely to behave in the process in the same way.

It is, therefore, the application which governs the selection of the most appropriate mean. Usually, enough is known about the process under study

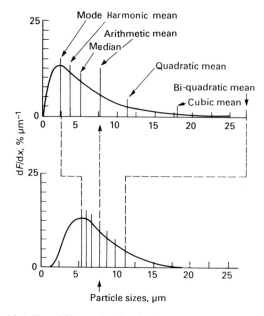

Figure 2. A1.1. Two different size distributions with the same arthmetic mean.

to be able to identify some fundamentals which can then be used as a starting point. The fundamental relations may be oversimplified or inadequate to describe the process fully, but it is still better to use these relations than to pick the mean definition out of a hat!

Two examples of some simple fundamentals used in the selection of the most relevant mean in solid–liquid separation are given below.

Example 2.A1.1

Application: flow through packed beds (e.g. cake resistance in filtration, cake washing or dewatering)

If the particle size distribution in *Table 2.6* (plotted in *Figure 2.8*) is by mass and the material forms a filtration cake, what is the value and definition of the mean to be used when correlating particle size with specific cake resistance?

Solution

Firstly, the decision about the definition of the mean has to be made. Equation 9.36 gives a fundamental relationship which can be used here

$$\alpha = \frac{K_o S_0{}^2}{\rho_s} \frac{(1 - \epsilon)}{\epsilon^3} \qquad (2.A1.1)$$

where α is the specific cake resistance, K_0 is a constant, S_0 is the volume specific area of the solids making up the cake and ϵ is the voidage of the cake.

We can relate the specific surface to particle size but no simple relationship exists for the voidage. All we can do here is to assume that ϵ is constant, independent of particle size; the task is then to represent the particle population by a mono-size system (mean size) which has the same S_0 as the real population.

It can be shown that, for a sphere, the specific surface is

$$S_0 = 6/\bar{x}$$

and this is to be equal to the total surface/total volume of the real distribution. The mean $\bar{x}$ is then to calculated as

$$\bar{x} = \frac{6 \cdot \text{total volume}}{\text{total surface}}$$

i.e.

$$\bar{x} = \frac{\Sigma x_i S_i}{\Sigma S_i}$$

where S_i is the surface area of a particle of size x_i.

Analytically, this can be written as

$$\bar{x} = \int_0^1 x \, dF_{\text{surface}}$$

which is the definition of the arithmetic mean of the distribution by surface $\bar{x}_a$ (equation 2.13 and *Table 2.4*).

The remaining task is to determine $\bar{x}_a$ from the distribution by surface; our size distribution is by mass, however, and a conversion has to be made. Rather than converting the whole distribution (from mass to surface) in this case, because the distribution is log–normal (see *Figure 2.8*), we can use the fact that the arithmetic mean by surface is equal to the harmonic mean by mass (see Appendix 2.2). The harmonic mean of the above distribution was determined in this example *(Example 2.1)* as 5.96 μm which is the mean size to be used in correlation with the specific cake resistance.

Example 2.A1.2

Application: mass recovery of solids in a dynamic separator such as a gravity settling tank or a sedimenting centrifuge

Which of the two particle size distributions given in *Table 2.A1.1.*, if suspended in a liquid, would you expect to give higher mass recovery in a gravity settling tank and why?

The two samples are of the same solid, i.e. are of the same solids density (greater than that of the liquid), suspended in the same liquid and the settling tank is run at identical operating conditions (same residence time, temperature, etc.) for both samples. The two distributions were determined by sedimentation and are given as per cent by mass.

Table 2.A1.1. TWO PARTICLE SIZE DISTRIBUTIONS DETERMINED
BY SEDIMENTATION (IN PER CENT BY MASS)

Particle size, μm	Sample 1, % undersize	Sample 2, % undersize
20	7.94	33.05
30	27.45	50.00
40	49.01	62.22
60	78.37	77.34
90	94.46	88.27
130	99.01	94.37
170	99.79	96.97
210	99.95	98.24
230	100.00	100.00

Solution

The above problem lies in determining which of the two distributions given is coarser, for the purpose of recovery in gravity settling tanks. Let us derive, from some simple first principles, the definition of a mean to be used in this problem as a criterion for the comparison.

Total recovery of any separator can be obtained by processing the feed size distribution (cumulative) $F(x)$ with the operating grade efficiency curve $G(x)$ of the separator. Mathematically, this can be written as

$$E_T = \int_0^1 G(x)\, dF \qquad (2.A1.2)$$

A simple, plug-flow model of the separation in a settling tank without flocculation gives the grade efficiency in the following form

$$G(x) = \frac{u_t\, A}{Q}$$

where A is the settling area, Q is the suspension flow rate and u_t is the terminal settling velocity of particle size x.

Assuming Stokes' law for the terminal settling velocity

$$u_t = \frac{x^2\, \Delta \rho\, g}{18\, \mu}$$

The above three equations, when combined, give the total recovery as

$$E_T = \frac{A\, \Delta \rho\, g}{18\, Q\, \mu} \cdot \int_0^1 x^2\, dF$$

where the integral is, of course, the definition of the quadratic mean, of the particle size distribution by mass because E_T is by mass (see equation 2.13 and *Table 2.4*).

The relevant mean to be used in the comparison of the two particle size distributions of this example is, therefore, the quadratic mean of the mass distribution. The graphical method outlined in section 2.4, when applied to the data given in the above example, yields the plots in *Figure 2.A1.2*. As the question does not require numerical answers, it is merely necessary to decide from the plots which of the two would give the greater area between the curve and the *y* axis.

As shown in *Figure 2.A1.2*, the deduction can be made from a simple

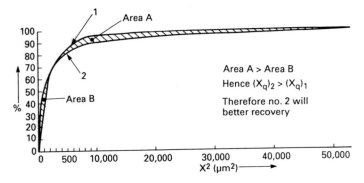

Figure 2.A1.2. *Plots of F against X² to determine the quadratic means*

comparison of the cross-hatched areas; as area A is greater than area B it follows that the quadratic mean of sample 2 is coarser than that of sample 1 and the recovery with sample 2 would, therefore, be greater.

This conclusion is, of course, subject to the assumptions made in the above derivation, but the example is merely used to illustrate the method and more realistic (and more complicated) assumptions may be used if necessary.

APPENDIX 2.2 CAN THE ARITHMETIC MEAN OF THE SURFACE DISTRIBUTION (SURFACE–VOLUME DIAMETER) BE DETERMINED FROM THE MASS DISTRIBUTION?

The distributions by surface and mass can be related by the following equation (from equations 2.2 and 2.3)

$$f_m(x) = k \times f_s(x)$$

This can be written explicitly for f_s, but by substituting dF for f

$$dF_s = dF_m / (k\ x)$$

By definition, the arithmetic mean of the surface distribution (see equation 2.13 and *Table 2.4*) is

$$(\bar{x}_a)_S = \frac{1}{k} \int_0^1 dF_M$$

which, after substitution, gives

$$(\bar{x}_a)_S = \int_0^1 x\ dF_S$$

i.e. the integral is equal to 1 and

$$(\bar{x}_a)_S = \frac{1}{k}$$

The constant k may be determined from the equation for dF_S, after integration of both sides from 0 to 1:

$$k = \int_0^1 \frac{1}{x}\ dF_M$$

which is the definition of the harmonic mean of the mass distribution. Therefore, $(\bar{x}_a)_S = (\bar{x}_h)_M$ for any shape of the distribution, but only if k is a constant and not a function of x (i.e. the shape factor is independent of particle size).

REFERENCES

1. Lloyd, P. J., and Ward, A. S., *Filtration and Separation*, **12**, 250 (1975)
2. Svarovsky, L., *Powder Technology*, **7**, 351–352 (1973)
3. Allen, T., *Particle Size Measurement*, 2nd edn, Chapman and Hall, London, (1975)
4. Aitchison, J. and Brown, J. A. C., *The Log-normal Distribution*, Cambridge Univ. Press, London, (1957)
5. Harris, C. C., *Powder Technology*, **5**, 39–42 (1971/72)
6. Harris, C. C., *Trans. SME*, **244**, No. 6, 187–190 (1969)
7. Svarovsky, L. and Friedova, J., *Powder Technology*, **5**, 273–277 (1971/72)
8. Harris, C. C., *Trans. SME*, **241**, 343–358 (1968)
9. Burt, M. W. G., Fewtrell, C. A. and Whatron, R. A., *Powder Technology*, **7**, No. 6 327–330 (1973)
10. Svarovsky, L., *Chemistry and Industry*, No. 15, 7 August 1976, 626–630 (1976)
11. Parker, G. J., *Atmospheric Environment*, **2**, 477–490 (1968)
12. Allen, T. and Khan, A. A., *Chem. Eng.* **238**, CE 108–112 (1970)
13. Svarovsky, L. and Svarovska, J., *Partikelmesstechnik*, Dechema-Monographien, Nr. 1589–1615, Band 79, 293–308, Dechema, Frankfurt-am-Main (1976) and Svarovsky, L. and Svarovska, J., *J. Phys. D: Appl. Phys.*, **8**, 181–190 (1975)
14. Allen, T. and Svarovsky, L., *Partikelmesstechnik*, Dechema-Monographien, Nr. 1589–1615, Band 79, 279–291, Dechema, Frankfurt-am-Main (1976)
15. Lloyd, P. J. (Ed.), *Particle Size Analysis*, Wiley, Chichester (1988)
16. Hinde, A. L. and Lloyd, P. J. D., *Powder Technology*, **12** 37–50 (1975)
17. Stanley-Wood, N. G., *Control & Instrumentation*, **7**, No 1, 30–35 (1975)
18. Schönert, K., Schwenk, W. and Steier, K., *J. Aufbereit. Verfahrenstech*, **7**, 368–372 (1974)
19. Lynch, A. J., Rao, T. C. and Whiten, N. J., *Proc. Australasian Inst. Mining Met.*, **223** 71–73 (1967)
20. Nakajima, Y., Gotoh, K. and Tanaka, T., *I&EC Fundamentals* **6**, No. 4, 587–592 (1967)
21. Svarovsky, L. and Hadi, R. S., 'A new on-line particle size analyser for fine powders', *Proc. Int. Symp. on Fine Particle Processing*, Vol. 1, Chap. 19, (Ed. P. Somasundaran), AIME, Las Vegas, 366–379 (1980)
22. Svarovsky, L., *Method and Apparatus for Monitoring Particle Size*, University Research Ltd., U.S. Patent No. 4179934, British Patent No. 1578157, German Patent DE 2932987 C 2
23. Svarovsky, L. and Hadi, R. S., 'A new, simple and sensitive on-line particle size analyser for fine powders suspended in gases', *Proc. 2nd European Symposium on Particle Characterization* (Nuremberg), 343–359 (1979) (also by R. S. Hadi)
24. *Bestobell Cyclometric Sizer*, Bestobell Mobrey, Slough
25. Svarovsky, L., *Hydrocyclones*, Holt, Rinehart and Winston, London (1984)
26. *The Y1308/RSM Slurry Sizer*, Carter Group Ltd., Slough
27. Holland-Batt. A. B., *Inst. Mining and Metall.*, **77**, C185 (1968)
28. *ASTM Manual on Presentation of Data and Control Chart Analysis*, ASTM Publication No. STP 15D, Philadelphia (1985)
29. Belanger, B., *Measurement Assurance Programs, Part I: General Introduction*, NBS Special Publ. No. 676-I (1984)
30. *Statistical Quality Control Handbook*, AT&T, Delmar Printing Co., Charlotte, NC (1984)
31. Feigenbaum, A. V., *Total Quality Control*, McGraw-Hill, New York (1983)
32. Speitel, K., 'Measurement Assurance', in *Handbook of Industrial Engineering*, Wiley, New York (1982)
33. *BCR Test Powders*, Community Bureau of Reference, Rue de la Loi 200, B-1049 Brussels, Belgium (also available from the National Physical Laboratory, Teddington)
34. Wood, B., Ph.D. Thesis, University of Bradford (1988)
35. *MINTEK/TOPIC 2/2 Monitoring wet mill performance with the Mintek/R.S.M/Slurry Sizer*, Cartner Group Ltd., Slough.
36. Putnam, R., 'Optimizing grinding mill loading by particle size analysis', *Mining Congress J.*, **Sept.**, 68–74 (1973)

3

Efficiency of Separation of Particles from Fluids

L. Svarovsky

Department of Chemical Engineering, University of Bradford

3.1 INTRODUCTION

Imperfection in the performance of any real separation equipment can be characterized by the separation efficiency. In this chapter basic definitions are given together with the relationships between the efficiency and particle size distributions of various combinations of the feed, underflow or overflow product streams. Practical considerations for grade efficiency testing and total efficiency prediction are given, together with worked examples.

The so-called 'grade efficiency concept' studied here is applicable to those principles of and equipment for solid–liquid separation whose separational performance does not change with time if all operational variables are kept constant. Hydrocyclones, sedimenting centrifuges or gravitational clarifiers are examples of such equipment; the concept is not widely used in filtration because there the efficiency changes with the amount of solids collected either on the face of the filter medium (in cake filtration) or within the medium itself (in deep bed filtration). Even in filtration, however, it is interesting to determine the grade efficiency of the clean medium which governs the initial retention characteristics of the filter and can be used for filter rating (see the example given in *Figure 3.4*).

As the efficiency of separation is very often particle-size dependent, some separational equipment can be, and often is, also used for the classification of solids. This is the area where the grade efficiency concept was first developed; it is now also widely used in gas cleaning.

Hence there are three major operations in which this concept is applicable:

1. solid–liquid separation;
2. solid–gas separation (gas cleaning and dust control);
3. solid–solid separation (often called classification of powders) with liquid or gas as a medium.

3.2 BASIC DEFINITIONS AND MASS BALANCE EQUATIONS

A single-stage separational apparatus can be schematically drawn as in *Figure 3.1*, where:

M is the mass flow rate of the feed (in kg s^{-1})
M_c is the mass flow rate of the coarse material in the underflow (in kg s^{-1})
M_f is the mass flow rate of the fine material in the overflow (in kg s^{-1})
$dF(x)/dx$ is the size distribution frequency of the feed
$dF_c(x)/dx$ is the size distribution frequency of the coarse material
$dF_f(x)/dx$ is the size distribution frequency of the fine material
x is the particle size
Q is the volumetric flow rate of the feed suspension (in m^3 s^{-1})
O is the volumetric flow rate of the overflow suspension (in m^3 s^{-1})
U is the volumetric flow rate of the underflow suspension (in m^3 s^{-1}).

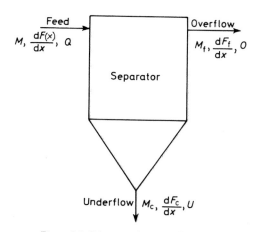

Figure 3.1. Schematic diagram of a separator

The total mass of the feed must be equal to the sum of the total masses of the products if there is no accumulation of material in the equipment, i.e.

$$M = M_c + M_f \qquad (3.1)$$

Mass balance must also apply to any size fractions present in the feed if there is no change in particle size of the solids inside the separator (no agglomeration or comminution). Hence for particles of size between x_1 and x_2:

$$(M)_{x_1/x_2} = (M_c)_{x_1/x_2} + (M_f)_{x_1/x_2} \qquad (3.2)$$

and also for each particle size x present in the feed:

$$(M)_x = (M_c)_x + (M_f)_x \qquad (3.3)$$

By definition, the particle size distribution frequency gives the fraction of particles of size x in the sample. The total mass of particles of size x in the feed for example is therefore the total mass of the feed M multiplied by the appropriate fraction dF/dx so that equation 3.3 becomes:

$$M\frac{dF}{dx} = M_c\frac{dF_c}{dx} + M_f\frac{dF_f}{dx} \tag{3.4}$$

3.2.1 Total efficiency

If a total (or overall) efficiency E_T is now defined as simply the ratio of the mass M_c of all particles separated to the mass M of all solids fed into the separator, i.e.

$$E_T = \frac{M_c}{M} \tag{3.5}$$

or, if mass balance in equation 3.1 applies:

$$E_T = 1 - \frac{M_f}{M} \tag{3.6}$$

equation 3.4 can be rewritten as:

$$\frac{dF}{dx} = E_T\frac{dF_c}{dx} + (1 - E_T)\frac{dF_f}{dx} \tag{3.7}$$

which relates the particle size distributions of the feed, the coarse product and the fine product. The same relationship holds for particle fractions between x_1 and x_2

$$F(x_2) - F(x_1) = E_T[F_c(x_2) - F_c(x_1)] + (1 - E_T)[F_f(x_2) - F_f(x_1)] \tag{3.8}$$

as well as for cumulative percentages corresponding to any size x

$$F(x) = E_TF_c(x) + (1 - E_T)F_f(x) \tag{3.9}$$

(Note that equations 3.8 and 3.9 are obtained by integration of equation 3.7.) The mass balance in equation 3.9 (or 3.7) allows the calculation of any one missing size distribution provided the other distributions and the total efficiency are known, i.e.

$F(x)$ from E_T, $F_c(x)$ and $F_f(x)$
$F_f(x)$ from E_T, $F_c(x)$ and $F(x)$
$F_c(x)$ from E_T, $F_f(x)$ and $F(x)$

or even E_T from $F(x)$, $F_c(x)$ and $F_f(x)$. The last combination is the basis of the analysis of errors in particle size distribution data (and of sampling) because equation 3.9 can be written in the form:

$$E_T = \frac{F(x) - F_f(x)}{F_c(x) - F_f(x)} \tag{3.10}$$

If the differences $F(x) - F_f(x)$ and $F_c(x) - F_f(x)$ are plotted against each other for different sizes x, a straight line with a slope of E_T should be obtained. Because of the errors in particle size measurement, however, there may be a considerable scatter in the results. *Figure 3.2* shows a practical example of a test with a hydrocyclone; the line is drawn at a slope of the actual (measured) total efficiency, which of course is also subject to measurement error.

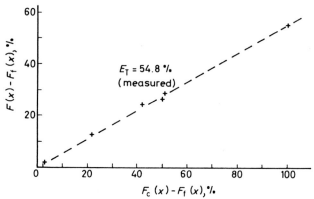

Figure 3.2. Example of scatter in the mass balance of particle size distribution data

Note that the cumulative percentages in equations 3.8, 3.9 and 3.10 can be either 'oversize' or 'undersize' as long as the same type is used for all the distributions in the given equation. As to the quantity considered, equation 3.7 applies to particle size distributions by mass, surface or number simply because, for example:

$$\left(\frac{dF}{dx}\right)_{by\ mass} = kx^3 \left(\frac{dF}{dx}\right)_{by\ number} \tag{3.11}$$

and as this is also true for the coarse and fine products (with the same constant k), x^3k will cancel out in equation 3.7. This does not necessarily apply to equations 3.8, 3.9 and 3.10 because the constant k in equation 3.11 may be particle-size dependent (when the particle shape factor varies with size) and may complicate the integration of equation 3.7. Only when the particle shape factor is constant throughout the size range in question, will equations 3.8, 3.9 and 3.10 also apply to particle size distributions other than those by mass.

3.2.2 Grade efficiency

As the performance of most available separational equipment is highly size dependent (and hence different sizes are separated with different efficiency), the total efficiency E_T defined in equation 3.5 (or 3.6) depends very much on the size distribution of the feed solids and is, therefore, unsuitable as a general criterion of efficiency. Thus values of total efficiency quoted in manufacturers' literature may result in misleading conclusions about the separational capability of equipment unless they are accompanied by the full particle size distribution of the feed solids (and the method of size analysis), the density of the solids and the operational data such as flow rate, temperature, type of fluid, solids input concentration, etc. A single value of the total efficiency cannot be used to deduce the separation capability of the equipment for any materials other than the actual test materials.

If, however, the mass efficiency is found for every particle size x, a curve referred to as the gravimetric grade efficiency function $G(x)$ is obtained which is normally independent of the solids size distribution and density and is constant for a particular set of operating conditions, e.g. fluid viscosity, flow rate and often also solids concentration. It is necessary, however, that the chosen characteristic particle size is a decisive factor in the principle of separation used in the equipment.

If, for instance, the separation effect is influenced only by the mobility of particles in fluids, the terminal settling velocity or Stokes' diameter could be used for the size x and the method of particle size analysis would be chosen accordingly (sedimentation or elutriation).

To make a given grade efficiency curve applicable to solid–fluid density differences $\Delta\rho_2$ and liquid viscosities μ_2 other than those quoted with the curve ($\Delta\rho_1$, μ_1), conversion of the particle size scale can be made assuming Stokes' law, from which

$$\frac{x_1}{x_2} = \sqrt{\left(\frac{\mu_1\Delta\rho_2}{\mu_2\Delta\rho_1}\right)} \tag{3.12}$$

This conversion, however, has to be made with caution and should be avoided wherever possible, not only because of the hidden assumptions in Stokes' law, but also because of the likely changes in the flow patterns in the separator under different viscosities.

Figure 3.3 shows a typical grade efficiency curve for a sedimenting (tubular) centrifuge, and *Figure 3.4* the curves for a depth and a surface-type cartridge

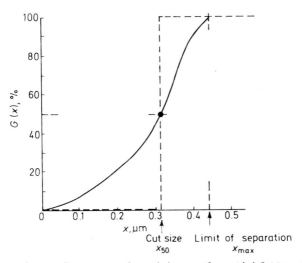

Figure 3.3. A typical grade efficiency curve for a tubular centrifuge, with definitions of x_{50} and x_{max}

filter (clean medium). The grade efficiency curves are usually S-shaped in devices that use either screening (cake filtration, strainers) or particle dynamics in which the body forces acting on the particles (which are proportional to x^3) such as inertia, gravity or centrifugal forces are opposed by drag forces (which are proportional to x^2).

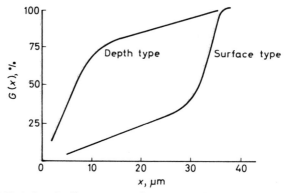

Figure 3.4. Typical grade efficiency curves for two types of cartridge filters (clean medium)[1]

Note that the S-shaped grade efficiency curves do not necessarily start from the origin—in applications with a considerable underflow to through-put ratio (by volume) R_f, the grade efficiency curves tend to the value $G(x) = R_f$ as $x \to 0$. This is a result of the splitting of the flow, or 'dead flux' that carries even the finest solids into the underflow in proportion to the volumetric split of the feed. Section 3.4 discusses possible modifications to the efficiency definitions which account for the volumetric split and illustrate only the net separation effect. Such 'reduced efficiencies' are widely used for hydrocyclones and nozzle-type disc centrifuges where large diluted underflows occur.

The value of the grade efficiency has the character of probability. This may be explained by considering the following: if only one particle of certain size x enters the separator, it will either be separated or it will pass through with the fluid. The grade efficiency will, therefore, be either 100% or 0%. If two particles of the same size enter the separator, the grade efficiency may be 100, 50 or 0% depending upon whether the separator will separate both, one or no particles. If a great number of particles of the same characteristic size enter the separator, a certain probable value of the number of separated particles will be reached.

This probability (not certainty) of the value of the grade efficiency occurs because different particles (of the same size) experience different conditions when passing through the separator. The finite dimensions of the input and output of the separator, uneven conditions for separation at different points in the separator and finally the different surface properties of particles (of the same size) influence the separation process to a great extent and are the reason for the probability character of the grade efficiency values.

It is for these same reasons that the grade efficiency value cannot be determined by a physically exact calculation. If, under certain simplifying assumptions, such a calculation is carried out, the results have to be corrected by coefficients determined experimentally, values of which may be very different from 100% according to the degree of the simplification adopted. Such a procedure can be used with separators that employ inertia principles, sedimentation, centrifugation, etc., where the particle trajectories in the separator can be estimated.

3.2.2.1 Important points on the grade efficiency curve

3.2.2.1.1 Cut size

A graph of the grade efficiency function is sometimes called the partition probability curve because it gives the probability with which any particular size in the feed will separate or leave with the fluid. The size corresponding to 50% probability is called the equiprobable size x_{50} and is often taken as the 'cut' size of the particular equipment (see *Figure 3.3*). Determination of this cut size (which is independent of the feed material) requires a knowledge of the whole grade efficiency curve. There are, however, two other simpler definitions of a cut size, but these give values not necessarily equal to x_{50}.

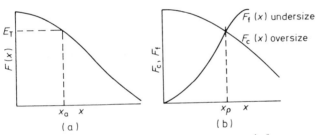

Figure 3.5. Definitions of the analytical cut size x_a and of x_p

The so-called *analytical cut size* x_a is the size such that ideally the feed solids would be split according to size (with no misplaced material) in the proportions given by the total efficiency E_T. In other words the analytical cut size corresponds to the percentage equal to E_T on the cumulative particle size distribution oversize $F(x)$ of the feed material (see *Figure 3.5a*), i.e.

$$F(x_a) = E_T \tag{3.13}$$

It is clear from the definition of x_a that it would be equal to x_{50} if the grade efficiency were a step function (the broken line in *Figure 3.3*) giving an ideally sharp classification of solids (the nearest to this is screening with a uniform aperture size); the two cut sizes would also be equal even for a non-ideal classification if both the coarse and fine products contained an equal quantity of misplaced material (i.e. material consisting of particles finer than x_a in the coarse product or particles coarser than x_a in the fine product). This of course leads to a practical conclusion that the analytical cut size x_a can be used as an estimate of x_{50} in those cases where the classification is very sharp with little misplaced material. The analytical cut size is widely used in particle size measurement.

In some applications, when E_T is unknown, another cut size x_p is sometimes used, defined as the size x at which the cumulative percentage undersize of the coarse fraction is equal to the cumulative frequency oversize of the fines (or vice versa), see *Figure 3.5b*. This cut size, however, is even more sensitive to the changes in the feed size distribution than the analytical cut size.

Further discussion of the cut size and its determination is given in section 3.3.1.5.

3.2.2.1.2 Limit of separation

There is always a value of particle size x above which the grade efficiency is 100% for all x. This is the size x_{max} of the largest particle remaining in the overflow after the separation (maximum particle size that would have a chance to escape) and will be called 'limit of separation' in the following (see *Figure 3.3*).

If the particle trajectories in the separator can be approximated, the most unfavourable conditions of separation are taken for determining this limit of separation. Examples of doing this may be found in theories of separation in centrifuges or settling tanks.

In practice, however, it is often difficult to determine the limit of separation accurately; in that case the size corresponding to 98% efficiency is measured thus giving a more easily defined point. This size x_{98}, sometimes called 'the approximate limit of separation' is widely used, for example in filter rating.

3.2.2.1.3 Sharpness of cut

When the principles of and equipment for the separation of solids from fluids are applied to solids classification, it is desirable to minimize the amount of misplaced material. This is related to the general slope of the grade efficiency curve which can be expressed in terms of a 'sharpness index' defined in many different ways: sometimes simply as the slope of the tangent to the curve at x_{50} or, more often, as a ratio of two sizes corresponding to two different percentages on the grade efficiency curve on either side of 50%, i.e. for example:

$$H_{25/75} = x_{25}/x_{75} \tag{3.14}$$

or

$$H_{10/90} = x_{10}/x_{90} \tag{3.14a}$$

or

$$H_{35/65} = x_{35}/x_{65} \tag{3.14b}$$

or alternatively, the reciprocal values of these.

3.3 BASIC RELATIONSHIPS BETWEEN E_T, $G(x)$ AND THE PARTICLE SIZE DISTRIBUTIONS OF THE PRODUCTS

From the definition in section 3.2.2, the grade efficiency is

$$G(x) = (M_c)_x/(M)_x \qquad (3.15)$$

which by the same argument as that following equation 3.3 in section 3.2 is

$$G(x) = \frac{M_c \, dF_c/dx}{M \, dF/dx} \qquad (3.16)$$

Using equation 3.5, which defines the total efficiency

$$G(x) = E_T \frac{dF_c}{dF} \qquad (3.17)$$

Equation 3.17 (or 3.16) shows how the grade efficiency can be obtained from the size distribution data of the feed and the coarse product, and the total efficiency E_T. It is apparent from equation 3.17 that the particle size distributions of the feed and the coarse product can be by mass, surface area or number as long as they are both by the same quantity as E_T. This is because the size-dependent shape factor would cancel out in the ratio (see equation 3.11). The evaluation of $G(x)$ is shown graphically in *Figure 3.6* where the curves of dF/dx and $E_T \, dF_c/dx$ are plotted in the same diagram. The grade efficiency $G(x)$ is then the ratio of the two values for any particle size x.

Most particle size analysis equipment, however, gives the cumulative size distribution $F(x)$ which is the integral function of the size frequency and has,

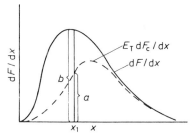

Figure 3.6. Determination of $G(x)$ from E_T, dF/dx and dF_c/dx. $G(x_1) = a/b$

therefore, to be differentiated in order to obtain the size frequency. This differentiation of two curves (feed and coarse product) can be avoided by using the cumulative distributions directly according to equation 3.17. This is shown graphically in *Figure 3.7* by plotting the values of $F(x)$ and $F_c(x)$ against each other for every particle size x and differentiating the curve.

The values of $dF_c(x)/dF(x)$ are then, according to equation 3.17, multiplied by E_T to obtain $G(x)$. Multiplication by E_T can be, of course, done before the differentiation:

$$G(x) = \frac{d[E_T F_c(x)]}{dF(x)} \qquad (3.18)$$

Note that it is of great help in practical evaluations to know the limiting

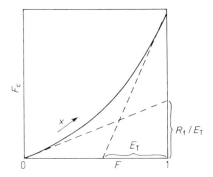

Figure 3.7. Determination of G(x) from E_T, F(x) and $F_c(x)$ (% undersize). ($G(x_1)$ is the slope of the line at x_1 multiplied by E_T)

values of the slope dF_c/dF; a maximum grade efficiency of 1 leads to (see equation 3.17)

$$dF_c/dF = 1/E_T$$

and the minimum grade efficiency equal to R_f leads to

$$dF_c/dF = R_f/E_T$$

If lines of slopes $1/E_T$ (for $x \to \infty$) and R_f/E_T (for $x \to 0$) are plotted in *Figure 3.7* through the diagonally opposite corners of the square corresponding to $x = \infty$ and $x = 0$, they provide the two limiting asymptotes and make it easier to draw a curve through the set of often scattered points (see the example in section 3.3.1.1).

The grade efficiency can also be obtained from the size distributions of the feed $F(x)$ and the fine product $F_f(x)$ or from the fine and coarse products, $F_f(x)$ and $F_c(x)$ from the following relationships, obtained by combining equations 3.17 and 3.7 (mass balance)

$$G(x) = 1 - (1 - E_T)\frac{dF_f(x)}{dF(x)} \qquad (3.19)$$

or

$$\frac{1}{G(x)} = 1 + \left(\frac{1}{E_T} - 1\right)\frac{dF_f(x)}{dF_c(x)} \qquad (3.20)$$

Equations 3.19 and 3.20 can be used with either the size frequencies or the cumulative percentages as is obvious from their mathematical form and is similar to the case of equation 3.17 described above. The following sections show the use of the basic relationships given above and are based entirely on the cumulative size distributions of products because this in practice leads to more accurate results.

3.3.1 Grade efficiency testing and evaluation

From the definition of the grade efficiency it is the efficiency of a separator if a mono-sized material is used as the feed. Such a mono-disperse powder can be prepared, but only with difficulty and in small quantities. If this material was used in the separator, the total efficiency obtained would give a single point on the grade efficiency curve corresponding to the given particle size.

This procedure could be repeated with several other mono-sized fractions of different sizes until enough points on the efficiency curve were obtained. This procedure would be time consuming and expensive. It is for this reason that, as an alternative, equations 3.17, 3.19 and 3.20 are used for grade efficiency testing, depending upon which two out of the three particle size distributions of the materials involved are available. Such an experiment, with a poly-disperse material as the feed, enables us to estimate the grade efficiency in one experiment.

Some experience is necessary in the choice of a suitable feed material. It is first of all necessary that the particles of the feed entering the separator do not agglomerate since the agglomerates would then be separated as larger particles and subsequently broken down to the original particles when the particle size analysis was carried out. The grade efficiency obtained would not correspond to the actual performance of the separator and would give higher values for smaller particles. The feed solids must therefore be well dispersed in the liquid and the concentration should be sufficiently low, say one or two per cent by volume at the most, for the number of impacts and interactions between particles to be small. Apart from particle interactions leading to hindered settling, higher concentrations of particles cause changes in the flow patterns in the separator thus affecting its performance. Tests at higher concentrations can, of course, also be carried out but the results cannot be reliably applied to materials and concentrations other than those used in the tests.

Depending on the actual arrangement of the testing rig, two of the three material streams involved (feed, coarse or fine product) are analysed for the grade efficiency determination and the appropriate equation 3.17, 3.19 or 3.20 used for evaluation. The most frequent combination is the feed and coarse product because of the convenience of their collection and size determination; it is not unusual to analyse all three materials because this provides a useful check on the mass balance and on the overall accuracy of the results (see section 3.2.1, equation 3.10).

The total efficiency can be determined from three different combinations of the material streams involved (equation 3.5 combined with the mass balance in equation 3.1); the mass flow rates of the solids are usually determined by measuring the total volumetric flow rates and the corresponding solids concentrations; the mass flow rate is, of course, the product of the two. The combination giving the smallest standard deviation of E_T (if all the variables are subject to random errors only) is that of the overflow and underflow streams. (The same conclusion applies to the testing of $G(x)$.)

The evaluation of the grade efficiency may be carried out graphically, in a table or using a computer. Tabular procedures usually do not give results of great accuracy often because only a relatively low number of points are available. The graphical methods are most versatile and instructive, and examples of these are given in the following sections. Simple computer programs can be written to carry out the task and these may save a great deal of time and effort in routine work. The author has successfully used a simple computing technique of fitting a second-order polynomial through three adjacent points in the appropriate square diagram and computing the differentiation for the middle point; this is done successively through the range of data, starting and finishing in diagonally opposite corners of the square diagram. This technique provides results which correspond very favourably with those obtained by graphical methods even if the number of data points available is as low as four or five.

Examples of grade efficiency evaluations from different combinations of the measured data are given in the following sub-sections 3.3.1.1–3.3.1.4. These are all based on a single test in which all three streams were analysed for particle size distribution and the total efficiency was also measured.

Example 3.1

A hydrocyclone was tested at certain operating conditions with a suspension of clay in water. Using the data given below and in *Table 3.1*, evaluate the

Table 3.1

Particle size of test powder, μm	Feed material $F(x)$, % oversize	Coarse product $F_c(x)$, % oversize	Fine product $F_f(x)$, % oversize
40	1	2	0.1
20	12	22	0.2
10	31	55	5
4	48	71	20
2	62	80	38

grade efficiency curve for the given operating conditions, and as a function of particle size for a density difference $(\rho_s - \rho_l)$ of 1000 kg m^{-3}.

Data: Density of solids $\rho_s = 2640$ kg m^{-3}
Density of water $\rho_l = 1000$ kg m^{-3}
Underflow-to-throughput ratio $R_f = 26.4\%$ v/v
Total efficiency $E_T = 54.8\%$

3.3.1.1 Evaluation of $G(x)$ from $F(x)$, $F_c(x)$ and E_T

Equation 3.17 is used to evaluate $G(x)$; $F_c(x)$ is plotted against $F(x)$ in a square diagram in *Figure 3.8* and the corresponding sizes marked beside each of the five points. Limiting lines are plotted, one at a slope of $R_f/E_T = 0.482$ through the point corresponding to $x = 0$ in the top right-hand corner, the other at a slope of $1/E_T = 1.825$ through the point of $x = \infty$ in the bottom left-hand corner of the square diagram. A curve is drawn through the data points so that it is asymptotic to the limiting lines. Tangents are then drawn to the curve at each data point in turn and their slopes measured. Following equation 3.17, the measured gradients are multiplied by E_T thus giving the values of $G(x)$ for the test powder—see *Table 3.2*, column 3 for results (section 3.3.1.3). For a particle size of 4 μm for example, the slope measured is 0.777, thus

$$G(4) = 0.777 \times 54.8 = 42.6\%$$

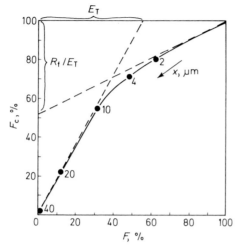

Figure 3.8. An example of G(x) determination from F_c, F (both oversize) and E_T

If the grade efficiency curve is now to be plotted against particle size for a material with a density difference of $(\rho_s - \rho_l) = 1000 \text{ kg m}^{-3}$, equation 3.12 is used to determine the particle size scale conversion. The size of particle which has the same grade efficiency as a particle of 4 μm in the test powder (assuming the viscosity remains the same) is given by

$$x_1 = x_2 \sqrt{\left(\frac{\Delta\rho_2}{\Delta\rho_1}\right)} = 4\sqrt{\frac{1640}{1000}} = 5.1 \text{ μm}$$

Figure 3.9 shows the resulting grade efficiency curve, which must go through the point of 26.4% $(= R_f)$ at $x = 0$.

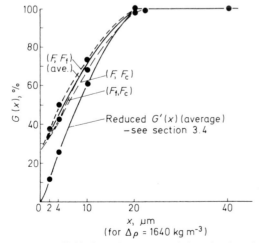

Figure 3.9. Plots of G(x) from Table 3.2, columns 3–6 and the reduced grade efficiency G'(x) from equation 3.34

Note that since particle size analysis data are subject to relatively large errors, the plot in *Figure 3.8* may show a considerable scatter. At best, a smooth curve is drawn through the scattered data points; the limiting lines provide useful guide-lines for this.

3.3.1.2 Evaluation of G(x) from F(x), $F_f(x)$ and E_T

If $F_f(x)$ is measured instead of $F_c(x)$, the following procedure can be adopted for the evaluation of $G(x)$. The evaluation is based on equation 3.19; $F_f(x)$ is plotted against $F(x)$ in a square diagram (*Figure 3.10*) in the same way as $F_c(x)$ in section 3.3.1.1. One limiting line is the x-axis ($F_f(x) = 0$) so that equation 3.19 gives $G(\infty) = 1$, the other, for the condition $G(0) = R_f$ at $x = 0$ has a slope of

$$\frac{(1 - R_f)}{(1 - E_T)}$$

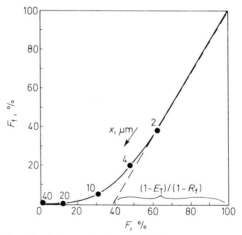

Figure 3.10. Determination of G(x) from E_T, F(x) and $F_f(x)$

and is drawn through the top right-hand corner of the diagram; the data line should again be asymptotic to both lines.

Tangents are then drawn to the curve at each data point in turn and their slopes measured. Following equation 3.19, the measured gradients are multiplied by $(1 - E_T)$ and then subtracted from 1 to give $G(x)$. For example, if the slope at 4 μm is 1.1

$$G(4) = 1 - (1 - 0.548)\,1.1 = 0.503 = 50.3\%$$

The resulting $G(x)$ (*Table 3.2*, column 4) is plotted in *Figure 3.9* and, as it applies to the same test as $G(x)$ from section 3.3.1.1, the two should be comparable.

The conversion of particle sizes is not given here as it is described in section 3.3.1.1.

3.3.1.3 Evaluation of G(x) from $F_c(x)$, $F_f(x)$ and E_T

If the size distributions of the products $F_c(x)$ and $F_f(x)$ are measured, equation 3.20 is used to evaluate $G(x)$. Values of dF_f/dF_c are again obtained from a square diagram—*Figure 3.11*—by drawing tangents to the curve plotted

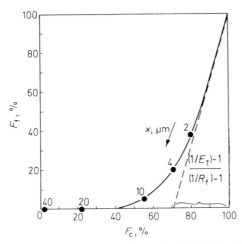

Figure 3.11. Determination of G(x) from E_T, $F_f(x)$ and $F_c(x)$

through the set of data points. The limiting lines in this case are the x-axis ($F_f(x) = 0$) for $x = \infty$ and a line at a slope of

$$\frac{(1/R_f) - 1}{(1/E_T) - 1}$$

drawn through the point corresponding to $x = 0$, in this case the top right-hand corner. The slopes measured, corresponding to given particle sizes, are then processed according to equation 3.20 so that for example, if the slope at 4 μm is 1.62, then

$$\frac{1}{G(x)} = 1 + \left(\frac{1}{0.548} - 1\right)1.62 = 2.336$$

hence

$$G(x) = 0.428 = 42.8\%$$

The rest of the results are shown in *Table 3.2*, column 5; the conversion of particle sizes is the same as in section 3.3.1.1. *Figure 3.9* shows the points from columns 3, 4 and 5 of *Table 3.2* all plotted in the same graph, and the full line represents the average $G(x)$ from column 6. The scatter of points in *Figure 3.9* gives some indication of the effect of errors in particle size analysis—doing all three analyses and evaluations as in this example, somewhat improves the accuracy of the resulting function. The best way of reducing random errors is of course repetition of the tests.

Table 3.2

Particle size of test powder $(\rho_s - \rho_l = 1640 \text{ kg m}^{-3})$, μm	Particle size $(\rho_s - \rho_l = 1000 \text{ kg m}^{-3})$, μm	G(x) from F, F_c, E_T, %	G(x) from F, F_f, E_T, %	G(x) from F_c, F_f, E_T %	G(x) average, %	G(x) from F_c, F_f, F, %
40	51.2	100.0	100.0	100.0	100.0	(91.2)
20	25.6	98.0	100.0	100.0	99.3	97.5
10	12.8	72.2	73.3	68.4	71.3	68.5
4	5.1	42.6	50.3	42.8	45.2	42.7
2	2.6	33.5	37.7	34.1	35.1	34.9
0	0	26.4	26.4	26.4	26.4	26.4

3.3.1.4 Evaluation of $G(x)$ from $F(x)$, $F_c(x)$ and $F_f(x)$

The grade efficiency curve can be determined from particle size analyses of the three streams alone, without the need to measure E_T

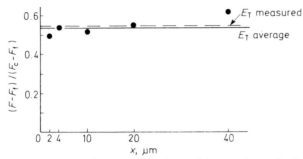

Figure 3.12. Variation in E_T derived from the mass balance of the particle size distributions of the three material streams

The basic equation for this may be derived from equation 3.9 by elimination of E_T using equation 3.17, this gives

$$G(x) = \left(\frac{F(x) - F_f(x)}{F_c(x) - F_f(x)}\right)\left(\frac{dF_c(x)}{dF(x)}\right) \tag{3.21}$$

which is in fact identical to equation 3.17 if equation 3.10 is substituted for E_T; as a result of the inaccuracy of particle size measurement and sampling, there is always a scatter in these 'derived' values of E_T if plotted against particle size x—see *Figure 3.12* for the example given previously (see also *Table 3.3*). The average value of E_T derived from the mass balance of the particle size analysis data using equation 3.10 is 53.7% and is in fact lower than the value of E_T measured directly, 54.8%; this indicates the errors in the measurement of the data for the evaluation of both of these values.

Equation 3.21 suggests $G(x)$ can be determined from the individual 'derived' values for the total efficiency together with the slopes dF_c/dF which can be determined graphically as described in section 3.3.1.1; column 7 in *Table 3.2* gives the results which are not very different from the results from the three previous sections (except for the point for $x = 40\,\mu m$ where the mass balance from the particle size distributions is at its worst—see *Figure 3.12*).

3.3.1.5 Simple evaluation of the cut size

If the complex separation performance of a size-dependent separator (best shown by a full grade efficiency curve) has for some reason to be described by a single number, in most applications it is best to use the concept of a cut size. In such a case, the cut size has to be unique and, if possible, independent of the size distribution of the feed solids, so that it characterizes only the machine run under the chosen conditions of operation. As the whole of the grade efficiency curve usually satisfies this condition, it is best to define the cut size as corresponding to a unique point on the curve. The equiprobable size x_{50} is probably best suited for this purpose and it is certainly the best of the three definitions of cut size reviewed here (section 3.2.2.1).

A simple way has recently been suggested[1] of determining x_{50} without the need for the whole grade efficiency curve. If $G(x)$ in equation 3.17 is equal to 0.5, it can be shown that the cut size x_{50} corresponds to the point on a plot of $(F-2E_\mathrm{T}F_\mathrm{c})$ at which:

$$\frac{\mathrm{d}}{\mathrm{d}x}(F-2E_\mathrm{T}F_\mathrm{c})=0 \qquad (3.22)$$

which corresponds to the maximum.

Alternative combinations of the material streams (equations 3.19 and 3.20) give:

$$\frac{\mathrm{d}}{\mathrm{d}x}[F-2(1-E_\mathrm{T})F_\mathrm{f}]=0 \qquad (3.23)$$

and

$$\frac{\mathrm{d}}{\mathrm{d}x}\left[F_\mathrm{c}-\frac{1-E_\mathrm{T}}{E_\mathrm{T}}F_\mathrm{f}\right]=0 \qquad (3.24)$$

Equations 3.22, 3.23 and 3.24 require plotting the composite functions of two solid size distributions and the total efficiency and finding the maximum (see the worked example elsewhere[2]), which requires no differentiation and can be performed easily in industrial production situations. Naturally, the test information necessary for this calculation is the same as for the grade efficiency curve itself: two particle size distributions and the total efficiency E_T.

The values of the other two commonly used cut sizes, the analytical cut size x_a and the 'curve intersection' cut size x_ρ (as defined in section 3.2.2.1.1) are always different from x_{50}, unless x_{50} is equal to the median of the feed size distribution x_g when all three coincide; the difference between x_{50}, and x_a and x_ρ increases as the median of the feed moves towards either end of the grade efficiency curve. This is best shown in the plot of $G(x)$ against $F(x)$ in *Figure 3.13*, from which it can be seen that both x_a and x_ρ are always on the same side of x_{50} as the median x_g of the feed, with x_a being always closest to x_{50}, hence:

if $x_\mathrm{g} < x_{50}$
$$x_\mathrm{g} < x_\rho < x_\mathrm{a} < x_{50}$$
if $x_\mathrm{g} = x_{50}$
$$x_\mathrm{g} = x_\rho = x_\mathrm{a} = x_{50}$$
and if $x_\mathrm{g} > x_{50}$
$$x_\mathrm{g} > x_\rho > x_\mathrm{a} > x_{50}$$

Theoretical conversion of the easily obtainable analytical cut size x_a into the equiprobable size x_{50} is possible if both the feed size distribution and the grade efficiency curve can be approximated by an analytical function. Thus for example if both of the above-mentioned functions are log–normal it can be shown that the total efficiency E_T can be determined analytically from the following expression[3]

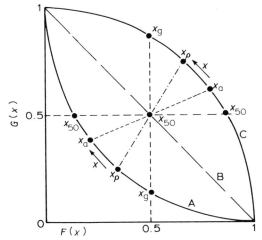

Figure 3.13. Plot of G(x) against F(x) for three different feed size distributions F(x), showing the relative position of the three cut sizes x_{50}, x_a and x_p with respect to the feed median x_g. Curve A—fine feed, i.e. $x_g < x_{50}$. Curve B—$x_g = x_{50}$. Curve C—coarse feed, i.e. $x_g > x_{50}$

$$E_T = \tfrac{1}{2} + \tfrac{1}{2} \operatorname{erf} \frac{\ln x_g - \ln x_{50}}{\sqrt{(2)}\sqrt{(\ln^2 \sigma_g + \ln^2 \sigma_{gs})}} \qquad (3.25)$$

where

x_g and σ_g are the median and geometric standard deviation of the feed size distribution respectively,

x_{50} and σ_{gs} are the equiprobable size (cut size) and the geometric standard deviation of the grade efficiency function respectively, and

the erf function is defined as

$$\operatorname{erf}(z) = \frac{2}{\sqrt{\pi}} \int_0^z e^{-t^2} \, dt \qquad (3.26)$$

Equation 3.25 can be solved using tables of the erf function. There are also some analytical approximations available which obviate the use of tables.

By definition of the analytical cut size x_a the cumulative percentage oversize of the feed at x_a is equal to the total efficiency E_T (equation 3.13) and hence, by integration of equation 2.21,

$$\tfrac{1}{2} - \tfrac{1}{2} \operatorname{erf} \frac{\ln x_a - \ln x_g}{\sqrt{2} \ln \sigma_g} = E_T \qquad (3.27)$$

Equation 3.27, in combination with equation 3.25, gives the final conversion formula

$$\frac{x_{50}}{x_g} = \left(\frac{x_a}{x_g}\right)^{\sqrt{\left[\left(1 + \frac{\ln \sigma_g}{\ln \sigma_s}\right)^2\right]}} \qquad (3.28)$$

The above equation can be used to calculate the true cut size x_{50} from the known size distribution of the feed defined by x_g and σ_g and from the simply measured analytical cut size x_a providing that an estimate of the geometric standard deviation σ_{gs} of the separator grade efficiency curve is available. Such information can be found in the literature. Gibson[4], for example, has recently given values of σ_{gs} for a disc centrifuge and a hydrocyclone under different operating conditions.

Better still, equation 3.25 can be used directly to calculate x_{50} from E_T but this involves the inverse of the error function.

Table 3.3 gives examples of three sets of E_T, x_{50}, x_a and x_p calculated for three different log–normal feed size distributions if processed by a log–normal grade efficiency. It can be clearly seen that, as predicted earlier in

Table 3.3. THREE DIFFERENT LOG–NORMAL SIZE DISTRIBUTIONS (FEED) WITH THE SAME $\sigma_g = 2.05$ AND MEDIANS 10, 15 AND 20 μm PROCESSED BY A LOG–NORMAL GRADE EFFICIENCY OF $\sigma_g = 1.65$ AND $x_{50} = 10\,\mu$m

Feed median x_g, μm	10	15	20
E_T, %	50.0	67.8	78.6
x_{50}	10.0	10.0	10.0
x_a	10.0	10.8	11.3
x_p	10.0	12.0	14.0

this section (*Figure 3.13*), the closer is the feed median to x_{50} the better is the agreement between x_{50}, x_a and x_p. x_a and x_p are always on the same side of x_{50} as the feed median x_g. The values of E_T were calculated from equation 3.25 such as, for example, for $x_g = 20$

$$E_T = \tfrac{1}{2} + \tfrac{1}{2}\,\mathrm{erf}\ \frac{\ln 20 - \ln 10}{\sqrt{(2)}\sqrt{(\ln^2 2.05 + \ln^2 1.65)}}$$

$$= \tfrac{1}{2} + \tfrac{1}{2}\,\mathrm{erf}\ 0.56$$

$$= 78.6\%$$

using standard tables of the erf function defined in equation 3.26. The analytical cut size x_a was calculated using equation 3.28, which for $x_g = 20$ again gives

$$x_a = 20 \times \left(\frac{10}{20}\right)\ \frac{1}{\sqrt{\left(1 + \left(\dfrac{\ln 1.65}{\ln 2.05}\right)^2\right)}}$$

$$= 11.33\ \mu\mathrm{m}$$

There is no simple way to determine the 'intersection' cut size x_p so this was done by calculating the size distributions of the products and finding their cross-over point as required by the definition shown in *Figure 3.5*.

Similar correction factors can be derived for other types of analytical functions that describe the feed and the grade efficiency curve.

3.3.2 Total efficiency determination from the grade efficiency and the size distribution of the feed

If the grade efficiency curve can be regarded as a characteristic parameter of a separator for particular conditions (flow rate, viscosity of liquid etc.), this curve can be used to determine the total efficiency that can be expected to be obtained with a particular feed material under the same conditions. The size distribution $dF(x)/dx$ of the feed material must be known, then, from equation 3.17:

$$G(x)\, dF = E_T\, dF_c \qquad (3.29)$$

which after integration in the given size range $0 - x_{max}$ gives

$$\int_0^1 G(x)\, dF = E_T \int_0^1 dF_c \qquad (3.30)$$

since the value of the total efficiency is a constant for a given application. From the definition of the particle size distribution

$$\int_0^1 dF_c = 1 \qquad (3.31)$$

which makes the integral on the right-hand side of equation 3.30 equal to 1. Hence the total efficiency is

$$E_T = \int_0^1 G(x)\, dF \qquad (3.32)$$

Evaluation of the integral can be carried out graphically, in a table or using a computer.

For graphical evaluation the values of $G(x)$ and F are plotted against each other (for every size x) in a square diagram—see *Figure 3.14*—and the area under the curve is found. The total efficiency is a ratio of this area to the area of the square (which would represent a total efficiency of 100% if $G(x)$ were 100% for all particle sizes).

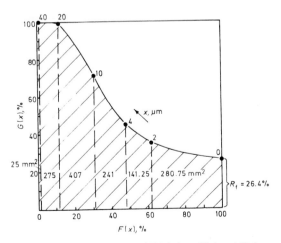

Figure 3.14. Prediction of E_T and $F_c(x)$ from $G(x)$ and $F(x)$
Total area (shaded) is 1370 mm², the area of the square is 2500 mm², hence $E_T = 1370/2500 = 0.548$
$F_c(2) = (280.75/2500)\,(1/0.548) = 0.205 = 20.5\%$ undersize or 79.5% oversize
$F_c(4) = (280.75 + 141.25)/(1/0.548) = 30.8\%$ undersize or 69.2% oversize etc.

If the size frequency curve dF/dx is to be used, the evaluation is carried out according to the modified equation 3.19:

$$E_T = \int_0^\infty G(x) \frac{dF(x)}{dx} dx \tag{3.33}$$

which represents multiplication of the values of $G(x)$ and dF/dx for every size x and then integration of the resulting curve.

If a digital computer is to be used, linear interpolation between the points in a plot of $G(x)$ against $F(x)$ may not give very accurate results because of the limited number of points on both the size distribution and the grade efficiency curve. The author has successfully used a technique for this integration that first finds parameters for a polynomial that can be fitted through every three adjacent points and then integrates the function obtained in this way step by step.

If both the feed size distribution $F(x)$ and the grade efficiency curve $G(x)$ can be approximated by an analytical function, the integration in equation 3.32 can be done analytically. Equation 3.25 can, for example, be used for log-normal functions—see section 3.3.1.5.

Example 3.2

Estimate the total efficiency that can be expected with a hydrocyclone which is identical with the one tested in section 3.3.1 and operated under the same conditions as in example 3.1, with feed material of a size distribution $F(x)$ as given in *Table 3.2*. Use the average grade efficiency curve obtained from the tests in section 3.3.1, column 6 in *Table 3.2*.

The solution is based on equation 3.32; *Figure 3.14* gives the required plot and yields, after integration of the curve (shaded area) $E_T = 54.8\%$ which is identical with the measured value given in section 3.3.1.

3.3.3 Determination of the size distribution of the products from the grade efficiency and the size distribution of the feed

Apart from the total efficiency of a separator, the quality of the products can also be predicted provided that the grade efficiency curve is known for the particular conditions of separation.

For determination of the size distribution of the coarse product equation 3.29 may be rearranged as

$$dF_c(x) = \frac{G(x)}{E_T} dF(x) \tag{3.34}$$

from which the value of $F_c(x_1)$ for any x_1 may be evaluated by integration

$$F_c(x_1) = \frac{1}{E_T} \int_0^{F(x_1)} G(x) \, dF(x) \tag{3.35}$$

provided that the total efficiency has already been found as in the previous section. Equation 3.35 may be evaluated graphically in a similar way to the total efficiency, but in this case the integration is carried out up to several different values of x until enough points on the cumulative size distribution of the coarse product are obtained—see *Figure 3.14* for an example.

If a size frequency curve dF/dx is to be used instead of the cumulative curve, the evaluation is simpler since, according to equation 3.31

$$dF_c/dx = \frac{1}{E_T} G(x) \frac{dF(x)}{dx} \tag{3.36}$$

i.e. the product of the grade efficiency and the size distribution for any x divided by the total efficiency gives the size frequency distribution of the coarse product.

For determination of the size distribution of the fine product, the procedure is very similar. From equations 3.11 and 3.16

$$dF_f(x) = \frac{1 - G(x)}{1 - E_T} dF(x) \tag{3.37}$$

which after integration gives

$$F_f(x_1) = \frac{1}{1 - E_T} \int_0^{F(x_1)} [1 - G(x)] \, dF(x) \tag{3.38}$$

or for dealing with size frequency curves

$$\frac{dF_f(x)}{dx} = \frac{1 - G(x)}{1 - E_T} \frac{dF(x)}{dx} \tag{3.39}$$

These two equations 3.38 and 3.39 may be evaluated in a similar manner to equations 3.35 and 3.36.

Example 3.3

Taking the data given in example 3.1, equation 3.35 is used for prediction of the particle size distribution of the separated coarse product. *Figure 3.14* shows the graphical method of solving the equation step by step where each integration is performed up to a given particle size x and the area obtained as a fraction of the total area under the curve gives the required cumulative per-centage $F_c(x_1)$.

For the size distribution of the fine product in the overflow, equation 3.38 is used directly; graphically, *Figure 3.14* can also be used for this except that in this case the areas above the curve are integrated. Alternatively, if the coarse product distribution has been worked out first, the mass balance in

Table 3.4

Particle size, μm	Coarse product $F_c(x)$, %	Fine product $F_f(x)$, %
40	1.8	0.0
20	22.0	0.0
10	51.6	6.0
4	69.2	22.2
2	79.5	40.7

equation 3.9 can be used to evaluate the fine product size distribution instead. *Table 3.4* gives the resulting size distribution worked out from the data given in example 3.1, which constitute the results of the tests in section 3.3.1. There are, of course, some discrepancies between the measured data in *Table 3.1* and the predicted values in *Table 3.4*; this is due to inaccurate mass balance in the experimental data (a result of measurement errors).

Note that as the total efficiency is needed for a prediction of the size distributions of the products, it must always be evaluated first using the method given in section 3.3.2.

3.4 MODIFICATIONS OF EFFICIENCY DEFINITIONS FOR APPLICATIONS WITH AN APPRECIABLE UNDERFLOW-TO-THROUGHPUT RATIO

3.4.1 Reduced efficiency

As was pointed out in section 3.2.2, in applications with an appreciable and diluted underflow, a 'reduced' efficiency concept is used if one wants to look at the net separation effect alone. The necessity for this modified efficiency springs from the simple fact that, as there is always some liquid accompanying the solids in the underflow, the total flow is split into two streams so that a certain 'guaranteed' efficiency is always achieved as a result of this split. In other words, the separator functions as a flow divider and divides the solids too, in at least as large a ratio as R_f the volumetric ratio of underflow to throughput. When looking at the performance of separators, it is then desirable to observe the net separation effect, i.e. to subtract the contribution of the dead flux. This gave rise to a number of possible new definitions of efficiency, reviewed by Tenbergen and Rietema[5]. The best and most widely used formula is one due to Kelsall[6] and also Mayer[7].

$$E'_T = \frac{E_T - R_f}{1 - R_f} \qquad (3.40)$$

where

E'_T is the so-called 'reduced' total efficiency

E_T is the total efficiency as defined by equation 3.5 (ratio of mass flow rates of solids) and

$R_f = U/Q$ is the underflow-to-throughput ratio (by volume) which is the minimum efficiency due to dead flux.

It can be seen that equation 3.40 satisfies the basic requirements for a definition of net efficiency in that it gives zero for conditions of no separation when $E_T = R_f$ and one for complete separation of solids when $E_T = 1$.

The grade efficiency curve $G(x)$, which in its unmodified form (defined by equation 3.15) also obscures the existence of a volumetric split of the flow, can be modified in the same way:

$$G'(x) = \frac{G(x) - R_f}{1 - R_f} \qquad (3.41)$$

which represents a net separation effect and, for inertial separation, goes through the origin since $G'(0) = 0$ (whereas $G(0) = R_f$, see *Figure 3.9*).

The size corresponding to $G'(x) = 50\%$ is used as the 'reduced' cut size x'_{50}.

This 'reduced' efficiency concept is widely used in hydrocyclones; the effect of this modification on the shape of the grade efficiency curve is shown in *Figure 3.9* which uses the average curve of $G(x)$ from *Table 3.2* (see section 3.3.1). It should be noted that the basic relationship between the total and grade efficiencies (equation 3.32) also holds for reduced efficiencies, so that:

$$E'_T = \int_0^1 G'(x)\,dF \qquad (3.42)$$

Note that the product size distributions cannot be calculated without the knowledge of R_f because equation 3.17 does not hold for $G'(x)$ and E'_T.

It is interesting to find that the often used clarification number[8] is in fact equal to the reduced total efficiency E'_T because it can be shown from the definition of E'_T that

$$E'_T = \frac{C - C_0}{C} \text{ (the clarification number)} \qquad (3.43)$$

(equations 3.42 and 3.43 indicate that C_0 is not affected by R_f if $G'(x)$ is not) or, alternatively:

$$E'_T = \left(\frac{C_u - C}{C}\right)\left(\frac{R_f}{1 - R_f}\right) \qquad (3.44)$$

or

$$E'_T = \frac{C_u - C_0}{C_u - C_0 + (C_0/R_f)} \qquad (3.45)$$

which offers three alternatives for convenient measurement of the reduced efficiency from the measurement of the concentration of solids C, C_u and C_0 in the feed, underflow and overflow respectively, and of the volumetric split R_f.

3.4.2 Other definitions of efficiency

As was explained in chapter 1, solid–liquid separation is often assessed by considering the solids recovery *and* the moisture content of the recovered solids simultaneously, with their relative importance varying with application. If a single parameter which combines both of these criteria together is required, some efficiency definitions originally proposed for powder classification may be used, as below. In this case it is advantageous to redefine the description of the three streams involved as follows:

M_F is the mass flow rate of the feed suspension
M_O is the mass flow rate of the overflow suspension
M_U is the mass flow rate of the underflow suspension
y_F is mass fraction of solids in the feed
y_O is mass fraction of solids in the overflow
y_U is mass fraction of solids in the underflow

In terms of the three mass fractions y_F, y_U and y_o, Newton[9] efficiency E_N is defined as:

$$E_N = \frac{(y_U - y_F)(y_F - y_o)}{y_F(1 - y_F)(y_U - y_o)} \tag{3.46}$$

which can be rewritten in another form:

$$E_N = \frac{M_U \cdot y_U}{M_F \cdot y_F} - \frac{M_U}{M_F} \frac{(1 - y_U)}{(1 - y_F)} \tag{3.47}$$

The latter form of Newton efficiency shows the physical significance because it is clearly the mass recovery of the solids E_T minus the undesirable 'recovery' of liquid, both in the underflow. This efficiency definition was selected as the best available by Tengbergen[5] but, more recently, Ogawa et al.[10] have pointed out that it gives uneven sensitivity to small changes in the mass fractions of solids in either the underflow or in the overflow (as can be seen from equation 3.46). In order to make the sensitivity more uniform over the whole range of possible values of y_U and y_o, Ogawa et al. derived another definition for separation efficiency, using the information theory, in the following form:

$$E_0 = 1 - \frac{(y_F - y_o)\phi(y_U) + (y_U - y_F)\phi(y_o)}{(y_U - y_o)\phi(y_F)} \tag{3.48}$$

where

$$\phi(y) = -y \ln y - (1 - y) \ln (1 - y) \tag{3.49}$$

Although this definition is an improvement on Newton efficiency with regard to the above-mentioned sensitivity, it is very complicated algebraically (it fails to satisfy one of the criteria laid down by Tengbergen[5] in that it should be simple) and it is therefore doubtful that it will be used in practice.

A far simpler solution to the sensitivity problem is in another possible definition in the form[11]:

$$E_s = \frac{(y_U - y_F)(y_F - y_o)}{(1 - y_F)y_F} \tag{3.50}$$

which gives a completely uniform sensitivity to changes in y_U and y_o over the whole range (as equation 3.50 is a straight line for either $E_s = f(y_U)$ or $E_s = f(y_o)$). The physical significance of E_s can be shown from equation 3.50: it is a product of

$$\frac{y_U - y_F}{y_F}$$

(equal to the relative improvement in the solids content of the underflow) and

$$\frac{(1 - y_o) - (1 - y_F)}{1 - y_F}$$

(equal to the relative improvement in water content of the overflow).

The three efficiency definitions quoted in this section suffer from the same disadvantage as for the total gravimetric efficiency in section 3.2.1 because their values for any specific equipment depend on the size distribution of the feed solids and they are therefore unsuitable as a general criterion of efficiency. This is, of course, unless they could be used in a differential form defined in the same way but for different particle sizes in turn; only Newton efficiency can be rearranged for that purpose[5].

It is fitting to end this review of efficiency definitions with a rather speculative but theoretically very basic definition. The thermodynamic efficiency may be defined as the basic energy of mixing of the solids in a fluid, in relation to the work done in a separator to 'unmix' the suspension. The former is as yet impossible to establish because little is known quantitatively about the thermodynamics of solid–fluid systems and the thermodynamic efficiency therefore remains a purely theoretical concept which might be worth studying in the future (see Chapter 23, Part 2).

3.5 THE USE OF SEPARATORS IN SERIES AND IN MULTIPLE PASS SYSTEMS

3.5.1 Two separators in series

It is often advantageous to use two or more separators in series; the first stage in the form of a low-cost, usually dynamic separator such a hydrocyclone or settling tank is used to remove the grit, and the second stage, a filter or a centrifuge is then used to remove the finer fractions. The first stage reduces the solids loading for the second, more efficient stage, with the additional benefit of the overall efficiency being higher than if the second stage were used on its own. *Figure 3.15* shows a schematic diagram of such an arrangement. The combined grade efficiency curve for the whole system is[12]:

$$G(x) = G_1(x) + G_2(x) - G_1(x)G_2(x) \tag{3.51}$$

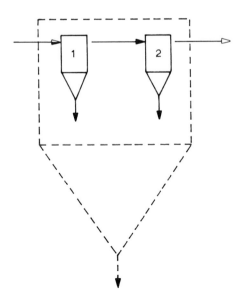

Figure 3.15. Schematic diagram of two separators in series

For n number of separators with identical grade efficiency curves $G_1(x)$ equation 3.51 becomes

$$G(x) = 1 - [1 - G_1(x)]^n \tag{3.52}$$

There is a law of diminishing returns in force here and it would in practice make little sense to use more than three separators with identical $G(x)$ (and therefore also the same cut size x_{50}) in series because the addition of another separator would increase the efficiency at 50% only by about 6%, which would not justify the additional capital and running costs.

For n greater than about 25, equation 3.52 can be modified[13] to

$$G(x) = 1 - \exp[-nG_1(x)]$$

3.5.2 A concentrator with a separator in series

Some separators are really used as concentrators, that is, the separated solids are not in the form of a cake but leave the separator still suspended in a liquid. In other words, a certain amount, say 5–10%, of the feed liquid flow rate leaves with the solids and a second stage separator has to be used to collect the solids from this 'underflow' stream.

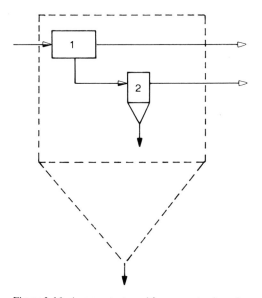

Figure 3.16. A concentrator with a separator in series

Thickeners and hydrocyclones are used in this way; *Figure 3.16* shows such a system schematically. The combined grade efficiency is then given by[12]:

$$G(x) = G_1(x) \cdot G_2(x) \tag{3.53}$$

and this cannot be better than $G_1(x)$; the overall performance can therefore only be as good and no better than the efficiency of the concentrator.

3.5.3 A concentrator with a separator in series, with feedback

If the efficiency of the separator (number 2 in *Figure 3.16*) is not sufficiently high, its overflow can be connected into the inlet of the concentrator, thus producing a feedback. *Figure 3.17* shows this schematically; the combined grade efficiency of the whole system is[12]:

$$G(x) = \frac{G_1(x)G_2(x)}{1 - G_1(x) + G_1(x)G_2(x)} \tag{3.54}$$

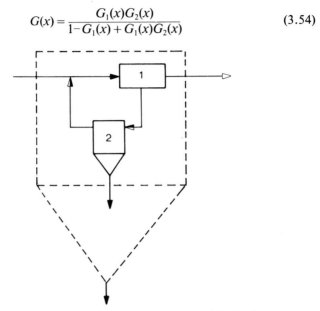

Figure 3.17. A concentrator with a separator in series, with feedback

This is always greater than $G_1(x) \cdot G_2(x)$ (from equation 3.53), hence the feedback always gives better overall performance, except for $G_2 = 100\%$ (when there is no point in having feedback), at the additional expense of a larger concentrator (and higher running costs) because it has to treat the extra recycle flow rate.

3.5.4 Multiple pass systems

The overall efficiency of a system can be increased using a multiple pass arrangement like the one shown in *Figure 3.18*. Part of the overflow is recycled back and mixed with the fresh feed before it enters the separator again. This mixing could in theory be done in a simple T-piece, but it is usual to use a sump or a reservoir in order to equalize the pressures and a pump is then needed to overcome the resistance of the separator—see *Figure 3.18*. If the fresh feed flow rate is q and the throughput of the separator is Q then the grade efficiency for the whole system can be shown[14] to approach an equilibrium value (neglecting the volume of the underflow)

$$G(x) = \frac{1}{1 + \frac{1 - G_s}{G_s} \frac{q}{Q}} \tag{3.55}$$

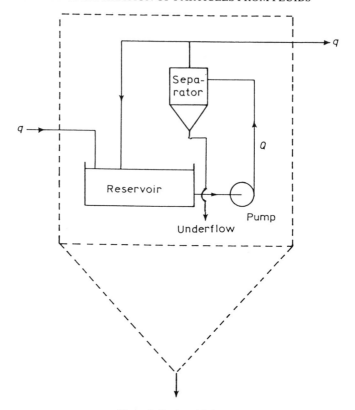

Figure 3.18. A multiple pass system

where G_s is the grade efficiency of the separator used. As q/Q is less than 1 it is clear from equation 3.54 that $G(x)$ is always greater than G_s. The separator, however, has to process greater flow rate at a lower feed concentration than it would in a one-pass system. The hydrocyclone is obviously the strongest contender here due to its ability to handle large flow rates and solid throughputs continuously and due to its increased efficiency with diluted feeds[14], but other types of separators or filters can be used.

REFERENCES

1. Trawinski, H., *Aufbereitungs-Technik,* **17** (5), 449–459 (1975)
2. Svarovsky, L., *'Measurement of efficiency of gas cleaning equipment',* Proc. Filtration Society's Conference on Filtration, Olympia, London, September 20–22 (1977) Also in: *Filtration and Separation,* **15** (4), 355–359 (1978)
3. Allander, C. G., *Staub-Reinhaltung Der Luft,* **18** (1), 15–17 (1958)
4. Gibson, K., *'Large scale tests on sedimenting centrifuges and hydrocyclones for mathematical modelling of efficiency',* Proc. of the Symposium on Solid–Liquid Separation Practice, Yorkshire Branch of the I. Chem. E., Leeds, England, March 27–29 (1979), pp. 1–10
5. Van Ebbenhorst Tengbergen, H. J. and Rietema, K., 'Efficiency of Phase Separations'. In

Cyclones in Industry, Rietema, K. and Verver, C. G., (Eds) Elsevier, Amsterdam (1961)
6. Kelsall, D. F., *'The theory and applications of the hydrocyclones'*. In Pool and Doyle: *Solid–liquid Separation,* H.M.S.O., London (1966)
7. Mayer, F. W., *Zement-Kalk-Gips* 19, No. 6, 259–268 (1966)
8. Fontein, F. J., Van Kooy, J. G. and Leniger, H. A., *Brit. Chem. Engng*, 7, 410 (1962)
9. Newton, H. W., *Rock Product*, 35 (26), 26–30 (1932)
10. Ogawa, K., Ito, S. and Kishino, H., *J. Chem. Engng of Japan*, 11 (1), 44–47 (1976)
11. Svarovsky, L., *The efficiency of separation processes*. In R. J. Wakeman, *Progress in Filtration and Separation*, Elsevier, Amsterdam (1979)
12. Crawford, M., *Air Pollution Control Theory*, McGraw-Hill, New York (1976)
13. Strauss, W., *Industrial Gas Cleaning*, p. 240, Pergamon Press, Oxford (1974)
14. Dudrey, D. J. and Riesberg, K. D., *'Hydrocyclones and their use in recirculating systems'*, *9th Annual Regional Symposium*. Twin Cities Chapter, A.I.Ch.E, Minneapolis, February 21 (1969)

4

Coagulation and Flocculation

Part I

M. A. Hughes

Department of Chemical Engineering, University of Bradford

4.1 INTRODUCTION

Solids which have to be separated from liquids vary both in size and morphology as well as chemical nature. The term 'colloid' is generally applied to those particles which are smaller than 1 μm and a dispersion of these particles in a fluid is called a 'sol'. Particles of less than 0.2 μm are called 'super' colloids and, although colloids are larger than molecules, they are too small to be seen under a microscope. Alexander and Johnson[1] classify colloid particles as those in the size range 10^{-6} to 10^{-9} m. Dispersions of larger particles are called 'suspensions'. The size limits mentioned above are only arbitrary and colloidal properties can be exhibited by suspensions. The separation of the very small particles in sols present more problems than do larger particles, hence techniques have been developed which cause agglomeration of the small particles and thus simplify the separation techniques.

Colloidal dispersions can be altered by treatment of the solid–liquid interface by the addition of either electrolytes or surface-active agents and, of course, adjustment of the physical conditions may lead to crystal growth and a change in the particle size and interfacial area. Procedures which lead to increased sol stability (i.e. which lead to uniform dispersion of the particles in the liquid) are called 'peptizing', 'stabilizing' or 'deflocculating' procedures whereas those which lead to breakdown of the sol (i.e. which cause the particles and liquid to separate out) are termed 'coagulating' or 'flocculating' procedures. The terms 'coagulation' and 'flocculation' are used widely to mean the same thing, but in the water treatment industry they refer to quite different processes (see later).

Dispersions can be classified into either lyophilic or lyophobic colloids. In the former the solid shows a marked affinity for water or some other dispersion medium so that sols are formed spontaneously on mixing. Lyophobic colloids exhibit a low affinity for the host medium. Examples of these colloids are clays, hydrated oxides etc. Examples of lyophilic colloids are macromolecules such as proteins, humic acids etc.

Lyophobic sols can be formed by chemical means or by mechanical mixing and it is this group of sols which is particularly sensitive to the addition of electrolytes to a bulk phase, being readily persuaded to flocculate upon such additions. Lyophilic colloids are less sensitive to the addition of electrolytes and very high concentrations of the electrolyte salts are necessary for precipitation.

A basic requirement for the formation of large particles is for smaller ones to 'come together'. The simplest picture of the approach of two charged spherical particles is given in *Figure 4.1*. We shall see that all particles carry a residual charge and a negatively charged particle is shown in *Figure 4.1* although positively charged particles are possible. Attracting forces such as the London and van der Waals forces are opposed by the interaction of like charges distributed over each particle. If the charge on the particle can be reduced then close approach is possible.

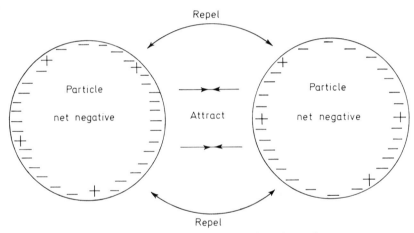

Figure 4.1. The approach of two like-charged particles

The total interaction between colloidal particles is discussed by Verwey and Overbeek[2] and Osipow[3]. A simplified approach is to consider the curves of energy against distance shown in *Figures 4.2a* and *4.2b* for repulsion and attraction respectively; these curves may be combined to give resultant interaction curves of the type shown in *Figure 4.3*.

The effect of thermal motion must also be included in the system since this will oppose any attraction which the particles may have for each other. Thus for a dispersion to be stable the potential energy maximum in the resultant curve must be considerably greater than kT (where k is the Boltzmann constant and T the absolute temperature). In *Figure 4.3* curves A, B, C and D therefore represent stable sol conditions whereas curves E, F and G represent instability. The height of the maximum on the potential energy curve is an indicator of the rate of flocculation and this height is determined by the potential drop on the double layer which surrounds the particle and extends into the bulk which contains the electrolyte.

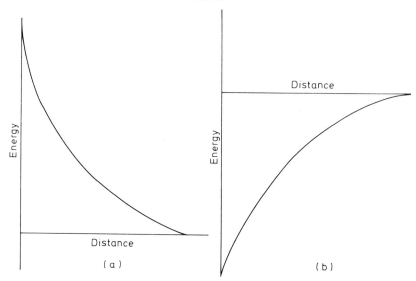

Figure 4.2. (a) Repulsion curve and (b) attraction curve for two like-charged particles approaching each other

Indeed, the charge surrounding a particle is a key to the flocculation process. It is now necessary to consider the origin of this charge and its distribution at and near to the solid–liquid interface.

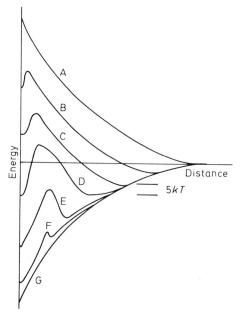

Figure 4.3. Resultant energy curves for particles approaching each other. Curves A, B, C and D demonstrate net repulsion, curves E, F and G demonstrate net attraction. Secondary minima occur in curves D, E and F; meta-instability may result at this separation distance

4.2 THE COLLOIDAL MODEL

4.2.1 The net charge on a solid

Most naturally occurring and man-made particles carry a residual charge on their solid surface, usually this is a net negative charge, as in the case of minerals and clays, although it can be net positive, as in the case of sewage sludge. There are three mechanisms which can cause this charge.

Mechanism 1

In crystalline materials the lattice is defective because of, e.g. Schottky defects[4] and thus a net excess of anions (negative) or cations (positive) exists at the surface. The net charge is compensated by an equivalent ionic charge at the surface and, on contact with water, the crystal releases the compensating ions to form a double layer. This behaviour is typical of ion-exchange materials such as zeolites, clays etc.

Mechanism 2

Some solids may be classified as sparingly soluble ionic crystals. When these are dispersed in water they exist in equilibrium with a concentration of the product ions, the concentration being determined from the solubility product. The potential of the solid (ψ_0) is determined from the Nernst equilibrium condition; for colloidal silver iodide for example, this gives

$$\psi_{0(AgI)} = A + \left(\frac{RT}{F}\right) \ln C_{Ag^+} \tag{4.1}$$

or

$$\psi_{0\,(AgI)} = B - \left(\frac{RT}{F}\right) \ln C_{I^-} \tag{4.2}$$

where A and B are constants, C is the concentration (activity) and F is Faraday's constant.

In general the potential of this type of solid is

$$\psi_0 = \left(\frac{RT}{vF}\right) \ln \left(\frac{C}{C_0}\right) \tag{4.3}$$

where v is the valency and C_0 is the zero point charge concentration. Of particular importance to the technologist is the behaviour of metal oxides and hydroxides which are notoriously difficult to separate from liquid phases, in the latter case the ions which determine the potential are H^+ and OH^-. We expect these systems to be especially sensitive to bulk electrolytes and to the pH of the bulk phase.

Mechanism 3

A third method by which net surface charge is generated is through the adsorption of specific ions from solution. In particular the adsorption can occur via a hydrogen-bonding mechanism where large organic molecules are adsorbed. Such agents are useful in reverting particles which are originally positively charged to negatively charged particles in order that the classical flocculation procedures devised for negatively charged solids may be used.

4.2.2 The double layer

The colloidal model for a net electronegative particle is shown in *Figure 4.4*. The layer of negative charge is surrounded by positive charges to give the double layer, the charge extends into the bulk but becomes more diffuse and random (a Boltzmann distribution is assumed). Certain potentials may be identified. A shear layer exists, which is the plane of slip between the double layer and the bulk media and the Nernst potential is the potential which exists between the shear plane and the solid surface. The zeta potential, ζ, is of especial importance since this is the potential which can be measured; it is the potential existing between the shear plane and the bulk phase.

A number of models exist which account for the double layer. The Gouy–Chapman model[4-6] was the first to indicate the way in which the potential, ψ, varies with the distance, x, into the bulk. The model is summarized in equation 4.4.

$$\frac{d\psi}{dx} = -\left(\frac{8\pi nkT}{\varepsilon}\right)^{1/2}\left[\exp\left(\frac{ve\psi}{kT}\right) - \exp\left(-\frac{ve\psi}{kT}\right)\right] \qquad (4.4)$$

where

k is the Boltzmann constant
v is the valency of the ion with opposite charge to that on the surface
n is the bulk concentration of this last ion
ε is the dielectric constant of the bulk liquid phase
ψ is the double layer potential at a distance x from the surface
e is the electronic charge.

Integration of equation 4.4 can lead to a simple statement in which the double layer potential varies with x, thus

$$\psi = \psi_0 \exp\left(-\varkappa x\right) \qquad (4.5)$$

where $\varkappa$ is the Debye–Huckel function

$$\varkappa = \sqrt{\left(\frac{8\pi e^2 N^2 I}{1000\varepsilon RT}\right)} \qquad (4.6)$$

in which N is Avogadro's number and I is ionic strength. Thus the potential of a charged surface falls away exponentially with increasing distance into the bulk phase. The potential is almost zero at a distance $x = 3/\varkappa$ from the surface

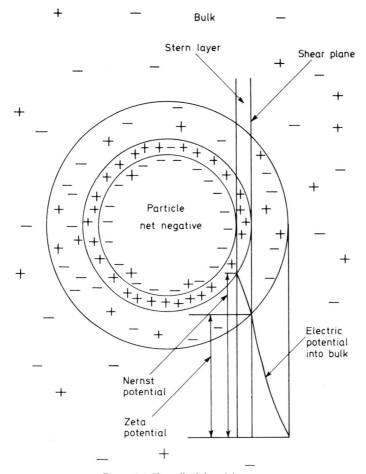

Figure 4.4. The colloidal model

and then beyond this distance the ions in solution experience no effect from the surface charge.

An expression for the charge density on the surface, σ, can be developed as in equation 4.7

$$\sigma = \left(\frac{\varepsilon \pi k T}{2}\right)^{1/2} \left[\exp\left(\frac{ve\psi_0}{2kT}\right) - \exp\left(-\frac{ve\psi_0}{2kT}\right)\right] \qquad (4.7)$$

Stern[7] improved on the Gouy–Chapman model by allowing for finite ionic size in the double layer. The total charge density then becomes,

$$\sigma = \sigma_s + \sigma_g \qquad (4.8)$$

where σ_g is the charge density as given by the Gouy model above, with ψ_0 replaced by ψ_δ which is the potential at the Stern layer (see Figure 4.4), and σ_s is the absorbed charge per unit area due to the Stern layer. The complete expression for the Stern model then becomes

$$\sigma = n\sigma_m \left[n + A \exp\left(-\frac{ve\psi_\delta}{kT} - \phi \right) \right]^{-1}$$

$$+ \left(\frac{\varepsilon nkT}{2\pi} \right)^{1/2} \left[\exp\left(\frac{ve\psi_\delta}{2kT} \right) - \exp\left(-\frac{ve\psi_\delta}{2kT} \right) \right] \qquad (4.9)$$

where

σ_m is a charge corresponding to a monolayer of counter ions
A is the frequency factor
ϕ is van der Waals energy.

The equation is complicated but represents the best model to date.

4.2.3 Compression of the double layer

It can be shown that for small colloid particles of low charge, and in the presence of electrolytes, the thickness of the double layer, d, is approximated by the reciprocal of the Debye–Huckle function, $\varkappa$ (equation 4.6). For water at 298 K, the thickness becomes

$$d = \frac{2.3 \times 10^{-9}}{\sqrt{I}} \text{ cm} \qquad (4.10)$$

It is seen to be dependent on the ionic strength and for some practical situations *Table 4.1* reflects this last fact. Consequently, the addition of various electrolytes at increasing concentrations compresses the double layer as shown in *Figure 4.5*. The double layer becomes very small on the addition of high valency salts in appreciable concentrations.

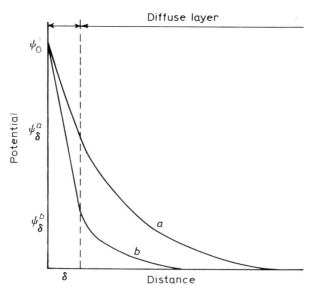

Figure 4.5. The Stern model for the electric double layer showing compression from (a) low to (b) high ionic strength

Table 4.1. ESTIMATES OF DOUBLE LAYER THICKNESS AROUND A PARTICLE IN VARIOUS MEDIA

Medium	$d = 1/\varkappa$, μm
Distilled water	900
10^{-4} M NaCl	31
10^{-4} M MgSO$_4$	15
River Thames water	4
Sea water	0.4

Thus one effect of the addition of indifferent electrolytes (that is, where electrolyte ions are not adsorbed on the surface) to the sol may be to reduce the double layer thickness and thus to promote closer approach of particles and consequent aggregation (see section 4.4).

4.2.4 The rate of aggregation

The process of aggregation is seen to require a low charge on each particle and a collision event. Assuming that electrical repulsion is absent, as a result of pretreatment with electrolyte, then the rate of aggregation depends on Brownian motion. In the assumed absence of velocity gradients, induced by e.g. stirring, we have the case of perikinetic aggregation or flocculation when Brownian motion alone dictates the rate.

According to the theory of von Smoluchowski[8,9] the most rapid aggregation will occur when every contact leads to the adherence of one particle to another. So the rate of perikinetic flocculation or aggregation is given by

$$-\frac{dn}{dt} = k_p n^2 \tag{4.11}$$

in which n is the number of particles present in a given volume and k_p is a specific rate constant. Hence on integration

$$n = \frac{n_0}{1 + k_p n_0 t} \tag{4.12}$$

then it follows that

$$t_{1/2} = \frac{1}{k_p n_0} \tag{4.13}$$

where $t_{1/2}$ is the half life.

The half life for particles, $t_{1/2}$, is the time taken for the initial number of particles to be reduced to one half and is related to the diffusion constant, D, the effective radius of the particles, r (assumed spherical) and n_0, the number of particles of radius r_0 initially present:

$$t_{1/2} = \frac{1}{8\pi D r n_0} \tag{4.14}$$

Taking the diffusion constant as

$$D = \frac{kT}{6\pi\eta r_0} \tag{4.15}$$

where η is the viscosity of the liquid bulk phase, then

$$t_{1/2} = \frac{3\eta}{4kTn_0} \tag{4.16}$$

Taking as an example water at $T = 298$ K

$$t_{1/2} \approx \frac{2.10^{11}}{n_0} \tag{4.17}$$

The more highly concentrated the sols, the faster the aggregation.

The rate constant k_p is also given by

$$k_p = \frac{4kT}{3\eta} \tag{4.18}$$

which demonstrates the effect of the thermal energy kT as reflected in the Brownian motion of the particles.

Equation 4.14 may be altered for the case of slow aggregation in which only a fraction β of the collisions leads to adherence between particles:

$$t_{1/2} = \frac{1}{8\pi Drn_0\beta} \tag{4.19}$$

Further analysis of slow coagulation is made in terms of the energy barrier, E^*, which exists for collisions. The coagulation is slowed down by a stability factor W which is given by

$$W = 2\int_2^\infty \frac{ds}{s^2} \exp\left(\frac{E^*}{kT}\right) \tag{4.20}$$

where $s = p/r$ and p is the distance between particles. Equation 4.20 approximates to

$$W \approx \frac{1}{2\kappa r} \exp\left(\frac{E^*_{max}}{kT}\right) \tag{4.21}$$

where κ is the Debye–Huckel function now given by

$$\kappa = \sqrt{\left(\frac{4\pi e^2 \Sigma n_{i0} z_i^2}{\varepsilon kT}\right)} \tag{4.22}$$

n_{i0} is the concentration of ions of type i in the bulk phase and z_i is the valency. Now since E^*_{max} can be evaluated, from the theory of Verwey and Overbeek[2], the dependence of the stability factor W on the electrolyte concentration and valency can be predicted. It can thus be shown that a 2:2 electrolyte provides less hindrance to the aggregation rate than does a 1:1 electrolyte at the same concentration.

Gregory[10] observes that Brownian motion alone is unlikely to produce aggregates of an acceptable size (say 1 mm) which leads to useful flocs in a reasonable time. In any case, if aggregates of many particles are formed by this mechanism the consequent reduction in particle numbers dramatically alters the rate to a very slow value.

Stirring a dispersion can cause an increase in flocculation rate — orthokinetic flocculation. At low energy inputs collisions of particles are induced through induced velocity gradients; obviously at high shear rates the aggregates break up again. Now, if every collision leads to an aggregate then equation 4.11 becomes, for orthokinetic flocculation,

$$-\frac{dn}{dt} = \tfrac{2}{3}Gd^3n^2 \qquad\qquad (4.23)$$

in which d is the diameter of the particle and G is the shear rate (s^{-1}).

Now as particles and aggregates collide, the probability of such collisions is increased through the energy imparted to the host fluid and so the reduced particle concentration (and hence reduced rate through equation 4.11) is partially offset. It also follows that

$$\frac{\text{orthokinetic rate}}{\text{perikinetic rate}} = \frac{G\eta d^3}{2kT} \qquad\qquad (4.24)$$

Gregory points out that if the shear rate, 10 s^{-1}, is used to describe gentle agitation, the two rates are equal when the particle diameter is around 1 μm but orthokinetic flocculation becomes more significant with larger particles and higher shear rates.

4.3   ELECTROKINETIC PHENOMENA AND THE ZETA POTENTIAL

It has already been said that a practical way to characterize the double layer is to measure the zeta potential, ζ. Several techniques are possible; these come within the category of electrokinetic phenomena.

In electrophoresis a potential gradient is applied to a dispersion and the movement of the charged particles relative to the stationary bulk phase is measured. Sedimentation potential is the reverse of the above; the potential which is developed when the charged particles precipitate through the stationary liquid is measured.

In electro-osmosis the charged solid is held stationary (possibly in the form of a plug), a potential gradient is applied, and the bulk liquid phase moves. Pressure is then applied until it just counters the flow, and this is the electro-osmotic pressure. Streaming potential is the opposite of electro-osmosis: the potential gradient generated when the bulk liquid phase is forced through the stationary charged solid is measured.

Commercial apparatus is available for the measurement of electrokinetic phenomena[11]. One such apparatus, to measure the electrophoretic effect, is shown in *Figure 4.6*. Observation cells are available in two forms, either cylindrical or flat; generally the flat cell is more robust. The potential is applied at platinized black electrodes and care must be taken to ensure that such electrodes do not become polarized during the measurements. The particles are observed through the microscope which must be focused into

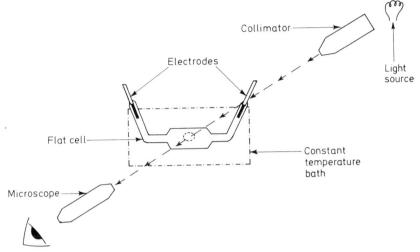

Figure 4.6. Schematic diagram of apparatus for measuring the electro-phoretic effect. The flat cell contains the dispersion of particles and these are subjected to an applied d.c. voltage across the electrodes. A light beam is focused into the flat cell at the predetermined stationary layer and the movement of the particles is observed through the microscope

the stationary layer of the bulk phase, otherwise 'cell wall effects' will enhance or slow down the true velocity, V_E, of the charged particles. The velocity of the particles is obtained by timing the passage of a number of individual particles across the calibrated grid set in the lens of the microscope and so an average V_E is obtained.

Kruyt[12] has related the electro-osmotic velocity of flow to the double layer potential at the shear plane and to the applied external field. The double layer must satisfy the relation

$$\nabla^2 \psi = -\frac{4\pi Z}{\varepsilon} \tag{4.25}$$

where

∇^2 is the Laplace operator
ψ is the double layer potential at a distance x from the surface
Z is the net space charge per unit volume at point x, and
ε is the dielectric constant of the medium.

When an external field E is applied, a volume of the bulk liquid of thickness dx is subject to a force

$$F = EZ \, dx \tag{4.26}$$

A number of layers adjacent to the shear layer will be moving at different velocities and will exert viscous drag on the latter. Consider a layer at position x from the surface; the drag force on this layer is

$$F_x = -\eta \left(\frac{\partial v}{\partial x}\right)_x \tag{4.27}$$

and at $x + $ dx the drag force is

$$F_{x+dx} = -\eta\left(\frac{\partial v}{\partial x}\right)_{x+dx} \tag{4.28}$$

Therefore the net frictional force on the layer is

$$F' = \eta\left(\frac{\partial v}{\partial x}\right)_{x+dx} - \eta\left(\frac{\partial v}{\partial x}\right)_{x} \tag{4.29}$$

In the steady state the total force on the layer is zero so that

$$EZ\,dx = \eta\left(\frac{\partial^2 v}{\partial x^2}\right)dx \tag{4.30}$$

Since at a plane surface

$$\nabla^2\psi = \left(\frac{\partial^2\psi}{\partial x^2}\right) \tag{4.31}$$

then

$$Z = -\frac{\nabla^2\psi\varepsilon}{4\pi} = -\frac{\varepsilon}{4\pi}\left(\frac{\partial^2\psi}{\partial x^2}\right) \tag{4.32}$$

and

$$-\frac{E\varepsilon}{4\pi}\left(\frac{\partial^2\psi}{\partial x^2}\right) = \eta\left(\frac{\partial^2 v}{\partial x^2}\right) \tag{4.33}$$

Equation 4.33 is now integrated over the whole liquid from the shear plane to infinity to give

$$-\frac{E\varepsilon}{4\pi}\frac{d\psi}{dx} = \eta\frac{dv}{dx} + C \tag{4.34}$$

where E is zero when

$$\left(\frac{\partial\psi}{\partial x}\right)_{x=\infty} = 0 \qquad \left(\frac{\partial v}{\partial x}\right)_{x=\infty} = 0$$

Further integration of equation 4.34 gives

$$-\left(\frac{E\varepsilon}{4\pi}\right)\psi = \eta v + C' \tag{4.35}$$

At $x = \infty$; $\psi = 0$ and $v = V_E$ where V_E is the true electro-osmotic velocity. Also at the shear plane $v = 0$ and $\psi = \zeta$ so that,

$$V_E = \frac{\varepsilon E\zeta}{4\pi\eta} \tag{4.36}$$

Equation 4.36 is used to determine ζ when V_E is measured (as described previously) and E, ε and η are known.

4.4 PRACTICAL APPLICATIONS OF THE ZETA POTENTIAL

If the zeta potential is an indicator of the electrical state of the double layer then when the value of ζ approaches zero the sols should become

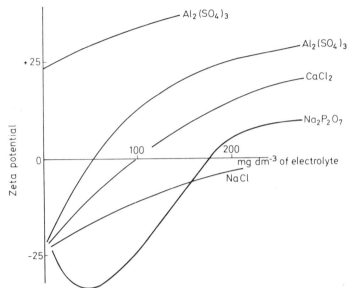

Figure 4.7. The changes in zeta potential when electrolytes are added. A 2.3 electrolyte. $Al_2(SO_4)_3$, is the most effective in reducing the initial negative zeta potential to zero.

unstable, and this should lead to clarification. This has often been demonstrated where inorganic electrolytes have been used to lower the zeta potential of industrial wastes, municipal wastes and paper processing slurries[13].

Riddick[14] has commented that the zeta potential is the most significant controlling factor in the achievement of coagulation, although it should not be regarded as the only parameter to be adjusted to produce the desired clarification. His remarks are pertinent to water clarification.

4.4.1 Coagulation

La Mer[15] has suggested that the term 'coagulation' should be restricted to the chemical destabilization of sols whereby electrolytes are added to a sol so as to reduce the charge on the particles and allow close approach and aggregation. This is also adopted by the industry concerned with water treatment. A 'primary coagulant' then becomes the salt which is added to achieve the effect.

The useful practical effect consequent upon the addition of indifferent electrolytes is now discussed.

Schulze[16,17] showed that certain lyophobic colloids became unstable and flocculated on the addition of electrolytes. The multivalent ions, e.g. Al^{3+} and Fe^{3+}, were very effective for net negatively charged particles. The Schulze–Hardy rule states that the coagulation effect is determined by the valency of the ion with opposite charge to that of the sol. *Figure 4.7* gives typical ζ measurements for a sol with an initial net charge of -25 mV dosed with small quantities (up to ~250 mg dm^{-3}) of various electrolytes (1:1, 1:2 and 2:3 types). An aluminium salt is seen to be the most effective and optimum dose is around 60 mg dm^{-3}. Note that the charge can be reversed from net negative to net positive. Of course, the addition of aluminium to already net positively charged sols only acts to increase the sol charge. A 1:3 electrolyte, e.g. $FeCl_3$, would be as effective as the aluminium salt.

It is now readily explained why primary coagulation is achieved when a sol is treated with either lime/alum (when net negatively charged sols coagulate) or with phosphates (when net positively charged sols coagulate).

The general dosage range for a useful primary coagulant is about 100–400 mg dm^{-3}.

Of particular interest is the effect of sodium metaphosphate (CALGON is a common coagulating agent and is made up from the sodium salts of meta- and pyro-phosphoric acids). In the case of *Figure 4.7*, the immediate effect of the metaphosphate is further to increase the negative ζ potential of the particles *then* the potential becomes more positive as the concentration of this electrolyte is increased. In this case the first additions lead to adsorption of $P_2O_7^{4-}$ ions on the surface but then the ionic strength effect takes over generating a reduction in the ζ potential towards zero as the ionic strength increases. The metaphosphate-type anions must be useful in treatment of a colloid with any initial positive ζ potential.

In fact, it can be shown that the critical coagulation concentration (c_f), which is the concentration of specifically adsorbing counter ions required to cause coagulation (usually a narrow range of concentrations), is inversely proportional to the sixth power of the charge on the ion. Lists of critical coagulation concentrations for electrolytes and sols exist[18]—see also *Table 4.2*. The c_f value depends to some extent on the nature of the sol and the usual quoted ratio of coagulating powers of ions of different charges, e.g. ~ 1:100:1000 for uni-: di-: tri-, is seen to be only approximate. Other factors besides the effect of the charge on the appropriate ion of the indifferent electrolyte dictate sol stability. Thus adsorption of the ion of the same charge

Table 4.2. CRITICAL COAGULATION CONCENTRATIONS FOR COLLOIDS, ACCORDING TO FREUNDLICH[18]

Electrolyte added	c_f value, mmol dm^{-3}		
NEGATIVELY CHARGED COLLOIDS			
	As_2S_3 sol	Au sol	Pt sol
NaCl	51.0	24.0	2.5
KCl	49.5	—	2.2
$CaCl_2$	0.65	0.41	—
$BaCl_2$	0.69	0.35	0.058
$Al(NO_3)_3$	0.095	—	—
$\frac{1}{2}Al_2(SO_4)_3$	0.096	0.009	0.013
POSITIVELY CHARGED COLLOIDS			
	Fe_2O_3 sol	Al_2O_3 sol	
NaCl	9.25	77.0	
KCl	9.0	80.0	
$\frac{1}{2}BaCl_2$	9.65	—	
K_2SO_4	—	0.28	
K_2CrO_4	—	0.60	
$K_3[Fe(CN)_6]$	—	0.10	
$K_4[Fe(CN)_6]$	—	0.08	

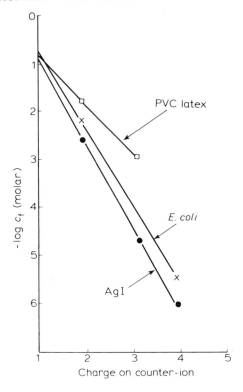

Figure 4.8. Variation of the critical flocculation concentration (c_f) with counter-ion charge

as the colloid, hydration of the ions in solution, and hydration of the surface of the particle may all influence stability of dispersions of colloids. An empirical relationship is usually found for the plots of log c_f against charge, and this is usually linear—see *Figure 4.8*.

Careful mixing of colloids of opposite charge can lead to coagulation and flocculation. However this is difficult to control, as the desired flocculation of say a positively charged colloid by addition of the negative one can easily be redispersed when the charge passes through zero only to become negative because of excess of the second colloid.

Natural clays, usually charged negative, might be used to coagulate a positively charged sol.

As stated above, the indifferent electrolyte effect is altered and possibly enhanced when the ions of that electrolyte undergo hydrolysis in the aqueous phase. We will now deal with cations in solution and the effect of the pH of the aqueous phase.

In the first place some cations and anions are not significantly hydrolysed at pHs in the range 1–14; these include common alkali metals such as Na^+ and K^+ and common anions like Cl^- and SO_4^{2-}. Of course, they will be hydrated and certain of these ions have a structuring or destructuring effect on the water. The Li^+ ion, in particular, structures water strongly and it would be of interest to study its effect on the double layer.

More important effects in practical coagulation are observed with ions which hydrolyse. The species and relevant stability constants can be pursued[19].

Of particular importance are the metal cations in solution and especially the species associated with aluminium, iron, silicon, magnesium and calcium. Of these, aluminium, iron and silicon hydrolyse the most readily (see *Table 4.3*), i.e. they give hydrolysis species even in acid media. Taking

Table 4.3. FIRST HYDROLYSIS CONSTANTS FOR METALS, AND THE pH AT WHICH PRECIPITATION OCCURS WITH APPROXIMATELY 0.02 mol dm^{-3} CONCENTRATION OF METAL[19]

Ion	$-log_{10} K_n$*	pH for precipitation
Fe^{3+}	2.22	2.0
Ce^{4+}	~0.7	2.7
Cu^{2+}	7.5	5.3
Cr^{3+}	3.82	5.3
Fe^{2+}	9.5	5.5
Co^{2+}	9.6	6.8
Mg^{2+}	11.4	10.5

* K_n refers to the first hydrolysis step of the type
$$M^{n+} + H_2O \rightleftharpoons MOH^{(n-1)+} + H^+.$$

aluminium as an example, a variety of species have been proposed ranging from hydrated Al^{3+} ions through $(Al_8(OH)_{20})^{4+}$, $(Al_6(OH)_{15})^{3+}$, $(Al(OH)_2)^+$ to $(Al(OH)_4)^-$ and the insoluble hydrated oxide $Al_2O_3 \cdot nH_2O$; the amounts of these species are dependent on pH and on the total aluminium concentration. Hall[20] treats the variation of ζ potential with pH in the system water/kaolinite dispersions to which aluminium has been added. The situation is complex but the kaolinite appears to form a layer of amorphous hydroxide on the surface even under conditions where coagulation is not possible, e.g. at pH less than 6–5. At higher pH values the amorphous oxide precipitates in the bulk of the solution and then the ζ potential of the aluminium hydroxide is a function of a polynuclear ion, e.g. $Al_6(OH)_{15}^{3+}$. The ζ potential is now at a maximum positive value and yet coagulation is efficient.

The relative ease of hydrolysis of certain metal species in increasingly acid media can be observed from a listing of the point of charge (p.z.c.) given by Gregory[10] for a series of hydrous oxides—oxide and p.z.c.—$SiO_2(2)$, $TiO_2(6)$, $Fe_2O_3(8.5)$, $Al_2O_3(9)$, $MgO(12)$. The p.z.c. is the characteristic pH value at which the oxide surface has no net charge, and it results from a series of reactions between the surface and the hydroxyl and hydrogen ions in solution, as illustrated in *Figure 4.9*.

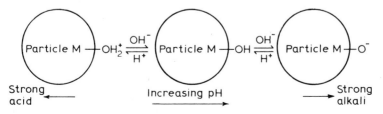

Figure 4.9. The changing character of the hydrous oxide surface as the bulk phase pH is altered

Another measure of the above effect is the well-known phenomenon that Fe^{3+} in solution begins to precipitate hydrated ferric oxides at fairly low pH, e.g. typically at a pH of 1.8.

Consequently, there must be a mechanism other than the effect of double layer compression whereby these hydrous oxides ($Al_2O_3nH_2O$, $Fe_2O_3nH_2O$ etc.) aid the growth of aggregates. The hydrous oxides proceed to gather aggregates, trapping aggregate particles in a network of hydrous oxide polymers. Riddick[14] considers that this progress is very fast involving times of up to a few seconds. Eventually this 'floc gathering' process, which can be aided by natural silica or added silica and by polyelectrolytes (see section 4.5) leads to efficient flocculation; the total process takes up to 15–30 minutes after application of the initial aluminium ion dose.

From the above, it is clear that the ageing of these flocs is important; a certain period is needed in a process for optimum floc size to be reached, thus leading to an easily separated solid. Further ageing may or may not have an adverse effect. In the latter case the floc becomes destabilized and might lead to some redispersion. Obviously care must be taken when flocs are separated out and then further treated; for example, a precipitate might be 'washed' in some instances and this may lead to alteration of the floc characteristics and some redispersion.

The chemistry of the aqueous phase and all the complexes which are possible is too detailed to be covered here and the reader should refer to standard texts on this subject.

4.4.2 Flotation

There is also a close relationship between the zeta potential and the efficient flotation of solids by purposely designed organic molecules which carry either positive or negative charge. The reader is referred to the work of Aplan and de Bruyn[21]. In general, anionic collectors are effective on positive surfaces and cationic collectors react with negative surfaces. If the material to be collected already has a charge close to zero then no collector is likely to be effective. Many applications exist in the minerals treatment industries.

4.5 FLOCCULATION BY POLYELECTROLYTES

The term 'flocculation' may be taken to cover those processes whereby small particles or small groups of particles form large aggregates. In the water industry the term is especially reserved for the formation of large flocs when the dispersion is stirred and aged in holding tanks. The term is also used for the dramatic effect when polyelectrolytes are added and a large stable floc is formed very quickly. 'Perikinetic flocculation' is a term reserved for floc formation brought about by Brownian motion alone whilst 'orthokinetic flocculation' is used to describe flocculation achieved by imparting velocity gradients to the dispersion through stirring. Orthokinetic flocculation is described in Part II by Ives. Perikinetic flocculation is touched upon in section 4.2.4.

One of the greatest advances in solid–liquid separation has been the development of polymers with remarkable abilities to flocculate sols when added in only trace quantities. Indeed, these polyelectrolytes may be used to supplement or replace the primary coagulants previously mentioned. Although they are considerably more expensive than primary coagulants the dose rate is much lower, typically 0.1–0.15 mg dm^{-3} of substrate to be treated. Already widely used in the mining industries, polyelectrolytes are now being considered for solid–liquid treatment applied to sewage.

4.5.1 The chemical nature of synthetic polyelectrolytes

The types of synthetic polymer which are considered useful are summarized in *Table 4.4.* It is observed that they fall into three main classes, non-ionic,

Table 4.4. MONOMERS AND POLYELECTROLYTES

MONOMERS

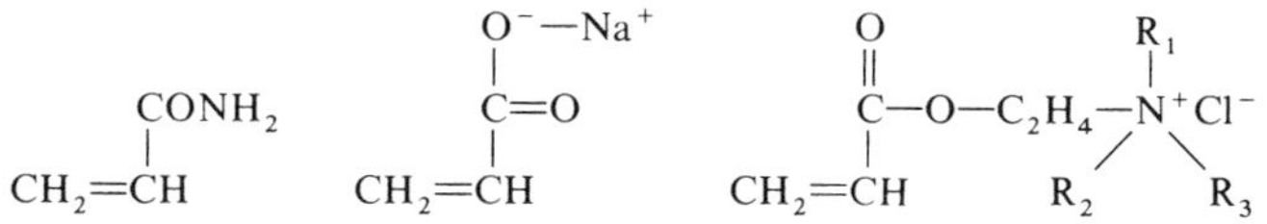

POLYMERS

Nonionic	Polyacrylamides	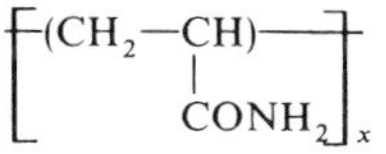
	Polyethyleneoxide	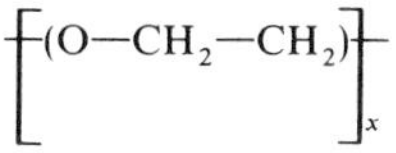
Anionic	Acrylamide co-polymer	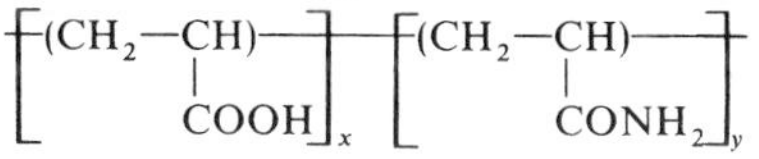
	Polyacrylics	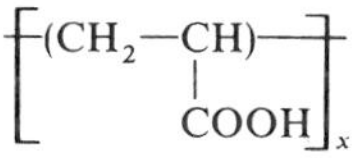
Cationics	Polyamines	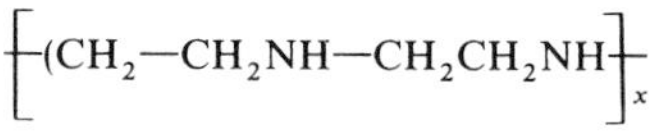
	Acrylamide co-polymers	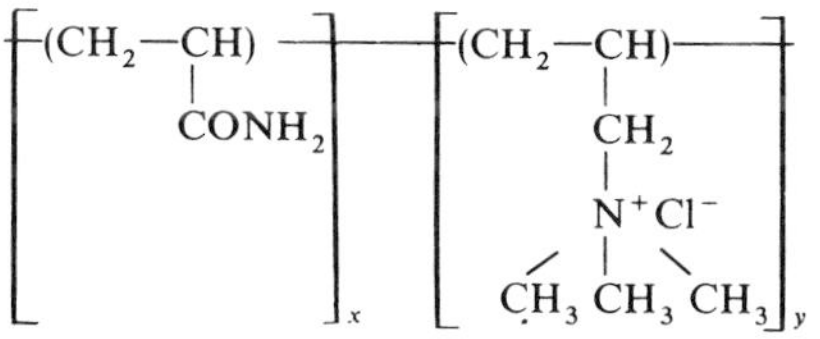

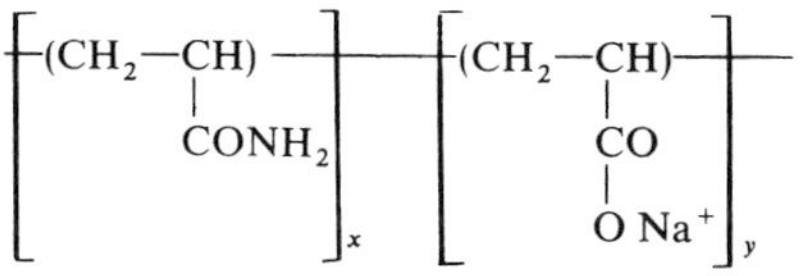

anionic and cationic. Obviously the anionic/cationic character can be altered at will by co-polymerization of the various monomers and any one product is then characterized by

1. its average molecular weight and
2. the charge density distribution within the polymer molecule.

Molecular weights can be classified according to high $\simeq 20 \times 10^6$, medium $\simeq 10 \times 10^6$, low $\simeq 5 \times 10^6$ and very low $< 1 \times 10^6$. A detailed description of the chemistry of polyelectrolytes has been given by Schwoyer[22].

4.5.2 Mode of action and application

The mechanism of flocculation by polyelectrolytes is considered to involve the two processes of surface charge neutralization and bridging. If the first mechanism is to be effective then one must take care to choose a polyelectrolyte whose charge is of the opposite sign to the charge on the particles. The charge density on the polymer will be an important measure of its capacity to flocculate. The process of charge neutralization and bridging then proceeds as shown in *Figure 4.10*.

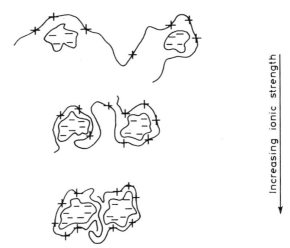

Figure 4.10. Flocculation by charge neutralization and bridging. (A cationic polymer is used to collect particles with a net negative charge)

The correct choice of polyelectrolyte is best made after laboratory trials on samples of the liquor to be clarified. Usually, for a given charge density, the polymer with highest molecular weight will give the fastest sedimentation rate. It will always be necessary to determine the optimum dosage for the best results, especially where an expensive chemical is in use.

Efficient use of the expensive agent is assured if a proper dosing system is adopted. The polymers are supplied either as solids or liquids and obviously, for efficient dosing at such low levels, the solid products must be taken up into some liquid to give a concentrate feed for the dosing plant. The Dow Chemical Company[23] have given diagrams of typical automatic dosing systems, where the polymer is a solid, and one such system is reproduced in *Figure 4.11*. Further useful practical points concerning efficient polyelectrolyte usage are listed:

1. add the polyelectrolyte to the main stream in a very dilute solution ($<0.1\%$);
2. add as near as possible to the point where flocculation is required;
3. add at points of local turbulence;
4. add in stages at different points;
5. add across the whole of the stream to be treated;
6. avoid turbulence subsequent to floc formation;
7. at high solids concentrations, add recycled or other dilution water to the system;
8. at low solids concentrations, recycle settled solids into the stream.

It will be gathered from the above that subsequent treatment of the flocs may lead to break up, indeed it is best to regard the flocculation of hydrophilic sols as reversible. Different techniques used for the physical separation of the flocs from the suspension liquid have shown that if belt presses are used then fragile flocs should be treated carefully, with the pressure being applied only gradually; even filter presses have provided problems in the past. Centrifuging would require a strong floc; high molecular weight polyelectrolytes are useful here, otherwise it is usual to employ polyelectrolytes of medium to low molecular weights but with high charge density.

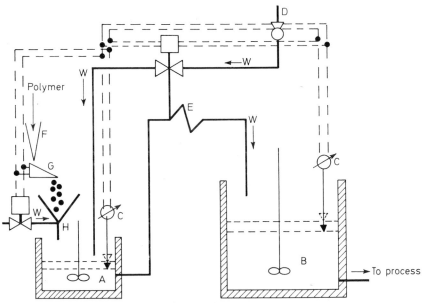

Figure 4.11. Schematic layout of a dilution system for a solid polyelectrolyte polymer. The polymer in hopper F is fed to a vibrator G which discharges into a dispenser H fed with water W. Tank A, (volume 1–2 m³) provides the first mixing and the product is fed through the mixing device E (fed with water from pump D) into tank B where the final mixing takes place. The solution is now fed to the process line. Automatic level controls are provided at C and are connected to valves and pumps as shown

4.5.3 Some examples of polyelectrolytes in action

The most important class of polymers are those anionic polyelectrolytes containing carboxyl groups. High molecular weight anionics ($> 1\,000\,000$), with low charge density, have many uses as flocculants for water and waste water.

Those cationic polymers in use for water treatment are nearly all based upon quaternized aminoesters or amino amides. Few non-ionic polyelectrolytes are used in water treatment and true non-ionic types are difficult to prepare. Polyacrylamides containing less than 1% hydrolysable groups are available. High molecular weight poly(ethyleneoxides) are used in flotation processes.

In recent years, work has been carried out on combinations of flocculants with other chemicals to achieve highly efficient solid/water separation.

Field[24] showed that a combined anionic flocculant/cationic coagulent system, at constant coagulent dose, performs better on coal-tailings than when flocculant alone is used. In general, it is the type of solid–liquid processes in use which dictate the *molecular weight* of the chosen electrolyte —see *Table 4.5*. On the other hand, the *chemical type* necessary generally depends on the slurry characteristics, i.e. solid chemical surface type, particle size distribution, the dissolved electrolytes and the pH.

Some of the major applications of polyelectrolytes are given in *Table 4.6*.

Table 4.5. CHOICE OF MOLECULAR WEIGHT FOR DIFFERENT PROCESSING TECHNIQUES

Process type	Molecular weight range
Sedimentation and centrifugation	High (in conjunction with highly cationic electrolyte for overflow clarity)
Vacuum filtration	Medium to high
Pressure filtration	Low to medium
Pressure belt filtration	Medium to high (in conjunction with low molecular weight, high cationic electrolyte)
Low solids clarification	Very low to high

4.6 OTHER CONSIDERATIONS

It cannot be claimed that this chapter is an exhaustive coverage of all those chemicals which are used to improve solid–liquid separations. For example, the potato starch which is commercially available has found wide use. This naturally occurring polymer with a molecular weight of about $1\,000\,000$ is made up of glucose units and operates in its neutral form through hydrogen bonding, and by suitable chemical pretreatment, starches bearing positive or negative charges can be made.

Other important water soluble polymers (biopolymers) include a variety of naturally occurring gums, e.g. Guar, Acacia (Gum Arabic), Tragacanth, Xanthan, etc. Guar gum, in particular, has found use as a food thickener as well as a flocculating agent in the minerals industry; it also has important industrial properties as an agent for friction reduction.

Flocculating agents may also be used in a selective manner to recover one solid which is in suspension with another unwanted material. In one sense this

Table 4.6. POLYELECTROLYTE USE IN SOLID–LIQUID SEPARATION

Substrate to be treated	Class of electrolyte	Typical dosage, per tonne of dry solids
Water (potable)	High molecular weight polyacrylamides of low toxicity with usage of non-ionic, cationic, anionic	Must not exceed 0.5 ppm w/v
	Modified starches (organic coagulents may be used in conjunction)	Average 0.1 ppm w/v
Effluents		
Textiles	Low molecular weight, highly cationic—followed by anionic	1–10 kg
Sewage sludge	High molecular weight polyacrylamides	1–4 kg
Paper pulp	High molecular weight cationic or anionic polyacrylamides for retention and drainage of paper stock	
	Dual component systems exist involving low molecular weight cationic first then high molecular weight anionic polyacrylamides to badge the floc	Up to 9.5 kg
	The new Hydrocol® system involves first the addition of cationic polyacrylamides then a modified bentonite to produce 'super coagulation'	
Mineral slurries		
Iron ore tailings	Medium molecular weight anionic polyacrylamide plus highly cationic coagulent	0.03 kg
Clayey coal tailings	Medium molecular weight anionic polyacrylamide plus highly cationic coagulent	0.03–0.05 kg
Acid leach liquors (Cu, U, Zn, Au, etc.)	Non-ionic or slightly anionic/cationic polyacrylamides also sulphonic acid polymers	0.03–0.15 kg
Alkali leach liquors (U, Bayer Process A1)	Highly anionic polyacrylamide	0.03–0.175 kg
Mine 'run-off' water, low solids	Highly cationic polyamine or anionic polyacrylamides plus highly cationic coagulents	1 ppm w/v
Coal slurries		
Frothed slimes	Medium to high anionic polyacrylamides	0.03–0.05 kg
Coal tailings	Low to medium/high anionic polyacrylamides	0.04–0.06 kg

Hydrocol® is a registered trademark of Allied Colloids Limited, Bradford.

is like the selective flotation practised in the mining industry and has great potential where there is now a continuing interest in the recycling of valuable materials from waste.

Finally, a comment must be made on the toxicity of the substances used for chemical pretreatment in solid–liquid separation. Some polyelectrolyte products may contain monomers as impurities and Croll et al.[25] have reported residues of acrylamide in water. Acrylamide has a high chronic toxicity, being a neurotoxin, and accumulation of the monomer in treated drinking water is guarded against. Primary coagulants in the form of inorganic salts are less toxic but nevertheless certain countries have standards for the final effluent from a mining complex which must not be exceeded. Hawley[26] reports Canadian regulations for aluminium sulphate, ferric sulphate and ferric chloride.

REFERENCES

1. Alexander, A. E. and Johnson, P., *Colloid Science*, p. 129, University Press, Oxford (1950)
2. Verwey, E. J. W. and Overbeck, J. Th. G., *Theory of the Stability of Lyophobic Colloids*, Elsevier, New York (1948)
3. Osipow, L. I., *Surface Chemistry*, Reinhold, New York (1963)
4. Chapman, D. D., *Phil. Mag.*, **11**, 425 (1906)
5. Gouy, G., *J. Physique*, **9**, 457 (1910)
6. Gouy, G., *Annls Phys.*, **7**, 129 (1917)
7. Stern, O., *Z. Elektrochem.*, **30**, 508 (1924)
8. von Smoluchowski, M. *Phys. Z.*, **17**, 557, 585 (1916)
9. von Smoluchowski, M., *Phys. Chem.*, **92**, 129 (1917)
10. Gregory, J., *Effluent and Water Treatment Journal*, **516**, (1977)
11. Szwarestajn, E. and Jain, S. C., *Indian Pulp and Paper*, **27**, 3 (1973)
12. Kryut, H. R., *Colloid Science*, Vol. 1, Elsevier, New York (1952)
13. Calver, J. V. M., *The Research Association for the Paper and Board, Printing and Packaging Industries, Bibliography No. 727* (1974)
14. Riddick, T. M., *Effluent and Water Treatment Journal*, **563**, (1964)
15. La Mer, V. K., *J. Colloid Science*, **19**, 291 (1964)
16. Schulze, H., *J. Prackt. Chem.*, **25**, 431 (1882)
17. Schulze, H., *J. Prakt. Chem.*, **27**, 320 (1883)
18. Freundlich, H., *Colloid and Capillary Chemistry*, Methuen (1926)
19. *Stability Constants Handbook*, Parts I and II, Publ. Chem. Soc., London (1957)
20. Hall, E. S., *J. Appl. Chem.*, **197**, (1965)
21. Fuerstenau, D. W. (Ed.), *Froth Flotation*, Vol. 170, p. 91, AIMM and Pet. Eng., New York (1962)
22. Schwoyer, W. L. K. (Ed.), *Polyelectrolytes for Water and Waste Water Treatment*, CRC Press, Boca Raton, FL (1981)
23. *Handbook on Separan*, Dow Chemical Co., Horgen, Switzerland (1976)
24. Field, R. J., *Coal Preparation*, **4**, 79 (1987)
25. Croll, B. T., Arkell, G. M. and Hodge, R. P. J., *Water Research*, **8**, 989 (1974)
26. Hawley, J. R., *The Use, Characteristics and Toxicity of Mine-Mill Reagents in the Province of Ontario*, Ministry of the Environment, Ontario, Canada (1972)

BIBLIOGRAPHY

Bikales, N. M. (Ed.), *Water-soluble polymers. Polymer Science and Technology*, Vol. 2, Plenum Press, New York (1973)

Finch, C. A. (Ed.), *The Chemistry and Technology of Water Soluble Polymers*, Plenum Press, New York (1983)

Molyneux, P., *Water-soluble Polymers: Properties and Behaviour*, Vol. 1, CRC Press, Boca Raton, FL (1985)

Coagulation and Flocculation
Part II—Orthokinetic Flocculation

K. J. Ives

Professor of Public Health Engineering, University College London

NOMENCLATURE

A	Area perpendicular to flow	m^2
A_p	Projected area of stirrer blade normal to motion	m^2
C	Floc volume concentration	—
C_D	Drag coefficient	—
C_s	Suspended solids concentration	$g\ m^{-3}$
c	Chezy coefficient	$m^{\frac{1}{2}}\ s^{-1}$
D	Grain or bubble diameter	m
d	Tube diameter	m
F	Filtrability number	—
f	Pipe friction factor	—
G	Mean velocity gradient	s^{-1}
H	Head loss	m
H_B	Baffle head loss	m
H_f	Friction head loss	m
h	Distance of sampling point below water level	m
i, j, k	Labels for i-, j- and k-particles, respectively	—
k_B	Boltzmann's constant	$J\ K^{-1}$
L	Length	m
N_i, N_j	Number concentration of i- and j-particles, respectively	m^{-3}
N_o, N_t	Number concentration of original and time t particles, respectively	m^{-3}
P	Power dissipated in fluid motion	W
p	Upper limit aggregate size label	—
Q	Volumetric flow rate of liquid	$m^3\ s^{-1}$
q	Sludge bleed flow	$m^3\ s^{-1}$
R	Hydraulic radius	m
r_i, r_j	Radius of i- and j-particles, respectively	m
T	Kelvin temperature	K

T_q	Torque	N m
t	Time	s
u	Velocity of liquid motion in the x-direction	m s^{-1}
V	Volume of liquid	m^3
V_a	Volume of air	m^3
v	Velocity of liquid being stirred, or flowing in flocculator	m s^{-1}
v_a	Apparent velocity of settling of particles	m s^{-1}
v_h	Hindered velocity of settling of particles	m s^{-1}
v_p	Velocity of stirrer blade	m s^{-1}
v_s	Settling velocity of single particle	m s^{-1}
x	Dimension in the direction of liquid motion	m
z	Dimension across the velocity gradient	m
β	Collision efficiency factor	—
Δp	Pressure drop	N m^{-2}
ε	Porosity of granular media	—
μ	Dynamic viscosity of liquid	kg m^{-1} s^{-1}
ρ	Density of liquid	kg m^{-3}
ρ_s	Density of particle, or floc	kg m^{-3}
ω	Angular velocity	rad s^{-1}

4.7 INTRODUCTION

It is commonly observed that gentle stirring promotes flocculation of particles which have been destabilized and which may have commenced to aggregate by Brownian motion (see Part I of this chapter). This is due to the velocity gradients which are induced in the liquid causing relative motion and therefore collisions between the particles which are present. Such flocculation caused by fluid motion is called 'orthokinetic', to differentiate it from that caused by Brownian motion, called 'perikinetic'.

A simple theory of flocculation kinetics can be derived for a constant velocity gradient (i.e. a uniform shear field, as for a Newtonian fluid shear stress, is proportional to velocity gradient). Such constant velocity gradients are difficult to achieve in practice, as they require parallel flat plates separated by a small gap moving at constant relative velocity. The closest experimental form has been in the annular gap between coaxial rotating cylinders (Couette apparatuses)[1,2]. Consequently, the theory has been extended to velocity gradients created in turbulent flow conditions, such as paddle stirrers, baffled channels, oscillating paddles, fluidized beds etc.

4.8 THEORY

4.8.1 Constant velocity gradient (uniform shear field)

As relative motion is the cause of particle collisions, it is useful to treat one particle (j) as a collector, being stationary in a constant velocity gradient du/dz (*Figure 4.12*). The liquid contains other particles (i), which will move relative to j if they are not on the centre line through j.

It is assumed that the uniform shear field is unperturbed by the presence of

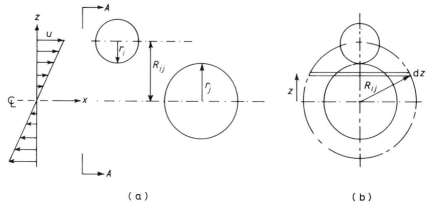

Figure 4.12. Definition sketch for orthokinetic flocculation. (a) Uniform shear field. (b) End view A–A

the particles and therefore the path of the i-particles is rectilinear. Although Arp and Mason[3] have shown that the curvature of the stream lines around the particles may appear to have a small effect, van de Ven and Mason[4] have shown that rolling of particles around one another does significantly reduce the collection efficiency of collisions. However, as a simplification each collision is assumed to result in an effective collection, i.e. flocculation with no electrical or hydrodynamic barriers. This simplification can be later adjusted by a collision efficiency factor $\beta \leqslant 1.0$ to take into account a proportion of ineffective collisions.

With reference to *Figure 4.12*, i-particles above the centre line (positive z) move in the direction x. If their centres lie within the hemi-cylinder radius $R_{ij}(=r_i+r_j)$, they will collide with the j-particle. Their velocities relative to the j-particle will depend on their distance z from the x-plane, and will equal $z(du/dz)$.

The liquid flowing through the upper hemi-cylinder, relative to the j-particle, is the sum of all flow through the elements of height dz.

$$\text{Element flow} = \text{area} \times \text{velocity}$$
$$dQ = 2[\sqrt{(R_{ij}^2-z^2)}]\, dz\, z(du/dz) \tag{4.37}$$

Total flow in the upper hemi-cylinder relative to j

$$Q_{\dagger} = 2(du/dz) \int_0^{R_{ij}} z[\sqrt{(R_{ij}^2-z^2)}]\, dz \tag{4.38}$$

There is an identical flow through the lower hemi-cylinder in the negative x-direction, so the total liquid flow relative to j is:

$$Q = 4(du/dz) \int_0^{R_{ij}} z[\sqrt{(R_{ij}^2-z^2)}]\, dz \tag{4.39}$$

which, integrated, becomes:

$$Q = \tfrac{4}{3}(du/dz)R_{ij}^3 \tag{4.40}$$

Q is the volumetric flow rate dV/dt, so for N_i particles per unit volume, the rate of collision of i-particles with the j-particle is:

$$N_i(dV/dt) = \tfrac{4}{3}N_i(du/dz)R_{ij}^3 \qquad (4.41)$$

If there are N_j j-particles per unit volume, then the total collision rate is:

$$dN_{ij}/dt = N_iN_j(dV/dt) = \tfrac{4}{3}N_iN_j(du/dz)R_{ij}^3 \qquad (4.42)$$

This is the simple Smoluchowski equation of orthokinetic flocculation, and shows that the rate of flocculation is second-order with respect to concentration, depends linearly on the velocity gradient and is proportional to the third power of the collision radius. Consequently, in these simple terms theory states that the rate of flocculation can be increased by:

1. increasing the collision radius of the particles; thus flocculation is self-enhancing as larger flocs are produced, also the presence of existing large particles, such as occurs in solids-contact flocculation, or floc-blanket clarifiers, will improve flocculation;
2. increasing the concentrations of particles present; thus flocculation is self-diminishing as particle numbers decrease due to aggregation; however, the provision of many new particles by precipitation from alum hydrolysis, for example, is advantageous; also the maintenance of a large number of flocs always to be present, as in solids-contact or floc-blankets is advantageous;
3. increasing the velocity gradient; however, there is a limit to the shear stress which flocs can withstand, so excessive gradients may cause floc break-up, particularly as the flocs grow in size; this can be overcome by decreasing the velocity gradient from an initially high value, when flocs are small, to lower values as the flocs grow: this is taper flocculation.

It must be stressed that equation 4.42 is a very simple form, because the aggregation of an i-particle with a j-particle creates a new particle (k), and the i- and j-particles disappear after aggregation. The new k-particle can itself interact with other i-, j- and k-particles, with appropriate concentrations and collision radii. The rate of change of aggregates of type $k(=i+j)$ is given by a two-term equation consisting of their appearance due to $i+j$ collisions, and their disappearance due to collisions with other particles,

$$dN_k/dt = \frac{1}{2}\sum_{\substack{i=1 \\ j=k-i}}^{i=k-1} \tfrac{4}{3}N_iN_j(du/dz)R_{ij}^3 - N_k\sum_{i=1}^{\infty} \tfrac{4}{3}N_i(du/dz)R_{ik}^3 \qquad (4.43)$$

The derivation of equation 4.43 is given in Ives[5]. It can be seen that it sums all the possible collisions from the original ($i=1$) particles up to an aggregate of infinite size. This upper limit is obviously impossible as it assumes an infinite supply of 1-particles, but more importantly it ignores the limit on floc size imposed by the disintegration of large flocs as a result of the shear stress provided by the velocity gradient. Therefore, equation 4.43 can be elaborated further with an upper limit aggregate size (p); it can also be written to replace the collision radii R_{ij} and R_{ik} by terms which include the radius of the original particles r_1 and the particle labels i, j and k.

Some simplifications are possible if the initial suspension is monodisperse at $t = 0$. Then $k = 1$, $N_k = N_o$ (original number concentration), and the initial rate of disappearance of 1-particles is:

$$-dN_1/dt = \tfrac{16}{3}(du/dz)r_1^3 N_o^2 \qquad (4.44)$$

At time t,

$$\sum_{k=1}^{p} N_k = N_t \qquad \text{(the total number concentration of all particles and aggregates)}$$

$$-dN_t/dt = \tfrac{16}{3}(du/dz)r_1^3 \sum_{k=1}^{p} N_k^2 k \qquad (4.45)$$

If the rate of flocculation is slowed by an energy barrier (see chapter 4, Part I), or hydrodynamic effects between approaching particles, then the right-hand side of equation 4.45 is multiplied by a collision efficiency β ($\leqslant 1.0$).

A fuller theoretical discussion of the orthokinetic flocculation rate equations is given in Ives[5]. Experimental tests of the kinetics of orthokinetic flocculation are described by Ives and Bhole[6, 7] and Ives and Al Dibouni[2].

4.8.2 Variable velocity gradient (non-uniform shear field)

In most practical cases, the flow in flocculators is turbulent, fluctuating rapidly in both position and time. Where the flow may be laminar in tube flocculators or lamella separators or deep bed filters, the velocity gradients are not uniform but parabolic or quasi-parabolic in form.

The velocity gradient (du/dz) is represented by an average value G, based on the power dissipated per unit liquid volume (P/V) by the non-uniform shear flow. Assuming that the liquid is Newtonian, the mean velocity gradient is given by:

$$G = \sqrt{\left(\frac{P}{V\mu}\right)} \qquad (4.46)$$

where μ is the dynamic viscosity.

The derivation of equation (4.46) is based on the shear force on an element cube of liquid, causing torsional work which is evaluated as work/time = power. Consequently, the value of G assumes that all elemental cubes of liquid in the volume V are being sheared at the same rate (on average) and that G is also a time-averaged value[5].

The power P can be calculated in terms of the head lost in flow through a pipe, a tube module flocculator, lamella separators or fixed (deep bed) or fluidized bed flocculators, whether in laminar or turbulent flow. It can also be estimated for paddle flocculators and stirred beakers (jar test apparatus) on the basis of paddle drag, or torque on the drive shaft. Such practical cases are dealt with in section 4.10.

4.8.3 Transition from perikinetic to orthokinetic flocculation

The simple orthokinetic rate equation 4.42 has an analogous equation for perikinetic flocculation, due to Brownian-motion collisions of particles (see Part I of this chapter, which deduces the half life of particles from such flocculation). The perikinetic rate equation is:

$$^p\left(\frac{dN_{ij}}{dt}\right) = 4\pi N_i N_j D_{ij} R_{ij} \tag{4.47}$$

where D_{ij} is the sum of the Stokes–Einstein diffusion coefficients for the i- and j-particles.

$$D_{ij} = \frac{2k_B T}{3\pi\mu R_{ij}} \tag{4.48}$$

where

$\quad k_B$ is Boltzmann's constant $(1.38 \times 10^{-23}$ J K$^{-1})$
$\quad T$ is Kelvin temperature $(K = °C + 273)$
$\quad \mu$ is dynamic viscosity of the liquid.

Combining equations 4.47 and 4.48:

$$^p\left(\frac{dN_{ij}}{dt}\right) = \frac{8N_i N_j k_B T}{3\mu} \tag{4.49}$$

The orthokinetic rate equation 4.42 is:

$$^o\left(\frac{dN_{ij}}{dt}\right) = \frac{4N_i N_j G R_{ij}^3}{3} \tag{4.42}*$$

$$\frac{^o(dN_{ij}/dt)}{^p(dN_{ij}/dt)} = \frac{\mu G R_{ij}^3}{2k_B T} \tag{4.50}$$

Initially for a monodisperse suspension $i = j = 1$ and

$$R_{ij} = 2r_1$$

When the orthokinetic rate equals the perikinetic rate, their ratio is one:

$$r_1 = \left(\frac{k_B T}{4G\mu}\right)^{\frac{1}{3}} \tag{4.51}$$

At 25°C, $T = 298$ K, $\mu = 0.895 \times 10^{-3}$ kg m^{-1} s^{-1}

$$\left(\frac{k_B T}{4\mu}\right)^{\frac{1}{3}} = 1.05 \times 10^{-6} \text{ m s}^{-\frac{1}{3}}$$

$$r_1 = 1.05 \times 10^{-6} G^{-\frac{1}{3}} \text{ m} \tag{4.52}$$
$$= 1.05 G^{-\frac{1}{3}} \mu\text{m}$$

Equation 4.52 can be used to calculate the collision radii of particles when perikinetic and orthokinetic rates of flocculation are equal, and flocculation is in transition from Brownian-motion (diffusion) dominated kinetics to fluid-motion kinetics. For example:

G	s^{-1}	1	10	20	50	100
r_1	μm	1.05	0.5	0.4	0.3	0.23

So for most cases of fluid motion caused by stirring or baffled flow the perikinetic rate dominates for particles less than 1 μm diameter, and the orthokinetic rate dominates for particles larger than 1 μm. For example, at $G = 10$ s^{-1} when the particle radius has increased to 5 μm (diameter 10 μm) orthokinetic flocculation will proceed 1000 times faster than perikinetic (equation 4.50).

Even when the particles are completely destabilized, with no repulsion energy barrier, there is evidence[4] that orthokinetic collisions will be reduced by particles rolling around one another without aggregating. This will multiply the orthokinetic rate (equation 4.42) by a factor of less than one, which will make the sizes of particles undergoing equal effective rates of perikinetic and orthokinetic flocculation smaller than those listed above (which assumed that all collisions were permanent and effective).

4.8.4 Taper flocculation

As mentioned in section 4.8.1, the magnitude of the velocity gradient G is limited by the possibility of large flocs being broken by the shear stress. Both the nature of the material constituting the flocs and the conditions under which they were formed affect their shear strength. For example, flocs formed with polymers are usually stronger; those with hydroxide flocs are usually weak; flocs formed rapidly in intense velocity gradients are usually more compact and, therefore, stronger than those formed in lower velocity gradients. As flocs become larger, they tend to become less dense and more susceptible to shear. Although studies have been made of floc break-up, no formal quantitative relationships are available because some flocs are broken by attrition of small particles and aggregates from their surfaces while others are sheared into two or more fragments of about equal sizes.

For aluminium or ferric hydroxide flocs, as formed in water treatment by paddle flocculators, the relationship illustrated schematically in *Figure 4.13* indicates the zones of aggregation and break-up. To maximize the rate of

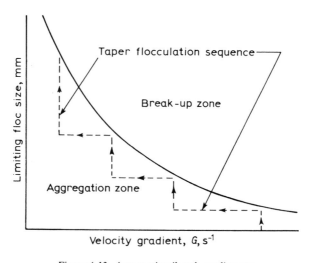

Figure 4.13. Aggregation/break-up diagram

flocculation without break-up as the flocs grow in size, a flocculation scheme could follow the broken line in *Figure 4.13*. That starts at a high value of G when flocs are small; as they grow, they approach the break-up zone and G is therefore reduced allowing further growth in the aggregation zone. This progressive reduction in G to keep the flocculation rate as high as possible, but below the break-up zone, is called 'taper flocculation'. It is usually achieved in practice by passing the flocculating suspension through a series of rotating paddles, which are sequentially rotating more and more slowly, until the flocculated suspension containing large flocs passes into a settling chamber. Care has to be taken in transferring flocs from paddle to paddle and chamber to chamber as weirs, orifices etc. may produce local velocity gradients which would cause the flocs to break up.

4.8.5 Optimum flocculation conditions

The best flocculation conditions are those which rapidly form large separable flocs, leaving no residual primary particles or small aggregates. However, it has already been shown that high rates of flocculation cannot be utilized if the velocity gradients are too high, thus causing floc break-up. So a balance has to be achieved between the velocity gradient G and the time of flocculation t, one compensating for the other.

In flow-through flocculation units (particularly paddle stirrers in series) where the concentration of suspended particles is low (typically 1000 p.p.m. $v/v = 10^{-3}$) the flocculation is characterized by the dimensionless product Gt (sometimes called the Camp number). Arising from observational data from waterworks in the USA, the optimum value of Gt is set between the limits 10^4 and 10^5 (see Camp[15]).

It has long been known (see Ives[5, 12]) that this is an insufficient criterion to cover all cases of flocculation, particularly those of floc blanket clarification, and that integration of equation 4.45 yields a dimensionless group GtC, where C is the floc volume concentration. Consequently, the very low G values (typically < 5 s^{-1}) in floc blanket clarifiers, are compensated by much higher C values (typically 0.1–0.2). It appears that the optimum GtC value lies between 100 and 500. This would agree with the value of $Gt = 10^5$ multiplied by $C = 10^{-3}$ ($GtC = 10^2$) for paddle-type flow-through flocculators.

4.9 LABORATORY TESTING

4.9.1 Jar tests

The traditional form of laboratory testing of stirred flocculating suspensions is to use the jar test apparatus (*Figure 4.14*). This consists of a set of beakers (usually 4 to 6) which are simultaneously stirred by a paddle in each so that conditions in each beaker (jar) are identical. The beakers are normally tall-form 1 litre capacity containing 0.8 litre of suspension, but sometimes larger units are also used. Unfortunately, no standard design exists so the shapes of beakers may vary, and the paddle type and dimensions may include vertically slatted (picket fence) stirrers extending through the whole depth, propeller

Figure 4.14. Jar test apparatus

stirrers at the bottom of the liquid, or flat blades. In some cases, fixed baffles (stators) are included in the beakers. A typical flat-bladed design is shown in *Figure 4.15.* The stirring speeds are also not standard, but usually an initial high-speed mixing of about one minute is followed by 10–20 minutes of

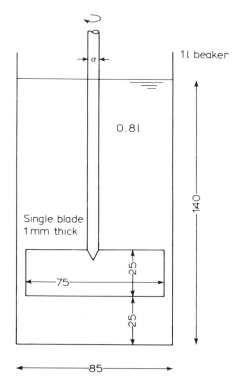

Figure 4.15. Typical jar test stirrer. All dimensions given in mm

slow stirring. These variations of geometry and stirring conditions make comparisons difficult between different jar test apparatuses.

Comparative testing in the beakers of a particular jar test apparatus is, however, valid due to the identity of stirring conditions. So the effects of various doses of flocculating chemicals, of different pH conditions, and of different types of chemicals (e.g. indifferent electrolytes, aluminium or ferric salts, various polyelectrolytes) can be determined. Usually, an optimum is required and this relates to the formation by flocculation of a separable suspension.

Frequently, the first assessment is visual, looking for the appearance of visible flocs (which normally is greater than about 100 μm) and their growth to sizes of about 1 mm with clear liquid between. As flocculation is a prior step to some solid–liquid separation processes, an assessment is required of their settling, filtration or flotation behaviour.

Settling is usually assessed in the beaker, by stopping the stirring and allowing some period of quiescence and measuring the turbidity of the supernatant liquid. If this is by sampling at a known depth, after a known time, then it can be assumed that all flocs with settling velocity greater than the depth/time are removed from the supernatant liquid. By taking samples at a given depth at sequential times, in a manner similar to the Andreasen pipette method (see chapter 2) a settling velocity frequency distribution can be determined for the flocculated particles. Alternatively, the flocculated suspension could be transferred to a sedimentation balance, Coulter counter or some other measuring apparatus, but the act of transfer may affect the flocs, possibly breaking the larger ones. Such a settling velocity frequency distribution would guide the design of a subsequent settling stage after flocculation.

Filtration of flocculated suspensions can be assessed using the filterability test method described in chapter 11. Using the filterability number F against the product Gt an optimum flocculation can be determined for filtration through a sample of porous granular material as exemplified in *Figure 4.16*. A variant of the filterability test apparatus replaces the sample funnel (see *Figure 11.10*) with a paddle-stirred beaker to allow direct testing of the flocculated suspension without transfer from the jar test to the filterability apparatus.

Flotation of flocs by air bubbles can be tested by using jar test beakers which have a false bottom diffuser plate so that after flocculation by stirring, diffused air can be released through the suspension to form a floc froth scum on the surface. Such a modified jar test apparatus has been designed by the UK Water Research Centre. Similarly, if dissolved air flotation is of interest then compressed air can be released into the bottom of the beaker to float the flocculated particles. Obviously in flotation it is the clarity of the bottom liquid which is important and sampling near the bottom is required for turbidity or particle count measurements.

Comparisons of different jar test apparatuses can be made by estimating the mean velocity gradient (G value) using equation 4.46:

$$G = \sqrt{\left(\frac{P}{V\mu}\right)}$$

where

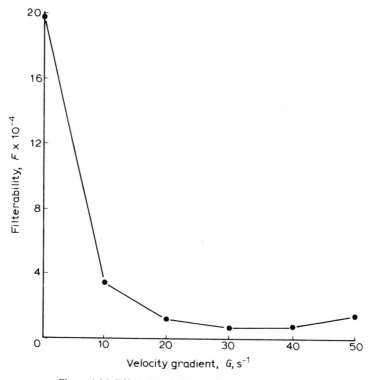

Figure 4.16. Effect of orthokinetic flocculation on filterability

V is the volume of the stirred beaker

P is the power transferred from the stirrer to the suspension.

This power can be measured by using a sensitive torquemeter (0.01–0.2 N mm) on the stirrer drive shaft. Then $P = T_q\omega$, where T_q is the measured torque and ω the rotation speed in rad s^{-1}. *Figure 4.17* shows values of G measured in this way, against rotational speed for the paddle stirrer in *Figure 4.15*. This shows that for a stirrer speed of 40 rev min^{-1} the G value is 25 s^{-1}, which is typical of jar test apparatuses. With a typical flocculation time of about 15 min, the product $Gt = 2.25 \times 10^4$, which is within the range 10^4–10^5 quoted for paddle type flow-through flocculators in water treatment[15].

An alternative method of determining the power P is to calculate the drag force on the paddle blade and multiply by the velocity of the blade relative to the suspension.

$$P = \text{drag force} \times (v_p - v) \qquad (4.53)$$

where $(v_p - v)$ is the paddle velocity minus the water velocity, giving the relative velocity.

$$\text{drag force} = C_D A_p \rho (v_p - v)^2 / 2 \qquad (4.54)$$

where

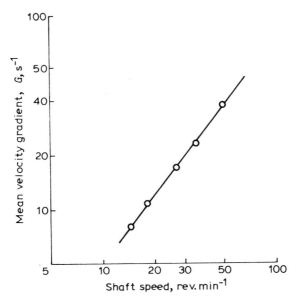

Figure 4.17. Velocity gradient produced by various shaft speeds in paddle flocculator of Figure 4.15

C_D is the drag coefficient, which depends on blade geometry and rotational speed

A_p is the projected area of the blade normal to the rotation direction

$\rho(v_p-v)^2/2$ is the Bernouilli dynamic pressure (the pressure energy equivalent of velocity energy).

It follows from equation 4.46 that G is given by

$$G = \sqrt{\left(\frac{C_D A_p \rho (v_p - v)^3}{2 V \mu}\right)} \tag{4.55}$$

It is difficult to determine C_D and (v_p-v) in a stirred beaker, but using special laboratory techniques Bhole[8] has determined for the stirrer geometry of *Figure 4.15* that C_D varies with G (and, therefore, with rotational velocity—*Figure 4.17*) as follows:

G s^{-1}	10	20	30	40	50
C_D	1.81	1.20	1.13	1.02	0.94

and that (v_p-v) was 0.52 times the blade tip velocity over this range of G values.

For example, taking the stirrer of *Figure 4.15*:

velocity of blade tip at 40 rev min^{-1} = $\pi \times 75 \times 10^{-3} \times \frac{40}{60}$ m s^{-1}
relative velocity $(v_p-v) = 0.52 \times \pi \times 75 \times 10^{-3} \times \frac{40}{60}$ m s^{-1}
          = 0.0816 m s^{-1}
drag coefficient at 40 rev min^{-1}, C_D = 1.16 by interpolation
projected area of blade, A_p = $75 \times 10^{-3} \times 25 \times 10^{-3}$ m^2
density of liquid (water), ρ = 10^3 kg m^{-3}
volume of liquid (0.8 litre), V = 0.8×10^{-3} m^3
dynamic viscosity (20°C), μ = 10^{-3} kg m^{-1} s^{-1}

$$G = \sqrt{\left(\frac{1.16 \times 75 \times 25 \times 10^{-6} \times 10^3 \times 0.0816^3}{2 \times 0.8 \times 10^{-3} \times 10^{-3}} \right)} \; \mathrm{s}^{-1}$$

$$= \sqrt{(738.6)} = 27 \; \mathrm{s}^{-1}$$

This agrees closely with the torquemeter value of 25 s^{-1}.

Assessment of flocculation in jar test apparatuses is useful in a comparative sense of finding the best chemical conditions, and showing that a separable floc can be formed under stirring conditions. The identification of the mean velocity gradient (G) is useful, but it cannot be assumed that conditions at full scale will be the same, even if the geometry of the beaker and stirrer were scaled up. Hydrodynamic conditions on the full scale are determined by the scale of the fluid motion and its spatial variation within the flocculator. Although a satisfactory G value may be determined in the jar test, the same average G value at full scale may lead to either better or poorer flocculation depending on the detailed geometry of the fluid motion, affected among other characteristics by the flow-through pattern in the full scale which is absent in the batch-type jar test.

4.9.2 Small-bore tubes

Bearing in mind the problem that scale-up is not meaningful from the jar test apparatus, an alternative test method is to avoid paddle stirrers (although they appear superficially like the full-scale paddle flocculators) and use small-bore tubes. These can be related to practice in the sense that they are flow-through devices, and their mean velocity gradients (G) can be easily measured or calculated.

Flow through small-bore tubes, of a few millimetres in diameter, is laminar and steady, not varying with time. The pressure drop along such a capillary can be measured manometrically, or calculated from Poiseuille's equation,

$$\Delta p = \frac{32 \mu v L}{d^2} \tag{4.56}$$

The power dissipated along the tube, in laminar flow, is the weight of liquid per unit time ($\rho g Q$) multiplied by the loss of pressure head (H)

$$P = \rho g Q H \tag{4.57}$$

But

$$\rho g H = \Delta p$$

So

$$P = Q \Delta p = \frac{32 Q \mu v L}{d^2} \tag{4.58}$$

The volume of liquid is the tube cross-sectional area (A) multiplied by the length (L).

$$\frac{P}{V} = \frac{32 Q \mu v L}{A L d^2} \tag{4.59}$$

But

$$Q/A = v \qquad \text{(mean velocity in the tube)}$$

$$\frac{P}{V} = \frac{32\mu v^2}{d^2} \qquad (4.60)$$

The mean velocity gradient (G) is given by equation 4.46

$$G = \sqrt{\left(\frac{P}{V\mu}\right)} \qquad (4.46)^*$$

$$G = \sqrt{\left(\frac{32v^2}{d^2}\right)} = 5.66\frac{v}{d} \qquad (4.61)$$

If the flocculation criterion Gt is calculated, as a tube is a flow-through flocculator, with no retention of particles:

$$t \text{ (mean residence time)} = \frac{AL}{Q} = \frac{L}{v} \qquad (4.62)$$

$$Gt = 5.66\frac{L}{d} \qquad (4.63)$$

This shows that the flocculation is dependent upon the tube geometry only, and independent of the flow rate, providing that G is within a reasonable range for aggregation without break-up.

There are refinements of the calculation of G and t to allow for the details of the paraboloid velocity distribution in a laminar flow tube. Among other aspects, this shows that particles which travel near the tube wall are subject to the highest velocity gradient, but the lowest velocity and hence the longest time. Therefore the local Gt near the tube wall is much higher than that near the centre of the tube. Such laboratory tube flocculators are usually coiled helically to make a more compact apparatus. The resultant curvature of the tube may induce a secondary circulation which is imposed on the primary laminar flow, therefore providing some lateral mixing of particles which would make the average values of G and t in equations 4.61 and 4.62 more realistic.

By taking certain values of G and t, it is possible to calculate the dimensions of laboratory tube flocculators. Assuming $G = 25$ s^{-1} and $Gt = 2.5 \times 10^4$ (values similar to the jar test—section 4.9.1), then $t = 10^3$ s. From equation 4.63

$$L = \frac{2.5 \times 10^4 d}{5.66}$$

The dimensions flow and volumes of tubes from 1 to 5 mm bore, are as given in *Table 4.7*. The highest Reynolds' number occurs for the largest tube; for water at 20°C, Re(max) = 110, i.e. well within the range of laminar flow.

The long tubes calculated above show the need to form them into helical coils in order to keep the apparatus compact. Such apparatuses have been used in several European countries but no published work has appeared up to 1981. Some references to microtube flocculation were made in 1979[9] but these were for specialized research and were below 0.1 mm in diameter.

Table 4.7

Tube bore d, mm	Tube length L, m	Tube volume, cm³	Flow rate, cm³ min⁻¹
1	4.4	3.5	0.21
2	8.8	28.0	1.66
3	13.2	93.0	5.60
4	17.6	·221.0	13.26
5	22.0	432.0	25.90

It is claimed that small-bore tube flocculators have advantages over the conventional jar test because there is a rapid response to changes in chemical conditions and so optimum conditions can be established fairly quickly using quite small sample volumes, typically about 20% of the jar test requirements.

The full-scale analogue of tube flocculation is in flocculation in pipes, although the flow would be turbulent, not laminar; this is discussed later in section 4.10.3. It may also be considered that flocculation in inclined tube modules is similar, as these are often in laminar or transitional flow.

4.9.3 Couette apparatus

Couette flocculators consist of coaxial rotating cylinders, with an annular gap between the cylinders comprising the flocculating volume. These apparatuses provide a velocity gradient which is almost uniform, having a very slight non-linearity due to the curvature of the annular gap.

Figure 4.18. Couette flocculator

Several designs have evolved, many with horizontal axes (i.e. both ends closed) with the outer cylinder rotating and the inner cylinder fixed, which provides best hydrodynamic stability. Some horizontal-axis Couette flocculators are flow-through designs, even with a tapered gap to provide taper flocculation[6, 7]. However, horizontal-axis apparatuses suffer from end effects which cause secondary circulation, so only a limited central zone (about one quarter of the length) is in defined laminar flow.

Vertical-axis Couette flocculators, with a free liquid surface, have only a bottom-end effect. The design shown in *Figure 4.18* is 150 mm high, with an inner cylinder diameter of 27 mm and a 3 mm annular gap filled with 40 ml of liquid. For mechanical convenience the inner cylinder rotates, the outer being fixed, although this is slightly less hydrodynamically stable than the converse. The fluid motion in this apparatus has been fully analysed experimentally and theoretically by Elson[1], and the lower-end effect is confined to the bottom 6 mm depth; all the rest of the annular volume is in almost uniform shear, with a nearly linear velocity gradient. The design of such Couette flocculators is described in Ives[5] and flocculation data is given by Ives and Al Dibouni[2].

Couette flocculators are basic research apparatuses of flocculation kinetics, and can be used with particle counters (e.g. Coulter counter) as a standard for other laboratory flocculators which have ill-defined velocity gradients.

4.10 PRACTICAL FLOCCULATORS

Flocculators used in practice can be classified into four groups:

1. *paddle:*   revolving on horizontal or vertical axes,
                oscillating, vertically or pendulum-like,
                propeller or turbine blades;
2. *baffles:* fixed in channels or tanks providing either 'over-under'
                or 'round-the-end' motion,
                labyrinth or honeycomb structures in modules for tanks
                or pipes (static mixers);
3. *pipes:*   turbulent flow provides the power dissipation;
4. *particles:* fixed beds (filters),
                fluidized beds (floc blanket, with possible solids recirculation),
                bubble swarms,
                differential settling.

Many, but not all, of these are reviewed and compared by Polasek[10].

4.10.1 Paddle flocculators

The power transferred from moving paddles to the liquid, for flocculation, is the drag force on the paddle multiplied by the distance moved per unit time (see section 4.9.1):

$$\text{power} = \text{force} \times \text{distance/time} = \text{force} \times \text{velocity}$$
$$\text{drag force} = C_\text{D}\rho(v_\text{p}-v)^2 A_\text{p}/2 \tag{4.54)*}$$

where

$\rho v^2/2$ is the Bernouilli dynamic pressure
$(v_\text{p}-v)$ is the relative velocity of the liquid to the paddle blade, assumed to be about 0.8 paddle blade mean velocity, and
A_p is the projected area of the blades normal to motion.

$$\text{power} = C_\text{D}\rho(v_\text{p}-v)^3 A_\text{p}/2$$

Mean velocity gradient $G = \sqrt{(\text{power/volume} \times \text{viscosity})}$

$$G = \sqrt{\left(\frac{C_\text{D}\rho(v_\text{p}-v)^3 A_\text{p}}{2V\mu}\right)} \tag{4.55)*}$$

The drag coefficient C_D depends on the length:width ratio of the paddle blade, at normal flocculator speeds:

length:width ratio	5	20	infinite
drag coefficient C_D	1.2	1.5	1.9

Example 4.1

Calculate the mean velocity gradient for the four-bladed paddle flocculator shown in *Figure 4.19*, in a tank containing 100 m³ of water at 20°C:

paddle length: width ratio $= 3/0.5 = 6$, therefore assume $C_\text{D} = 1.2$
water viscosity at 20°C $= 10^{-3}$ kg m⁻¹ s⁻¹

paddle velocity $v_\text{p} = \dfrac{4\pi \times 2}{60}$ m s⁻¹   (mean)

relative velocity of water to paddle $(v_\text{p}-v) = 0.8\, v_\text{p}$

$$(v_\text{p}-v) = \frac{0.8 \times 4\pi \times 2}{60} = 0.33 \text{ m s}^{-1}$$

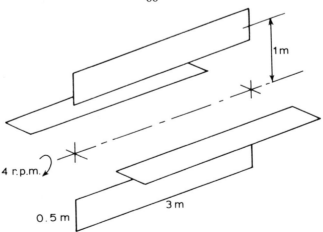

1m

4 r.p.m.

3 m

0.5 m

Figure 4.19. Four-blade paddle flocculator

mean velocity gradient G from equation 4.55:

$$G = \sqrt{\left(\frac{C_D\rho(v_p-v)^3 A_p}{2V\mu}\right)}$$

$$= \sqrt{\left(\frac{1.2\times 10^3 \times 0.33^3 \times 4 \times 1.5}{2\times 100 \times 10^{-3}}\right)}$$

$$= 36 \text{ s}^{-1}$$

Generally, the G value should be between 20 and 75 s^{-1}.

Figure 4.20 shows typical arrangements of horizontal- and vertical-axis paddle flocculators. The axes should rotate in alternate directions along the tank, e.g. in *Figure 4.20* the first and third paddles rotate clockwise, but the

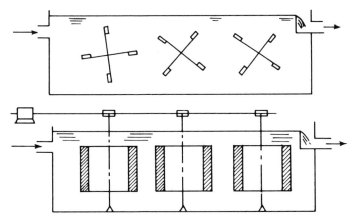

Figure 4.20. Paddle flocculators

second rotates anticlockwise. This rolls the liquid containing the flocs from paddle to paddle without excessive shear as the paddle blades pass each other. If taper flocculation is desired, each sequential axis will rotate more slowly. The choice between horizontal and vertical axes depends primarily on mechanical considerations, such as the power drive system and the maintenance of bearings and support frames.

Detention times for such flocculation tanks should be 1200–1800 s. For the example above, $G = 36$ s^{-1} and, assuming $t = 1500$ s, the product $Gt = 5.4 \times 10^4$ which is within the optimum range 10^4–10^5 for flow through flocculators with no solids retention.

The horizontal velocity (mean) of the liquid should be 0.25–0.4 m s^{-1}. Paddle tip velocities should not exceed 0.8 m s^{-1}, and the total paddle blade area should not be greater than 20% of the cross-sectional area of the tank, otherwise the liquid will roll with the blades without the desirable velocity gradient.

Oscillating or reciprocating paddle flocculators may be on shafts mounted vertically, which move vertically up and down in the direction of the shaft axis, or they may be on frames swinging to and fro in a pendulum-like motion. Both types impart a harmonic motion to the blades.

For the vertical agitator, the distance moved by the blades is the same for all and the relative velocity is expressed by the harmonic mean velocity; a drag coefficient of $C_D = 3.0$ has been suggested. With this information the velocity gradient is calculated from the power dissipated, using equation 4.55.

For the pendulum-type agitator (*Figure 4.21*, after Polasek[10]) the motion of the blades is through small arcs, increasing with distance down the pendular frame. To maintain a uniform intensity of agitation over the entire depth, it has been proposed that the blades should be distributed logarithmically along the frame depth as indicated in principle in *Figure 4.21*.

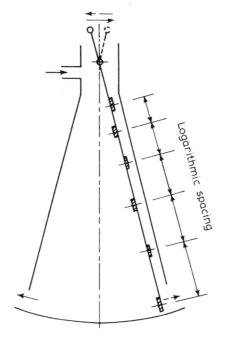

Figure 4.21. Pendulum flocculator

Power can be calculated from the mean harmonic velocity and drag coefficient from paddle shape, summed for all the paddles. Further details are given by Polasek[10].

Propeller or turbine blade flocculators are difficult to assess from a theoretical standpoint as they vary greatly in form and efficiency. An example of successful taper flocculation using turbine agitators is described by Bernhardt and Schell[11], for the formation of ferric hydroxide flocs for phosphorus removal of the water supplying the Wahnbachtal reservoir. The practical approach to such propeller/turbine flocculators is to make torque measurements on the shaft of a prototype in the liquid, using the relationship $P = T_q \omega$ (where T_q is torque and ω is angular velocity).

4.10.2 Baffle flocculators

The flows around baffles in a channel, whether 'over-under' or 'round-the-end', or through slatted baffles, or through labyrinths in channels or pipes, all dissipate energy as head loss, which per unit time gives the power. As in flow-through tubes (section 4.9.2) the power dissipated is the weight of liquid per unit time $(\rho g Q)$ multiplied by the head loss (H),

$$P = \rho g Q H \qquad (4.57)^*$$

The head loss is greatest at each change of direction around a baffle H_B, and least in straight stretches of flow H_f. The direction-change head loss H_B is given by equation 4.64

$$H_B = n C_D \frac{v^2}{2g} \qquad (4.64)$$

where

n is the number of direction changes, and
v is the mean velocity (Q divided by cross-sectional area).

The drag coefficient C_D is about 3.0 for 180 degree changes in direction, and 2.0–2.5 for 90 degree changes.

The head loss along the straight channels is given by the Chezy equation 4.65,

$$H_f = \frac{L v^2}{c^2 R} \qquad (4.65)$$

where

c is the Chezy coefficient, and
R is the hydraulic radius of the channel.

The Chezy coefficient depends on the channel wall roughness and the hydraulic radius, and can be calculated from formulae in standard books on fluid mechanics. In practical terms H_f will be much less than H_B, so as a first approximation $H \simeq H_B$.

Velocities in baffled channels are usually between 0.25 and 0.5 m s^{-1}, and retention time varies between 10 and 60 min. However, a disadvantage of baffled channels is that their performance varies with flow rate, whereas mechanical flocculators can be operated independently of flow rate.

Example 4.2

Calculate the number of baffles required in a round-the-end baffled channel under average conditions of water velocity, $v = 0.3$ m s^{-1}, $t = 30$ min (1800 s) and Gt for flow-through flocculation $= 5 \times 10^4$:

$$G = \frac{5 \times 10^4}{1.8 \times 10^3} = 28 \text{ s}^{-1}$$

volume of channel $= LA$

From equations 4.46 and 4.57

$$G = \sqrt{\left(\frac{\rho g Q H}{L A \mu} \right)}$$

But $Q/A = v$ and $v/L = 1/t = 1/1800$ and, substituting equation 4.64 for H,

$$G = \sqrt{\left(\frac{\rho g n C_D v^2}{t\mu 2g}\right)}$$

$$n = \frac{2G^2\mu t}{\rho C_D v^2}$$

At 20°C, water viscosity $\mu = 10^{-3}$ kg m^{-1} s^{-1}

$$n = \frac{2 \times 28^2 \times 10^{-3} \times 1800}{10^3 \times 3.0 \times 0.3^2}$$

$$= 10.45, \text{ say } 11 \text{ baffles}$$

For slatted baffles the head loss can be assessed using the flow past fixed blades where the relative velocity is the liquid flow-through velocity through the slat gaps, and the drag coefficient is given by the baffle geometry, as in section 4.10.1.

In labyrinth or honeycomb flocculators (static mixers) the head loss has to be measured experimentally with a prototype module in a test rig. Then equation 4.57 can be used to calculate the power dissipation and, hence, the mean velocity gradient G.

4.10.3 Pipe flocculators

The natural turbulence of flow through a pipe can create velocity gradients leading to flocculation. In the laboratory, small-bore tubes operate in laminar flow (section 4.9.2) with Reynolds' numbers less than 2000. In practical pipe flow the situation is turbulent and head loss is given by the Darcy–Weisbach equation 4.66.

$$H = \frac{4fL}{d} \times \frac{v^2}{2g} \tag{4.66}$$

where f is the pipe friction factor, which depends on the pipe roughness and Reynolds' number, and is usually given in graphical form in standard books on fluid mechanics.

The mean velocity gradient is calculated from equations 4.46 and 4.57

$$G = \sqrt{\left(\frac{\rho g Q H}{V\mu}\right)} \tag{4.67}$$

The pipe volume is LA, and the criterion $Gt = 10^4 - 10^5$ for flow-through flocculators is used.

Such pipe flocculators tend to create rather high velocity gradients at practical flow rates and need to be very long to give the required detention times. Therefore it may be useful to utilize an existing pipe as a flocculator, but it is not economic to build one for the purpose, as a concrete channel with baffles would be more compact and cost less.

Example 4.3

Calculate the length of 600 mm diameter pipe required to produce a Gt value of 5×10^4 at a water flow of 20 Ml d^{-1}:

friction factor $f = 0.005$
water temperature 10°C
kinematic viscosity $\mu/\rho = 1.31 \times 10^{-6}$ m² s⁻¹

From equation 4.67

$$G = \sqrt{\left(\frac{\rho g Q H}{L A \mu}\right)} = \sqrt{\left(\frac{\rho g v H}{L \mu}\right)}$$

$$v = \frac{Q}{A} = \frac{20 \times 10^3 \times 4}{3600 \times 24 \times \pi \times 0.6^2} = 0.82 \text{ m s}^{-1}$$

$$\frac{H}{L} = \frac{4f}{d} \frac{v^2}{2g} = \frac{4 \times 0.005 \times 0.82^2}{0.6 \times 2g} = \frac{0.011}{g}$$

$$G = \sqrt{\left(\frac{\rho g \times 0.82 \times 0.011}{\mu g}\right)} = \sqrt{\left(\frac{0.82 \times 0.011}{1.31 \times 10^{-6}}\right)} = 83 \text{ s}^{-1}$$

$$t = \frac{5 \times 10^4}{83} = 0.6 \times 10^3 \text{ s}$$

$$L = vt = 0.82 \times 0.6 \times 10^3 = 492 \text{ m, say } 500 \text{ m}$$

4.10.4 Particle flocculators

With the exception of sedimentation where differential settling velocities of flocs create relative motion, hence orthokinetic flocculation, particle flocculators rely on the liquid drag past the particles to create the velocity gradients.

4.10.4.1 Fixed bed (deep bed filters)

In the case of fixed bed (deep bed granular filters) flocculators the power is dissipated as laminar flow through granular media, with local velocity gradients in the pores formed by quasi-parabolic velocity distributions. The system is analogous to small-bore tubes (section 4.9.2) bearing in mind the capillary model concept of the Kozeny–Carman equation 4.68 for head loss through a fixed bed,

$$\frac{H}{L} = \frac{5\mu v(1-\varepsilon)^2}{\rho g \varepsilon^3}\left(\frac{6}{D}\right)^2 \qquad (4.68)$$

where

ε is the porosity of the fixed bed (pore volume/total volume), and
D is the grain diameter of the particles of the fixed bed.

Combining equations 4.46 and 4.57 to give the mean velocity gradient, and substituting for the liquid volume in the pores $V = \varepsilon A L$

$$G = \sqrt{\left(\frac{\rho g Q H}{\varepsilon A L \mu}\right)} \qquad (4.69)$$

Substituting for H/L from equation 4.68 and noting that $v = Q/A$

$$G = \sqrt{\left(\frac{180\ v^2(1-\varepsilon)^2}{\varepsilon^4 D^2}\right)} = \frac{13.4v(1-\varepsilon)}{\varepsilon^2 D} \tag{4.70}$$

The residence time t is the pore volume εAL divided by the volumetric flow rate Q,

$$t = \frac{\varepsilon AL}{Q} = \frac{\varepsilon L}{v} \tag{4.71}$$

The product Gt, being a criterion of flocculation for flow-through flocculators, is given by the product of equations 4.70 and 4.71

$$Gt = \frac{13.4(1-\varepsilon)L}{\varepsilon D} \tag{4.72}$$

Consequently, flocculation is dependent on the geometry of the fixed bed, and not on the flow rate.

Example 4.4

Calculate the value of Gt for a fixed bed, being a sand water filter, where the grain size $D = 0.5$ mm, porosity is 0.4 and thickness $L = 600$ mm:

$$Gt = \frac{13.4(1-0.4)0.6}{0.4 \times 0.5 \times 10^{-3}} = 2.4 \times 10^4$$

that is within the range 10^4–10^5.

The value of G requires the flow rate. For normal rapid sand filtration $v = 5$ m h^{-1} = 5/3600 m s^{-1},

$$G = \frac{13.4v(1-\varepsilon)}{\varepsilon^2 D} = \frac{13.4 \times 5 \times (1-0.4)}{3600 \times 0.4^2 \times 0.5 \times 10^{-3}}$$
$$= 140 \text{ s}^{-1}$$

This is a high value, so large flocs do not form, which is suitable as they would clog the small filter pores. Since filtration velocities may be used up to 15 m h^{-1} in water and waste water treatment practice, even higher G values may be experienced.

Such fixed bed flocculation is normally accompanied by simultaneous filtration of the flocs which are formed. This causes clogging and so the accumulated flocs must be flushed away periodically (typically, every day) by backwash fluidization, as in normal deep bed filtration practice.

Flocculation prior to deep bed filtration is normally used, either in a previous flocculator and settling (or flotation) tank, or by direct flocculation in the filter chamber above the filter media. Flocculation within the fixed bed is sometimes called contact flocculation, or flocculating filtration.

In studying flocculation in such fixed beds (and also for fluidized beds) it is difficult to follow the kinetics of flocculation, or even to study the reduction in the numbers of primary particles, due to the removal of so many flocs by deposition on the fixed bed grains (filter effect). However, as the kinetics of removal of particles by filtration is first-order with respect to particle concentration (see chapter 11), but the kinetics of flocculation are second-order (equation 4.44), it is possible to separate the two effects by making observations at different inlet particle concentrations.

4.10.4.2 Fluidized bed (floc blanket clarifiers)

In the case of fluidized bed flocculators the power dissipated is the energy of drag past the fluidized bed particles per unit time. This occurs in the floc blanket clarifier where the previously formed floc particles comprise the fluidized bed. For a fluidized bed the drag force past the floc particles is equal to their weight in the liquid.

For such flocculator-clarifiers the hydraulic flow must be steady and maintain a steady fluidization of the existing floc particles; the incoming flocculating particles must aggregate to a size equal to the existing flocs or, more likely, be collected on them; and there must be a balance between the incoming solids and the withdrawal of excess floc to maintain a steady state.

The steady fluidization requires that the upward flow velocity v ($= Q/A$ where A is the tank plan area) be equal to the hindered settling velocity of the fluidized bed v_h:

$$v = v_h = v_s(1-C)^{4.5} \tag{4.73}$$

where the hindered settling is expressed in terms of the single particle settling velocity (v_s, often Stokes' velocity) and the fluidized particle volume/volume concentration (C) in the liquid; equation 4.73 is the well-known Richardson–Zaki relationship (see chapter 5). Typically, C is maintained between 0.1 and 0.2 vol./vol., and the upflow velocity is about 2 m h^{-1} = 0.55 mm s^{-1}.

From equation 4.73, taking $C = 0.15$,

$$0.55 = v_s(1-0.15)^{4.5}$$

from which $v_s = 1.14$ mm s^{-1}.

Using Stokes' law:

$$v_s = \frac{g(\rho_s-\rho)D^2}{18\mu} \tag{4.74}$$

diameter of flocs

$$D = \sqrt{\left(\frac{18\mu v_s}{g(\rho_s-\rho)}\right)}$$

In floc blanket clarifiers the density of the existing flocs is about 1005 kg m^{-3}, and at 20°C viscosity $\mu = 10^{-3}$ kg m^{-1} s^{-1},

$$D = \sqrt{\left(\frac{18 \times 10^{-3} \times 1.14 \times 10^{-3}}{9.81\,(1005-1000)}\right)} \quad \text{m}$$

$$= 0.65 \times 10^{-3} \text{ m} = 650\ \mu\text{m}$$

In constant plan (flat-bottom) clarifiers the upflow velocity should be uniform at all levels and the floc blanket (fluidized bed) should have a uniform concentration and uniform particle size. In expanding upflow clarifiers (conical, or hopper-bottom) the lower levels should have greater upflow velocities and, therefore, lower floc volume concentrations or particles. In practice, such expanding upflow is highly unstable and massive vertical eddies cause recirculation, with almost uniform concentrations and particle sizes throughout the fluidized bed. A definite zone of separation forms at the surface with a clean water zone above it, relatively free from primary particles and flocs.

The kinetics of flocculation in a fluidized bed are described by the equations in section 4.8. The simplified Smoluchowski equation 4.42 can be modified by the special bimodal distribution of the fluidized bed (floc blanket) particles:

$$dN_{ij}/dt = \tfrac{4}{3} N_i N_j G R_{ij}^3 \qquad (4.42)*$$

The incoming primary particles $i = 1$ are very small (< 1 μm), and it is the reduction of these particles N_1 which is important. The existing particles j are large (100–1000 μm) and form the fluidized floc particles. In steady state $N_j = $ constant, and the size of the aggregated particles $R_{ij} \simeq R_j$ as the addition of primary particles scarcely increases their radii. Equation 4.42 therefore becomes:

$$\frac{dN_{ij}}{dt} = -\frac{dN_1}{dt} = \tfrac{4}{3} N_1 N_j G R_j^3 \qquad (4.75)$$

The floc volume concentration, in the fluidized bed, is the number of j-particles per unit volume (N_j) multiplied by the volume of one j-particle

$$C = N_j \tfrac{4}{3} \pi R_j^3 \qquad (4.76)$$

So the rate of disappearance of primary particles is

$$-\frac{dN_1}{dt} = \frac{GCN_1}{\pi} \qquad (4.77)$$

Integrating from $t = 0$, $N_1 = N_0$ to $t = t$, $N_1 = N_t$:

$$\frac{N_t}{N_0} = \exp\left(-\frac{GCt}{\pi}\right) \qquad (4.78)$$

So the kinetics of fluidized bed flocculation predicts an exponential decline in the number of primary particles as the liquid passes through the fluidized bed (residence time t). This has been confirmed in experimental laboratory models using 1 μm polystyrene latex suspension destabilized in water, flocculated by passage through a fluidized bed of PVC/PVA co-polymer spheres 125–150 μm in diameter. These experiments, at University College London, have also confirmed that the flocculation is second-order with respect to concentration (equation 4.42) and removal of particles in the fluidized bed is first-order with concentration (filter effect—see chapter 11).

The mean velocity gradient is calculated from the power dissipated as drag past the fluidized bed particles. This exceeds by far any power dissipated in inlet turbulence or flow conditions (Ives[12]):

drag force = weight (fluidized bed equilibrium)
weight = $g(\rho_s - \rho)CAL$

where

A is the plan area of the tank (assumed constant), and
L is the depth of the fluidized bed.

$$\text{power} = \text{drag force} \times \text{velocity} = g(\rho_s - \rho) CAL \frac{Q}{A}$$

$$\frac{\text{power}}{\text{volume}} = \frac{P}{V} = \frac{g(\rho_s - \rho) CAL}{AL} \times \frac{Q}{A} = g(\rho_s - \rho) Cv \qquad (4.79)$$

From equation 4.46

$$G = \sqrt{\left(\frac{P}{V\mu}\right)} = \sqrt{\left(\frac{g(\rho_s - \rho) Cv}{\mu}\right)} \qquad (4.80)$$

Using the previous values for equations 4.73 and 4.74

$C = 0.15$, $v = 0.55$ mm s^{-1}, $\rho_s = 1005$ kg m^{-3}, $\mu = 10^{-3}$ kg m^{-1} s^{-1}

$$G = \sqrt{\left(\frac{9.81 \times (1005 - 1000) \times 0.15 \times 0.55 \times 10^{-3}}{10^{-3}}\right)}$$

$= 2$ s^{-1} (it is normally < 5 s^{-1} in floc blanket clarifiers)

This is a very low value compared with conventional paddle flocculators ($G = 20$–75 s^{-1}). The mean residence time in practical fluidized bed flocculators (t) is usually 20–30 min (1200–1800 s). So the product $Gt = (0.24$ to $0.36) \times 10^4$. This is less than the range for flow-through flocculators with no solids retention (10^4–10^5). But the low value of Gt is compensated by the high value of solids concentration C, which results from the criterion GCt of the theory (equation 4.78). Taking typical values:

$$GCt = 2 \times 0.15 \times 1500 = 450$$

Starting with the assumption that in a flow-through paddle flocculator all the primary particles are identical ($i = j = 1$) equation 4.44 can be integrated to a form similar to equation 4.48 but with a different numerical constant ($4/\pi$) because the appearance of a k-particle is accompanied by the disappearance of two i-particles (similar to equation 4.43). However, the term GCt appears relevant even for flow-through paddle flocculators. In water treatment a typical coagulant dose would be 50 mg l^{-1} as $Al_2(SO_4)_3 18 H_2O$, being 4.05 mg l^{-1} as Al ion. It has been estimated by Hudson[13] that 1 g of Al ion produces 240 ml of floc particles. Therefore, floc volume concentration is:

$$C = 4.05 \times 10^{-3} \times 240 \text{ ml l}^{-1}$$
$$= 4.05 \times 240 \times 10^{-6} \text{ vol/vol}$$
$$= 972 \times 10^{-6} \simeq 10^{-3} \text{ vol/vol}$$

Therefore

$$GCt = 10^{-3} \times 10^5 = 100 \text{ when } Gt = 10^5$$

It seems, therefore, that the values of GCt are comparable even in very different types of flocculation systems, such as the floc blanket and the flow-through paddle flocculators.

The balance of incoming and outgoing solids to maintain a steady volume fluidized bed is readily calculated from the volumetric concentrations of the incoming suspension and the fluidized bed. Solids are removed by withdrawing excess fluidized bed material (e.g. floc blanket) over a slurry weir or into a concentrator cone at the surface of the fluidized bed (zone of separation into the clear liquid zone). In the examples calculated above:

incoming solids from 50 mg l⁻¹ aluminium sulphate

$$C_i = 10^{-3} \text{ vol/vol}$$

outgoing solids (so-called sludge bleed) of floc blanket

$$C_o = 0.15 \text{ vol/vol}$$

Therefore ratio of sludge bleed suspension flow (q) to inflow (Q) is C_o/C_i because solids flux balance

$$C_i q = C_o Q$$

$$q = \frac{C_o Q}{C_i} = \frac{10^{-3}Q}{0.15} = 0.007Q$$

Therefore sludge bleed flow rate $= 0.7\%\ Q$.

For high alum doses the percentage increases proportionately. Because of inefficient slurry weirs or cones, some clear liquid may also be withdrawn, increasing q to values up to 3% Q. This is a high value, leading to excess of sludge to be disposed. Values of 5% or more are normally unacceptable.

This discussion and calculations of fluidized bed flocculators are based on a parallel-sided tank. Hopper (inverted pyramid) and conical shapes are quite common and the values of GCt, and fluidized bed stability, involve integrations throughout the fluidized bed depth (Ives[12]). In water treatment practice there is a growing interest in the development of flat-bottom, parallel-sided floc blanket clarifiers.

Solids recirculation with pumps or impellers takes advantage of the increase in the concentration term (C) in the criterion GCt, to enhance flocculation. There are many designs and these, together with normal floc blanket tanks, are presented by Sontheimer in Ives[5]. Also some discussion of such flocculators and the methods of model testing are described by Stevenson in Purchas[14].

4.10.4.3 Bubble flocculation

Bubbles rising through a suspension create velocity gradients due to their motion. These velocity gradients could be used for flocculation in principle, but early assessments of bubble flocculation indicated that the diffusers which existed at that time created bubbles that were too large to be effective (Camp[15]).

With the advent of flotation as a useful treatment process in water and waste water treatment, made possible by the creation of finer bubbles by dissolved air precipitation, and electrolysis, the concept of bubble flocculation can be reconsidered. In the flotation process in water treatment, flocculation takes place prior to bubble flotation, usually by mechanical (paddle) flocculatior with a coagulant salt (such as alum). It seems possible that the flocculation and flotation processes could be combined with the bubbles providing the flocculating power.

As for other particle flocculators, the power is derived from the drag force on the rising bubbles. Small bubbles will rise in a laminar regime (Stokes).

Stokes' drag force per bubble = force = $3\pi\mu v_s D$

power per bubble = force × velocity = $3\pi\mu v_s^2 D$

Stokes' law $v_s = \dfrac{g(\rho_s - \rho)D^2}{18\mu}$ (rising velocity of bubble)

Assume density of air in bubble is negligible (1.2 kg m^{-3}) and that no hindered rising or mass lift will occur, then

$v_s = \dfrac{g\rho D^2}{18\mu}$ (negative sign is dropped as v is opposite vectorially to g)

$$\text{power per bubble} = \frac{\pi g^2 \rho^2 D^5}{108\mu} \qquad (4.81)$$

From equation 4.46

$$G = \sqrt{\left(\frac{P}{V\mu}\right)} \qquad (4.46)*$$

$$\frac{P}{V} = \frac{\text{power}}{\text{volume}} = \frac{\text{power}}{\text{bubble}} \times \frac{\text{bubbles}}{\text{volume}}$$

$$= \frac{\text{power}}{\text{bubble}} \times \frac{\text{air volume}}{\text{volume}} \times \frac{1}{1 \text{ bubble vol}}$$

$$= \frac{\pi g^2 \rho^2 D^5}{108\mu} \times \frac{V_A}{V} \times \frac{6}{\pi D^3}$$

$$\frac{P}{V} = \frac{g^2 \rho^2 D^2 V_A}{18\mu V} \qquad (4.82)$$

$$G = 0.236 \frac{g\rho D}{\mu} \sqrt{\left(\frac{V_A}{V}\right)} \qquad (4.83)$$

where V_A/V = volume of air supplied per unit tank volume.

In dissolved air flotation bubbles are formed as microbubbles which grow with the release of pressure to values typically of 0.1–0.5 mm. The air is dissolved in recycled water, being about 10% of the flow, with values of 5–7 g air m^{-3}. This amounts to a free air volume per unit water volume (V_A/V) of about 5×10^{-4}.

Taking $D = 0.1$ mm and $\mu = 10^{-3}$ kg m^{-1} s^{-1} (20°C) for water, from equation 4.83:

$$G = \frac{0.236 \times 9.81 \times 10^3 \times 0.1 \times 10^{-3}}{10^{-3}} \sqrt{(5 \times 10^{-4})}$$

$$G = 5 \text{ s}^{-1}$$

If $D = 0.5$ mm

$$G = 25 \text{ s}^{-1}$$

Consequently, if the bubbles can be kept fine enough, useful G values can be generated. The use of bubble flocculators is still in the experimental stage, but there are indications that they may be applicable in practice.

4.10.4.4 Differential settling

Natural suspensions may flocculate, or previously flocculated suspensions may flocculate further, if the particles or flocs are settling at different velocities. Faster settling particles may collide with slower settling particles leading to aggregation if they are destabilized. The aggregates will then settle faster due to their increased mass, and possibly experience further collisions and aggregations.

The orthokinetic definition sketch of *Figure 4.12* will still apply, with relative velocity u replaced by the differential settling velocity $(v_i - v_j)$ in the direction of gravity (represented by the x-direction). The flux of i-particles through the cylinder radius R_{ij} $(= r_i + r_j)$, centred on a j-particle, is

$$N_i\frac{\mathrm{d}V}{\mathrm{d}t} = N_i\pi(r_i + r_j)^2(v_i - v_j) \tag{4.84}$$

As there are N_j j-particles present, the collision rate is

$$N_iN_j\frac{\mathrm{d}V}{\mathrm{d}t} = \frac{\mathrm{d}N_{ij}}{\mathrm{d}t} = N_iN_j\pi(r_i + r_j)^2(v_i - v_j) \tag{4.85}$$

Assuming that the particles settle according to Stokes' law, and that $\rho_i = \rho_j = \rho_s$ (constant particle density):

$$v_i - v_j = \frac{2g}{9}\frac{(\rho_s - \rho)}{\mu}(r_i^2 - r_j^2) \tag{4.86}$$

Noting that $(r_i^2 - r_j^2) = (r_i + r_j)(r_i - r_j)$

$$\frac{\mathrm{d}N_{ij}}{\mathrm{d}t} = N_iN_j\frac{2\pi g}{9\mu}(\rho_s - \rho)(r_i + r_j)^3(r_i - r_j) \tag{4.87}$$

Equation 4.87 can be developed further to allow for the creation of k-particles and their possible collisions with i- and j-particles, and so on. This is set out by Ives[5].

The equations cannot be treated exactly like other orthokinetic equations because of the limitation $i \neq j$. If $i = j$ there would be no differential settlement and, therefore, no collisions and aggregation. Similarly, it cannot be applied to an initially monodisperse suspension because all particles would settle at the same rate. If the suspension were subject initially to significant perikinetic (Brownian diffusion) flocculation then it would become heterodisperse and flocculation by differential settling would follow.

If a sample of suspension to be settled is available it is not difficult to determine whether flocculation by differential settling will occur. An apparatus similar in principle to the Andreasen pipette method (chapter 2) is used, but with sampling at more than one depth, such as the settling tube with multiple sampling ports shown in *Figure 4.22*. Samples are taken at the

127

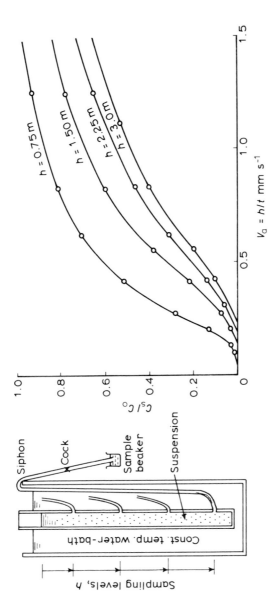

Figure 4.22. Multi-port settling tube and settling data (flocculating suspension)

various depths h_1, h_2, h_3 etc., below the surface level of the suspension, at successive time intervals t. These samples are analysed for suspended solids concentration C_s, and reported as a fraction of the original concentration C_o. Such data are given in *Table 4.8* and plotted in *Figure 4.22*.

Table 4.8. SETTLING ANALYSIS DATA

$h_1 = 750\,mm$ time $t(s)$	$= 0$	600	900	1200	1800	2700	3600	5400	7200
h_1/t (mm s^{-1})	=	1.25	0.83	0.63	0.42	0.28	0.21	0.14	0.10
C_s (g m^{-3})	= 86	80	70	61	44	24	12	3	1.3
$100\,C_s/C_o$	= 100	93	81	70.5	51.5	28	13.5	3	1.5
$h_2 = 1500\,mm$									
h_2/t (mm s^{-1})	=	2.50	1.67	1.25	0.83	0.56	0.42	0.28	0.21
C_s (g m^{-3})	= 86	83	74	67	52	33	19	7	3
$100\,C_s/C_o$	= 100	96	86	77.5	60	38	22	8	3
$h_3 = 2250\,mm$									
h_3/t (mm s^{-1})	=	3.75	2.50	1.90	1.25	0.83	0.63	0.42	0.31
C_s (g m^{-3})	= 86	84	76	70	56	40	27	12	5
$100\,C_s/C_o$	= 100	98	88.5	81	65	46.5	31	13.5	6
$h_4 = 3000\,mm$									
$h_4 t$ (mm s^{-1})	=	5.00	3.33	2.50	1.67	1.12	0.83	0.56	0.42
C_s (g m^{-3})	= 86	85	77	71	59	46	34	17	8
$100\,C_s/C_o$	= 100	99	89.5	83	69	53	40	20	9.5

If these values all lay on one single curve of $100\,C_s/C_o$ against h/t, the suspension would be discrete with particles retaining their settling velocities at all depths. As the values for different depths give different curves on *Figure 4.22*, the particles are flocculating as they settle, thus having increasing settling velocities at increasing depths.

Note that the settling velocity h/t is only an apparent settling velocity (v_a) because the particles have been growing in size due to flocculation, and accelerating during their descent through h and time t. So the true instantaneous settling velocity (v_s) is a function of h:

$$t = \int_0^h \frac{dh}{v_s} \qquad (4.88)$$

$$v_a = \frac{h}{t} = h \bigg/ \int_0^h \frac{dh}{v_s}$$

$$v_a \int_0^h \frac{dh}{v_s} = h \qquad (4.89)$$

Differentiating with respect to h,

$$\frac{d}{dh}\left(v_a \int_0^h \frac{dh}{v_s}\right) = 1 \qquad (4.90)$$

$$\frac{v_a}{v_s} + \frac{dv_a}{dh}\int_0^h \frac{dh}{v_s} = 1$$

$$\frac{v_a}{v_s} + \frac{dv_a}{dh}\frac{h}{v_a} = 1$$

$$v_s = \frac{v_a}{1 - \left(\dfrac{dv_a}{dh} \times \dfrac{h}{v_a}\right)} \qquad (4.91)$$

So it is possible to calculate the true settling velocities of the particles from their apparent settling velocities. This, however, is rarely required and *Figure 4.22* gives sufficient information for the design of a settling tank, providing that the observations in the settling tube cover the full range of depth expected in the tank.

4.11 CURRENT DEVELOPMENTS

Compared with the rapid development of new synthetic organic polymer flocculants, the developments in orthokinetic flocculations will be slower, as quite substantial hardware is usually involved. Certain trends which are apparent now may be developed further, or become more widely adopted.

4.11.1 Tubular flocculators

Tube settler modules are commonplace in settling tanks, but their basic function is to increase the total plan area for settlement, as required by settling tank theory. However, these tube modules also act as orthokinetic flocculators, mainly in laminar flow, with Reynolds' numbers of about 500.

In pipes at higher Reynolds' numbers, ranging from 2000 to 25 000, and therefore in turbulent flow, Döll[16] has demonstrated the mixing and flocculation conditions in laboratory systems. These employed silica particle suspensions, flocculated with three different cationic polymers, and the turbulent root mean square velocity gradient (G) characterized the influence of flow rate on reaction rate. In West Berlin, a practical pipe flocculator has been used prior to the entry of treated waste water into a sedimentation tank, based on the concepts set out in section 4.10.3, with the pipe set out in a snake-like pattern on the ground surface.

4.11.2 Fixed beds

The use of deep bed filters as flocculators is quite well established; some were developed in the 1950s as 'contact clarifiers' (USSR). If the suspended solids loading is light (e.g. below 150 mg l⁻¹) they may function without undue rapid clogging; but if the filtration effect dominates, the advantages of flocculating filtration may be lost.

At the other extreme, the static mixer, comprising labyrinth baffles in a tubular module, can only provide flocculation with no retention.

There is a possibility of a development between these two, with a packed bed providing flocculation with a non-clogging retention. An example of this is the 'Banks' pebble bed clarifier, where sewage that has been previously biologically treated flows upwards through a fixed bed of gravel (pebbles). The resulting flocculation produces biological flocs which settle back loosely on top of the pebble bed, not impeding the flow. The accumulated flocs can be periodically flushed away from the top surface.

4.11.3 Fluidized beds

Floc blanket clarification uses the flocs themselves to form the fluidized bed, to enhance flocculation. In laboratory models a particulate suspension (such as polymer spherical beads) may be used to form the fluidized bed. In any case, the upflow rate is limited by the settling velocity of the fluidized suspension. This is affected linearly by the density of the suspension particles, and greater upflows (therefore more production per unit plan area of tank) can be passed through more dense particle suspensions.

Advantage has been taken of this effect in fluidized bed flocculators in practice, which use 'micro-sand' on which flocculation takes place. This development, which was pioneered in Hungary, now appears as the 'Cyclofloc' process. The sand is coated (activated) with an alginate flocculant, and incoming suspensions of flocculent aluminium or ferric salts are flocculated by the fluidized bed and retained on the sand. A continuous discharge of some floc-coated sand, recycles it through an external washing process which separates the floc and the sand. The cleaned sand is continuously fed into the tank inflow to sustain a steady-state process.

Further development of this concept of solid particle nuclei for fluidized bed flocculation has been described by Sibony[17].

There are also processes developing in which the particles are ion-exchangers, charged with Al^{3+} ions which exchange with Ca^{2+} in hard waters. The released aluminium hydrolyses to form alum floc, which can be retained on the particles in a fluidized bed. Such particles can be withdrawn, washed clean, and recharged with Al^{3+}.

The combination of a paddle flocculator within a fluidized bed produces much denser flocs by the turbulent eddies which provide pressure gradients on the flocs, squeezing out their interstitial water. The resultant particles are dense, with a layered onion-like structure. The high density enables high upflow rates through such modified floc blanket clarifiers[18].

REFERENCES

1. Elson, T. P., 'Velocity profiles of concentric flow between coaxial, rotating cylinders with a stationary low boundary'. *Chem. Eng. Sci.,* **34**, 373–377 (1979)
2. Ives, K. J. and Al Dibouni, M., 'Orthokinetic flocculation of latex microspheres'. *Chem. Eng. Sci.,* **34**, 983–991 (1979)
3. Arp, P. A. and Mason, S. G., 'Orthokinetic collisions of hard spheres in simple shear flow'. *Can. J. Chem.,* **54**, 3769 (1976)
4. van de Ven, T. G. and Mason, S. G., 'The microrheology of colloidal suspensions. VII. Orthokinetic doublet formation of spheres'. *Colloid and Polymer Sci.,* **255**, 468–479 (1977)
5. Ives, K. J., *The Scientific Basis of Flocculation,* Sijthoff and Noordhoff, Alphen aan den Rijn, Netherlands (1978)
6. Ives, K. J. and Bhole, A. G., 'Study of flowthrough Couette flocculators. I. Design for

uniform and tapered flocculation'. *Wat. Res.,* **9,** 1085–1092 (1975)

7. Ives, K. J. and Bhole, A. G., 'Study of flowthrough Couette flocculators. II. Laboratory study of flocculation kinetics'. *Wat. Res.,* **11,** 209–215 (1977)

8. Bhole, A. G., 'Measuring the velocity of water in a paddle flocculator'. *J. Amer. Wat. Wks. Ass.,* **72,** 109–115 (1980)

9. Vadas, E. B., Goldsmith, H. L. and Mason, S. G., 'The microrheology of colloidal dispersions. 1. The microtube technique'. *J. Coll. Interfac. Sci.,* **43,** 630–648 (1973)

10. Polasek, P., 'The significance of mean velocity gradient and its calculation in devices for water treatment'. *Water S.A.,* **5,** 196–207 (1979)

11. Bernhardt, H. and Schell, H., 'The technical concept of phosphorus elimination at the Wahnbach estuary using floc-filtration (Wahnbach system)' [in German]. *Wasser u. Abwass Forsch.,* **12** (3/4), 123–133 (1979)

12. Ives, K. J., 'Theory of operation of sludge blanket clarifiers'. *Proc. Inst. Civ. Engrs,* **39,** 243–260 (1968)

13. Hudson, H. É., 'Physical aspects of flocculation'. *J. Amer. Wat. Wks. Ass.,* **57,** 885–892 (1965)

14. Purchas, D. B., *Solid/Liquid Separation Equipment Scale-up,* Uplands Press, Croydon (1977)

15. Camp, T. R., 'Flocculation and floccuation basins'. *Trans. Am. Soc. Civ. Engrs,* **120,** 1–16 (1955)

16. Döll, B., 'Particle destabilization in turbulent pipe flow', *Proc. 14th IAWPRC Biennial International Conference* (Brighton, 1988), Book 2, pp. 435–442, IAWPRC, London (1988)

17. Sibony, J., 'Clarification with microsand seeding. A state of the art', *Water Res.,* **15,** 1281–1290 (1981)

18. Ide, T. and Kataoka, K., 'A technical innovation in sludge blanket clarifiers'. *Proc. Second World Filtration Congress (London),* pp. 377–385, Filtration Society, Uplands Press, Croydon (1979)

5

Gravity Thickening

D. G. Osborne

PT Kaltim Prima Coal, Jakarta, Indonesia

–NOMENCLATURE

a, A	Cross-sectional area of settling	m^2
C	Concentration of solids. Weight solids per weight liquid	$kg\ kg^{-1}$
C_f	Concentration of feed	—
C_o	Concentration of overflow	—
C_u	Concentration of underflow	—
d	Circular tank diameter	m
G	Solids handling capacity (SHC)	$kg\ m^{-2}\ h^{-1}$
h	Height of interface from base	m
K_G	Geometric constant	—
κ_l	Fluid density	$kg\ m^{-3}$
$s\kappa_s$	Solids density	$kg\ m^{-3}$
P_r	Raking power	kW
P_t	Trench power	kW
Q	Flow rate	—
Q_f	Feed rate	$m_3\ h^{-1}$
Q_o	Overflow rate	$m^3\ h^{-1}$
Q_u	Underflow rate	$m^3\ h^{-1}$
t	Time	s
t_t	Time of retention	s
T	Torque	Nm
T_r	Raking torque	Nm
v	Peripheral velocity of rakes	$m\ s^{-1}$
V_t	Sum of solids and liquid volume in the compression zone	m^3
u, U	Settling velocity of solids in the concentration zone of minimum settling rate	$cm\ s^{-1}$
μ_s	Friction coefficient (solids on solids)	—
w	Upward velocity of propagation of concentration zone of minimum settling rate	$cm\ s^{-1}$
W	Solids handling rate	h^{-1}

5.1 INTRODUCTION

The separation of the solid and liquid phases of a suspension is very often a two-stage process. The first stage, known as thickening, is physicochemical and involves the conversion of discrete, unflocculated particles of the suspension into a thickened underflow with a clear overflow, often by the addition of a coagulant or flocculant. This step is capable of a high degree of control, and is efficient, irrespective of the scale of the operation. This reliability is, however, of vital importance in terms of establishing the conditions required for a similar effectiveness of the second stage. The second stage, predominantly an engineering operation, is to reduce the remaining water content of the thickened underflow by a suitable solid–liquid separation process, thereby transforming it into a compact, rigid solid containing only small quantities of water.

Perhaps the most troublesome section of a treatment plant flowsheet is that set aside to perform the initial, two-fold function of providing an acceptable clarified water product and a thickened slurry product suitably controlled for effective subsequent dewatering. Very often, even a single element of disregard in designing this section rapidly becomes apparent by resultant poor operation culminating in further expenditure for remedial steps. These, all too commonly, do no more than provide only marginal improvement and often result in higher operating costs, e.g. increased flocculant cost.

Two distinct forms of sedimentation vessels are in common usage. The clarifier is used, as the name suggests, for the clarification of a dilute suspension to obtain an overflow containing minimal suspended solids. The thickener is used to thicken a suspension to obtain an underflow with a high solids concentration whilst also producing a clarified overflow.

In both the cases the assessment of the sedimentation behaviour of the solids within the fluids will allow the correct size of vessel to be determined. It is therefore important to know a little about the way in which settling solids behave during sedimentation.

5.2 THE SEDIMENTATION CONCEPT (BATCH SETTLING)

Particle size, particle density and fluid viscosity are readily recognized factors to be considered in any sedimentation process. Less obvious are particle shape and orientation, the distortion of a deformable particulate, the interference of one particle with another (particularly when concentrations are high), the nearness of the wall of the container, convection currents and the like. These also have a significant and yet varying influence on the separation process.

Materials with particle diameters of the order of a few microns settle too slowly for most practical operations. Wherever possible, such particles are agglomerated, or flocculated, into relatively large clumps called 'flocs', that settle out more rapidly. In such a context flocculation can be considered to include both the effect of coagulation and the effect of polymer flocculation.

The settling of discrete particulate masses based initially upon modifications of Stokes' law has featured in the research of many workers with as many different solid–liquid systems. Early research work in this field considered the settling behaviour of spherical particles for a wide range of concentrations, from dilute to concentrated suspensions. A similar approach to that used for spheres was adopted for a study of the sedimentation of small uniform particles by Steinour[1] and later (1950) by Hawskley[2], who both correctly assumed that the upthrust acting on the particles is determined by the density of the suspension rather than that of the fluid alone. For the sedimentation of uniform particles, the increased drag is probably more attributable to a steepening of the velocity gradients than to a change in viscosity as a result of solids concentration. In practical terms the rate of sedimentation of a suspension of discrete fine particles is extremely difficult to predict because of the large number of factors involved. However, a number of empirical equations has been obtained for the rate of sedimentation of suspensions, as a result of tests carried out in measuring jars or cylinders. For a given solid and liquid the chief factors which affect the process are the height of suspension, the diameter of the containing vessel and its shape, and the solids concentration. Wallis[3] attempted to bring together results obtained from a variety of conditions to categorize the behaviour of settling sediments of uniform particles ($>100\,\mu$m), sufficiently large for anomalous viscosity effects and flocculation to be regarded as negligible.

The settling of flocculated particulate masses is a complicated process involving rearrangements in the sediment long after the flocs themselves have settled. Bottom-lying flocs are compressed by the weight of the others that settle upon them, since flocs usually are bundles of particles held together by weak forces, and which have entrapped within their structure considerable quantities of the liquid medium. This produces a sediment with varying degrees of density. A simplified history of the batch sedimentation of a flocculated suspension is shown in *Figure 5.1*. The upper section is a continuous plot of the four zones that have been observed to develop during sedimentation and the lower sections show the containers at various stages of settling, according to Comings et al.[4].

Cylinder A contains a uniformly mixed but flocculated suspension. Cylinder B shows the situation shortly after when, at the very base of the cylinder, there is a zone (zone 4 in *Figure 5.1*) which consists of a mixture of flocs and comparatively large grains resting at the bottom upon one another. This zone has formed from flocs initially very close to the container's bottom. Immediately above the lowest zone is a transition layer that is intermediate in density between the deposit and the suspension. This is the zone in which liquid is being squeezed out from among a network of flocs—or a compression zone (zone 3 in *Figure 5.1*). Some authors have preferred to break the compression zone into two; notable among these are Coe and Clevenger[5], whose work in the field of sedimentation provided the basis for a considerable amount of work by others who accepted their original assumptions. The work of Coe and Clevenger, however, concerned the settlement of non-colloidal particles as initially discrete units. A deliberately flocculated suspension does not appear to exhibit quite so clearly the characteristics which would allow two sub-zones to be distinguished within the compression zone, and cylinders B and C contain a layer of flocs still in suspension. Sometimes this is referred to as the 'free settling zone', but a

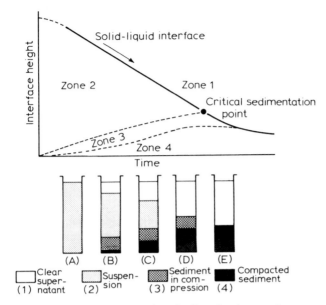

Figure 5.1. Sedimentation of a flocculated suspension

more apt term is that suggested by Anderson and Sparkman[6]—'the zone of collective subsidence'—since, substantially every particulate group, though settling, is retaining its position relative to its neighbours. Zone 2 maintains the concentration of the original suspension. The upper zone (zone 1 in *Figure 5.1*) is a zone of liquid which, ideally, is clear if the suspension is well flocculated, and turbid or discoloured if not, thereby still containing fine suspended particles. When well flocculated, the boundary between the liquid and the suspension is normally sharp and easily located almost immediately after mixing. Other zonal boundaries are less sharply defined and are often difficult to detect because of the opacity of the suspension. However, X-ray adsorption techniques have been used to locate them (Gaudin and Fuerstenau[7]). Cylinder C illustrates the continued settling effect when the upper and lower zones increase in volume, the suspension zone decreases, and the zone of sediment undergoing compression remains essentially the same in volume but moves upward. This continues until (cylinder D) the suspension zone disappears and all the solids exist in the form of a sediment. This condition is known as the 'critical sedimentation point'. As the graphical representation shows, the solids–liquid interface follows an approximately linear relationship with time until this point is approached. Following a short transitional phase, the sedimentation continues at an even slower rate, attaining eventually the final condition represented by cylinder E. The compression of the sediment occurs between D and E, and the time involved in this compression phase constitutes the major part of the time involved for the total process. The liquid which accompanies the flocs into the deposit is slowly expelled under the weight of the sediment above. This continues until an equilibrium is established between the weight of the flocs and their mechanical strength.

Overall process time often depends very much on the type of settling aid used. Coagulants such as lime may give process times of several hours whereas acrylic-based polymer flocculants can achieve process times of a fraction of this. Process requirements and cost are therefore very important considerations in the selection of a settling aid between these two extremes.

Spherical and quite compact flocs can often be achieved by using combinations of electrolytes and polyelectrolytes and mixing in a specialized fashion (Hamza[8]). Such flocs are not only denser and faster settling, but also entrain less of the suspending liquid and hence produce a more distinct separation.

5.3 FACTORS AFFECTING SEDIMENTATION

The most important factors affecting the rate and the mode of settling of particulate suspensions in a fluid of relatively low viscosity are:

1. the nature of the particles: size distribution, shape, specific gravity, mineralogical and chemical properties etc;
2. the proportion of solids to liquid forming the suspension, and concentration effects;
3. the type of pretreatment: chemical conditioning, flocculation, heating/cooling etc;
4. the type of containing vessel: size, shape, wall effects etc.

Each of these factors will now be discussed in a little more detail.

5.3.1 The nature of the particles

Spherical or near-spherical particles or agglomerates settle considerably more rapidly than non-spherical plate or needle-like particles of similar weight. Flocculation has an obvious benefit in that it may convert a group of irregularly sized and shaped particles into a fairly well-rounded aggregate, which greatly improves the settling characteristics of the suspension. Autocoagulation may occur if the particles are mineralogically or chemically suited to utilize an inherent ionic environment even without the addition of a flocculant.

5.3.2 Concentration effects

Increasing the number of evenly distributed particles in a fluid decreases the rate of fall of each individual particle. This is readily appreciated if it is considered that the particles, instead of falling through a stationary liquid, are themselves static and the fluid is rising. Since the fluid is unable to pass through positions occupied by the particles, its velocity must increase to compensate for this partial blocking of the flow channel. The relative velocity effect, however, does not account completely for the slowing of sedimentation with increases in solids concentration. Particle interference by collision and coagulation are other factors. Highly concentrated suspensions exhibit severe curtailment of individual particle settling as a result of a phenomenon usually referred to as hindered settling, which causes the descent rate of the particulate mass to occur at a more or less uniform rate. Their motion is said to be hindered because settling takes place by a massive subsidence rather than by individual particle fall.

Under batch settling conditions, the behaviour of flocculated suspensions with variation in initial solids concentration has been observed to exhibit three quite distinct modes[9,10]. These distinctly different modes of settling have been defined to coincide with three separate slurry concentration types: dilute, intermediate and concentrated, illustrated in *Figure 5.2* by means of settling time/descent curves. In dilute suspensions, the individual grains or floccules behave discretely and descend freely in the return fluid which is displaced upward between them. Intermediate suspensions, in which the floccules are in loose mutual contact, settle by channelling, provided that the suspension height is sufficiently great. The channels are of the same order of diameter as the flocs and are developed during an induction period in which an increasing quantity of return fluid forces its way through the pulp mass.

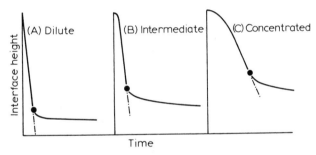

Figure 5.2. Effect of concentration on sedimentation[11]

In concentrated pulps, or in zones of intermediate concentration formed during the subsidence of more dilute pulps, the formation of return flow channels is not possible, either because of lack of sufficient height in the zone or because of the relatively small quantity of fluid remaining near the base of the container. The only fluid flow possible is therefore through the minute voids between the primary particles, and this results in relatively low compaction rates. In this region the degree of compaction depends upon the weight of overlying solid and it is suggested that flocculation results in the formation of a particle–particle structure, which contains a relatively large quantity of voidage water and which can resist collapse because of friction between the composite primary particles at the points of contact.

The critical sedimentation point, mentioned earlier, is distinctly noticeable for all three types of settling behaviour marked by a sudden decrease in settling rate, the formation of 'volcanoes' in the surface and a sharpening of the interface. Very concentrated suspensions require fairly large suspension heights for a useful assessment of the settling behaviour to be achieved.

All three types of settling curve demonstrate commonality in behaviour in that all three have three phases:

1. zone settling: most evident in dilute suspensions;

2. channelling: most evident in intermediate suspensions;
3. compression: most evident in concentrated slurries.

Looking at these phases collectively, as in *Figure 5.3*, each type is clearly obstructive and requires quantifying under different conditions of feed solids concentration in order that the impact of each on thickener capacity design can be assessed.

Scott's work[10,11] includes mathematical equations which appear to describe satisfactorily the three types of subsidence and the experimental effect of these modes of behaviour on the settling performance in large sedimentation units.

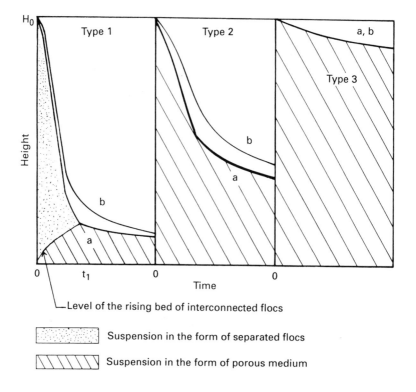

Figure 5.3. Initial concentration from concentration ranges of: type 1, zone settling; type 2, channelling (phase settling); type 3, compression

5.3.3 Flocculation effects

In flocculated particles a number of smaller particles are clustered together. Flocs or floccules, though they settle at rates considerably higher than the fastest of the individual particles, may contain large amounts of entrapped water within their voidages. Their shapes and density bear little or no relation to the nature of the original particles. For mathematical prediction of the settling rate, which is extremely complex, new shape factors and density values, quite different to those of the actual particles, must be determined and this is achieved by methods that are still largely empirical. Because floc size, and to a lesser extent shape, largely depends on the type of flocculant used, the rate of settling achieved will depend not only upon flocculant type but also on the extent to which dispersion and ultimately adsorption have occurred.

Fitch[12], combining observations made in the earlier section regarding solids concentration with those made for flocculated suspensions, related solids concentration with interparticle interaction. *Figure 5.4* shows the differing settling regimes. The upper line approximates an indefinite transition between clarification and zone settling, where clarification is characterized by a very low solids concentration. When the particles are discrete (always separated) the behaviour is often called hydroseparation or classification and obeys Stokes' law. Truer clarification occurs for flocculated solids which settle out as individual flocs. With increasing solids concentration, the solids contact each other more and more frequently and settle together under zone settling conditions. During zone settling, flocs may cohere and begin to link into a plastic structure, thereby settling all at the same rate.

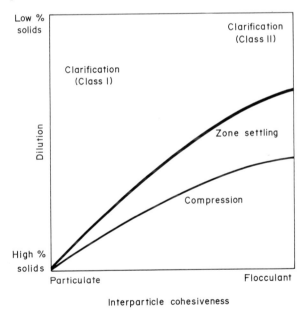

Figure 5.4. Sedimentation regime for thickeners and clarifiers

At some even higher concentration, the pulp structure becomes so firm that it develops compressive strength, enabling it to support further sedimentary layers deposited above it. This compression regime is therefore created by a squeezing effect of the upper layers which are each approaching the end of the previous zone settling regime. Flocs which initially form as loose structures are transformed into compact aggregates containing a minimum amount of interstitial fluid at the termination of compression.

5.3.4 Sedimentation vessel

The presence of a stationary wall or boundary near a settling particle upsets the normal flow pattern about the particle, thereby reducing its settling rate. If the ratio of the diameter or mean diameter of the vessel to the diameter of the particle is greater than about 100, the walls of the container appear to have no effect on the settling rate of the particle[3].

The height of suspension afforded by the vessel does not generally affect either the rate of sedimentation or the consistency of the sediment ultimately obtained. If, however, the solids concentration is high, sufficient height must be afforded by the vessel to allow the sedimentation process time to include a period of free settlement.

Provided that the walls of the vessel are vertical and the cross-sectional area does not vary with depth, the vessel shape has little effect on the sedimentation rate. However, if changes occur in either cross-sectional area or wall inclination the effect on the sedimentation process may be considerable. This is particularly true in the case of so-called high capacity thickeners, which are described later.

5.4 THICKENER DESIGN

5.4.1 Fundamental similarities

The thickener, or clarifier, is the industrial plant in which the concentration of a suspension is increased purely by sedimentation, with the formation of a clear liquid. In some cases the concentration of the feed suspension is in excess of 10% by weight and hindered settling takes place more or less from the onset of sedimentation. Thickening devices may be designed to operate as batch or continuous units, and consist largely of relatively shallow tanks from which the clear liquid is taken off at the top and the thickened liquor at the bottom. *Figure 5.5* shows a simplified diagram of a typical gravity thickener.

In order to obtain the largest possible throughput from a thickener of given size and type, the rate of sedimentation should be as high as possible. In most cases the rate is artificially increased by the addition of small quantities of either a coagulant or a flocculant, which are usually selected following a thorough examination of various types recommended for a fairly specific application. Flocculation agents, in particular, can produce vastly differing performances in similar applications and because of the wide variety available and their relatively high cost, wise selection is usually ensured by thorough testing. Hamza[8] has suggested selection procedures and demonstrated variation in application.

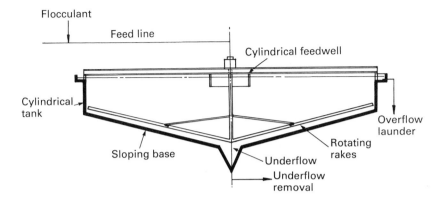

Figure 5.5. Simplified diagram of a typical gravity thickener

5.4.2 Settling capacity

The settling capacity of sedimentation units is directly proportional to the area of the vessel or tank and is still usually determined by the free or zone settling rate of the solids in suspension, a rate which is independent of the depth of the liquid. We have seen that the pulp, in settling, passes through zones of variable concentration between that of the feed and that of the final discharge. Consequently in the intermediate zones, which exist between these limits, each particle will encounter different settling rates and the zone that exhibits the lowest settling rate in proportion to the liquid separated will control the size of the unit, since all solids must eventually pass through this zone.

The capacity of continuous sedimentation units is usually based on their ability to perform the two-fold function of both thickening and clarifying at the required rate. The plan area of the unit controls the time allowed for the settlement of solids through the liquid for a given feed rate of liquid and is important in determining the clarification capacity. Likewise, the depth of the unit controls the time allowed for thickening the sludge for a given rate of feed of solid and is important in determining the thickening capacity.

General design of sedimentation tanks does not, for the most part, follow strict geometric proportions. The relationship between depth and diameter is important only to the extent that the tank volume will provide sufficient retention time, taking into consideration factors such as operating efficiency and mechanical design.

5.4.3 Thickening capacity for flocculated solids

So far, in reviewing sedimentation characteristics, the discussion has been confined to batch settling conditions. To satisfy the design requirements of continuously operating thickeners these characteristics must be translated to a dynamic situation. *Figure 5.6* shows a typical concentration profile for a continuous gravity thickener. The size of vessel required to thicken a given quantity of dilute slurry to a desired discharge concentration depends upon the 'actual' settling rate/concentration relationship which applies to the solids under the real conditions present within a continuously operating thickener. The reliability of any method for sizing thickeners therefore depends upon the accuracy with which this relationship can be evaluated.

If the settling rate was a unique function of concentration, static batch settling tests such as those proposed by Coe and Clevenger[5], Shannon and Tory[13] and Fitch[12] would be perfectly satisfactory. However, later work[14] indicated that other variables affect the settling rate for a given concentration. The depth of descent of the test pulp in batch settling tests with jars of varying height has been shown to affect the settling rate while the weight of solids per unit area and the presence of slow raking, in units which incorporate such a device, affect the settling behaviour and consequently the final sediment density achieved[15]. Furthermore, Kynch[16] showed elaborately that if settling rates were a unique function of concentration, the maximum concentration would be reached immediately at the base of the column with the various concentration levels being propagated upwards at constant rates. This was not, however, observed for flocculated clay suspensions by Gaudin and Fuerstenau[17], suggesting that for such slurries additional factors besides solids concentration must affect the settling rate.

Because of these additional factors it is essential to discover how closely the settling rate obtained in batch tests approximates to that in a continuous thickener. This is particularly significant in view of the relative differences in pulp height, the presence of rakes in the latter and the fact that differences in flocculant dispersion can create widely varying conditions. Some idea of thickener capacity is essential before the unit is installed and usually the only effective means of obtaining this is by batch settling tests with representative samples of slurry and simulated conditions for flocculation and dispersion. Ideally, continuous pilot testing would provide the most reliable data, but unless the thickener is a retrofitted unit or an additional unit to be installed in an existing plant, this form of testing is usually not possible. Perhaps the most reliable approach for determining thickener area requirements from batch settling test data for flocculated suspensions is that suggested originally by Yoshioka et al.[18] and later by others[19] for specific types of suspensions. Batch settling tests, because of their simplicity have been the most widely used approach to designing thickeners for many years. However, methods like those of Coe and Clevenger[5], Yoshioka et al.[18] and Talmage and Fitch[19], are based on sedimentation behaviour and can give a reasonably accurate indication of the area and capacity required for a thickener handling particulate suspensions. If the suspension is flocculated this accuracy may be affected, depending on the degree of flocculation. When sizing a thickener for handling slurries of this type[20], both the sedimentation process in the sedimentation zone and the consolidation process (compression) in the thickening zone (shown in *Figure 5.6*) must be evaluated. The sedimentation zone can be simulated by means of the batch settling tests referred to earlier because zone settling of individual flocs permits solids handling capacity of this zone to be determined.

The compression process in the thickening zone is a different matter. There has been only limited success in devising a simple and reliable test procedure for determining thickening zone area and depth. It has therefore been common practice to apply empirical rules, despite extensive research which has been carried out during the past two decades to develop a more analytical approach.

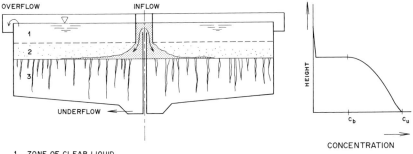

1. ZONE OF CLEAR LIQUID
2. SEDIMENTATION ZONE
3. THICKENING ZONE

Figure 5.6. Typical concentration profile in a continuous gravity thickener. 1, Zone of clear liquid; 2, sedimentation zone; 3, thickening zone

Because the compression process for flocculent slurries behaves more like a filtration process, i.e. liquid flowing through a porous medium, rather than true sedimentation, this approach has been favoured by researchers.

Using this approach, Kos[21] has described a batch test employing the fluid upflow effect to evaluate compression zone behaviour which has been used successfully for highly flocculent non-segregating suspensions.

5.4.4 Power and torque

One of the most important parameters used for specifying a thickener mechanism is the torque requirement for the drive system. Despite this, very little design data exist other than an almost totally empirical approach.

For many years 'rule-of-thumb' formulae of the following type have been employed[22]

$$T = K_G d^2 \tag{5.1}$$

where T is the total torque requirement, K_G is a constant combining specific rake mechanism design with vessel geometry, and d is the tank diameter.

Empirically determined values of K_G are substituted to suit the anticipated duty of the unit—see *Table 5.1*. Charts like the one shown in *Figure 5.7* are also used.

More recently, attempts have been made to identify a more definite approach towards design. Two 'typical' models have been defined as being representative of common sediment types: neither is a true extreme[23]. The sediment types are:

1. those containing a range of solids from fine, flocculated particles to coarse grains—such sediments are typical of those treated by thickeners employed in coal and mineral processing plants;

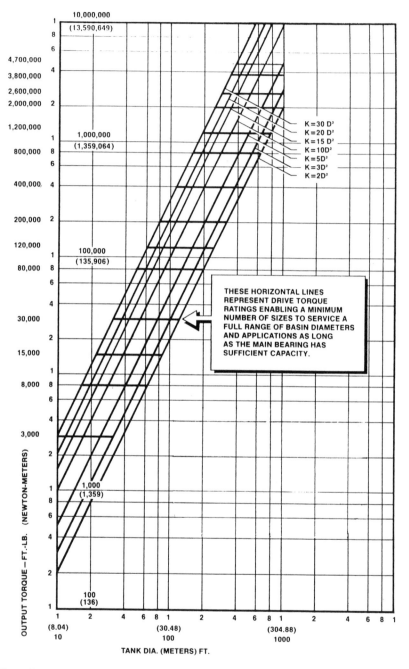

Figure 5.7. Available torque versus tank diameter for the normal range of K values. The standard Eimco drive torque ratings are noted on the figure. The K values apply only to units of feet and feet-pounds, for units of metres and Newton-metres divide K by 0.1263. For most industrial and minerals service K ≥ 12.5; for most waste water and water-treatment service K = 3 to 12; for water-softening service K = 5 to 15, except in Florida where K = 20 to 30

2. those containing fine, flocculent material forming a uniform sediment having consistent fluid properties.

Table 5.1. PRELIMINARY TORQUE SELECTION CHART FOR INWARD RAKING THICKENER MECHANISMS[25]

Item	Duty classification*			
	Light	Standard	Heavy	Extra heavy
Solids loading, $m^2 \, ton^{-1} \, day^{-1}$	> 4.7	1.4 – 4.7	0.5 – 1.4	< 0.5
Underflow concentration, % dry solids	⩽ 5	5 – 30	30 – 50	> 50
% Total solids < 74 μm, % minus 200 mesh	100	85 – 100	50 – 85	< 50
% Total solids > 210 μm, % plus 65 mesh	0	0 – 5	5 – 15	> 15
Specific gravity of dry solids	1.0 – 1.25	1.25 – 3.0	3.0 – 4.0	> 4.0
Torque determination where torque = kd^2				
K for Newton-metre torque where d = tank diameter in meters	15 – 58	73 – 131	146 – 292	> 292

*Some examples of duty classification as applied to sedimentation equipment. Light duty: river, or lake water clarification, metallic oxides, brine clarification. Standard duty: magnesium oxide, lime softening, brine softening. Heavy duty: copper tails, iron tails, coal refuse tank, coal, zinc or lead concentrates, clay, titanium oxide and phosphate tails. Extra heavy duty: uranium countercurrent decantation (CCD), iron ore concentrate, iron pellet feed, titanium ilmenite.

The first type requires strong rake mechanisms capable of moving a hard, packed coarse layer which forms at the base of the vessel. The rakes must be capable of directing and ploughing all solids to the underflow discharge via a trench. If they become buried, the torque will probably exceed any operable value. In practice, this basal sediment usually represents the maximum torque achievable during normal operation. The second type, being a uniform sediment, need only be stirred by a relatively light-duty rake system to maintain flow to the discharge point. Hence, the torque conditions will be very low in comparison to the first type.

Completely dispersed free-settling pulps, representing the extreme case, are rarely handled in rake-type gravity thickeners. Pulps of this type, often referred to as hydroseparating, or classifying pulps settle freely and torque calculation is readily determined from data obtained by simple test methods. A good example of this type of hydroseparator is a rake-type or scroll-type classifier as shown in *Figure 5.8*. Size and capacity determination is based on the specification of the following

1. solids handling capacity of the settling unit;
2. diameter or width of the unit;
3. angle of slope of the base;
4. angle of inclination of the rake blades;
5. velocity of the rakes;
6. solids and liquid densities.

A test is conducted to determine the static coefficient of friction for solids–plate and solids–solids and these data are included in empirically

Figure 5.8. Scroll-type classifiers

based formulae of the type given earlier to determine the torque value and power requirement.

A similar type of batch test is employed for slurries of type 1 but it should be kept in mind that the presence of even a small amount of flocculated slimes greatly decreases the torque value. Most slurries handled in thickeners are highly flocculated. A more complex model is therefore required for type-2 slurries, taking into account the concentration gradient, i.e. the effect of variation in pulp concentration from the top to the bottom of the thickening zone, and the effect of non-Newtonian fluid behaviour, of which the most common example is thixotropy. These effects can be measured by means of specially designed test apparatus. Rotating blade devices or specialized types of viscometer may prove suitable for obtaining a value of yield for the pulp under examination. The torque value is computed by adding together the various elements, i.e. rake blades, truss structure etc., as determined for the model for a specific type of design. Each value is usually corrected by factors that are often proprietary to a particular manufacturer.

All of the foregoing has direct impact on thickener power requirements. The power values given in *Figure 5.9*[24] give an indication of how the total power requirement varies in accordance with duty and vessel diameter. This kind of graphical approach can, however, only be used to obtain an indication because, as we have seen, the nature and type of sediment formed is commonly a major factor in determining the power requirement.

Raking power[23] is defined as the sum of the individual power requirement for the two principal functions, i.e.

1. raking solids into the underflow trough, P_r;
2. moving solids around the trough into the discharge point, P_t.

Hence $P = P_r + P_t$ for which theoretical determinations assume that solids

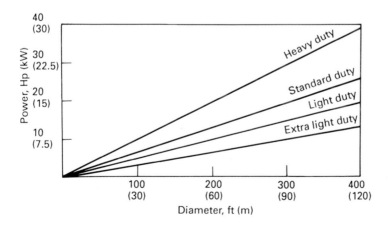

Figure 5.9. A plot of power requirement versus diameter for thickeners

settle uniformly over the entire vessel area. Approximate values may be obtained using formulae which take into account the solids handling capacity, vessel diameter, solids and liquid density and frictional effects, i.e.

$$P_r = \frac{Wd}{1000} \left[\frac{P_s - P_l}{P_s} \right] K_G \tag{5.2}$$

$$P_t = \frac{Wd_t U_s}{234} \left[\frac{P_s - P_l}{P_s} \right] \tag{5.3}$$

5.4.5 Thickener design by means of batch settling tests

The previous sections have provided a review of the properties and behaviour of settling suspensions and the way in which these influence the performance of a gravity thickener. The classic work on the capacity of a continuously operating thickener was that provided by Coe and Clevenger[5] in 1916 and was based upon the assumption that the rate of settlement within the zone of settling is a function only of the solids concentration. If so, then it should have the same value in a batch test as in continuous operation. They devised a formula to give the required area of thickener vessel in terms of the various concentrations, using batch settling tests conducted at different initial pulp concentrations to indicate corresponding settling rates at various horizons in the vessel.

Once a thickener has achieved steady operating conditions, the total solids flow rate at any level between the feed and underflow points remains the same and the overflow is hopefully clear. Then, by solids volumetric balance, and using the nomenclature given on page 132,

$$\frac{Q_f C_f}{\rho_s} = \frac{QC}{\rho_s} = \frac{Q_u C_u}{\rho_s} \tag{5.4}$$

and by liquid volumetric balance,

$$Q_o = Q \left(1 - \frac{C}{\rho_s} \right) - Q_u \left(1 - \frac{C_u}{\rho_s} \right)$$

which, by substitution, gives

$$Q_o = Q \left(1 - \frac{C}{\rho_s} \right) - \frac{QC}{C_u} \left(1 - \frac{C_u}{\rho_s} \right)$$

or

$$Q_o = Q \left(1 - \frac{C}{C_u} \right) = QC \left(\frac{1}{C} - \frac{1}{C_u} \right)$$

Substituting for QC and dividing by A (the thickener area) gives

$$\frac{Q_o}{A} = \frac{Q_f C_f}{A}\left(\frac{1}{C} - \frac{1}{C_u}\right)$$

(5.5)

where Q_0/A represents the superficial velocity of the liquid being displaced by settling solids; if the thickener is to discharge a clear overflow this must not exceed the settling rate of solids at concentration C. Thus, by substituting the velocity by U, the equation becomes

$$A = \frac{Q_f C_f}{U}\left(\frac{1}{C} - \frac{1}{C_u}\right)$$

(5.6)

which is the classic Coe and Clevenger equation[5]. This is often expressed in terms of solids flux G (the mass rate of solids flow per unit of area), i.e. $G = Q_f C_f/A$,

$$G = \frac{U}{\left(\frac{1}{C} - \frac{1}{C_u}\right)}$$

(5.7)

Coe and Clevenger's method then consisted of determining corresponding values of U and C between C_f and C_u by means of individual batch settling tests, calculating each thickener area and selecting the maximum value for design purposes.

Application of the Coe and Clevenger method *per se* can often lead to values of thickener area which are inadequate, as described by Scott[10,11], Yoshioka *et al.*[18], Talmage and Fitch[19], and by others. The main source of error suggested for this tendency to underestimate is the behaviour of the suspensions in batch tests, which often leads to overestimation of thickener fluxes. Such behaviour may occur as a result of channelling or short-circuiting of fluid through the higher concentrations and may possibly be partly attributable to segregation of the particulate during compression in the case of flocculated suspensions. If the cylinder diameter used is small, wall effects may cause irregularities in settling behaviour. The Coe–Clevenger relationship may still be valid in compression but no good theoretical model has been developed which relates interface subsidence rates in batch tests to settling rates in continuous operation.

In a batch test, settling commences with a uniform initial concentration of solids—*see Figure 5.10*. The concentration in zone 3 must range between the original concentration in zone 2 and that of the final condition in zone 4. If the solids handling capacity per unit area (SHC) is lowest at some intermediate concentration a zone of such concentration must start to build up, since the rate at which solids enter this zone will be less than the rate at which they will leave. The mathematician Kynch[16] showed how to determine the concentrations and fluxes in these zones by constructions on the transition section of a single batch settling curve. Talmage and Fitch[19] used Kynch's approach to determine the critical zone, or limiting flux, for a continuous thickener directly from a Kynch-type construction.

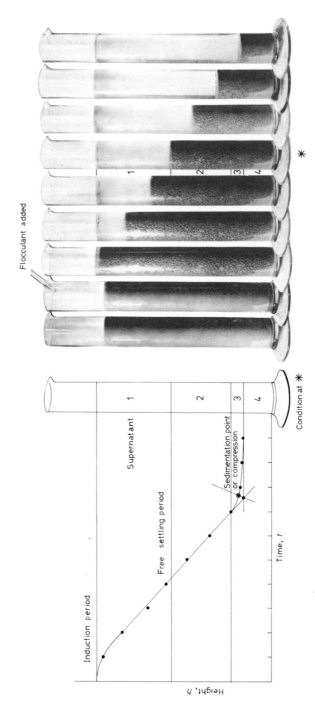

Figure 5.10. Batch settling testing

Combining Kynch's analysis with the Coe–Clevenger formula gives the graphical construction from which unit area can be obtained (unit area being the area of thickener required to handle unit mass of solids in unit time)—*see Figure 5.11a*. The total mass of solids in the batch test is $C_0 h_0 a$. When any capacity-limiting concentration layer reaches the interface, all solids in the column must have passed through it since it was propagated up from the base of the column. If the concentration of this layer is C_j at time t_j,

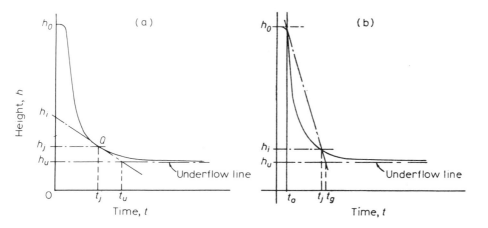

Figure 5.11. (a) Talmage–Fitch construction. (b) Oltman construction

$$C_j a t_j (u_j + w_j) = C_0 h_0 a \qquad (5.8)$$

If h_j, is the height of the interface at time t_j, and since it has been shown by Kynch that the upward velocity of any specific layer is constant,

$$w_j = \frac{h_j}{t_j}$$

which, when substituted in equation 5.8, gives

$$C_j a t_j \left(u_j + \frac{h_j}{t_j} \right) = C_0 h_0 a$$

which simplifies to

$$C_j = \frac{C_0 h_0}{(h_j + u_j t_j)} \qquad (5.9)$$

Referring again to *Figure 5.11a*, the value of u_j is the slope of tangent h_i to Q at $t = t_j$

$$u_j = \frac{h_j - h_i}{t_j - 0}$$

(5.10)

which simplifies to

$$h_i = h_j + t_j u_j$$

(5.11)

Combining equations 5.7 and 5.8 gives

$$C_j h_i = C_0 h_0$$

This means that h_i is the height of uniform slurry of concentration C_j which contains the same amount of solids as the initial slurry.

The settling velocity as a function of concentration may be developed from a single settling test by use of the above relationship. Using arbitrarily chosen values of concentration, C_j, the corresponding value of h_i can be calculated. u_j can then be determined as the slope of the line drawn through the point h_i and the tangent to the settling curve and a complete set of data showing u_j as a function of C_j can therefore be developed from one settling test. These data can be represented graphically as shown in *Figure 5.12*.

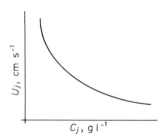

Figure 5.12. Settling rate as a function of slurry concentration

In order to specify the area requirement of a thickener the concentration layer requiring the maximum area to pass a unit mass of solids must be determined. This may be done by calculating the unit area required for a series of concentrations using the data showing u as $f(C)$, developed previously, and substituting in the Coe–Clevenger formula (rearranged from equation 5.7)

$$\text{unit area} = \frac{A}{Q_f C_f} = \frac{\left(\dfrac{1}{C_j} - \dfrac{1}{C_u} \right)}{u}$$

Whichever concentration layer gives the largest unit area is then used as a

Whichever concentration layer gives the largest unit area is then used as a design base. With *Figure 5.11a*, a simple geometrical construction may be used to obtain these areas directly.

In Kynch's equation

$$C_0 h_0 = C_j h_i = C_u h_u \qquad \text{(i.e. equation 5.11)}$$

where C_u is the desired underflow concentration and h_u is the corresponding height:

$$\frac{1}{C_j} = \frac{h_j}{C_0 h_0}$$

and

$$\frac{1}{C_u} = \frac{h_u}{C_0 h_0}$$

hence

$$\frac{1}{C_j} - \frac{1}{C_u} = \frac{h_j - h_u}{C_0 h_0}$$

If u_j is represented by slope h_j at Q

$$u_i = \frac{h_u - h_i}{t_u}$$

so that

$$\left(\frac{1}{C_j} - \frac{1}{C_u}\right) = \left(\frac{h_u - h_i}{C_0 h_0}\right) \bigg/ \left(\frac{h_u - h_i}{t_u}\right)$$

thus

$$\text{unit area} = \frac{t_u}{C_0 h_0} \qquad (5.12)$$

The Talmage and Fitch method for determining the unit area corresponding to any pulp concentration C_j in the free settling range is therefore as follows:

1. determine h_j and h_u from the following material balances; point h_i corresponding to an arbitrarily selected concentration, C_j,

$$C_0 h_0 = C_j h_i = C_u h_u$$

2. draw an 'underflow' line parallel to the time axis at height h_u on the settling curve, as shown in *Figure 5.11b*;
3. draw a tangent to the settling curve through point h_i;
4. read t_u at the intersection of the tangent and the underflow line;
5. calculate

$$\text{unit area} = \frac{t_u}{C_0 h_0}$$

In some cases, where the induction period is very marked, a correction to t_u can be made by subtracting a small time interval to compensate for induction. This would need to be estimated from the settling curve. Because it is only the hindered settling zone and not the part where the interfacial layer goes into 'compression' that determines thickener area, it is important to note whether the underflow line through h_u cuts the settling curve above or below the compression point. If this line is below, then the tangent to the curve must pass through the compression point as for *Figure 5.11a*. If, however, the underflow line intersects the curve above the compression point then the tangent required, and hence t_u, is the one that corresponds to the intersection of the underflow line and the curve. The objective in both cases remains the same, to obtain the maximum value of t_u for the hindered settling zone.

If Kynch's model was completely valid, flux pots determined from the constant-rate sections of batch settling curves at various initial concentrations (Coe and Clevenger zone settling tests) should give identical values to those determined by Kynch constructions in the initial period of descent. They rarely do and discrepancies between the two may occasionally be quite large. Empirical data from operating thickeners have shown that, in general, Coe and Clevenger tests overestimate thickener fluxes and lead to underdesign of thickener area. The Kynch-based Talmage–Fitch approach underestimates critical fluxes and leads to overdesign. Fitch's colleague at Dorr Oliver, H. Oltmann, introduced another procedure[14] which has found considerable favour because it appears to fall between the two. The Oltmann procedure, like that of Talmage and Fitch, requires a single batch settling test and relies upon identifying a compression point. It is therefore quite dependent upon this point being fairly clearly evident from the curve. In cases where difficulty is experienced it may prove worthwhile to plot the height-time curve on a log–log scale or to plot $\log (h - h_\infty)$ where h_∞ is the height of settled solids at infinite time against time on a linear scale. (This construction is included in example 5.2 later in this section.) In the latter graph the compression leg is usually linear. The most commonly applied and perhaps most dubious approach to locating the compression point in the more 'obvious' cases is by bisecting the angle between the projected settling linear section and the projected compaction section. The point where the bisection intersects the curve is the compression point, but if discontinuities exist the intersection will usually 'look' wrong.

Once the compression point (corresponding to h_i) has been located on the curve, the construction shown in *Figure 5.11b* should be made. If an induction period occurs, the free settling line is extended back to its intercept with the original pulp height at point h_0. The underflow concentration line h_u, equal to $h_0 C_0/C_u$, is then drawn in, where C_u has been obtained from a detention test.

The time t_g that corresponds to the value of h_u is read off, and the intersection with the induction line is noted. The value of t_u is then,

$$t_u = t_g - t_0$$

The corresponding critical thickener unit area will then be

$\dfrac{(F)t_u}{C_0 h_0}$ usually expressed in square metres per tonne per hour $(m^2 t^{-1} h^{-1})$ (or day)

where F is a completely empirical correction known as the 'metric rule' (the old three-foot rule), which is applied if the depth of the compression zone is determined as being over one metre. A common allowance of 20% is made in such cases, which means that F becomes 1.2. Critical thickener flux is therefore equal to

$G_c = \dfrac{C_0 h_0}{F t_u}$ usually expressed in tonnes per square metre per hour $(t\ m^{-2} h^{-1})$ solids loading

Example 5.1

The settling curve shown in *Figure 5.13* was obtained for a coal tailings slurry batch settling test. The solids concentration of the test slurry was 5% g l^{-1} solids by weight and the data shown in *Table 5.2.* were recorded during the test. Final concentration is 30% g l^{-1}. A starch-type flocculant was used for the test at a dosage of 1.0 kg per tonne dry solids.

The Talmage–Fitch construction described earlier has been added to the curve:

$$C_0 h_0 = C_j h_j = C_u h_u$$

Then

$$5 \times 17 = C_j h_j = 30 \times h_u$$
$$2.8 = h_u$$

and

$$5 \times 17 = \quad 10\ h_j$$
$$8.5 = \quad h_j$$

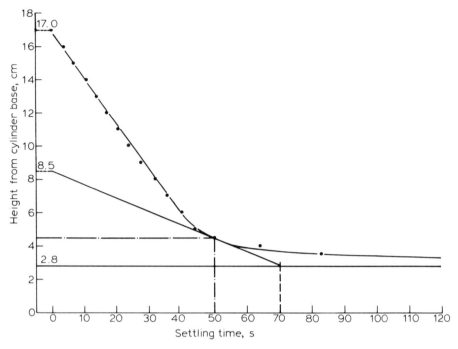

Figure 5.13. Settling curve. Material: coal tailings 5% weight concentration flocculated with starch at 1.0 mg per g solids

From the construction of the tangent, t_u is 70 s, or 1.167 h, and h_0 is 17/100 m. Thus

$$G = \frac{C_0 h_0}{F t_u} = \frac{5 \times 17}{1.2 \times 100 \times 1.167}$$

where

        compression zone exceeds 1 m,
        $F = 1.2$,
        $G = 0.6$ t h^{-1} m^{-2},
        unit area is $1/G = 1.667$ m^2 t^{-1} h^{-1}.

For a thickener required to handle 100 t h^{-1} the required diameter would be

$$d = \left[\frac{4 \times 1.667 \times 100}{\pi} \right]^{\frac{1}{2}}$$

$d = 14.6$ m or 15 m, corresponding to an area of 167 m^2 (177 m^2 design).

    Once the area is established, the depth of the compression zone can be determined from the retention time in this zone. This time is controlled by the rate of discharge of the underflow, which in turn depends on the desired concentration in the underflow. As soon as this is established, the approximate volume of the compression zone can be estimated and this is usually taken from the settling rate–concentration curve—*see Figure 5.12.*

Table 5.2.

Height from base, cm	Time of settling, s
17.0	0
16.0	4.0
15.0	7.0
14.0	11.0
13.0	14.0
12.0	17.0
11.0	20.5
10.0	23.5
9.0	27.5
8.0	32.0
7.0	35.5
6.0	40.0
5.0	44.0
4.5	50.0
4.0	64.0
3.5	83.0
3.0	131.0

If C_{ave} is the average concentration expressed as a mass fraction, in the compression zone, then $(1 - C_{ave})/C_{ave}$ is the average liquid–solid mass ratio in this zone. If t_r is retention time then the total volume V_c is the sum of the solids and liquid volumes in the zone thus:

$$V_c = \left(\frac{Q_f C_f t_r}{\rho_s}\right) + \left(\frac{Q_f C_f}{\rho_1} \cdot \frac{1-C_{ave}}{C_{ave}}\right)$$

(5.13)

If h_c is the required depth of the compression zone for a flat-bottomed tank the depth will become:

$$h_c = \frac{V_c}{A}$$

and for a circular tank of diameter d

$$h_c = \frac{4V_c}{\pi d^2}$$

(5.14)

which can also be suitably modified to consider the conical section of the thickener. The total depth of a thickener is usually estimated by allowing for clarification and settling zones as well and these usually lie in the cylindrical section. Between $\frac{3}{4}$ and $1\frac{1}{2}$ metres are usually allowed for each to provide for size capacity when this technique is applied.

Example 5.2

The second example deals with the design of a thickener to be used in a countercurrent decantation circuit in a gold ore treatment plant. Such slimes are usually ultra-fine in size and require treatment by flocculants to achieve acceptable overflow clarity. The example includes concluding comments on the final selection and shows how important it is to consider the *entire* treatment circuit.

Example 5.3

Thickening test on the slime fraction of two composite gold ore samples A-01 and A-02.

Purpose

To estimate thickener size for the hydrocyclone CCD washing circuit over-flow slimes.

Method

The settling tests were run in the following manner. Two 600-g samples of the composite were rod-milled (separately) at 63% solids using 2.4 kg of slaked lime per ton ore to a grind size of about 80% passing mesh. The ground pulp was deslimed at 200 mesh by wet screening. The -200 mesh fractions of both ore samples were combined in a 5-in i.d. settling column, mixed with a perforated stirrer and flocculated with Wonderfloc 123 Pt1201 at a dosage of 0.035 kg ton^{-1} ore (i.e. slime fraction). The height of the pulp–supernatant interface was measured to generate the settling curve.

A Kynch analysis of the settling curve was made to determine the unit area for thickening at the end of the first falling rate settling period. The unit area (UA) was calculated from

$$UA = 0.063 \frac{t_u}{C_0 H_0} \text{ m}^2 \text{ ton}^{-1} \text{ day}^{-1}$$

where C_0 is the initial solids concentration (gcm^{-3}), H_0 is the initial pulp height (cm), and t_u is the settling time to reach desired underflow solids concentration (min).

The 'compression' point was determined by plotting H versus t on log–log paper where

$$H = \frac{H_i - H_u}{H_0 - H_u} \text{ dimensionless}$$

where H_u is the ultimate interface height (cm) and H_i is the interface height (cm) at time t.

Since the overflow stream from the washing circuit is expected to run from 7 to 12% solids, the settling tests were run at 10–11% initial solids concentration.

Data

Work sheets for the two settling tests are included in *Tables 5.3* and *5.4*. Settling curves for A-01 and A-02 are shown in *Figures 5.14* and *5.15* respectively, while curves of *H* versus *t* are plotted in *Figure 5.16*.

Table 5.3. WORKSHEET FOR THE SETTLING TEST ON SAMPLE A-01

Thickening data sheet
Test: DOSB   Sample: A-01   Weight: 586.2 g, 614.8 g   Date: 1 June 1984

Grind
MOG = 65%   Solids: 63%   Water: 350 g   Time: 10 min 9 s $Ca(OH)_2$: 1.6 g

Thickening
Initial % solids: 11.1%   Pulp volume: 3529 cm³   Pulp weight: 3778 g
Initial pulp height: 28.3 cm   Material weight: 422.8 g, −200 mesh
Flocculant: Wonderfloc 123   Addition: 0.03 kg ton⁻¹, 7.0 ml
Vessel: 12.6 cm diameter   Initial pH: 10.5   Temperature: 17°C

Settling data

Time	Height, cm	Average % solids	H	Remarks
0	28.3	11.1	1.000	
0.5	23.85		0.812	
1.0	19.65		0.635	
1.5	16.15		0.487	
2.0	14.40		0.414	
2.5	13.25	22.0	0.365	
3.0	12.35		0.327	
3.5	11.65		0.297	
4.0	11.10		0.274	
4.5	10.65		0.255	
5.0	10.30	27.2	0.241	
6	9.70	28.6	0.215	
7	9.25	29.7	0.196	
8	8.85	30.8	0.179	
9	8.55	31.7	0.167	
10	8.25	32.6	0.154	
12	7.80		0.135	
14	7.45		0.120	
16	7.15		0.108	
18	6.85		0.095	
20	6.65	38.5	0.086	
25	6.25		0.069	
30	5.95		0.057	
35	5.72	43.0	0.047	
40	5.55		0.040	
45	5.40	44.9	0.034	
60	5.10	46.7	0.021	
90	4.80		0.008	
120	4.70		0.004	
180	4.65	49.8	0.002	
ULT	4.65	49.9		Supernatant slightly turbid

Table 5.4. WORKSHEET FOR THE SETTLING TEST ON SAMPLE A-02

Thickening data sheet
Test: DOSB   Sample: A-02   Weight: 601.6 g, 595.6 g   Date: 1 June 1984

Grind
MOG = 65   % Solids: 63%   Water: 350 g   Time: 7 min 30 s $Ca(OH)_2$: 1.6 g

Thickening
Initial % solids: 9.9%   Pulp volume: 3691 cm^3 Pulp weight: 3940 g
Initial pulp height: 29.60 cm   Material weight: 391.8 g, −200 mesh
Flocculant: Wonderfloc 123   Addition: 0.03 kg ton^{-1}, 8.0 ml
Vessel: 12.6 cm diameter   Initial pH: 10.6   Temperature: 18°C

Settling data

Time	Height, cm	Average % solids	H	Remarks
0	29.6	9.9	1.000	
0.5	23.45	12.4	0.755	
1.0	18.5	15.3	0.559	
1.5	15.0	18.5	0.419	
2.0	13.1	20.8	0.344	
2.5	11.8	22.8	0.292	
3.0	10.96	24.2	0.259	
3.5	10.4	25.3	0.237	
4.0	9.9	26.4	0.217	
4.5	9.5	27.3	0.201	
5.0	9.2	28.1	0.189	
6	8.67	29.5	0.168	
7	8.20	30.8	0.149	
8	7.9	31.7	0.137	
9	7.6	32.7	0.125	
10	7.38	33.5	0.117	
11	7.13	34.4	0.107	
12	7.0	34.9	0.101	
13	6.8	35.7	0.093	
14	6.62	36.5	0.086	
15	6.39	37.5	0.077	
16	6.31	37.8	0.074	
17	6.19	38.4	0.069	
18	6.11	38.7	0.066	
19	6.00	39.3	0.062	
20	5.95	39.5	0.060	
25	5.68	40.9	0.049	
30	5.48	42.0	0.041	Supernatant turbid
40	5.11	44.2	0.026	
50				
60	4.8	46.2	0.014	Supernatant turbid
360	4.80	48.7	—	Supernatant clear

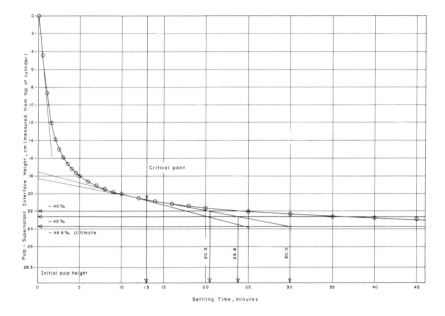

Figure 5.14. Settling curve for A-01; − 200 mesh slimes from 65 MOG

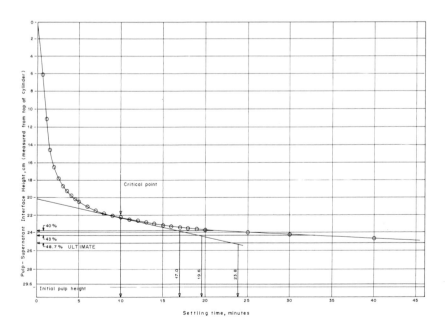

Figure 5.15. Settling curve for A-02; − 200 mesh slimes from 65 MOG

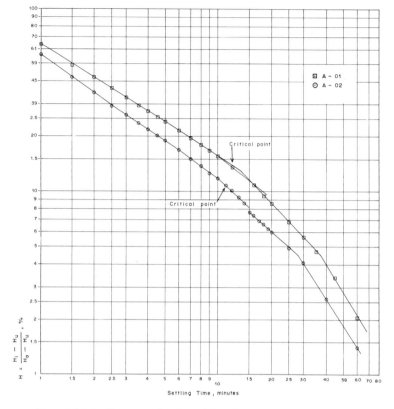

Figure 5.16. Settling curves for determining the transition point (example 2)

Results

Sedimentation for both ore types is very rapid, requiring less than 3 h to complete the settling cycle. The final percentage solids varied from 48.7% for A-02 to 49.9% for A-01. These values compare with an average of 54% solids for previous tests run on ORE-1, ORE-2 and ORE-3 (at -200 mesh grinds). Neither material showed a significant constant rate settling period (see *Figures 5.14* and *5.15*). Location of the critical points from the curves in *Figure 5.16* are not precisely defined, but the selected points are conservative.

Unit area calculations

The curves in *Figures 5.14* and *5.15* indicate underflow lines corresponding to the final percentage solids, 43% solids and 40% solids. For design purposes, an underflow of 43% is selected. To reach this underflow concentration, A-01 required a sedimentation time of 23.8 min and A-02 required 19.6 min. It is interesting to note that A-02 reaches its ultimate solids concentration in 23.8 min. Using these data, the following unit areas were calculated.

For A-01, at 43% solids: $C_0 = 0.11982$ g cm^{-3}, $h_0 = 28.3$ cm,

$$\text{UA} = 0.063 \times \frac{23.8}{0.11982 \times 28.3} = 0.44 \text{ m}^2 \text{ TPD}^{-1} \text{ at } 43\% \text{ solids}$$

(TPD = t day^{-1}, i.e. 24-h operation; TPH = t h^{-1}).
For A-01, at 40% solids,

$$\text{UA} = 0.063 \times \frac{20.5}{0.11982 \times 28.3} = 0.38 \text{ m}^2 \text{ TPD}^{-1} \text{ at } 40\% \text{ solids}$$

$$\text{Design UA} = 0.46 \text{ m}^2 \text{ TPD}^{-1}$$

Thickener size

The projected feed stream tonnage from the CCD circuit is 62 TPH of solids.
Allowing for a 50% increase in capacity to 95 TPH, the thickener size is

$$D = \left(\frac{4 \times 0.46 \times 95 \times 24}{\pi} \right)^{0.5} = 36.72 \text{ diameter } (1060 \text{ m}^2)$$

Select a 38 m diameter basin.

Discussion

A 38 m diameter thickener provides a unit area of about 0.65 m^2 TPD^{-1} (assuming that at a 65 mesh grind 35 weight% is -200 mesh at a daily throughput of 5000 ton of mill feed). At a required unit area of 0.46 m^2 TPD^{-1}, the thickener is capable of handling 102 TPH of CCD slimes. If the required unit area decreases to 0.38 m^2 TPD^{-1}, the thickening capacity increases to 125 TPH (3000 TPD) which would correspond to a grind of about 120 mesh.

The capacity of the thickener to settle solids is limited by the velocity of upward flowing water to the overflow. For a feed stream containing 95 TPH at 8% solids and for an underflow at 43% solids, the upward water velocity is 1.28 cm min^{-1} (at 70 TPH at 8% solids, the velocity is 0.95 cm min^{-1}). Thus, particles with a settling velocity of less than 1.28 cm min^{-1} will be carried to the overflow. Unflocculated particles smaller than about 20 μm will not settle.

Previous settling tests have demonstrated that higher flocculant dosages (up to 0.05 kg t^{-1}) reduce the thickening area requirements. Consequently, a design unit area of 0.46 m^2 TPD^{-1} at an underflow of 43% solids obtained with a flocculant dosage of 0.03 kg t^{-1} would appear to be a safe design factor for handling CCD slimes on the transition ores. Inasmuch as the downstream unit operations (carbon columns and carbon-in-pulp) are being designed for an underflow of 38% solids, the 38 m diameter thickener should handle any flow situations likely to be encountered under normal operating conditions.

5.4.6 Clarifier design by means of long tube data

No attempt has so far been made to consider the special requirements for the design of vessels for use as clarifiers rather than thickeners. The method which follows is one of a number which are used but this method, although perhaps the most commonly applied, still relies heavily on empirical data and often necessitates adjustment by a generous safety factor where the behavioural characteristics of the suspension are uncertain.

In the first place the clarification zone is determined by:

u_0—the overflow velocity, and
t—the detention time

Thus pool area is related to both depth and flow rate, i.e.

$$u_o = \frac{Q_f}{A} \quad \text{and} \quad t = \frac{Ah}{Q_o}$$

where

Q_0 is the volume of overflow per unit time,
A is the plan area of the pool, and
h is the perpendicular depth.

This simple relationship should be borne in mind for later on.

The long tube test is used to determine the clarification zone requirements for a clarifier. In concept, it represents the cylindrical element extending from top to bottom of the unit. This is stimulated by a long glass or plastic cylindrical tube of such a length that will allow determination of the maximum overflow rates for a certain detention time. Possibly the tube could be as long as 3 or even 4 m. Sampling taps are usually located at regular intervals (200–300 cm pitch) down the tube each normally comprising a glass or plastic tube with a pinched rubber closing tap.

Slurry is introduced to the tube following flocculation and an upper level is established at the topmost outlet. The timer is then started and the column is allowed to settle until a visually acceptable clarity is achieved. After flushing, each successive outlet is quickly sampled from top to bottom. The test is repeated for several detention times.

The final data tabulated for each set of tests will therefore relate:

cumulative depth (m) or h with corresponding cumulative average concentration (mg dm^{-3}) or c, from which overflow rate in m h^{-1} is calculated by $u_0 = h + t$.

The value of c is normally corrected to compensate for dissolved or colloidal solids. In the case of the latter, the final 'real' clarity will be a measure of the effectiveness of either a coagulant or a polymeric flocculant. Nevertheless, the ultra-fine or colloidal particles remaining in suspension must be considered as being in permanent suspension.

Assuming that a required value is suggested for Q_0 the following example should serve to indicate the normal procedure for establishing a design basis for the clarifier following a long tube test:

overflow rate required	$= 133 \text{ m}^3 \text{ h}^{-1}$	Q_0
feed solids concentration	$= 366 \text{ mg dm}^{-3}$	C
overflow solids concentration,		
maximum tolerable	$= 10 \text{ mg dm}^{-3}$	C_0
ideal overflow	$= 6.75 \text{ m h}^{-1}$	u_0
ideal detention time	$= 1.10 \text{ h}$	t
suspensoid factor	$= 0.70$	f_s

Thus

$$
\begin{aligned}
\text{pool volume, } V \quad &= Q_0 \cdot t \\
&= 133 \times 1.1 \\
&= 146.3 \text{ m}^3
\end{aligned}
$$

$$
\begin{aligned}
\text{pool area, } A \quad &= \frac{Q_0}{u_0} \cdot \frac{1}{f_s} \\[2mm]
&= \frac{133}{6.75} \times \frac{1}{0.7} \\[2mm]
&= 28.1 \text{ m}^2
\end{aligned}
$$

$$
\begin{aligned}
\text{pool depth, } h \quad &= \frac{V}{A} \\[2mm]
&= \frac{146.3}{28.1} \\[2mm]
&= 5.2 \text{ m}
\end{aligned}
$$

$$
\begin{aligned}
\text{pool diameter, } d \quad &= \sqrt{\left(\frac{4A}{\pi}\right)} \\[2mm]
&= \sqrt{\left(\frac{4 \times 28.1}{\pi}\right)} \\[2mm]
&= 5.8 \text{ m}
\end{aligned}
$$

The long tube method suffers the limitation of requiring constant refilling of the tube for each test, which leads to unavoidable inconsistencies. One approach used to overcome this limitation is to employ a much larger diameter tube therefore reducing the incremental effects of removing the successive samples insignificantly small compared with the original amount of sample. Obviously, there is a practical limit to which this remedy can be taken and alternatives are usually employed to provide another estimate of pool dimensions for comparison purposes. A common alternative, used specifically for flocculated suspensions, is known as the detention method. This method involves conventional batch settling tests and relates to clarifiers requiring almost total solids removal.

5.4.7 Thickener design by the solids flux concept

Although the Coe–Clevenger and Talmage–Fitch methods for designing thickener dimensions from batch settling tests provide a useful tool for obtaining acceptable estimates, they are not nearly concise or precise enough for continuous thickener design[20].

Solids flux is defined as the mass rate of solids flow per unit of area. In a batch settling test, the solids flux will have a single component—that due to the settling velocity. Thus, if G_s is the batch solids flux, it will represent the product of settlement velocity U and concentration C. However, in a continuous thickener with symmetrical dimensions the solids flux consists of two components:

1. a bulk transport component by virtue of the underflow rate, G_B;
2. a component due to settlement relative to the slurry, which is equivalent to the batch flux, thus:

$$G = G_B + G_s$$
$$= CV_u + G_s$$
$$= CV_u + CU$$
$$= C(V_u + U)$$

where V_u is the transport velocity, and
U is the settling velocity.

Once the underflow rate Q_u has been determined for a given thickener, the V_u value may be calculated and a graph of total flux against solids concentration can be constructed as shown in *Figure 5.17*. This graph relates to one underflow rate only, as a different graph is obtained for each rate. This makes the practical value of such an approach questionable, although the graph is useful in demonstrating the behaviour of solids in a continuous thickener.

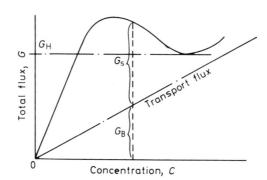

Figure 5.17. Total flux against solids concentration

Note that if $G < G_H$, i.e. relatively light loading, there is only one corresponding value of concentration. Up to G_H there exists a zone of constant concentration for any particular underflow rate. For $G \geqslant G_H$ there exist two zones of constant concentration but the flux corresponding to G_H

indicates the maximum practical value for stable thickener operation and thus G_H is called the 'solids handling capacity' (SHC) of the thickener at a given underflow rate.

For a thickener operating at steady state and with a clear overflow the mass of solids entering the thickener in unit time equals the mass of solids leaving in the underflow, i.e.

$$Q_F C_F = Q_u C_u$$

$$Q_u = Q_F C_F \cdot \frac{1}{C_u}$$

but

$$V_u = \frac{Q_u}{A} = \frac{Q_F C_F}{A} \cdot \frac{1}{C_u}$$

and by definition

$$G = \frac{Q_F C_F}{A}$$

Thus

$$V_u = G \cdot \frac{1}{C_u}$$

and therefore substitution in

$$G = CV_u + G_s$$

becomes

$$G = \frac{GC}{C_u} + G_s$$

or

$$G_s = G - \frac{G}{C_u} \cdot C$$

$$G_s = G \left(1 - \frac{C}{C_u} \right)$$

This equation may also be derived from the Coe–Clevenger equation. Normally, in designing a thickener or considering an existing thickener capacity, both G and C_u are known or fixed, and therefore only G_s and C are unknown. Thus the flux equation itself does not give an explicit method for determining thickener area but the equation does represent bulk settling rate values obtained from hindered settling tests.

When C_s is plotted against C the result is linear with intercept G and slope G/C_u. The line produced is called 'an operating line'. The batch flux relationship is also plotted with axes G_s and C and the intersection thus obtained of the operating line with the batch flux curve gives the values of C which satisfy both relationships.

Figure 5.18 indicates the flux plot for a thickener operating well within its SHC. In the thickener only a single zone at concentration C_t exists corresponding to the intersection of the line with the flux curve. If the flux remains at value G, the underflow concentration becomes C_{ui} thereby creating a second zone of constant concentration at C_a. This value of C_a corresponds to the tangent to the flux curve and the G value is called limiting. This defines the solids handling capacity for the thickener at a *given rate of underflow.*

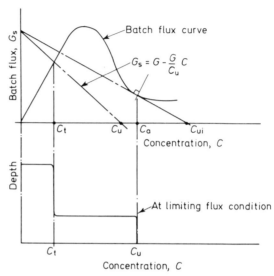

Figure 5.18. Flux plot for a thickener operating well within its solids handling capacity (SCH)

Practical use of this method is usually made by obtaining several operating lines by means of tests. A combination of altering the applied flux, G, and underflow concentration, C_u, will eventually provide an idea of the limiting flux. Caution is obviously required in increasing the practical value of C_u too much for fear of causing the thickener to slime-up, particularly in the case of flocculated feed slimes. Once a practical maximum value of C_u is obtained, it can be marked on the flux plot and a tangent drawn to determine the limiting flux. Tank area may then be obtained by dividing the maximum rate of solids loading by the limiting flux. Experience has indicated that the actual thickener throughput will be about 90% of the value obtained. Example:

total flow to thickener	$= 2350$ m³ h⁻¹
at a feed concentration	$= 4$ g l⁻¹
maximum underflow concentration	$= 10$ g l⁻¹
	$= 10$ kg m⁻³
underflow rate	$= \dfrac{2350 \times 4}{10}$
	$= 940$ m³ h⁻¹
limiting flux (from tangent construction)	$= 4.7$ kg m² h⁻¹

tank area

$$= \frac{9400}{4.7}$$

$$= 2000 \text{ m}^2$$

tank diameter

$$= \sqrt{\left(\frac{4A}{\pi}\right)}$$

$$= \sqrt{\left(\frac{8000}{\pi}\right)}$$

$$\simeq 50 \text{ m}$$

5.4.8 Impact of thickening zone on design

The compression process which occurs in the thickening zone has been discussed in the previous section and earlier in section 5.3.3. In order to evaluate the depth of the thickening zone for a given loading, it would be necessary to carry out an extensive series of steady-state runs on continuously operated laboratory-scale or pilot-scale thickeners. This does not represent a very practical approach, which explains why the most common approach still employed is largely empirically based on previous installations and using zone settling determined capacity. Evaluation of the sedimentation zone, as we have seen, is carried out by using the relationship between settling velocity and concentration. Kos[26] pointed out that the sediment forming in the thickening zone is in a compression state and its superficial velocity can take on values from close to zero (minimum) up to some maximum. If this maximum can be determined as a function of concentration, then by means of Kynch based methods the thickening zone area can be determined.

Fitch[14] suggested another steady-state approach which could be applied to determine the critical concentration layer corresponding to the point of overloading. This test approach is known as the upflow test and the concept is illustrated in *Figure 5.19*. During the upflow test the superficial velocity can be maintained constant by means of pumping a constant flow of liquid into the test column. The suspension in the column settles until it achieves an equilibrium concentration profile which is the same as at the top of the compression zone in a slightly overloaded thickener. Upflow testwork on alum sludge[21] produced a concentration distribution as shown in *Figure 5.20*.

The concentration at the base of the column is that which corresponds to the maximum superficial velocity (i.e. the velocity used during the test). By carrying out the same experiment at various flow rates the corresponding concentrations can be measured and the relationship between velocity and concentration thereby established. From these data, corresponding solids loading (flux) and solids concentration can be determined and shown graphically as in *Figure 5.21a*. The value of this approach can be seen in *Figure 5.21b* where the Kynch based methods and upflow methods are compared with actual results obtained from a laboratory continuous thickener. This comparison suggests that upflow testing can prove valuable especially for sludges.

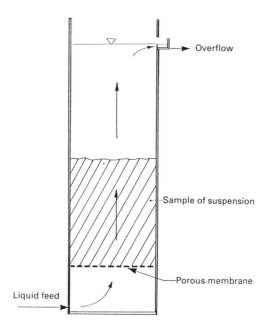

Figure 5.19. Schematic diagram of the upflow test apparatus

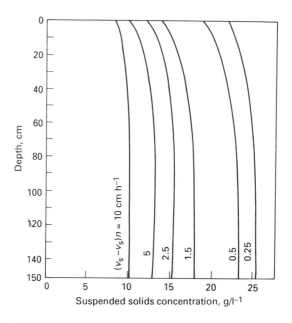

Figure 5.20. Theoretical shape of the concentration profiles in the upflow test

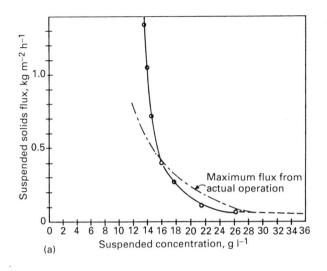

(a)

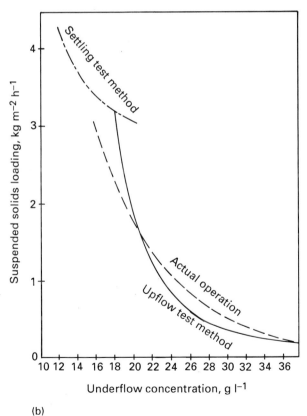

(b)

Figure 5.21. (a) Suspended solids flux as a function of concentration. (b) Comparison of settling test, upflow test and actual loading of a thickener

5.5 THICKENER TYPES

Circular thickeners, generally regarded as the conventional type, are normally provided with a mechanism consisting of a drive head, shaft or cage, and rotating arms. The latter serves two purposes: to convey the settled solids to the point of discharge, and to provide channels within the compressing sediment to permit the release of water in order to obtain a relatively dense and manageable underflow concentration. While the design of this mechanism is of primary importance, the cost of the thickener tank usually far exceeds that of the mechanism. Careful consideration must therefore be given to the overall design to minimize tank costs[27] (*Figure 5.22*).

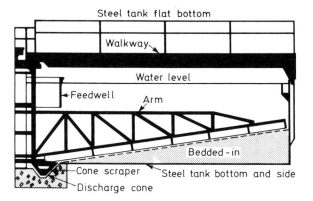

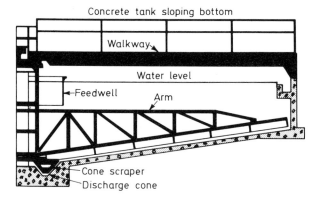

Figure 5.22. Thickener tank designs (two common types)

Thickener tanks (representing 50 or 60% of the total installed cost) are constructed of steel, concrete or a combination of both. The steel tank is usually the more economical alternative in sizes of less than 25 m in diameter. The tank bottom is often flat while the mechanism arms are sloped towards the central discharge area. With this design, settled solids must 'bed-in' to form a false sloping floor. Steel floors are rarely sloped to conform with the arms because of expense. Concrete bases and sides become more common in the larger sized tanks. In many cases the settled solids, because of particle size, tend to slump and will not form a solid false bottom. In these cases, the floor selected should be a concrete one and poured to match the slope of the arms. The tanks may also be constructed with sloping concrete floors and steel cylindrical shells which are welded to a steel ring cast into the concrete floor perimeter.

The method of supporting the mechanism depends primarily on the tank diameter. In small sizes the drive head is supported overhead on a super-structure or bridge spanning the tank, with the arms being fixed to a drive shaft. Above approximately 25 m, a stationary centre column, of either steel or concrete, supports the drive head and mechanism. Raking arms are attached to a rotating driving cage which is in turn bolted to the main drive gear. *Figure 5.23* shows these two alternatives.

(a)

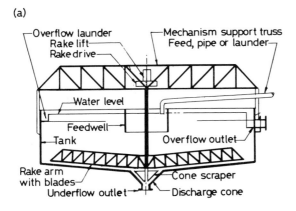

(b)

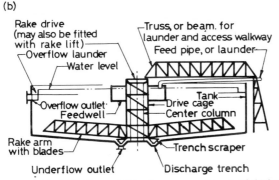

Figure 5.23. (a) Bridge and (b) column support types of thickener

A variation of the column support-type thickener is the caisson thickener, which is constructed by enlarging the central pier sufficiently so that discharge pumps can be located internally in the bottom of the centre column. The sedimented underflow is then raked to a trench surrounding the caisson and the solids enter the suction side of the underflow pumps, which pump vertically through the column and then along the access walkway. This form of construction provides the means for creating the very largest single unit thickeners, some with diameters of over 200 m. In such cases, the vessel is created by the excavation of an earthen bowl.

Feedwell designs have altered very little in conventional thickener designs although one or two new systems, developed with higher handling capacities in mind, have emerged. One such type is the Fitch feedwell shown in *Figure 5.24*. This consists of three horizontal shelves extending around the inside of a cylinder. The feed is split into two equal tangentially fed streams. The impingement of these opposing streams upon each other destroys random feed stream velocities and simultaneously promotes flocculation, thus improving the resultant settling rate and detention efficiency. Inflowing feed cannot plug this feedwell because the shelves are completely open to the inside and present no upstream edges to trap debris.

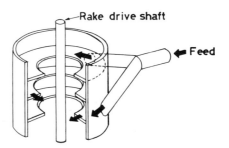

Figure 5.24. The Fitch feedwell

5.5.1 Raking arm mechanism construction

The construction must be strong enough so that the required torque to rake the settled solids to the discharge point can be applied. The necessary design also depends upon the nature of the compressed solids and must therefore cater for exceptional loading for emergency situations caused by, for example, temporary stoppage under load.

There are many different types of construction varying from a single rigid member with blades bolted directly on (limited to a maximum of 15 m diameter tanks) to many truss-type arm designs in the larger tanks (above 35 m diameter). To eliminate large structural members in motion close to the settled sediment on larger tanks, a form of extension arm can be incorporated. These are known as 'thixo' arms since they were specifically developed for sediments with thixotropic tendencies. Pipe construction of the arms can further reduce sealing effects and for this reason some fairly large diameter mechanisms are constructed with round steel piping. *Figure 5.25* shows examples of raking arm types.

In most conventional thickeners the rake arms are connected to a central shaft (bridge-type thickeners) or to a cage arrangements (centre pier thickeners). An alternative arrangement, shown in *Figure 5.26a*, consists of a hinged rake arm which is fastened to the base of the drive cage or central column. The hinge is designed to provide simultaneous vertical and horizon-

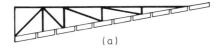

(a)

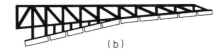

(b)

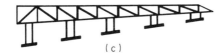

(c)

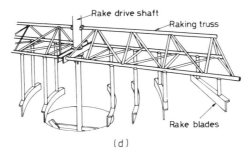

(d)

Figure 5.25. Alternative rake-arm designs: (a) conventional design; (b) design used for large diameter tanks with double sloping bottoms and thixo arms in inner area; (c) thixo arms used for thixotropic slimes to prevent 'doughnut' formations and attain maximum underflow densities; (d) Dorr–Oliver pipe construction

tal movements of the rake arm, which is hauled around the column by means of cables connected to a torque arm structure. The torque arm is rigidly connected to the central column below the water level. This type of rake is designed to lift automatically when the torque, developed as the rake moves through the sludge bed, counter-balances the weight of the rake. As the torque demand is reduced, the rake will move downwards while continuing to transport sediment towards the centre. This type of thickener finds wide application for solids which are thixotropic or very finely divided and partially flocculated.

Another type of thickener is the traction thickener, in which the central pier not only partially supports the rake mechanism but also serves as a pivot around which the rake mechanism rotates. Movement is provided by a drive mechanism or trolley, which is connected to a single long arm extending from the central pier to the tank periphery. The drive trolley imparts motion to the rake arm by means of a rail-mounted drive, the rail extending around the entire circular periphery of the tank. Usually an additional two or more shorter rake arms are located symmetrically around the central column.

(a)

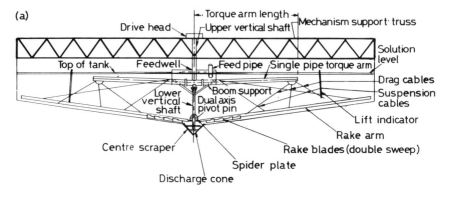

(b)

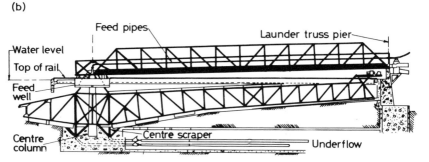

Figure 5.26. (a) Cable support-type thickener. (b) Traction-type thickener

Traction thickeners are manufactured in size from 50 to 150 m diameter and are easily recognizable by the peripheral rail. *Figure 5.26b* shows a traction thickener.

Most conventional thickeners are installed with some type of lifting device for the arms, particularly in applications which involve thixotropic materials. The fine clays etc. tend to gel, retarding flow to the withdrawal point and causing a ring or 'dough-nut' formation. If the arms can be raised and lowered, the ring can usually be removed. *Figure 5.27* shows a simple lifting device commonly applied to tailings thickeners. Manual or motorized lifting devices with automatic or semi-automatic controls are available for lifting and lowering the rake mechanism while it continues to rotate.

5.5.2 Drive types

Bridge type thickeners of up to 15 m diameter are usually driven by a single worm-spur gear combination, as shown in *Figure 5.28*. Larger units up to 50 m in diameter usually have drive heads with a double worm and spur gear arrangement, sometimes manufactured with secondary gearing as shown in *Figure 5.29*. Central pier-type thickeners are fitted with drives which are so designed as to provide drivage via a worm gear set to a large diameter ring gear through a pinion wheel. The ring gear itself provides a locus to which the

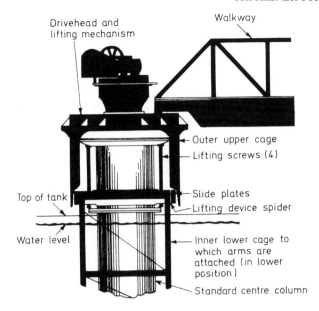

Figure 5.27. Platform-type lifting device as used on centre column supported thickeners

drive cage is connected. As thickener diameters become larger towards the caisson type of thickener, this gearing becomes more complex and the load-carrying capability and rotational stability combined with the very large torque load become limiting factors in design. The most powerful drives are therefore fitted with hydrostatic bearing systems to ensure the necessary protection against metal wear. These bearing systems have been installed with diameters of greater than 7 m. They comprise two basic parts: a runner

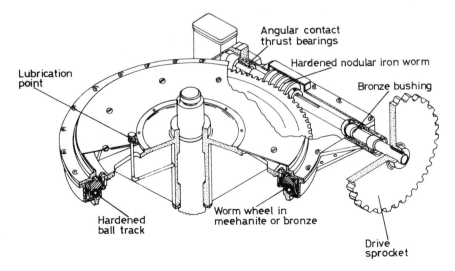

Figure 5.28. Single worm-spur gear drive. (Courtesy of Dorr–Oliver)

178

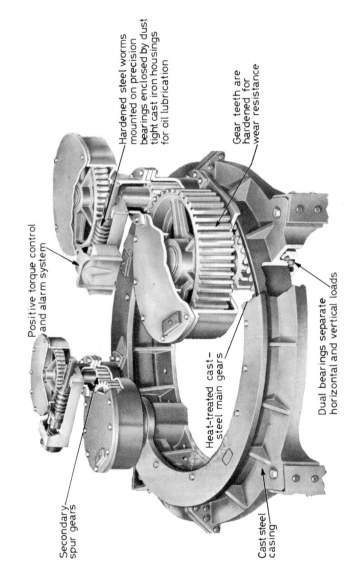

Hardened steel worms mounted on precision bearings enclosed by dust tight cast iron housings for oil lubrication

Gear teeth are hardened for wear resistance

Positive torque control and alarm system

Dual bearings separate horizontal and vertical loads

Heat-treated cast-steel main gears

Secondary spur gears

Cast steel casing

Figure 5.29. Double worm and spur gear arrangement. (Courtesy of Eimco)

and thrust pads. An oil film capable of supporting the load is maintained between the runner and each pad, with oil being constantly replenished under pressure by a controlled hydrostatic pumping system, thus ensuring very high thrust and radial load-carrying capacity.

Common accessories to conventional thickeners include overload protecting devices, torque indication of various types and, as mentioned earlier, lifting mechanisms with varying degrees of automation in operation. Most recently, level-sensing devices have been introduced which serve to provide further warning of emergency conditions and, under normal flow conditions, greater operational control. Remote indication in central control rooms now provides operators with information regarding rake position and loading, sludge line level and flow conditions in and out of the thickener, thereby ensuring safe and efficient operation and encouraging the application of the larger diameter single unit thickeners.

5.5.3 Pumping systems

Pumps which withdraw the underflow through tunnel systems have had wide application, the advantage being that high underflow concentrations can be handled without fear of blockage. The main disadvantage is the cost of constructing the tunnel beneath the main body of the thickener and many alternatives have been suggested for eliminating this, the most popular being the submergible pump system, an example of which is shown in *Figure 5.30*. However, with the larger diameters of the caisson type, with pumps mounted next to the discharge trough, tunnels can be eliminated entirely and cavitation and clogging are prevented with short pump suction lines while the pumps are still located in a readily accessible, and thus easily maintainable, position.

Centrifugal slurry pumps are most commonly used for underflows of all types with all sizes of thickener but quite often diaphragm (positive displacement) pumps are used. An example of a mechanical version of this type is shown in *Figure 5.31*. The main advantage of such pumps is that if excessive densification occurs, the positive displacement pump will continue to operate, whereas a centrifugal pump would be prone to blockage. Spring-assisted air-operated diaphragm pumps are used for pumping dense and digested sludges. An example of this type of pump is shown in *Figure 5.32*. Electrical impulses from an adjustable timer alternately open and close the solenoid-controlled air valves which admit and exhaust air from the diaphragm cavity.

5.5.4 Equipment cost

Obtaining reliable up-to-date indication of equipment cost is always difficult but occasionally useful data are published. Mular[28] suggests that cost = $a(x)^b$, where x is the tank diameter. This is also presented in graphical form in *Figure 5.33* as a means of obtaining reasonable cost estimates for thickener mechanisms. For thickeners the values of a and b are given as 147 and 1.38 respectively. The range of x for which these values are effective is 10–225 ft. Given the required handling capacity, e.g. 100 tons of dry solids per day, the thickener area can be approximated by using an appropriate value from the

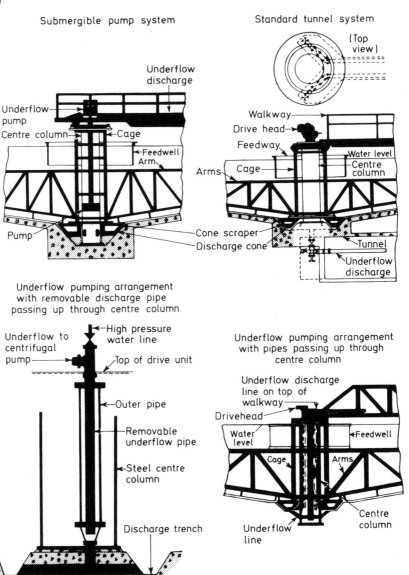

Figure 5.30. Different types of pumping systems

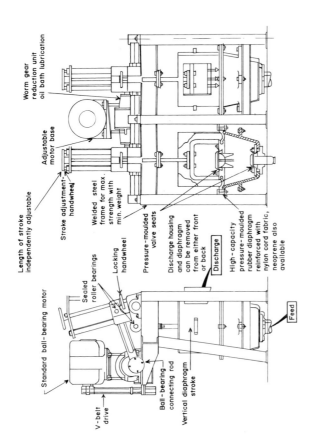

Length of stroke
independently adjustable

Worm gear
reduction unit
oil bath lubrication

Adjustable
motor base

Stroke adjustment
handwheel

Welded steel
frame for max.
strength with
min. weight

Pressure-moulded
valve seats

Discharge housing
and diaphragm
can be removed
from either front
or back

Discharge

High-capacity
pressure-moulded
rubber diaphragm
reinforced with
nylon cord fabric,
neoprene also
available

Standard ball-bearing motor

Sealed
roller bearings

Locking
handwheel

V-belt
drive

Ball-bearing
connecting rod

Vertical diaphragm
stroke

Feed

Figure 5.31. Diaphragm slurry pump. (Courtesy of Dorr–Oliver)

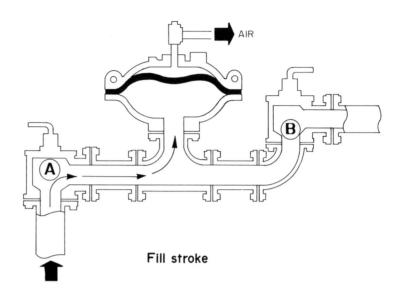

Fill stroke

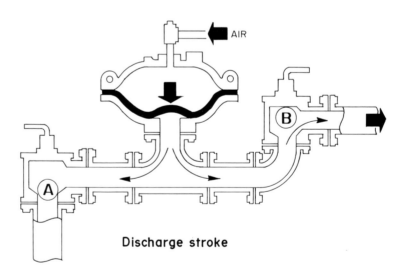

Discharge stroke

Figure 5.32. ODS° multipurpose diaphragm pump. A, B, ball valves. (Courtesy of Dorr–Oliver)

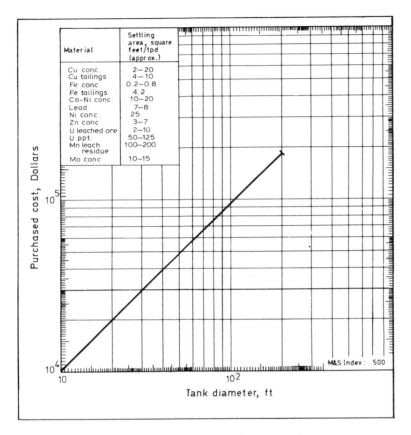

Material	Settling area, square feet/tpd (approx.)
Cu conc.	2—20
Cu tailings	4—10
Fe conc.	0.2—0.8
Fe tailings	4.2
Co-Ni conc.	10—20
Lead	7—8
Ni conc.	25
Zn conc.	3—7
U leached ore	2—10
U ppt	50—125
Mn leach residue	100—200
Mo conc	10—15

Figure 5.33. Thickener mechanism costs[29]

table (*Figure 5.33* inset). If copper tailings are to be thickened, the value of area required per ton per day has been established from experience as ranging from 4 to 10. Choosing 8, the total area becomes 800 ft² and the corresponding diameter 32 ft, which from the graph suggests a cost of $31 500. If it were suggested that the approach described above be applied to determining area requirements, it would be a contradiction of all that is implied by the previous section on thickener design. Design area must be obtained by means of the appropriate test work using the material to be thickened. The graph does, however, provide reasonable indication of cost for estimation purposes.

Substantial savings in installation costs can be realized with large thickeners, where tank wall lengths for equivalent settling areas are lower, and tunnel lengths (where required for gravity underflow) are reduced. When a caisson is used, with pumps mounted next to the discharge trough, tunnels can be eliminated entirely.

Table 5.5 illustrates savings potentials for three typical installations. Case A compares two 91.5 m diameter units with an equivalent 130 m diameter

Table 5.5 (Courtesy of Dorr–Oliver)

Smaller diameter thickener

Case	No. off	Size	Vol. concrete in wall, m³	Length of tunnel, m
A	2	91.5	801	91.5
B	2	106.75	934	106.75
C	2	122.0	1068	122.0

Equivalent large thickener

Case	No. off	Size	Vol. concrete in wall, m³	Length of tunnel, m
A	1	130.0	567	65.0
B	1	152.5	667	76.25
C	1	173.75	761	86.875

Percentage saving for large thickener

Case	Vol. concrete in wall, m³	% savings	Length of tunnel (no caisson), m	% savings	% savings with caisson (no tunnel)
A	234	29	26.5	29	100
B	267	29	30.5	29	100
C	307	29	35.125	29	100

unit. Case B compares two 106.75 m diameter units with an equivalent 152.5 m diameter unit. Case C compares two 122 m diameter units with an equivalent 173.75 m diameter unit. The examples involve a wall section of 3.05 m by 0.8 m thick, with a 0.3 m by 0.9 m section and a cross-sectional area of 0.2 m² for the launder, giving a total section area of 1.4 m².

5.6 HIGH CAPACITY THICKENING SYSTEMS

What has been described so far in this chapter has been the 'conventional' thickener, which at present is by far the most familiar sedimentation-clarification unit. The trend in the use of flocculation agents, which have simultaneously developed in capability as their function and application have become more widely understood, initially brought with it a greater handling capacity potential for the conventional thickener. Many existing installations that had struggled as a result of marginal practical capacities were transformed by the selection of effective flocculants. New installations were then designed, with flocculant application featured in the design calculations. Inevitably new types of thickener emerged, offering high solids handling capacities from relatively small areas but depending for their effectiveness on the selection of very effective flocculants and subsequent efficient mixing and dispersion of the flocculant solutions. High-capacity thickeners in this category are the deep cone type, the so-called Enviroclear thickener and the Eimco Hi-capacity thickener. Only one high-capacity thickener can be regarded as a totally different concept, this being the lamella thickener.

5.6.1 The deep cone thickener

The use of thickening tanks with deep, steep-sided conical bottoms is by no means a new approach (as it dates back to the turn of the century in the treatment of coal and metallurgical ores). With the introduction of polyelectrolyte flocculants came the reutilization of deep cone vessels, and the emergence of a variety of alternative designs largely characterized by the presence or absence of a slow stirring type of mechanism. The most widely used deep cone is that developed by British Coal as described by Abbott *et al.*[30], which is principally used to prepare froth flotation tailings for disposal without the necessity for further dewatering. The shell of this type of cone comprises an upper cylindrical launder section located above the main conical section, as shown in *Figure 5.34*. The stirrer mechanism is in three parts: the lower section of the shaft to which is attached the stirrer paddle, the mid-section main drive shaft, and the coupling to the gear box which provides a stirring speed of 2 rev. min^{-1}. The discharge mechanism consists of a

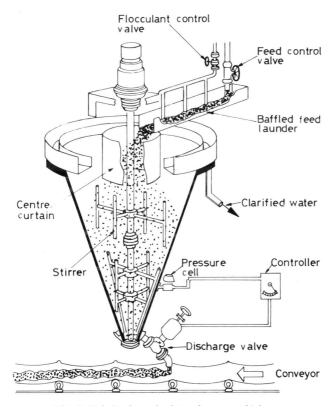

5.34. British Coal standard 4 m deep cone thickener

pneumatically operated 150 mm Saunders valve, which automatically receives a signal to open from a pressure transducer located in the wall of the cone apex. Tailings are fed to the cone from a flocculant mixing tank, or via a feed head box, to the cone feed launder.

Cones of this design range from 3 to 3.7 m in diameter, 4.3 m deep and one capable of handling 70 m^3 h^{-1} of feed at 6% solids by weight to produce a final discharge which ranges in moisture content from 25 to 35% by weight. The discharged tailings have a sausage-like appearance and are therefore fairly readily conveyed for disposal.

Recently, a further stage in the predisposal treatment of this material has been the addition of cement to solidify further the underflow in order to produce a more stabilized product for subsequent dumping.

5.6.2 The Eimco Hi-capacity thickener

This unit is shown in *Figure 5.35*. The feed enters through a hollow drive shaft into which flocculants can be added in total or in parts to the inflowing slurry and dispersed by an internal mechanical mixer. The feed then exits from the mixing chamber and is directed into a formed sludge blanket, at the base of which a conventional raking system is located to draw the denser sediment to the centre cone. Fast settling flocs are therefore encouraged to pass to the sludge zone as quickly as possible in moderately quiescent conditions, with the clear supernatant passing upwards away from the blanket. The height of the slurry blanket is automatically controlled by means of a level sensor. Eimco thickeners of this type are especially suitable

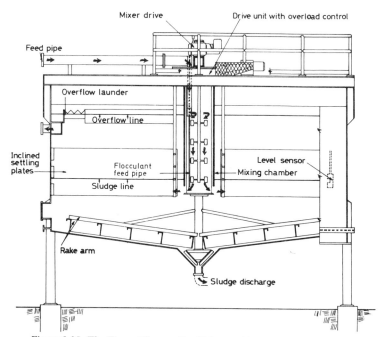

Figure 5.35. The Eimco Hi-capacity thickener. (Courtesy of Eimco)

where space is restricted and they have been installed in numerous existing plants to provide extra fines handling capacity or to support increases in handling capacity of the process circuits.

5.6.3 The Enviro-Clear thickener

This unit is shown in *Figure 5.36*, although a number of other alternative arrangements do exist. The pulp feed is introduced in the centre of the unit. A vertical feed pipe extends into the unit and is faced by a baffle plate which forces the incoming feed into an initial horizontal direction. The gap between

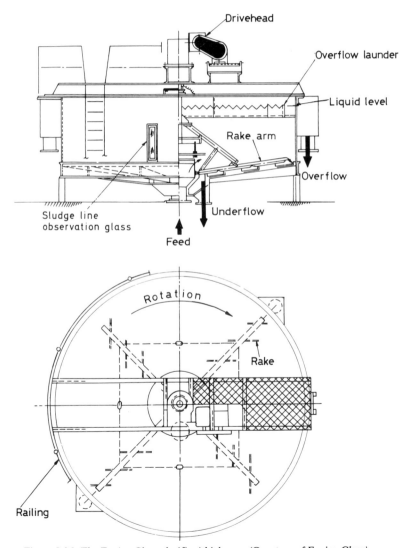

Figure 5.36. The Enviro-Clear clarifier/thickener. (Courtesy of Enviro-Clear)

the baffle plate and the end of the feed pipe determines the velocity with which the feed is introduced into the sludge bed. The feed inlet pipe can enter the unit in two alternative ways: bottom inlet with the feed pipe located in the centre of the sludge boot, or side inlet with the feed pipe entering from above through a centre well surrounding the rake drive shaft. The concentrated solids collect in the bottom of the unit where rotating rakes move the solids to the centrally located mud boot from which they are discharged. The discharge can be by gravity flow or by pump suction. The clarified liquid is discharged at the top over either peripheral or radial overflow weirs. The shell portion of the unit contains a vertical sight glass for visual inspection of the operation and for observation of the bed level. Ultrasonic level-sensing can also be used for automatic monitoring of the sludge level. Mechanical or hydraulic rake drives are used for fixed or variable raking speeds. The Enviro-Clear thickener is available with diameters from 3.8 m handling 170 m^3 h^{-1} to 18 m handling up to 3850 m^3 h^{-1}, and it has already found wide application in pollution control in the minerals industry, the food and paper industry and a variety of other special applications.

5.7 CLARIFIER TYPES

5.7.1 Introduction

Depending upon the application, clarifiers can range from simple rectangular or circular settling basins to multitray or inclined plate units. All are characterized by being capable of removing small concentrations of suspended solids from a fluid to produce a clarified overflow of a specified quality.

So far, this chapter has placed emphasis on gravity thickeners making the distinction that clarifiers normally treat suspensions having less than 1% suspended solids content (by volume) whilst thickeners are used for higher concentrations. There is of course some overlap depending largely upon the nature of the solids, i.e. colloidal, precipitates, flocculant, etc. Some mention of clarifier design was included in section 5.4.6. In this section we examine the various common types of clarifier in current use.

5.7.2 Rectangular and circular basins

Perhaps the simplest form of clarifier is the rectangular basin type shown in *Figure 5.37* which comprises a concrete rectangular basin with a sloping base from which settled solids are removed by means of some form of rake or paddled system. Most clarifying units are, however, circular tanks with a central raking assembly which are really light-duty thickener units. Varieties offer enhanced flocculant mixing or recirculation capabilities as demonstrated in *Figure 5.38*.

Multitray versions of the circular clarifier are quite common where a single high-walled tank accommodates up to four levels or trays, all served by a common raking shaft fitted with rakes at each level.

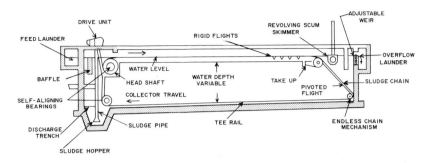

Figure 5.37. Rectangular clarifier basin. (Courtesy of Dorr–Oliver)

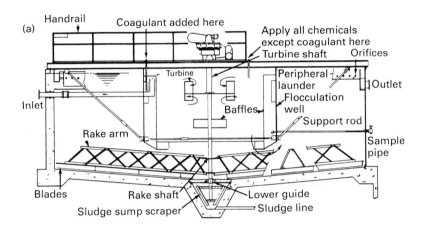

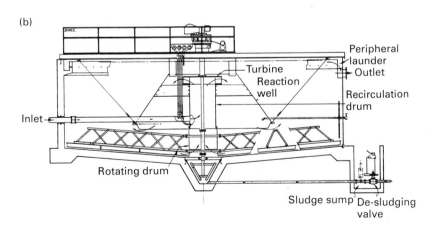

Figure 5.38. (a) Flocculating clarifier with mixing turbine. (b) Solids recirculating clarifier

5.7.3 Inclined plate or lamella clarifiers

Another type of clarifier is the inclined plate or lamella type which is commonly employed both as a clarifier and as a thickener depending upon the application. Its appeal as a potential thickener is its compactness and potential space saving.

This unit is shown in *Figure 5.39* and was originally a development of the Axel Johnson Institute for Industrial Research in Sweden. It consists of a series of inclined plates in close proximity to one another so that the effective area becomes the horizontal projected area of each plate. The incoming pulp is introduced either directly into the feed box or into a flash mix and flocculation tank. The slurry is presented to the lamella plates through a

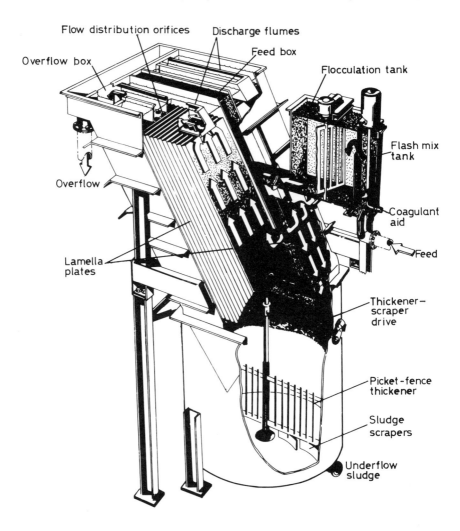

Figure 5.39. The Axel Johnson 'lamella' thickener/clarifier

bottomless rectangular feed curtain from where it flows to the plates, as shown in *Figure 5.39*, exiting at the top of the tank through flow distribution orifices. The solids settle out against the inclined surfaces of the plates and slide downwards to the sludge hopper, whereupon additional densifying is achieved by action of a low amplitude vibrator located inside the hopper.

Other varieties of the lamella unit have been developed. One is the tube settler, marketed in addition to the lamella, by Parkson Corporation of Florida. Tubes, instead of plates, are used but the principle is the same. The tubes are generally about 50 mm^2 in cross-section and about 950 mm long. However, this unit can only be applied for specially selected applications because it is vulnerable to uneven solids distribution of the feed. Many lamella units are now used and applications include the treatment of gas scrubber effluents, mill scale solutions and fly-ash suspensions. Lamella types are also used in coal and mineral treatment applications where their role is often combined clarification and thickening.

Another variety of inclined plate is the Serpac separator designed for water clarification applications such as surface-water clarification, lime softening and root crop washwater clarifications. Flow rates of between 30 and 80 m^3 h^{-1} m^{-2} are quoted for these applications. This could represent a factor of up to ten times the rate possible with a circular-basin type clarifier. The unit is shown in *Figure 5.40*.

5.8 FLOCCULATION FEED SYSTEMS FOR THICKENERS AND CONES

5.8.1 Mixing the flocculant

High molecular-weight flocculants require a definite period of time to go into solution, since the water must hydrate the long molecules before they will uncoil to form a uniform solution. The time necessary for this to take place is appreciably reduced when the powder particles are uniformly distributed throughout the water. Such uniform distribution is often still accomplished with one of the two dispersing systems shown in *Figure 5.41*, but both are vulnerable to the formation of glutinous conglomerates which cause flocculant wastage unless their operation is carefully observed. The Mining Research and Development Establishment of the British Coal Corporation (MRDE) developed the Bretby Autex Unit shown in *Figure 5.42*. This dispenser can be adapted for either hand or automatic feeding and has been demonstrated as being a reliable mixer for all polyelectrolyte flocculants. In general, it is advisable to employ slow mechanical paddle mixing to complete dissolution in 20–30 minutes. Depending upon the flocculant, polyelectrolytes are usually prepared in stock solutions which are diluted to 0.01–0.1% wt before introduction to the slurry. Starch flocculants are normally used as 0.5–1% wt solutions.

5.8.2 Adding the flocculant

Most manufacturers agree that multi-nozzle feeding systems for the intro-

192

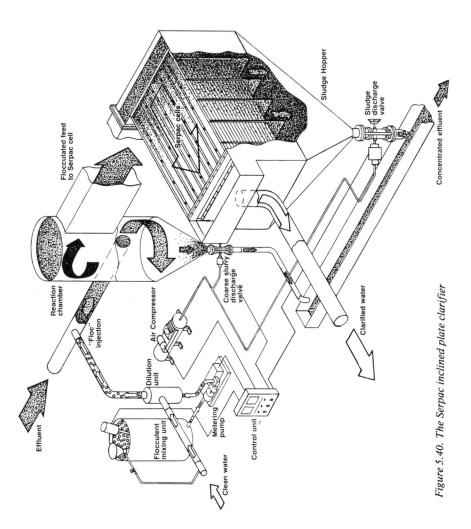

Figure 5.40. The Serpac inclined plate clarifier

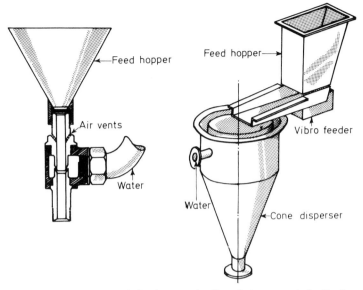

Figure 5.41. Mixing aids for dry granular flocculation reagents (wetters)

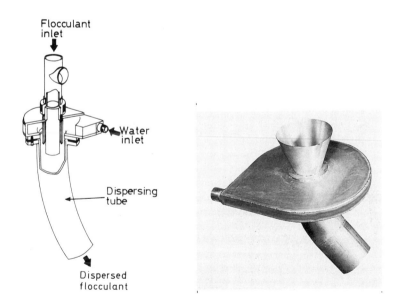

Figure 5.42. The Bretby Autex disperser. (Courtesy of British Coal)

duction to the feed produces the most effective flocculation and reduces localized overflocculation to a minimum. *Figure 5.43* illustrates two alternative methods for accomplishing this effect.

Many treatment plants still use inadequate methods for both mixing and adding flocculants. It is therefore well worth considering installing a fully automatic system which mixes the flocculant, meters its flow to the addition point and disperses a predetermined amount throughout the feed slurry.

MRDE originally developed a flocculant-mixing system which is completely automatic and which incorporates the Autex disperser. This is shown in *Figure 5.44* and the standard unit comprises a mixing tank fitted with a mixing impeller. This unit, known as the Bretby Mark II automatic flocculant mixer, can be incorporated in a total flocculation control system which

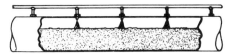

Multi-nozzle feeding into pipe

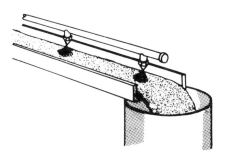

Figure 5.43. Multi-nozzle feeding systems

utilizes another MRDE development, the Clarometer, to give total regulation of flocculant usage. It is now widely used and various commercially produced units have developed the concept further[31]. *Figure 5.45* shows the total circuit. The clarometer receives a sample of the thickener feed and tests the settling rate, which is then compared by a controller with a desired time. If flocculant dosage requires adjustment following this measurement, the controller causes the flocculant control valve to be repositioned. This type of control system is discussed in further detail in the following section under the umbrella of general thickener control systems.

5.9 CONTROL SYSTEMS

Thickener performance is often assured by means of an integrated central system which can react to changes in feed conditions, cause regulation in flocculant dosage and react to a rising or falling sludge line.

Figure 5.46 shows a commonly applied control configuration. In this sytem underflow solids concentration (density) is measured by means of a nuclear gauge and controlled by adjusting the outflow of slurry. An override on the controller output ensures that the flow rate does not fall below a preset critical level. The underflow rate may be adjusted by either a pinch valve or a variable drive on the underflow pump. In the case of a diaphragm pump this may be in the form of a two-speed drive. Minimum flow may be set by measuring flow using a magnetic flow meter, calibrating pump speed, or

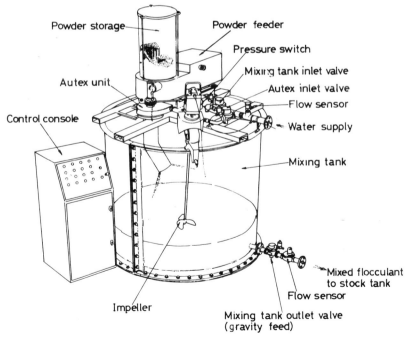

Figure 5.44. The Bretby Mark II automatic flocculant mixer. (Courtesy of British Coal)

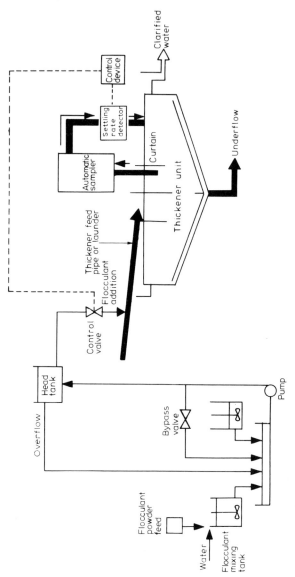

Figure 5.45. MRDE flocculation control system incorporating the Clarometer and Bretby Mark II automatic flocculant mixer. (Courtesy of British Coal)

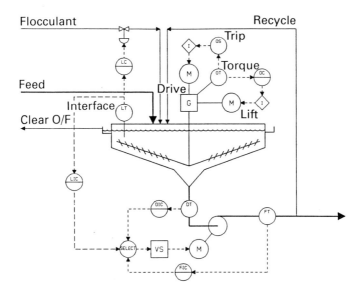

Figure 5.46. Typical thickener control system incorporating a recycle line

setting the pinch valve to a minimum stop. *Figure 5.46* shows a system with a recycle line to allow the thickener to increase the density by raising the sludge line. This is useful during start-up but is often also used in day-to-day operation.

As we have already seen, thickener rake control becomes crucially important with increasing diameter of the thickener. If rake drive torque exceeds the critical design value the ensuing damage to the drive and mechanism can be extensive. Rake control is normally achieved by means of limit switches operated by a sliding shaft on the worm gear. If remote indication is required, the shaft can also operate a position transmitter such as an electrical potentiometer. If torque increases, the rakes are usually automatically raised until a safe value is achieved; the rakes then lower again. To achieve the most efficient effect may be a manual operation but can be initiated by a timing device set to a specified torque value.

Sludge line level control systems have been the subject of several research endeavours. The level is usually measured in a point level basis using a series of ultrasonic or optical turbidity switches. A tracking probe device is also available which uses photometric cells but depends upon the formation of a clear interface to be effective. Bed level is usually adjusted by one or more of the following:

1. variable speed drive on underflow pump;
2. flocculant dosage adjustment; and/or
3. partial re-cycling.

Adjustments are usually referenced against changes in overflow clarity and in turn this can be measured by some form of clarometer. There are a number of commercially available units often with microprocessor based control for automatic adjustment of flocculant addition. This type of unit is most useful for thickeners handling low density, colloidal or precipitated solids or for high-capacity thickeners discussed earlier in section 5.6.

When reviewing overall thickener control systems and their applications in the various metallurgical, chemical and municipal fields, one of the most complex examples is that of the countercurrent decantation systems. As described earlier, these systems are used for washing and dewatering leached mineral products. *Figure 5.47* shows an example of a partial system (showing three units which could, in fact, be extended to as many as eight). The key functions in the system[32] shown are:

1. feed conductivity control: a continuous sample is tapped from the feed line and passed to a conductivity probe. The signal obtained is used to control automatically the addition of acid to the feed tank;
2. flocculant addition: a rotameter gives a continuous flow-rate measurement;
3. thickener rake torque: an axial displacement of the worm drive shaft in the drive head gives a measure of torque loading and activates a rake height control system;
4. sludge level: an ultrasonic detector provides sludge level indication;
5. underflow density: a nuclear gauge provides continuous measurement and controls a hydraulically actuated apex valve to correct to a set point;
6. pump box level: a float-actuated high-level switch(es) actuates alarms;
7. leach pump interlock: the leach pumps are interlocked with each of the thickener drives to facilitate by-passing if required;
8. countercurrent and decantation washwater temperatures: these are measured by temperature transmitters, and adjusted by balancing valves.

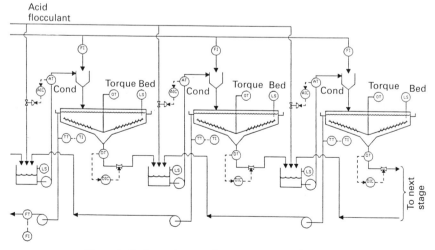

Figure 5.47. Control system for countercurrent decantation

5.10 PROCESS MODELLING

As the sedimentation theory has been developed and thickener behaviour has become more fully understood, various researchers have investigated thickener modelling for off-line (simulation) and on-line state prediction and estimation. As might be expected, the initial work concentrated on the dynamic behaviour of clarifiers. Work on dynamic behaviour of thickeners has been limited and is often not supported by experimental verification, being based solely upon sedimentation theory. A recently developed model[33] uses a continuous thickener as the experimental system to obtain direct on-line comparison of the predictive equations. The final model is therefore semi-empirical with adaptive coefficients based on proven sedimentation concepts. The model thickener is divided into four operating zones: clear liquid, settling, compaction and removal. Each zone is then modelled by an equation based upon sedimentation theory, described earlier. Modelling the total system, taking into account the physical properties involved, is achieved using the 'extended Kalman filtering' technique. By comparing selected measured values with predicted ones, the model parameters and constants can be constantly adapted. The measurement of the basic disturbances to the thickener were found to be most important to the performance of the Kalman filter. Hence measured variables included:

> underflow density (nuclear density gauge);
> underflow velocity (volumetric flow rate sensor);
> solids content (X-ray analysis);
> sludge level (ultrasonic sensor);
> torque (torque meter);
> rake position (limit switch);
> valve position (position sensor).

Estimated variables included:

> zone locations (depths);
> zone concentrations (per cent solids);
> viscosity values.

Calculated variables included:

> sludge level;
> volume flow rates of feed, underflow and overflow;
> mass balance;
> mass inventory (solids stored in the thickener);
> settling rates.

REFERENCES

1. Steinour, H. H., 'Rate of sedimentation—non-flocculated suspensions of uniform spheres—suspensions of uniform size angular particles', *Ind. Eng. Chem.*, **36**, 618–624, 804–847 (1944)
2. Hawksley, P. G. W., 'The effect of concentration on the settling of suspensions and flow through porous media', *Inst. Phys. Symp. 1950,* p. 114 (1950)
3. Wallis, G. B., 'A simplified one-dimensional two-component vertical flow', *Inst. Chem. Eng. Symp. on Interaction between Fluids and Particles,* **19**, 9–16 (1962)

4. Comings, E. W., Pruiss, C. E. and De Bord, C., 'Continuous settling and thickening', *Ind. Eng. Chem.*, **46**, 1164-1172 (1954)

5. Coe, H. S. and Clevenger, G. H., 'Methods for determining the capacities of slime thickening tanks', *Trans. AIME*, **55**, 356, 384 (1916)

6. Anderson, A. A. and Sparkman, J. E., 'Review of sedimentation theory', *Chem. Eng.*, **66**, 75-80 (1959)

7. Gaudin, A. M. and Fuerstenau, M. C., 'Experimental and mathematical model of thickening', *Trans. Soc. Min. Eng.*, **223**, 122-129 (1962)

8. Hamza, H. A., 'Least cost flocculation of clay minerals by polyelectrolytes', *Trans. IMM*, Sect. C, **87**, C212 (1978)

9. Michaels, A. S. and Bolger, J. C., *I. and E. C. Fundamentals*, **1**, 24 (1962)

10. Scott, K. J., 'Experimental study of continuous thickening of a flocculated silica slurry', *Ind. Eng. Chem. Fund.*, **7**, 582 (1968)

11. Scott, K. J., *Trans. I.M.M.*, **77**, 185 (1968)

12. Fitch, E. B., 'Current theory and thickener design', *Ind. Eng. Chem.*, **58**, 18 (1966)

13. Shannon, P. T. and Tory, E. M., *Ind. Eng. Chem.*, **57**, 18 (1965)

14. Fitch, E. B., 'Current theory and thickener design', *Filtration Separation*, 335-359, 480-488, 636-638 (1975)

15. Dell, C. C. and Sinha, J., *Trans. AIME*, **235**, 375 (1966)

16. Kynch, G. J., *Trans. Faraday Soc.*, **48**, 166 (1952)

17. Gaudin, A. M. and Fuerstenau, M. C., 'Experimental and mathematical model of thickening', *Trans. Soc. Min. Eng.*, **223**, 122-129 (1962)

18. Yoshioka, N., Hotta, Y., Tanaka, S., Naito, S. and Tongami, S., 'Continuous thickening of homogenous slurries', *Chemical Engineering, Tokyo*, **21**, 66-74 (1957)

19. Talmage, W. P. and Fitch, E. B., 'Determining thickener unit areas', *Ind. Chem.*, **47**, 38-41 (1955)

20. Wilhelm, J. H. and Naide, Y., 'Sizing and operating continuous thickeners', *Mining Engineering,* **Dec.**, 1710-1718 (1981)

21. Kos, P., 'Theory of gravity thickening of flocculent suspension and a new method of thickener sizing', *2nd World Filtration Congress*, pp. 595-603 (1979)

22. Seifert, J. A., 'Selecting thickeners and clarifiers', *Chem. Eng.,* **Oct.**, (1987)

23. Willus, C. A., and Fitch, B., *Determining Thickener Torque Requirements: Technical Report*, Dorr-Oliver Publication (1979)

24. Bradie, M. N., 'Power requirements for concentration and dewatering circuits. *In* (A. L. Mular and M. A. Anderson, Eds) *Design and Installation of Concentration and Dewatering Circuits*, Chap. 25, pp. 376-387, Soc. Mining Engineers SME AIME (1986)

25. King, D. L. and Baczek, F. A., 'Characteristics of sedimentation based equipment'. *In* (A. L. Mular and M. A. Anderson, Eds) *Design and Installation of Concentration and Dewatering Circuits*, Chap. 7, Soc. Mining Engineers, SME AIME (1986)

26. Kos, P., 'Fundamentals of gravity thickening', *Chem. Eng. Prog.,* **Nov.**, pp. 99-105 (1977)

27. Dale, L. A. and Dahlstrom, D. A., 'Design and operation of thickening equipment for closed water circuits in coal preparation', *Trans. AIME, Coal Preparation, Mudd Series, Am. Inst. Min. Pet. Eng., New York*, **12**, 38-50 (1965)

28. Mular, A. L., 'The estimation of preliminary capital costs'. *In* (A. L Mular and R. B. Bhappu, Eds) *Mineral Processing Plant Design*, 2nd edition, Chap. 3, pp. 52-70, Soc. Mining Engineers, SME AIME (1980)

29. Mular, A. L., 'Mineral processing equipment costs and preliminary capital cost estimates', *Can. Inst. Min. Metal.*, **18** (1978)

30. Abbott, J. *et al.*, 'Coal preparation plant effluent disposal by means of deep cone thickeners', *VI International Coal Preparation Congress*, (Paris), Paper 20E (1972)

31. Machling, K. L., 'Automatic flocculant control improves thickener performance', *Coal Mining* **August**, pp. 48-50 (1985)

32. Carriere, K. C., 'Control strategies for solid–liquid separation'. *In* (A. L Mular and M. A. Anderson, Eds) *Design and Installation of Concentration and Dewatering Circuits*, Chap. 35, pp. 534-547, Soc. Mining Engineers, SME AIME (1986)

33. Basur, O. A. and Herbst, J. A., 'Improved thickener performance through the use of an extended Kalman filter'. *In* (A. L. Mular and M. A. Anderson, Eds), *Design and Installation of Concentration and Dewatering Circuits*, Chap. 57, pp. 835-842, Soc. Mining Engineers, SME AIME (1986)

Further reading

Emmett, R. C., 'Selection and sizing of sedimentation–based equipment'. *In* (A. L. Mular and M. A. Anderson., Eds), *Design and Installation of Concentration and Dewatering Circuits*, Chap. 7, Soc. Mining Engineers, SME AIME, (1986)
Handley, J., 'Sedimentation: an introduction to solids flux theory', *Water Pollut. Control*, Paper 5, 230–240 (1974)
Richardson, J. F. and Zaki, W. N., 'Sedimentation and fluidisation', *Trans. Inst. Chem. Eng.*, **32**, 35 (1954)
Savage, P., 'NCB refines waste dumping', *Coal Age,* **May,** 184–188 (1980)
Keane, J. M., 'Sedimentation: theory, equipment and methods', *World Mining,* **Nov.**, (1979)
King, D. L., 'Thickeners'. *In* (A. L. Mular and R. B. Bhappu, Eds) Chap. 27, pp. 541–577, Soc. Mining Engineers, SME AIME, (1980)

6

Hydrocyclones

L. Svarovsky

Department of Chemical Engineering, University of Bradford

NOMENCLATURE

c	Feed solids concentration (fraction by volume)
c_o	Solids concentration in the overflow (by volume)
c_u	Solids concentration in the underflow (by volume)
Cy	Characteristic cyclone number
D	Hydrocyclone body diameter
D_i	Inlet equivalent diameter (by area)
D_o	Overflow diameter
D_u	Underflow diameter
E_T	Total mass recovery
E'_T	Reduced total recovery defined as $E'_T = [E_T - R_f] / (1 - E_T)$
Eu	Euler number
$G(x)$	Grade efficiency (efficiency of separation for particle size x)
$G'(x)$	Reduced grade efficiency defined as $G'(x) = [G(x) - R_f] / (1 - R_f)$
l	Vortex finder length
K	Constant
L	Overall length of hydrocyclone body
n	Exponent
n_p	Exponent
N	Number of hydrocyclones in parallel
m	Constant
p	Constant
Q	Suspension feed flow rate in each hydrocyclone
Q_T	Total suspension feed flow rate
r	Radius or constant
R_f	Underflow-to-throughput ratio (flow ratio)
Re	Reynolds number
Stk_{50}	Stokes' number
Stk'_{50}	Reduced Stokes' number (based on x'_{50})
Ss	Geometric standard deviation of $G'(x)$
T	Residence time

v_a	Axial velocity
v	Characteristic velocity
v_t	Tangential velocity
v_r	Radial velocity
x	Particle size
x_g	Solids mass median size
x_{50}	Cut size (particle size at $G'(x) = 50\%$)
x'_{50}	Reduced cut size (at reduced grade efficiency $G'(x) = 50\%$)
Δp	Static pressure drop
ϵ	Turbulent viscosity
λ	Dimensionless parameter
μ	Liquid viscosity
ρ	Liquid density
ρ_s	Solids density

6.1 INTRODUCTION AND DESCRIPTION

Cyclones have found wide application in various fields of technology, such as gas cleaning, burning, spraying, atomizing, powder classification etc. They are also used for solid–liquid separation; the cyclones specially designed for liquids are referred to as hydrocyclones, hydraulic cyclones or hydroclones. The basic separation principle employed in cyclones is centrifugal sedimentation, i.e. the suspended particles are subjected to centrifugal acceleration, which makes them separate from the fluid. Unlike centrifuges (which use the very same principle), cyclones have no moving parts and the necessary vortex motion is performed by the fluid itself.

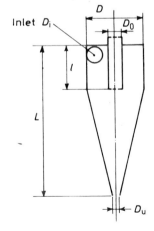

Figure 6.1. Schematic diagram of a typical hydrocyclone

Figure 6.1 shows a cross-section of a hydrocyclone of conventional design. It consists of a cylindrical section joined to a conical portion. The suspension of particles in a liquid is injected tangentially through the inlet opening in the upper part of the cylindrical section and, as a result of the tangential entry, a strong swirling motion is developed within the cyclone. A portion of the liquid containing the fine fraction of particles is discharged through a cylindrical tube fixed in the centre of the top and projecting some distance into the cyclone; the outlet tube is called the overflow pipe or vortex finder. The remaining liquid and the coarse fraction of the material leave through a circular opening at the apex of the cone, called the underflow orifice.

As with all separation principles involving particle dynamics, a knowledge of the flow pattern in the hydrocyclone is essential for understanding its function and subsequently for the optimum design and evaluation of the particle trajectories, which in turn allow prediction of the separation efficiency. A short account of the flow pattern within typical hydrocyclones and the known or probable behaviour of solid particles in the flow is given in the following section. This is done for the case of low viscosity liquids under conditions in which the particles cause little or no interference to the flow patterns (i.e. for low solid concentrations).

The operating characteristics, optimum design criteria and applications of hydrocyclones are dealt with in the remaining sections.

6.2 LIQUID FLOW PATTERNS

The flow pattern in a hydrocyclone has circular symmetry, with the exception of the region in and just around the tangential inlet duct. The velocity of flow at any point within the cyclone can be resolved into three components: the tangential velocity v_t, the radial velocity v_r and the vertical or axial velocity v_a, and these can be investigated separately.

From the experimental data in the literature, values obtained by Kelsall[1] are shown in *Figures 6.2, 6.3 and 6.4*; they were obtained with an optical device which did not interfere with the flow. In spite of their limitations, because of the special conditions of measurement and the cyclone dimensions, these data have formed the basis of some theoretical correlations devised to account for cyclone performance.

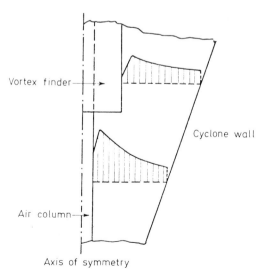

Figure 6.2. Tangential velocity distribution in a hydrocyclone

6.2.1 Tangential velocity

At levels below the rim of the vortex finder, the tangential velocity v_t increases considerably with decreasing radius down to a given radius, which is smaller than the exit radius of the vortex finder (*see Figure 6.2*). This can be described by the relationship

$$v_t r^n = \text{constant} \qquad \text{(where } n \text{ is normally } 0.6 \leqslant n \leqslant 0.9) \qquad (6.1)$$

As the radius is further increased, the tangential velocity decreases and is proportional to r, this relationship holds until the cylindrical air column (which normally forms in a hydrocyclone discharging at atmospheric pressure) is reached. At levels above the rim of the vortex finder, the break in the rise of v_t occurs at a larger radius as can be seen in *Figure 6.2*. Apart from this phenomenon and the wall effects, v_t is independent of the vertical position so that envelopes of constant tangential velocity are cylinders coaxial with the cyclone.

6.2.2 Axial velocity

As can be seen from *Figure 6.3*, there is a strong downward flow along the outer walls of both the cylindrical and conical portions. This flow is essential for cyclone operation since it removes the particles that have been separated into the underflow orifice; it is for this reason that it is not essential to build cyclones with the apex pointing downwards and the cyclone efficiency is only very little influenced by its position relative to the gravity field.

The downward current is partially counterbalanced by an upward flow in the core region, depending on the underflow-to-throughput ratio. There is a

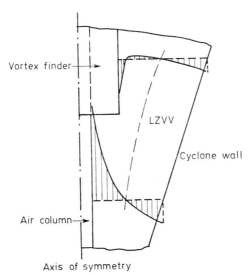

Figure 6.3. Vertical (axial) velocity distribution in a hydrocyclone. LZVV, the locus of zero vertical velocity

well defined locus of zero vertical velocities (LZVV) which follows the profile of the cyclone.

Above the rim of the vortex finder, the largest downward velocities again occur near the cyclone wall. At radii between the cyclone wall and the vortex finder, the axial velocity becomes upward. Around the vortex finder strong downward flow may be observed. This is due to wall-induced flow which runs inward along the top of the cyclone.

6.2.3 Radial velocity

The radial velocity components are normally much smaller than the other two components and, as such, they are difficult to measure accurately. As can be seen from *Figure 6.4*, the radial velocity is inward and its magnitude decreases with decreasing radius. The radial position of zero radial velocity is not known.

At levels above the rim of the vortex finder, there may be outward recirculatory flows, and near the flat top of the cyclone there are strong inward radial velocities directed towards the root of the vortex finder, thus causing the above mentioned short circuit flow down the outside wall of the vortex finder.

It should be pointed out here that this short account of velocity profiles in a hydrocyclone is only qualitative; the flow patterns are highly complex even for water with a low specific gravity and viscosity, and it may be incorrect to assume that precisely similar profiles occur in cyclones with a considerably different geometry or with liquids of high viscosity.

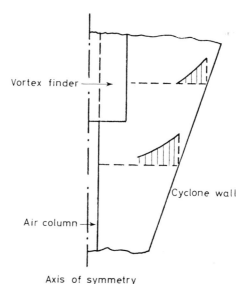

Figure 6.4. Radial velocity distribution in a hydrocyclone

6.3 MOTION OF SUSPENDED PARTICLES

Several authors have attempted to calculate particle trajectories in the cyclone and to derive formulae at least for the equiprobable size x_{50} if not for the whole grade efficiency curve. Some of these theories are discussed in section 6.6; certain observations are given here and refer to the probable behaviour of solid particles in sufficiently dilute suspensions.

When the solid particles enter near the cylindrical wall they can be dispersed radially inwards because of the intensive turbulent mixing in the feed sections. There is, however, very little information about the behaviour of the liquid in the cylindrical section: this portion of the cyclone is usually treated as a preliminary separating zone and the precise separations are thought to be performed in the conical section.

As was suggested by Kelsall, any particles present in the downward flows near the conical wall (*see Figure 6.3*), can move radially inwards only if the liquid moves inwards. It is, therefore, obvious that if a fraction R_f of the feed liquid goes into the underflow, then the same fraction R_f of all particles, independent of their settling rate, must also go with the liquid, together with the particles separated from the remaining fraction of the liquid $1 - R_f$ leaving in the underflow. This is an important phenomenon peculiar to hydrocyclones and, consequently, a correction has to be made in evaluation of the true grade efficiency curve—see section 3.4.

A particle at any point within the flow in a hydrocyclone is basically subjected to two forces: one from both the external and internal fields of acceleration (gravity and centrifugal forces) and the other from the drag exerted on the particle by the flow. The gravity effect is normally neglected in hydrocylones, so that only centrifugal and drag forces are taken into account. The movement of a particle in both the tangential and vertical (axial) directions is unopposed by any forces, so that its velocity components in those directions can be taken to be equal to the corresponding flow velocity components v_t and v_a. Since the centrifugal force acts in the radial direction, it prevents the particle following the inward radial flow—see *Figure 6.4*—and the particle is subjected to 'centrifugal elutriation'. If the centrifugal force acting on a particle exceeds the drag, the particle moves radially outwards and, if the drag is greater, the particle is carried inwards.

Since the drag force and the centrifugal force are determined by the values of v_r and v_t respectively (for a given particle), the relative values of v_r and v_t at all positions within the separation zone are decisive for the overall performance of the cyclone.

Our aim in developing the theory of the separation in hydrocyclones is to have a model that describes the process so closely that any need for test work is obviated. This is an enormous task, however, because the process is extremely complex.

Consider the complexity of the flow patterns with clean liquid (see later in this section) before we even put any particles in the flow! As our knowledge of particle–particle interaction and of particle presence on swirling, turbulent flow is still inadequate for this application, there is really no such model in existence yet. Most theories, including the most up-to-date analytical and/or numerical flow simulations, only apply to dilute systems which are rare in industry.

The problem is that a hydrocyclone may be presented with a whole range of feed slurries: from nearly clean liquids to concentrations of solids of up to 30 or 40% by volume, and from very coarse, fast settling slurries to colloidal, slow settling ones.

Let us consider the following two extreme cases:

1. if the feed is very coarse, the particles will settle readily on entry into a ribbon of solids on the wall, swirling into the underflow (this can be observed in transparent cyclones), and the main bulk of the flow inside will be essentially clean liquid;
2. if the feed is very fine, say sub-micron in size, the centrifugal field inside will be insufficient to cause any separation and the concentration anywhere in the cyclone will be essentially the same as in the feed.

Clearly, most real cases will be somewhere in between, when the particle concentration varies with position throughout the cyclone body anywhere between that of the clean liquid and the underflow concentration. Accordingly, the suspension density and apparent viscosity will vary, and different feed solids particle size distributions will result in different spatial distributions of these two variables.

Both the pressure drop and the separation efficiency (as all the other performance criteria we may be interested in) depend strongly on the cumulative effect of the density and viscosity distributions within the cyclone flow and the precise knowledge of these distributions (for any distribution of particle size in the feed) is essential for any model to work. To predict these spatial distributions, together with the complex effects of underflow orifice crowding, internal flow eddies and turbulence, short circuit flow, turbulent pick-up of solids from the boundary layer, non-Newtonian behaviour of concentrated slurries etc., is an enormous task, in my view still well beyond our capabilities today.

It is not surprising, therefore, that research workers and scientists searching for the ultimate model have had to make assumptions of one kind or another in order to circumvent this problem. Many will claim, however, universal validity of their model and some, like mathematicians for example, even suggest that we may no longer need actual tests because their computer simulation does the same job and much more cheaply!

6.4 PRESSURE DISTRIBUTION WITHIN THE FLOW, STATIC PRESSURE DROP

Due to the vortex flow in the hydrocyclone, the static pressure in the flow increases radially outward. This 'centrifugal static head' is primarily determined by the distribution of both, the tangential fluid velocities and the suspension densities, within the flow and it constitutes the major contribution to the total pressure loss across an operating hydrocyclone.

It follows, therefore, that tangential velocity distributions at low solids concentrations can be estimated from simple measurements of radial static pressure. This was the idea behind the early studies of tangential velocity distributions in clean liquid flow. Driessen[2] was the first to derive an expression relating the tangential velocity, v_t, to the radial pressure distribution by assuming the radial velocity component negligible in relation to the tangential component, so that

$$v_t^2/r = dp/(\rho \cdot dr) \qquad (6.2)$$

This relationship was then used to calculate the tangential velocities from static pressure measurements in different places within hydrocyclones run with clean liquids. Driessen[2] and many others following him thus deduced the general expression for tangential velocity profiles in the outer vortex given previously in equation 6.1, where n is an empirical exponent, usually from 0.6 to 0.9. Note that for a free vortex in inviscid flow $n = 1$, while in a forced vortex (solid body rotation) $n = -1$.

It is a common misconception that the pressure loss in the hydrocyclone is due to friction; the friction component actually plays a minor part and the centrifugal head dominates. This can be demonstrated in analogy with the static head under gravity—see *Figure 6.5*. If the feed point to a gravity tank is at the bottom of the vessel, as indicated in the figure, the static head has to be overcome in order to pump in fluid and overflow it at the top. In a working hydrocyclone, as shown in the lower part of *Figure 6.5*, the spin of the fluid causes increasing pressure radially outwards and, because the inlet is in the

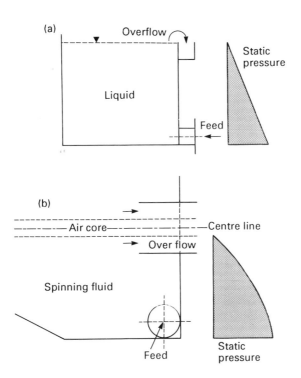

Figure 6.5. The analogy between the static head in a gravity tank (a) and the centrifugal head in a hydrocyclone (b) (note that the hydrocyclone is shown with its axis horizontal)

region of high pressure and the overflow at a point of low pressure, the static head must again be overcome in order to pump in additional fluid.

When describing the pressure loss in an operating hydrocyclone, it is common practice to relate pressure drop and flow rate in the same way as for any other flow devices, using a dimensionless pressure loss coefficient. This is defined later in this chapter (equation 6.9) as the Euler number, Eu, and, for the reasons indicated above, it must not be seen as an equivalent of the friction factor in pipes because it has very little to do with friction.

6.5 HYDROCYCLONE FUNCTION, DESIGN AND MERITS

The separation efficiency of a hydrocyclone has a character of probability. This is to do with the probability of the position of the different particles in the entrance to the cyclone, their chances of separation into the boundary layer flow and the general probability character of turbulent flow. Coarse particles are always more likely to be separated than fine particles. Effectively, the hydrocyclone processes the feed solids by an efficiency curve called 'grade efficiency', which is a percentage increasing with particle size (see chapter 3 for more details about grade efficiency). *Figure 6.6* shows the

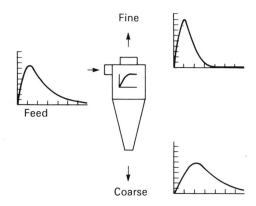

Figure 6.6. A schematic diagram of the results of the classification process in a hydrocyclone

process schematically; the solids in the feed enter the cyclone and are processed by the grade efficiency curve. There are two products of the separation: the coarse product (i.e. the solids in the underflow) and the fine product (i.e. the solids in the overflow). Every hydrocyclone is therefore primarily a classifier, although we may use it as a separator by setting the cut size as low as possible to recover as much of the solids as possible into the 'coarse product'. Chapter 3 describes the classification process in more detail.

6.5.1 The effect of cyclone proportions

There has been a lot published about the relative proportions of the cyclone dimensions and their effect on separation efficiency and pressure drop. A user need not be concerned with this aspect except for two observations. Firstly, there is a rule by which every measure which increases resistance to flow improves solids recovery, and vice versa. This applies to all proportions of the cyclone body, within certain reasonable limits, except for the length of the cyclone. Thus for example, a cyclone with relatively small inlet and outlet openings is expected to give higher mass recovery but will offer higher resistance to flow and therefore have lower capacity. Several 'optimum' or recommended designs have been published and, as these have been well tested and enough is known about their performance, they may be adopted if needed.

Secondly, one dimension of a cyclone should be made variable and that is the underflow orifice diameter. Correct adjustment of this dimension is vital for the best operation of the cyclone because the optimum size of the opening cannot be predicted reliably. It is for this reason that the underflow orifice diameter is often regarded as an operating (rather than design) variable. The orifice diameter is best adjusted after start-up of the plant and also during operation whenever some operating conditions change or when the orifice itself gradually wears out.

6.5.2 The effect of operating variables

There is a whole host of operating conditions that affect the performance of hydrocyclones. Perhaps the most important are the operating pressure drop and the feed concentration. With increasing pressure drop the efficiency of separation increases but the law of diminishing returns applies. There is little point in increasing the pressure beyond 5 or 6 bar, the typical operating pressures for larger cyclones being between 1 and 2 bar. With increasing feed concentration the efficiency of separation rapidly falls off and hydrocyclones are therefore operated with dilute feeds whenever high total mass recoveries are sought.

6.5.3 Typical sizes and performance ranges

The diameters of individual cyclones range from 10 mm to 2.5 m, cut sizes for most solids range from 2 to 250 μm (for the definition of the cut size consult chapter 3), flow rates (capacities) of single units range from 0.1 to 7200 m^3 h^{-1}. The operating pressure drops vary from 0.34 to 6 bar, with smaller units usually operated at higher pressures than the large ones. The underflow solids concentrations that can be achieved with hydrocyclones rarely exceed 45 or 50% by volume, depending on the size and design of the unit, operating conditions and the nature of the solids being separated.

In order to make full use of the advantages of the hydrocyclone it is often best to use multiple units, connected either in series or in parallel. In clarification duties for example, the parallel connections allow the more efficient, smaller diameter units to be used to treat high flow rates. The series connections on the other hand, are used to improve overall recoveries in clarification, to produce thicker underflows and clearer overflows simultaneously, to wash solids or to sharpen the classification or sorting.

6.5.4 Merits and disadvantages

The relative merits of hydrocyclones can be summarized as follows:

1. they are extremely versatile in application in that they can be used to clarify liquids, concentrate slurries, classify solids, wash solids, separate two immiscible liquids, degas liquids or sort solids according to density or shape;
2. they are simple, cheap to purchase, install, and run, and require little in the way of maintenance and support structures;
3. they are small relative to other separators, thus saving space and also giving low residence times, which gives them an advantage in terms of the speed of control over the sedimentation classifiers for example;
4. the existence of high shear forces in the flow is an advantage in classification of solids because it breaks any agglomerates, and also in the treatment of thixotropic and Bingham plastic slurries.

The disadvantages of hydrocyclones may be listed as follows:

1. they are somewhat inflexible once installed and operated, giving low turn-down ratios due to the strong dependence of their separation performance on flow rate and feed concentration; they are also inflexible due to their general sensitivity to instabilities in feed flow rate and solids concentration;
2. there are limitations on their separation performance in terms of the sharpness of cut, the range of operating cut size, dewatering performance or the clarification power; some of these characteristics may be improved in multistage arrangements, but at additional costs of power and investment;
3. they are susceptible to abrasion but steps can be taken to reduce abrasive effects;
4. the existence of shear may sometimes turn into a disadvantage because flocculation cannot be used to enhance the separation as in the case of gravity thickeners (as most flocs do not survive the shear).

6.5.5 Two basic design types

The conventional, cono-cylindrical hydrocyclone with a single, tangential entry comes in two basic shapes, depending on the included angle of the cone—see *Figure 6.7*. The narrow-angle design, with angles of up to about 25°, is used more widely than the wide-angle design with angles being anywhere from 25° to 180°. The cone angle has a fundamental effect on the existence of circulating flows in the cone: at narrow angles such flows are supressed and this makes the cyclone efficient for separation of fine particles. Such cyclones are then used whenever the required cut size is relatively low

Narrow angle
design

Wide angle
design

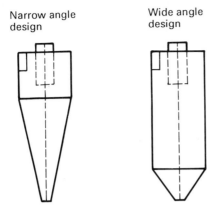

Figure 6.7. The two conventional hydrocyclone designs

such as in clarification applications, in thickening and in classification of fine materials.

The washing action of the circulating flows in cyclones with wide angles, however, is beneficial in the following cases:

1. in classification, the sharpness of cut is improved (bearing in mind, however, that the wide-angle cyclones are able to achieve only relatively coarse cuts);
2. the cut size of wide-angle cyclones may be as high as 400 μm or more, i.e. appreciably higher than is possible with the narrow-angle designs;
3. the circulating flows in wide-angle designs act like circulating fluidized beds and can sort materials according to particle density or particle shape; the best known application here is that of 'water only' cyclones in sorting minerals by density.

6.5.6 Categories of applications

Applications of hydrocyclones in industry fall into eight broad categories of two-phase separation with the liquid being the suspending medium:

1. liquid clarification;
2. slurry thickening;
3. solids washing;
4. solids classification by particle size;
5. solids sorting according to density or particle shape;
6. particle size measurement (off-line or on-line);
7. degassing of liquids;
8. separation of two immiscible liquids (the dispersed phase may be either lighter or heavier than the continuous phase).

Recent developments in the last category show that hydrocyclones can separate oil from water, dewater light oils and produce highly concentrated samples of a lighter dispersed phase.

Each application listed above has its particular requirements and goals, and it calls for changes in the design and operation of the cyclone to make the cyclone most suitable for each case. It is therefore necessary when discussing the design and operation of hydrocyclones to refer to the above-mentioned categories of application. In principle, however, any hydrocyclone separates particles (solids, droplets or gas bubbles) of the dispersed phase from the liquid (continuous phase) on the basis of the density difference between the phases, and the separation depends heavily on particle size (or on particle density if the system is not homogeneous).

Particle size which separates at 50% efficiency is referred to as the 'cut size' and is commonly used to characterize the performance of a hydrocyclone. This may be understood to be the aperture size of an ideal screen which would give similar recovery as the cyclone: smaller cut size leads to higher recovery. In the first three categories of use listed above, the aim is to set the cut size sufficiently low to obtain high solids recovery. In classification and sorting applications, however, sharpness of cut is another important factor in assessing the hydrocyclone performance as it controls the amount of mis-placed material in the two products.

6.6 THEORIES OF SEPARATION

Before giving a short review of the current theories I warn the reader to view any theory or model with an open mind and only as an aid to our efforts in hydrocyclone design and modelling, and not as a complete solution or a substitute for real test results.

A vast amount of work has been published concerned with modelling the flow and the separation process in a hydrocyclone. The different approaches to the problem can be classified into seven categories (*Figure 6.8*) as follows.

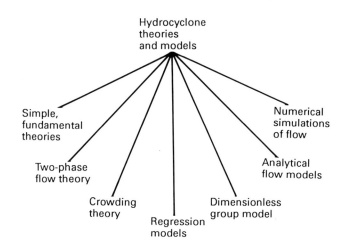

Figure 6.8. Hydrocyclone theories and models

1. Simple, fundamental theories which take no or little account of the effect of the flow ratio (or the size of the underflow orifice), of the feed concentration and of the feed size distribution. The influence of some cyclone dimensions on the cyclone performance is included.
2. A more sophisticated, two-phase flow theory which includes the effects of feed concentration and the mean particle size in the feed.
3. The crowding theory which explains the strong effect of the size of the underflow orifice on cyclone performance in some cases.
4. The all-embracing empirical models, based mostly on regression analysis of the measured data, applicable only to the specific systems tested (e.g. copper ore in Krebs cyclones etc.).
5. The chemical engineering approach based on dimensionless groups, combining 1, 2, 3 and 4 above.
6. The analytical mathematical models of the flow patterns inside the hydrocyclone and of the particle trajectories, including the boundary layer flow, the short circuit flow and the internal eddies but at low feed concentrations only.
7. The numerical simulations of fluid flows assuming axial symmetry within the main flow. Low feed concentrations are again often assumed and particle trajectories can be studied.

A short account of the above categories of theories is given in the following; a full review may be found in Svarovsky[3].

6.6.1 The simple, fundamental theories

Each theory in this category offers a relatively simple correlation for the static pressure drop and the cut size of a hydrocyclone described by a few (but often not all) dimensions. The theories fall into two main groups: the equilibrium orbit theory and the residence time theory.

6.6.1.1 The equilibrium orbit theory

The equilibrium orbit theory is based on the concept of the equilibrium radius, originally proposed by Driessen[4] and Criner[5]. According to this concept, particles of a given size attain an equilibrium radial orbit position in the cyclone where their terminal settling velocity is equal to the radial velocity of the liquid. Particles are therefore 'elutriated' by the inward radial flow according to the balance of the centrifugal and drag forces, and Stokes' law is usually assumed.

The fine particles reach equilibrium on small radii where the flow is moving upwards (and into the overflow), while the coarse particles will stay on large radii, in the downward flow, and finish in the underflow. The dividing line (or better, the surface) is the locus of zero vertical velocity (LZVV). The particle size, the equilibrium radius of which is coincident with LZVV must then be the cut size, x_{50}, which has an equal chance of reporting to either underflow or overflow. Detailed knowledge is required of the inward radial velocities which determine the drag force, as well as of the tangential velocities which cause the centrifugal force. Most authors assume some simplified shape for LZVV and then work out the average radial velocity through the surface. Most also assume Stokes' law but some consider a more general relationship for the drag coefficient. A detailed review of the different approaches is given by Bradley[6] and, more recently, by Svarovsky[3].

Probably the best known and most credible approach to the equilibrium orbit theory is that due to Bradley and Pulling[7]. This is based on the discovery of the 'mantle' by the same authors, i.e. an area in the region immediately below the vortex finder where there is no inward radial velocity. Consequently, the authors only used a conical surface below the mantle, as shown in *Figure 6.9* in the derivation of an expression for the cut size. The theoretically obtained constants were slightly adjusted by comparison with experimental results with a 38 mm diameter cyclone.

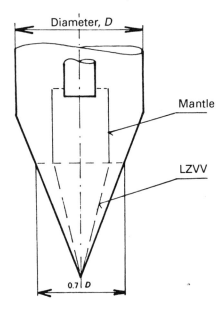

Figure 6.9. The conical surface and the mantle according to Bradley and Pulling[7]

The equilibrium orbit theory in all its various forms suggested by various authors, can be criticized on the grounds that it takes no account of the residence time of the particles in the cyclone. Not all particles may be able to attain equilibrium orbits within their residence time. The theory also takes no account of turbulence as it might affect particle separation. Despite the above disadvantages, many of the various forms of the equilibrium orbit theory (as reviewed more fully by Svarovsky[3]) give reasonable predictions of cyclone performance at low feed solids concentrations, particularly if used under similar conditions and with similar cyclone designs and sizes as in the original work of their respective proposers.

6.6.1.2 The residence-time theory

The residence-time theory assumes non-equilibrium conditions and considers whether a particle will reach the cyclone wall in the residence time available. Rietema[8] first proposed this theory and assumed homogeneous distribution of all particles across the inlet. The cut size will then be the size of the particle which, if entering precisely in the centre of the inlet pipe will just reach the wall in residence time T. In mathematical terms, this means that the particle radial settling velocity integrated with time should therefore be equal to half the inlet diameter. Rietema proposed a 'characteristic cyclone number', Cy, and suggested that it should be as small as possible considering the variables in it.

A series of large-scale experiments[8] with a hydrocyclone 76 mm in diameter of variable proportions yielded a minimum value of Cy = 3.5 for a set of proportions which has since become one of the recommended 'standard' designs (refer to a later section for further discussion). Recent experimental investigations at Bradford University[9] carried out with three sizes of Rietema's optimum design (22, 44 and 88 mm in diameter) at 1% volume feed concentration have reproduced Rietema's results, but only with very dilute underflows: values of Cy about twice as large have been found under more practical operating conditions.

Rietema's theory does not take into account the radial fluid flow, it neglects any effects of inertia, it takes no account of hindered settling at higher concentrations and it assumes any influence of turbulence to be negligible. A more recent version of the residence-time theory, the so-called 'bulk model' due to Holland-Batt[10], does take into account the radial fluid flow. He simply used the hold-up time of the liquid in the cyclone (flow rate per cyclone volume) as the residence time, average radial fluid velocity (flow rate per wall area of the cyclone) and a general continuity equation for two-dimensional flow to derive an expression for the cut size.

The original equation for the cut size does not include the important effects of the inlet and vortex finder diameters. Holland-Batt suggested[10] that their omission ". . . will be compensated to some extent by specification of the pressure drop and volumetric capacity of the hydrocyclone as input variables", but it still represents a serious omission. Holland-Batt did not suggest any pressure drop – flow rate relationship to be used in conjunction with the above theory. One important contribution made by Holland-Batt was his introduction of hindered settling. He simply adopted a hindrance factor from gravity-hindered settling theory and multiplied by it the radial particle settling velocity. Strangely, the effect of the underflow-to-throughput ratio was not included. Holland-Batt also modified his 'bulk model' for the case when non-spherical particles are not characterized by the equivalent Stokes' diameter, and a shape factor must therefore be used. He also considered the case of particle motion outside the Stokes' region.

Another theoretical approach to cut size prediction that can be classified as another version of the residence-time theory is that of Trawinski[11]. In direct analogy with gravity settling Trawinski used Stokes' law, an effective clarification area and an average acceleration in a hydrocyclone to derive an expression for the cut size. The same author also proposed[11] a rather simplistic correlation for the pressure drop–flow rate relationship.

There has also been a Russian theoretical development that can be loosely classified as a version of the residence time theory[12,13]. This is primarily concerned with overall separation of solids into the overflow and underflow and is based on a stochastic theory of separation processes. Stokes' law is used in describing resistance to particle motion but, curiously, angular velocity is employed to characterize the spin flow in the cyclone and not the tangential velocity as is more usual. The separation in either of the outgoing streams is expressed as an integral of radial gradients in the probability density with dimensionless particle residence time and it is shown to tend to a limiting value for a given set of operating conditions and feed material. The theory requires the knowledge of a coefficient described as 'intensity of random effects' and this can be found from experiments. Kutepov et al.[12] gave an empirical correlation for this, as obtained from a series of tests, expressed as a function of the cyclone design and inlet velocity. The prediction of efficiency using this theory is not particularly good: predicted overall penetrations of 9% and 3.5% in two different examples given by Kutepov et al.[12] were measured experimentally as 11% and 5% respectively. As the Russian theory does not use the concept of grade efficiency or cut size, it does not lend itself easily to direct comparison with other theories.

The residence-time theory, despite its very different approach and assumptions, often leads to correlations of very similar form to those from the equilibrium orbit theory. Either of the two theories will work better for the respective geometries to which they were 'tailored' and applied by their authors[3].

6.6.2 The turbulent two-phase flow theory

The effect of turbulence on the separation in hydrocyclones has been of concern to researchers ever since the early work of Driessen[4]. One aspect of interest is how turbulence modifies the tangential velocity profiles, i.e. its effect on the exponent n in the exponential equation for the tangential velocity in equation 2.2. Rietema[14] made a detailed investigation into this problem and estimated the turbulent viscosity with the aid of the tangential velocity profiles measured by Kelsall[1]. This was done using a dimensionless parameter λ:

$$\lambda = v_r \cdot v / \epsilon \qquad (6.3)$$

where V_r is the radial velocity in the cyclone, r is the radius of the cyclone and ϵ is the turbulent viscosity.

A more rigorous viscous turbulent model of single-phase flow, based on a Prandtl mixing length theory was published by Bloor and Ingham[15]. Like Rietema, these authors obtained theoretical velocity profiles, but they used variable radial velocity profiles calculated from a simple mathematical theory. The turbulent viscosity was then related to the rate of strain in the main flow and the distribution of eddy viscosity with radial distance at various levels in the cyclone was derived.

Most recently, Duggins and Frith[16] have challenged the concept of isotropic eddy viscosity and showed that not only does the eddy viscosity vary with position in the cyclone but its value is different in two mutually perpendicular directions and the ratio of the two values is not constant. They proposed a new method for calculating velocities where the eddy viscosity for the axial and radial momentum equations is calculated from the conventional turbulent model (k–e model) and the viscosity for the tangential momentum equation from a mixing length expression. This method, therefore, allows for anisotropy of turbulence in the swirling flow.

Schubert and Neesse[17] have recently proposed a separation model based on turbulent two-phase flow. They assumed a homogeneous, stationary turbulence field, with particles moving under Stokes' law. They also assumed that the particles are smaller than the smallest eddies and that their concentration is 'sufficiently low'. According to Schubert and Neesse particle transport consists of sedimentation flux superimposed on the turbulent diffusion flux. For mathematical convenience, they also assumed the centrifugal force field to be homogeneous. Schubert and Neesse derived two general models for turbulent cross-flow classification: depending on the way the underflow and overflow streams are removed, they distinguished between the suspension partition model where the flow is divided between the overflow and underflow without any change in total cross-section, and the suspension tapping model where the discharge streams are 'tapped' from the main flow through small outlet openings. The latter model was then applied to the separation in a cyclone, with the resulting equation for the cut size. This correlation takes no account of the spread of the feed distribution but it is the first one reported here which allows for the effect of the feed solids on the separation in a hydrocyclone, with turbulence damping effect increasing with increasing fineness of the solids.

The theory of Schubert and Neesse points to small cone angle and curved feed inlets for clarification duties where high mass recoveries are required. The authors compared the above correlation with many published experimental results and found the correspondence better for umbrella (or spray) discharge. No flow rate–pressure drop relationship was proposed.

The latest contribution from Neesse and Schubert[18] is concerned with the case of higher solids concentrations where the underflow orifice may restrict the free discharge of the solids. They propose the existence of a limit in the solids loading and propose an empirical correlation for the discharge capacity of the underflow orifice. This brings their work into the realms of the crowding theory considered below.

6.6.3 Crowding theory

This theory was first suggested by Fahlstrom[19] who proposed that the cut size is primarily a function of the capacity of the underflow orifice and of the particle size analysis of the feed. He argued that the crowding effect, or hindered discharge through the apex, can swamp the primary interaction to the extent that the cut size can be estimated from the mass recovery to the underflow. Fahlstrom quoted the discharge capacity to be from 1.5 to t3 $cm^{-2} h^{-1}$ in grinding circuits, but this depends on the absolute size of the underflow orifice and it is not realistic for small cyclones. In any case, any crowding effect must surely depend on the physical proximity of the solid particles and this depends on volume rather than mass. Fahlstrom's crowding theory seems feasible in principle but his original justification is false. He attributed the differences between the analytical cut size and the equiprobable size to the crowding effect. Furthermore, he found the difference to be a function of particle size. This difference is now known to be due to the non-ideal shape of the grade efficiency curve and it exists whether or not underflow crowding takes place.

A much more scientific proof of the crowding theory has recently emerged from some mathematical modelling work by Bloor et al.[20]. This is a hydrodynamic model of the flow both in the cyclone body and in the boundary layer (the cyclone design is that of Kelsall[1]), but it makes no predictions of conditions at underflow because it breaks down at the vertex. The model assumes inviscid flow in the main body and viscous flow in the boundary layer. It allows plots of particle trajectories, assuming their homogeneous distribution on entry to the cyclone cone.

Using a correlation for the slurry viscosity as a function of particle concentration and using concentration averaged across the boundary layer, Bloor et al.[20,21] then modelled the flow in the boundary layer and this represents the most interesting part of their work, later also published by Laverack[22]. The model allows plots to be made of particle concentrations, volume fluxes and layer thickness along the boundary layer. This in turn allows the effect of increased feed concentration on the conditions in the boundary layer to be studied: the increase was found to thicken the layer as well as to increase the flux, and the underflow would have to be increased to accommodate this. Furthermore, the authors found that the conditions at the cut-off point were decisive. The technique cannot deal with a double structured boundary layer near the vertex.

Unlike the theories reviewed in the previous sections, the model by Bloor et al. does not lead to any simple correlations for the cut size or the grade efficiency curve, but it is the first time that any direct proof of the crowding theory is given. On the basis of an actual set of conditions studied, the authors give a quantitative example: if a cyclone operates satisfactorily at 5% solids concentration and this is increased to 15%, the underflow rate must be increased by a factor of 1.6 in order to prevent overcrowding and possible blocking of the underflow orifice.

Another theoretical development, originating from Freiberg and also reported in the previous section[17], has recently been adapted to include the crowding of the underflow orifice[18]. An empirical equation for the discharge capacity of the underflow orifice is proposed.

In conclusion to this section on the crowding theory, this theory is certain to be valid in principle. Anyone involved in hydrocyclone testing will have observed the strong influence of underflow orifice control on the cut size. The question is how to describe the effect quantitatively: ultimately, the effect may be simply related to the underflow concentration by volume, in conjunction perhaps with the absolute size of the orifice and the size distribution of the feed. Further development of this theory is expected.

6.6.4 Comparison of the simple theories reviewed so far

Direct comparison of the different theories is not really fair because the respective authors always tailored their theory to a set of experimental results which were never sufficiently comprehensive or general to allow them to include the effect of all design variables. Rietema's residence-time theory, for example, does not really offer a correlation for any design other than the 'optimum design' of his own obtained by optimization from his tests.

It can be concluded that none of the simple theories give a completely general correlation that can be applied to any combination of the relative cyclone proportions. The vast majority of them, however, show that for a family of geometrically similar cyclones, there is a dimensionless group (we shall call it the 'cyclone number') which should be a constant. This constant can be obtained from experiment, rather than from the correlations given by the various theories, and this approach leads to much more reliable performance prediction. This is the approach adopted in the chemical engineering model for hydrocyclone scale-up.

Design variables are not the only ones to affect the cyclone number. While the effects of pressure, flow rate, viscosity and the densities of the fluid and particles are included, the strong influence of the feed concentration and of underflow orifice control are not. Most of the theories only apply to low solids concentrations, some (Holland-Batt, Schubert and Neesse) offer a correction for hindered settling and others consider to some extent the effect of the underflow-to-throughput ratio (Bradley) or the limited capacity of the underflow orifice (Schubert and Neesse). Once again, the effect of solids concentration for the purpose of scale-up is best described by dimensionless correlations derived from pilot tests and this is shown in some detail later.

At higher concentrations, the size distribution of the feed solids, and perhaps particle shape too, also affect the cyclone performance (because the spatial distributions of suspension viscosity and density within the hydrocyclone depend on what is being fed to it). Only the turbulence theory of Schubert and Neesse[17], and the Russian stochastic approach[12,13] include the effect of the average particle size of the feed solids.

As was pointed out at the beginning of this section, there are several other models available for predicting cyclone performance, but those were obtained from curve-fitting exercises and, as such, do not represent a true physical model of the separation process in hydrocyclones. These models are reported in the next section.

6.6.5 The regression models

This is the group of so-called mathematical models which are in fact based almost entirely on regression analysis of test data. They are concerned with the two main performance characteristics of hydrocyclones, the capacity (or pressure drop) and the separation efficiency in the form of the cut size. These two aspects are dealt with separately here.

The presence of higher concentrations of particles in the flow modifies the velocity profiles in the cyclone and increases capacity. Some authors take the view that the effect is the same as of higher liquid viscosity and density, and recommend the use of the apparent suspension viscosity of the feed slurry and density of the slurry rather that of the suspending liquid. While this approach oversimplifies the real process in the cyclone, the increase in cyclone capacity with concentration is small anyway and the errors may be acceptable (in practice, clean liquid data are often used regardless of concentration and the capacity may then be adjusted in plant by throttling or by other means).

The effect of really high feed concentrations on hydrocyclone capacity, in concentration of minerals, was studied by Lynch and co-workers[23-26] and also by Plitt[27]. Hundreds of experiments were carried out with various cyclone designs and operating conditions, leading to empirical 'models' suitable for mathematical modelling of closed circuit mineral grinding circuits or similar applications. Unfortunately, only aqueous systems were used and the authors did not seek conclusions in dimensionless form. The resulting equations are purely empirical and dimensionally inconsistent, requiring empirical constants with a dimension, and specified units to be used for each variable. There is little point in reproducing their equations in this report and the reader is referred to either the original references or the book[3].

The correlations for the pressure drop–flow rate relationship change from paper to paper in the publications by Lynch and co-workers[23-26] depending on the test system. Cyclones of 500 mm diameter were tested with silica and copper ore at concentrations of 15–65% by weight, whilst in another series of experiments[24,25], cyclones from 100 to 380 mm in diameter were tested with limestone at concentrations of 15–70% by weight. Finally, another series of tests[26] at concentrations of limestone from 40 to 70% by weight are reported.

The correlations usually take the following form:

$$Q = K D_o^{n} \cdot D_i^{m} \cdot D_u^{p} \cdot (\Delta p)^r \tag{6.4}$$

where K, n, m, p and r are all empirical constants having specific values for each system tested. Some correlations also include a concentration term.

Plitt[27] took the data of Lynch and Rao[23] and added his own, with smaller cyclones up to 150 mm diameter tested with silica flour, and used the data, 297 individual tests in total, to derive by regression analysis yet another correlation for the pressure drop – flow rate relationship, not dissimilar to the above equation.

The formulae proposed by Lynch et al. and Plitt have been applied to other cyclone sizes and slurries (see, for example, Apling et al.[28]), and neither model was found to be entirely successful: it was necessary to change the constants to fit the predicted results to the experimental data.

The variety of equations given by Lynch and co-workers for different studies is in itself indicative of this problem. As the equations are not based on any physical model of the actual process and represent merely a curve-fitting exercise, it is only to be expected that they cannot be used for reliable predictions unless the constants used are measured for each system under consideration.

As to the effects of operating variables like pressure drop and the solids concentration on the cut size, most authors agree that it is an exponential relationship. The work of Lynch quoted above resulted in several correlations while Plitt[27] suggested another.

The above-mentioned correlations are subject to the same criticism as the pressure drop–flow rate correlations discussed previously. They were obtained by curve-fitting a set of experimental data and do not represent a real model of the separation process. In situations where materials different to those used in the original tests are being handled, it is necessary to obtain a new set of constants to make the correlations fit the test data.

A further note must be added here about the grade efficiency curve. All the above-quoted correlations use the cut size as the representative of the separation efficiency. The full grade-efficiency curve is the best way to characterize the separation efficiency because only if this is known can an accurate estimate of the total mass recoveries be made. The conventional way round this problem in practice is to make a prediction of the operating cut size and then generate the full grade-efficiency curve by using a generalized grade-efficiency curve which is plotted against a dimensionless size x/x_{50}. Such generalized functions were reviewed by Bradley[6], but since then more detailed studies have been made of this such as the work by Lynch and Rao, and a full review can be found in a later book by Svarovsky[3].

One of the functions successfully fitted to the reduced grade-efficiency curve is the log–normal law (consult chapter 2 for the definition of log–normal law). Gibson[29] tested small-diameter Mozley hydrocyclones (25 and 50 mm diameter) with china clay at concentrations up to 35% by weight, and found geometric standard deviations from 1.77 to 2.0, depending on the size of the vortex finder. Larger vortex finders yielded lower standard deviations, thus giving sharper classification.

On the basis of the available experimental evidence it can be concluded that for a given cyclone design and low feed solids concentrations (say below 1% by volume) the shape of the reduced grade-efficiency curve is reasonably constant. At higher solids concentrations, it becomes dependent on the feed material and experimental measurement may be necessary. Alternatively, with smaller diameter cyclones, the log–normal law with a measured geometric standard deviation may also be used. Knowledge of the cut size is of course necessary in any of these cases to obtain the full curve.

6.6.6 The dimensionless group model

The model developed and tested at Bradford recently[9,30,31] is an example of a chemical engineering approach to the design and scale-up of single or multiple cyclone installations. It is based on fundamental theory combined with dimensional analysis to produce the necessary correlations, and, in keeping with the usual practice in chemical engineering, the required constants are derived from tests rather than from theory.

The procedure centres around the three dimensionless equations 6.5 to 6.7 (below) which fully describe the function of a hydrocyclone within its operational limits. If a given cut size and underflow concentration are specified, simultaneous solution of the three equations yields the cyclone diameter, the number of cyclones to be used in parallel and the size of the underflow orifice which, in turn, gives the required performance and capacity (under the given pressure drop). If the operating pressure drop is not specified, several design options are obtained and the final design is selected either from some operating constraints or from economic optimization. In such optimization, the operating pressure drop can be traded off, for the same cyclone, against the flow ratio. The geometrical similarity normally concerns all internal dimensions of the individual cyclones except the size of the underflow orifice which is regarded as an operational variable.

Table 6.1. TWO WELL-KNOWN HYDROCYCLONE GEOMETRIES AND RESPECTIVE
PERFORMANCE CONSTANTS (NARROW-ANGLE DESIGNS)

Cyclone type	D_i/D	D_o/D	D_u/D	l/D	L/D	Cone angle degrees	$Stk_{50} \cdot Eu$	K	n_p
Rietema[8]	0.28	0.34	0.20	0.4	5	20	0.0611	316	0.134
Bradley[7]	0.133	0.20	0.07	0.33	6.85	9	0.1111	446.5	0.323

In the model presented here, all design proportions were according to
Rietema's optimum design as per *Table 6.1*. The model was tested with two
materials: chalk of density 2780 kg m^{-3} and alumina hydrate of density 2420
kg m^{-3}. The suspending liquid used in the hydrocyclone tests was water with
0.1% Calgon as dispersing agent. The feed solids concentration was varied
from 1 to 10% by volume.

The correlations derived in the previous study[9] were as follows:

$$Stk'_{50} \, Eu = 0.0474 \, [\ln(1/R_f)]^{0.742} \exp(8.96c) \tag{6.5}$$

$$Eu = 371.5 \, Re^{0.116} \exp(-2.12c) \tag{6.6}$$

$$R_f = 1218 \, (D_u/D)^{4.75} \, Eu^{-0.30} \tag{6.7}$$

where the various dimensionless groups are defined as follows.

The Reynolds number, Re, to define the flow:

$$Re = v \, D \, \rho/\mu \tag{6.8}$$

The Euler number is a pressure loss factor based on the static pressure drop
across the cyclone:

$$Eu = \Delta p/(\rho v^2/2) \tag{6.9}$$

Stk_{50} is the Stokes number (for cut size x_{50}):

$$Stk_{50} = x_{50}^2 \, (\rho_s - \rho) \, v/(18 \, \mu \, D) \tag{6.10}$$

where ρ_s and ρ are the densities of the solids and of the liquid respectively, μ is
liquid viscosity and D is the cyclone diameter. If the reduced cut size x'_{50} is
used in the above equation in place of x_{50}, the reduced Stokes number Stk'_{50} is
obtained (for definitions of the parameters see the Nomenclature given at the
beginning of this chapter).

All the above equations use the superficial velocity in the cyclone body as
the characteristic velocity, i.e.

$$v = 4Q/(\pi \, D^2) \tag{6.11}$$

Equations 6.5 to 6.11 allow reliable cyclone design and scale-up at moderate
concentrations up to about 10% by volume. If the full data on the operation
and performance of a hydrocyclone are known then the dimensionless groups
necessary for scale-up may be calculated directly from equations 6.8 to 6.10.

Conversely, the knowledge of the scale-up constants as given in equations 6.5 to 6.7 may be used to predict the performance of a hydrocyclone of a given design or the design of a hydrocyclone installation to meet a particular performance requirement.

Looking at equations 6.5 to 6.7, one can make some interesting observations. Firstly, the resistance coefficient, Eu, in equation 6.6 is not affected by the size of the underflow orifice D_u/D (or by the flow ratio R_f). Secondly, something that every hydrocyclone operator will know well, the flow ratio R_f is very strongly affected by the setting of the underflow orifice D_u/D.

Equations 6.5 to 6.7 primarily represent the test results of the study published previously but, unlike some other models in the literature, they are based on a physical model of hydrocyclone operation and particle behaviour and, as such, have a much better chance of success when applied to conditions outside those tested. Another research project at Bradford is directed at the effects of solids concentrations greater than 10% when many practical slurries show non-Newtonian behaviour.

6.6.6.1 Examples of design choices

Equations 6.5 to 6.7 (with the dimensionless group definitions in equations 6.8 to 6.11) represent a full model necessary for hydrocyclone design or scale-up. Since its development, we have used the model for the design and optimization of industrial hydrocyclone systems. The model can be easily programmed and used on a microcomputer, and thus provides valuable insight into the effects of any of the variables involved. *Figures 6.10 to 6.12* show three-dimensional representations of the three relationships.

In order to demonstrate the usefulness of the model, it was used to provide operating alternatives for an existing hydrocyclone of internal diameter $D = 88$ mm, Rietema's proportions and operated with chalk in water at feed solids concentrations up to 10% by volume. *Figure 6.13* gives the plot of the operating conditions if the overflow clarity is to be constant (reduced cut size of 10.6 μm). As can be seen from this plot, at any given feed concentration, the choice is between low pressure drops at high flow ratios, R_f, and high pressure drops at low flow ratios. The operating pressure drop can, therefore, be traded off against the concentration of solids in the underflow. A lower pressure drop (i.e. specific energy) requires greater amounts of water to be accepted with the solids, and vice versa. Depending on the particle size of the feed solids and the cost of further dewatering of the solids in underflow, the two running costs of pressure drop and of dewatering of the solids in the underflow can be weighed against each other and the operating conditions optimized accordingly.

The scale-up model based on dimensionless groups simplifies hydrocyclone design and allows the designer to select the best compromise between hydrocyclone performance and the design and operating conditions likely to achieve it. The model is particularly suitable for computer simulations where many different alternatives can be computed and the interaction of the hydrocyclone with other plant items can be described for overall optimization—see the example later in this section. As to which particular alternative is adopted depends on the pressure drop which is available or suitable. If this is not determined by some other consideration, optimization of the total cost of the system (running costs and capital charges) may lead to the best design; Gerrard and Liddle[32] have proposed a procedure designed for precisely this choice.

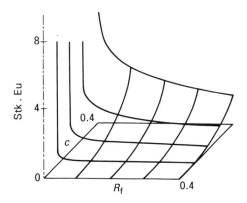

Figure 6.10. Graphical representation of equation 6.5

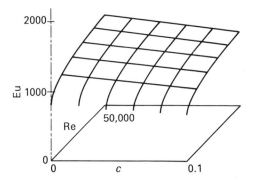

Figure 6.11. Graphical representation of equation 6.6

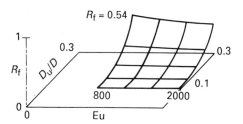

Figure 6.12. Graphical representation of equation 6.7

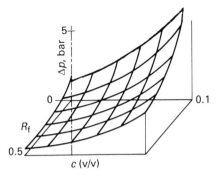

Figure 6.13. Plot of operating variables for a particular cyclone operated to give constant overflow clarity. $D = 88$ mm, chalk in water, $x'_{50} = 10.6$ μm, Rietema's design

6.6.7 Analytical flow models

The analytical flow models are based on mathematical solutions of the basic flow mechanics. Reference has already been made to the work of Bloor and Ingham[20-22] who have persued this direction for many years. Their model assumes inviscid flow in the main body of the flow and is capable of producing particle trajectories for efficiency calculations. It also considers flows in the top boundary layer, the side boundary layers and around the air core. They found that if the flow is assumed to be irrotational (i.e. having no vorticity) then unreasonable results are obtained.

In developing their model, Bloor and Ingham[15] first assumed the fluid to enter the hydrocyclone with uniform momentum and used the form of vorticity distribution determined by the Polhausen method. Bloor[33] has recently shown that the momentum cannot be distributed uniformly in the entering fluid and that the secondary motions produced from realistic entry conditions are of sufficient strength to generate the required levels of vorticity in the cyclone.

Bloor and Ingham published a whole series of papers when developing their main body flow model. At one stage[20] they introduced the effect of the leakage of particles to the overflow through the boundary layer under the top cover, calculated grade efficiency curves and found them to be in broad agreement with Kelsall's experiments[1]. The same authors also investigated the central region of solid-body rotation, which incorporates the air core, by employing a viscous turbulent model based on a Prandtl mixing length theory[15] as was discussed above.

The above reviewed theoretical work at Leeds University is still in progress and gaining in recognition as the results get closer to the reality of the processes in the cyclone. The model will not be fully appreciated by engineers, however, until it comes up with some easily understandable, practical correlations for the cut size or the grade efficiency curves, in a form suitable for technical practice. It is undoubtedly the model most likely to get closest to the description of the actual flow and particle separation in a hydrocyclone; in its final form, it may not require any experimental test constants.

6.6.8 Numerical simulations of flow

Instead of solving the equations of flow analytically, the methods of computational fluid mechanics can be used to develop numerical simulations of flow. There are now in existence several commercial general fluid flow computer software packages which can be used for mapping the streamlines of the flow inside a hydrocyclone.

Numerical calculations of the complete flow field, using axially symmetric flow models and solving the full viscous equations of motion, have been carried out by Boyson et al.[34] and by Rhodes et al.[35]. These simulations are suitable for assessing parameters in design optimization provided that they are based on realistic boundary conditions and a comprehensive flow model. Most recently, Bloor et al.[36] have produced a numerical simulation for viscous flow in a hydrocyclone at unrealistically low Reynolds numbers. Unlike other authors, however, they did not ignore the three-dimensional character of the flow around the entry point and used three different models to simulate the entry flow.

Numerical simulations are more flexible for design purposes than either analytical models or a full experimental investigation. They are, however, no substitute for experimental tests as yet, particularly for high solids concentrations when particle–particle interactions and possible non-Newtonian effects are at play. When considering a new geometry for a hydrocyclone, a numerical simulation of the flow can be very useful.

6.7 HYDROCYCLONE SELECTION AND SCALE-UP

6.7.1 General Strategy

The very first consideration in the process of hydrocyclone selection and sizing is the choice of narrow-angle or wide-angle designs. The choice is made according to the application: wide-angle designs are only beneficial in the cases of classification duties (and when the cut sizes are to be larger than about 50 μm) and of sorting by particle shape or density. For the other possible applications, namely in clarification, thickening and classification of fine particles, narrow-angle designs are the only choice. The remainder of this section is, therefore, devoted entirely to narrow-angle designs and the reader is referred to Svarovsky[3] for further reading on wide-angle types.

There are two somewhat different schools of thought with regard to hydrocyclone geometry. Most manufacturers, due to commercial considerations, produce only a limited range of cyclone diameters and, in order to be able to cover a full range of flow rates and cut sizes, each cyclone size can be altered in design proportions to vary its performance. In marketing their units, manufacturers emphasize that their cyclones are always designed specifically for each application, almost custom-built, and that this is the only way to obtain an optimum design. Such an approach requires reliable quantitative know-how as to the effect of design variables and, as the equipment user normally does not have this information, he or she has to seek assistance from the supplier. The claim of custom-building is true of course, but it is a case of assembly from the range of standard parts as a commercial necessity rather than a scientific virtue.

An alternative approach open to users or suppliers who are in a position to build a hydrocyclone of any diameter, is to use geometrically similar families of cyclones of 'standard' design. All the cyclone proportions are related to body diameter and the only design variable is then the cyclone size. This simplifies the cyclone selection remarkably and makes it accessible to people who do not necessarily have access to reliable scale-up information on the effects of cyclone proportions. The above-mentioned cyclone designs are either geometries obtained by careful optimization by scientists like Rietema[8] or Bradley and Pulling[7] or simply any practical or commercial cyclone geometries that have been well tested and can therefore be reliably scaled-up. *Table 6.1* lists two such cyclone geometries and the test data (see *Figure 6.1* for cyclone dimensions and section 6.6.6 for definitions of the dimensionless groups in *Table 6.1*), others can be found elsewhere[3]. Hydrocyclones designed in this way have a better chance of being close to an optimum design than do those selected from only a few sizes and assembled from kits to suit a particular application. Commercial considerations force most manufacturers to follow the latter route, but many also adopt geometric similarity in their range, thus combining the two approaches.

The approach used in this chapter is based on families of geometrically similar cyclones so that all design variables are omitted from the scale-up correlations (except the size of the underflow orifice which should be variable and is considered here to be an operating variable).

The significance of the dimensionless groups and constants in *Table 6.1* will be explained in the following section. When faced with the problem of selecting the most suitable hydrocyclone or set of hydrocyclones for a particular task, one can consult manufacturers' literature and consider some/one size from their range. A preliminary selection is usually based on capacity and the separation performance is then checked using additional information available from the manufacturer or using a model or theory. The following section considers this route.

The simple theories or simplified charts from manufacturers are usually satisfactory for a preliminary selection; when more detailed and reliable predictions of performance are required a model or theory must be used which includes the effects of feed solids concentration and of the underflow orifice setting. The choice is then narrowed to the empirical models or the semi-empirical scale-up using dimensionless groups because the other theories reviewed in section 6.6 are not yet useful for this.

As discussed in section 6.6.5, there have been some attempts to construct empirical 'models' based on large-scale tests, most notably those of Lynch and Rao[25] and Plitt[27] which include both design and operating variables in the correlations. The models are based on linear-regression analysis and are merely an analytical representation of tests with specific slurries on a range of cyclone sizes and designs. The apparent all-embracing nature of Plitt's and Lynch's equations is deceptive: the constants must usually be recalibrated for any particular slurry and hydrocyclone because the original correlations do not take into account all the effects which are important at high concentrations, like particle shape or surface properties. Should Plitt's or Lynch's correlations be used for hydrocyclone design, one would have to choose a particular set of design proportions to be used in the correlations, thus effectively choosing a hydrocyclone design and, when doing so, one may just as well use a known and well-tested design together with a simpler model, and test the effects of material properties at higher concentrations on a small cyclone of the same family. The correlations of Plitt and Lynch are very useful, however, for qualitative illustrations of the effect of changes of any particular variable.

It is, therefore, considered more appropriate in this book to use the 'chemical engineering' approach of dimensionless groups and their semi-empirical relationships, as given in section 6.6.6.

6.7.2 Preliminary design charts

As a very preliminary step in the process of hydrocyclone selection, the low concentration performance data is often used even when the actual feed concentrations are higher than 1% v/v. In clarification duties, however, this data would be correct for final design providing that high flow ratios are to be used.

At low solids concentrations of below 1 or 2% by volume, the flow pattern in hydrocyclones is unaffected by the presence of particles, and particle–particle interaction is negligible. The volume of the particles that separate into the underflow is small and the underflow-to-throughput ratio, R_f, is usually assumed to have no effect on the cut size, x_{50}, except for the effect of flow splitting which can be easily accounted for by using the reduced efficiency concept (section 3.4.1). Dimensional analysis coupled with the conclusion of most of the simple theories of separation in hydrocyclones (see section 6.6.1, and Svarovsky[3] and Bradley[6] for full reviews) give two relationships between three dimensionless groups:

$$Stk_{50} \cdot Eu = constant \qquad (6.12)$$

and

$$Eu = K \cdot (Re)n_p \qquad (6.13)$$

where K and n_p are empirical constants for a family of geometrically similar cyclones; the dimensionless groups in the above equation have all been defined in section 6.6.6. *Table 6.1* gives the constants for two well-known and tested cyclone geometries. Our own test work[30] with Rietema's geometry confirmed the validity of the constants derived in the original work[8], but only for high R_f ratios, i.e. those greater than about 23%; at lower ratios the separation performance becomes worse than predicted.

Other geometries give other values of the constants in the above equations; the group $Stk_{50} \cdot Eu$ varies from 0.06 to 0.33 for most cyclones, and the exponent n_p is usually between 0 and 0.4. The constants may be computed from full experimental test data or, conversely, knowledge of the constants allows the prediction of performance (or design to match a required performance) of a hydrocyclone of a given size and geometry. Design charts can be constructed for a particular geometry to aid size selection or establish limits in performance. *Figure 6.14* gives an example of such a design chart for Rietema's geometry specified in *Table 6.1*.

Similar design charts exist for some commercial cyclones, but as only a limited range of cyclone sizes are available from a given manufacturer, such charts are not continuous but merely show areas of cut size/capacity covered by the range of cyclone diameters.

The selection chart in *Figure 6.14* is very simple to use: a hydrocyclone is selected on the basis of the required capacity and available pressure drop. If the reduced cut size is too high then the flow rate is divided by 2, 3, 4 etc., and the procedure repeated with the lower value of Q until the cyclones (to be used in parallel) become small enough to give the required value of reduced cut size. The chart shows that reduced cut sizes down to almost 2 μm should be possible with 10 mm hydrocyclones. This has recently been confirmed in another investigation[37], but the units had to be run at high underflow dilutions.

6.7.3 Design options

The conventional hydrocyclone design procedures have been based on a rather simplistic view of the hydrocyclone function: the cyclone size is selected from the capacity and available pressure drop requirements, with the cut size not being a free choice but fixed by the former two requirements (reduction in cut size can only be achieved by using a greater number of smaller cyclones in parallel). This approach ignores completely the effect of the underflow orifice size on the cut size, and also on the solids concentration in the underflow. The procedure based on the model in section 6.6.6 centres around the three dimensionless equations 6.5, 6.6 and 6.7 which fully describe the function of a hydrocyclone within its operational limits. If a given cut size and underflow concentration are specified, simultaneous solution of the three equations will yield the cyclone diameter, the number of cyclones to be used in parallel and the size of the underflow orifice which will produce the required capacity (under the given pressure drop) and performance. If the operating pressure drop is not specified, several design options are obtained and the final design is selected either from some operating constraints or from economic optimization, see the example below.

It should be pointed out here that the scale-up procedure described here is specifically designed for separation or classification of fine particles. These are the particles predominantly used in the chemical industry and the approach, using dimensionless groups, is in common with the treatment of other unit operations in chemical engineering. The geometrical similarity concerns all internal dimensions of the individual cyclones except the size of the underflow orifice which is regarded here as an operational variable.

Another point worth making concerns the fluid density and viscosity used in this model. Some people believe that, at high solids concentrations, one should use the apparent viscosity of the suspension and the actual density of the feed suspension. This is an unrealistic approach for the following reasons. At the relatively moderate concentrations for which the model was designed (0–10% by volume), fine particles are surrounded by the suspending fluid and the drag on them depends on the viscosity of the fluid. Further-

232

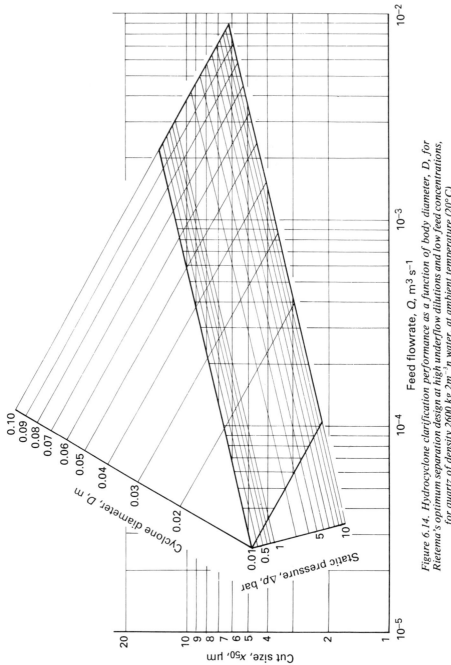

Figure 6.14. Hydrocyclone clarification performance as a function of body diameter, D, for Rietema's optimum separation design at high underflow dilutions and low feed concentrations, for quartz of density 2600 kg 2m^{-3}n water, at ambient temperature (20°C)

Feed flowrate, Q, m^3 s^{-1}

Cyclone diameter, D, m

Static pressure, Δp, bar

Cut size, x$_{50}$, µm

more, the critical zone which determines whether or not a particle will separate is not in the entry but further down stream in the cyclone body, where the flow is depleted of the coarser, easily separated solids, and the concentration there is much less than that in the feed.

A similar case may be made for the use of density: in Stokes' law, the buoyancy of particles in the separation zone must be taken into account. The fine particles displace the continuous phase and hence it is the density of the liquid that is used in the model. In any case, the suspension density in the zone is not known but is likely to be much less than that of the feed. The second use of fluid density is in the resistance coefficient, Eu. The density to be used there depends on how we define the Euler number; the dynamic pressure in the denominator (equation 6.9) is simply a yardstick against which we measure the pressure loss through a cyclone. We have used the clean liquid density in the dynamic pressure; alternatively, the feed suspension density may be used. It is immaterial which of the two densities is used (they are both equally unrealistic) provided the case is clearly defined: conversion from one to the other is a simple matter.

Example 6.1 Hydrocyclone selection, Rietema's geometry

The design procedure described here is best carried out with a computer programme. The example below is calculated with one such commercially available programme[38]. The programme sizes a hydrocyclone, or a number of units in parallel, for a given duty. The first problem to be solved is to establish the task required from the cyclone and relate it to the reduced cut size which can then be used by the programme. In order to facilitate relatively easy conversion of performance data such as total recovery, overflow concentration and all the reduced values, both the size distribution of the feed solids and the reduced grade-efficiency of the hydrocyclone are assumed to follow the log–normal law. The data input needed for running the programme, therefore, includes the median size and the geometric standard deviation of the feed solids, and the geometric standard deviation of the reduced grade-efficiency curve.

The programme first converts the task given, via three different input options, into the operating cut size required. The three options are for classification, clarification or thickening duties. It then calculates, from the given operating conditions and the required cut size, the design options available (number of cyclones in parallel, their diameter and the operating pressure drop necessary) and allows the user to choose a suitable set. If the task specified leads to unreasonable design options, the user may select to go back and change the task. If one of the design options is selected as acceptable, the final design proportions of the hydrocyclones may also be obtained. A print option allows the printing of the final table of results.

The operating conditions used in this example are listed in *Table 6.2*. As can be seen, the same performance may be obtained either with fewer, larger cyclones operated at high pressures or with more smaller units at lower pressure drops. The designer must select from the options which is the most appropriate or most economical; if, for example, two hydrocyclones are selected to be used in parallel, then the operating pressure will be 4.7819 bar and the cyclone diameter $D = 0.0849$ m. All the other dimensions may be calculated from the proportions given in *Table 6.1* (except for D_u which is determined from equation 6.7), i.e. $L = 0.4247$ m, $D_i = 0.0238$ m, $D_u = 0.0184$ m, $D_o = 0.0289$ m and $l = 0.0380$ m.

Finally, the whole of the above example can, of course, be calculated manually from the equations in section 6.6.6, but a computer programme allows a faster computation of the alternatives and a better appreciation of the effect of the different variables on the final design solution.

Table 6.2. SUMMARY OF THE OPERATING CONDITIONS USED IN EXAMPLE 6.1

Feed flow rate, Q	0.005 m^3 s^{-1}
Liquid viscosity	0.001 Ns m^{-2}
Liquid density	1000 kg m^{-3}
Solids density	2600 kg m^{-3}
Solids concentration, c	0.05 v/v
Solids median	20 μm
Geometric standard deviation of the feed solids size distribution	3
Geometric standard deviation of the reduced grade efficiency curve of the cyclone	2
Performance criteria	
Total coarse recovery by mass, E_T	70.92%
Total reduced mass recovery, E_T	67.69%
Actual cut size, x_{50}	10 μm
Reduced cut size, x'_{50}	11 μm
Overflow concentration, c_o	0.0162 v/v
Underflow concentration, c_u	0.3546 v/v
Flow ratio, R_f	0.1

Design options to meet the above performance

No. of cyclones in parallel	Diameter, D, m	Pressure drop, bar
1	0.1089	7.4620
2	0.0849	4.7819
3	0.0735	3.6860
4	0.0063	3.0645
5	0.0612	2.6555
6	0.0573	2.3622
7	0.0542	2.1396
8	0.0517	1.9638
9	0.0496	1.8208
10	0.0477	1.7017
11	0.0461	1.6007
12	0.0447	1.5138

6.7.4 Multiple cyclones in series

Series connections of separators are a common way of improving the performance of single units. In the case of hydrocyclones, due to their low capital and running costs, multiple series arrangements are quite frequently used. As to which particular layout and arrangement should be used depends on the actual application in question.

If clarification of the liquid is the ultimate purpose and it cannot be achieved in a single-pass arrangement, several stages in series in the direction of the overflow may be considered, such as the arrangement shown in *Figure 6.15*. Alternatively, a multiple-pass arrangement may be considered, such as that shown schematically in *Figure 6.16*; hydrocyclones are particularly suited for this. An additional benefit from this arrangement is that it dilutes the feed for the cyclone which improves the cyclone efficiency.

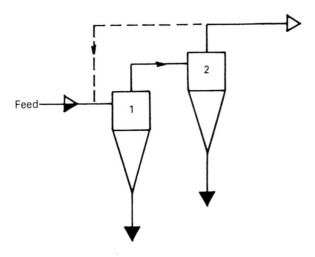

Figure 6.15. *Two hydrocyclones in series for clarification, with a possible partial recycle*

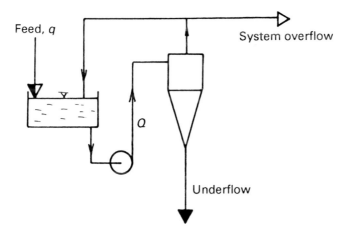

Figure 6.16. *A multiple-pass arrangement for clarification*

In thickening duties, if the required degree of thickening cannot be achieved with one cyclone, two or more in series may be used, this time in the direction of the underflow—see *Figure 6.17*. The overflow from the second

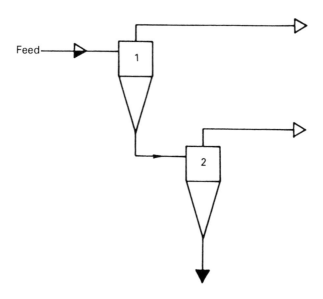

Figure 6.17. Two hydrocyclones in series for thickening

or subsequent stages may be partially recycled back to the feed of the first stage.

If both thickening and clarification are required simultaneously, a two- or three-stage arrangement is required where some cyclones are used as thickeners and others as clarifiers. *Figures 6.18* and *6.19* show examples of such arrangements, depending on whether one or two thickening cyclones are needed (i.e. depending on the concentration of the feed: if the feed is very dilute, two thickeners will be needed). Whether a hydrocyclone is used as a thickener or as a clarifier is shown in *Figures 6.18 and 6.19* by the nature of the underflow stream: rope discharge indicates a thickening duty whilst a spray discharge is for clarification.

In one arrangement, shown schematically in *Figure 6.18* as an example, the thickener is in the first stage, followed by one (or more) clarification stage, the underflow from which is returned to the feed. The overall recovery of the whole plant is better than the recovery of any of the individual cyclones used. The recycle should be more dilute than the feed, so that the feed to the thickening cyclone is diluted, because divergent performance of the plant could result otherwise. If the feed is very dilute (less than say 1%) and the thickening cyclone might not produce sufficiently thick underflow in one stage, another arrangement may be used (*Figure 6.19*) where the first stage is clarification, with the thickener treating the underflow of the first stage. Another clarifier is then used to clean the overflow from the thickener. Such a plant is relatively small and achieves good thickening, but the recovery is not as good as in the arrangement shown in *Figure 6.18*.

In classification duties, hydrocyclones give a relatively poor sharpness of

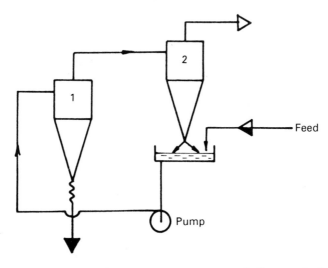

Figure 6.18. Two hydrocyclones in series with a second-stage underflow recycle, for simultaneous thickening and clarification

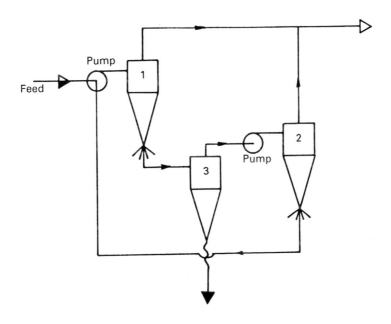

Figure 6.19. A three-stage arrangement for clarification and thickening of dilute feeds

cut in one pass and, consequently, in order to minimize the amount of misplaced material, the classification may be done in two or more stages. *Figure 6.20* shows how the classification can be sharpened by using another stage on both the overflow and the underflow. If the cut size of all three cyclones is the same, the process will cut at the same particle size as the front cyclone but the grade efficiency will be significantly steeper. Typically, a sharpness index (x_{75}/x_{25}) of 2.4 for a single cyclone can be reduced to 1.6 using this arrangement. Note from *Figure 6.20* that if the underflow from the previous stages is to be classified further, it must be diluted in order to obtain good sharpness of cut and the required cut size.

Finally, multiple-series arrangements are also extensively used in counter-current washing of solids. The reader is referred to chapter 15 for the description and design of the washing trains used for this purpose.

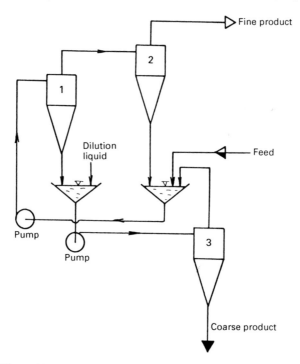

Figure 6.20. A reclassification arrangement designed to sharpen the cut

6.8 DESIGN VARIATIONS, OTHER DESIGN FEATURES

The conventional hydrocyclone has been subjected to considerable development to improve its characteristics. A selection of various design features is given below.

In the pulp and paper industry, water is often added in the underflow 'rejects' chamber to reduce losses of fibre with the rejected granular dirt. Similarly, the sharpness of cut in the classification of granular solids can be improved by injecting clean liquid into the flow near the apex of the cone. The liquid displaces the slime fraction that would normally end up in the underflow; commercial systems exist for this (e.g. Krebs Engineers' Cyclowash). The quantity of liquid added is usually equal to or greater than the underflow rate without liquid injection. The benefits of this depend on the cost of clean liquid needed for the injection: reclassification may be a cheaper alternative.

Another way to sharpen the classification is to operate the cyclone upside down so that the solids are separated into a container above the cyclone and subsequently fall back into the cyclone to be washed again. In this way, the container contents recirculate until the fines entrained first time around are disentrained to go to the overflow. A multistage operation using cyclones of different cut at each stage can then separate the feed into several closely sized fractions depending on the number of cyclones used.

If magnetic particles like finely divided magnetite are to be recovered with hydrocyclones, the recovery may be enhanced with an external magnetic field suitably shaped so as to produce a radial field perpendicular to the cyclone axis. A commercial 'magnetic cyclone' is based on this principle.

Some hydrocyclones used in the pulp and paper industry use cones which steepen gradually towards the lower end, as shown in *Figure 6.21*. This steepening reduces the centrifugal force which presses the separated particles onto the wall and it is supposed to reduce the abrasion in the lower part of the cone. An alternative to a curved cone may be a stepped cone, also shown in *Figure 6.21*, which is found to be particularly beneficial in gas cyclones where it reduces re-entrainment.

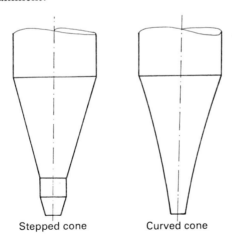

Stepped cone   Curved cone

Figure 6.21. The curved cone and the stepped cone as alternative cone shapes

The cleaning efficiency and the blocking characteristics of cyclones applied to pulps (saw dust, newsprint, sulphite, etc.) can be improved by using spiral cones. These have a single or double spiral cast into the cone, with two sides inclined at a negative angle which represents no or negligible extra cost with plastic cones. However, there is no improvement for granular solids.

6.8.1 The underflow orifice design and control

When operating a hydrocyclone with a free underflow orifice, discharging into atmospheric pressure, the discharging slurry can be observed to have different shapes depending on the feed size distribution, separation efficiency and the size of the underflow orifice. At relatively low underflow concentrations, the discharge has an 'umbrella' or spray-type shape because the outgoing slurry is still spinning quickly. Under such conditions, the air core inside the cyclone is vented to the atmosphere through the centre of the umbrella discharge. When more solids report to the underflow, because of an increase in the feed concentration, in the separation efficiency or because of a decrease in the orifice size, the boundary layer flow carrying the particles into the underflow joins in the middle and the underflow comes out as 'rope' discharge, no longer spinning out in a spray. The underflow orifice becomes overloaded and significant quantities of solids will start overflowing, resulting in loss of efficiency. On further throttling of the underflow orifice the discharge becomes 'lazy', in the form of a snake, and there is an acute danger of blocking (in fact blocking sometimes occurs before this stage is reached).

The simplest way to operate a hydrocyclone is to open both the underflow and overflow to the atmosphere because this ensures correct hydraulic balance between the two outlets, independent of what happens downstream of either of the outlets. Virtually all test data available on different cyclone designs, as well as theoretical or empirical formulae for performance characteristics, are for such conditions of equal back pressure on both discharge streams.

If for some reason the cyclone discharge must be against back pressure, it may still be worthwhile to maintain the balance between the underflow and overflow. In some applications such as when two or more cyclones are used in series, the back pressure may not be the same on both outlets (in the case of the liquid–liquid separation cyclones for example), and this is quite acceptable provided that the pressures are maintained in the same ratio because this has a profound effect on the flow split in the cyclone and the operating cut size.

Underflow rate can be controlled by back pressure with a valve downstream but this is generally not recommended for individual, small-diameter cyclones because in such cases the throttling valve itself is likely to block. It is, however, the only option for multiple assemblies of small-diameter cyclones because the underflow opening of the individual units cannot be reduced below a certain size for the increased danger of blocking. In any case, it is not feasible with a multi-unit arrangement to change the underflow orifices of the constituent cyclones as the cyclones are numerous and often cast in blocks of steel or plastic, and are, therefore, inaccessible anyway.

Only small-diameter or specialized hydrocyclones have a fixed underflow orifice, but most commercial units are supplied with a variable one. This is because the optimum size of the opening cannot be reliably predicted and the correct adjustment of it is vital for the best operation of the cyclone. The size of the underflow orifice affects directly the underflow-to-throughput ratio, the underflow concentration and the cut size. The underflow orifice is best adjusted after start-up of the plant and during operation whenever any operating conditions change. Several possible designs are available for this adjustment: replaceable nozzles, mechanically adjustable openings, pneumatically or hydraulically controlled orifices, or even self-adjusting devices which maintain a constant underflow density or concentration. Whatever the method of controlling the size of the underflow opening, the orifice should always be kept circular because the underflow slurry is still rotating when it comes out and blocking would be more likely with an orifice of any other shape.

It is also important to mention here that a hydrocyclone by no means necessarily requires continuous underflow: it can just as well be operated with a closed 'grit pot' under the discharge orifice and this can then be intermittently discharged by a manually or automatically operated purge valve. The separation efficiency with a grit pot is always lower than with continuous underflow (in other words the use of the grit pot coarsens the separation) but this can often be offset by using a multiple-pass system. The grit pot improves the sharpness of cut, due to the exchange of liquid between the pot and the cyclone, and the accompanied washing effect (with no net flow in either direction).

6.8.2 Types and shapes of inlet

Another important design feature is the method of introducing the feed slurry to the hydrocyclone. A single tangential inlet is most common: there is little advantage in using multiple entry which would complicate the design of the manifolding. One commercial design of multicyclone systems for separating organic matter from seawater or other classification duties, however, features multiple entry (3–6 inlets evenly distributed around the periphery) and an annular overflow orifice. A dual inlet is used in one of the designs developed for liquid–liquid separation duties.

The shape of the cross-section of the inlet is another design variable. It might be circular or rectangular (with the longer side parallel with the cyclone axis), the latter being slightly better because, for the same flow area, it brings the particles a little closer to the wall on entry. Rectangular inlets do not represent a manufacturing complication because nowadays most cyclones are cast.

The form of the entry along the flow and the shape of the top cover are probably more important to the cyclone performance than the cross-sectional shape of the inlet, and these parameters are subject to many proprietary variations. At least one manufacturer uses an involute entry where, instead of a simple tangential inlet, the feed enters via a spiral. This is supposed to minimize the intersecting angle between the incoming feed and the already rotating fluid inside the cyclone, and thus reduce turbulence and energy requirements. This is bound to have some effect on separation efficiency but whether the gain is significant enough to justify the design complication remains to be seen. Other manufacturers reduce the turbulence on entry by making the top cover in the shape of a helix so that the tangentially introduced feed does not impinge on the flow inside and it also gets some downward momentum. Helical inlets are also known to increase abrasion resistance of gas cyclones but no direct evidence exists for hydrocyclones.

The turn-down ratio and/or the cut size of a hydrocyclone can be altered by changing the size of the inlet. Some small-diameter cyclones are available with replaceable cylindrical sections which are cast together with the vortex finder and inlet section, and two or more alternative inlet sizes may be used. Another alternative is to use inserts in the inlet or shrouds which can be set without disassembly of the inlet section. Our own development work at Bradford University has produced a piston-like inlet control device which allows inlet size to be changed without interrupting the flow, but this would probably prove too expensive to be produced commercially.

6.8.3 Size and shape of the vortex finder, pressure recovery

The inside diameter of the vortex finder is also often used to control the capacity and cut size. This is done by using the above-mentioned replaceable cyclone tops or by inserting different sized nozzles into the vortex finder. The shape of the vortex finder is not subject to much variation; the only significant feature is the occasional use of skirts or discs at the bottom edge to counter the short circuit flow down the outside of the vortex finder. This is relatively rare, however, as the benefit is doubtful and possibly not worthwhile in view of the manufacturing or casting complications.

Cylindrical overflow boxes are sometimes used, with a tangential off-take, on top of some larger cyclones. This is supposed to recover some of the energy of the spinning overflow but, once again, the actual benefit is small in view of the total pressure drops involved in hydrocyclones. Such pressure recovery devices are much more common and beneficial with gas cyclones where the operating pressure drops are about 100 times lower than those normally used with hydrocyclones.

6.8.4 Multicyclone arrangements

Both theory and practice show that smaller diameter hydrocyclones give for the same pressure drop better separation efficiency. In clarification duties or when classifications at low cut sizes are required, banks of small-diameter hydrocyclones are used in parallel to treat high flow rates. This may be done either by manifolding several single cyclones in parallel or, when a large number of really small units is to be used, by enclosing the cyclones in a single housing, with common discharge chambers. Such multicyclone units may themselves be manifolded in parallel in multiple arrangements, or they might be stackable to produce compact columns capable of treating very high flow rates.

The available arrangements can be divided into four basic types (as shown schematically in *Figure 6.22*), with small variations within each type. The first two types, (a) and (b), are linear arrangements while the other two, (c) and (d), are circular. Circular arrangements lend themselves much better to an even distribution of the feed because each unit can have an identical length of piping between it and the centrally positioned header.

Arrangement (c) in *Figure 6.22* is probably most common for manifolding a multiplicity of single cyclones together. It is sometimes referred to as the

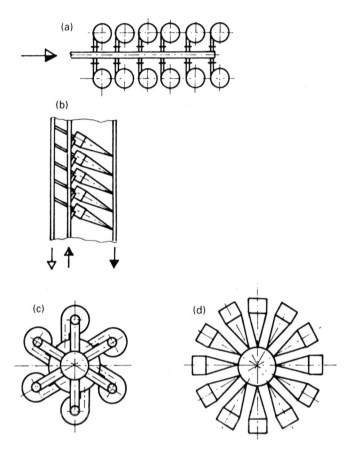

Figure 6.22. Types of multiple hydrocyclone arrangements. (a) Plan view. (b) Elevation. (c) Plan view. (d) Plan view

'spider arrangement' due to the shape of the necessary pipework, particularly the sweeping 180° turns of the overflow pipes from each unit to the central well. Like in the horizontal linear arrangement, the cyclones here are also in vertical positions or slightly inclined. The underflow discharge may be arranged to discharge freely into a common launder around the central feed well and the operator is therefore able to see the discharges of the individual units and spot any blockages.

The vertical circular arrangement is also highly suitable for use in the compact multiple cyclone units which in this case take the shape of circular boxes, often stackable. The cyclones are either mounted inside as individual units, using simple clamping systems, or cast into plastic or metal, often together with a prestrainer and the associated ducting.

6.9 APPLICATIONS

Applications of hydrocyclones in industry fall into several broad categories: clarification, thickening (or both simultaneously), classification, washing, sorting, liquid–liquid separation, liquid degassing and particle size measurement. Here, we shall concentrate on the first three categories, and for examples of the other categories the reader is referred to Svarovsky[3].

6.9.1 Clarification

The aim of clarification is to produce clear overflow, which is the same as maximizing the mass recovery of solids from the feed. The feed liquid has only small amounts of solids and it is the clarification of the liquid which is of primary interest, not the concentration of the solids in the underflow.

The operating variables for clarification are: dilute feed, relatively open underflow orifice (i.e. dilute underflow, typically below 12% of solids by volume but the less the better), and high pressure drop. The design variables are: small-diameter, narrow-angle cyclones (there is a limit in the body diameter here due to the liability to blocking) nested in parallel in multicyclone arrangements, high efficiency design (i.e. small inlet and overflow orifices, etc.). In single-pass installations, hydrocyclones can give minimum cut sizes of around 2 μm but this can be further improved by using multiple-pass or recirculating systems.

Applications of hydrocyclones in clarification include, for example:

1. recovery of catalysts in the oil and chemical industries;
2. separation of sand from sugar cane juice;
3. removal of scale particles from the jet water and cooling water of the rollers in steel rolling plants;
4. separation of corrosion products in circulating systems in the nuclear power industry;
5. removal of solids (steel or metal swarf) from metal working coolants, cutting fluids or circulating washwater in the engineering industry;
6. clean-up of washwater in washing machines for motor parts;
7. removal of silt from well water;
8. clean-up of water from gas scrubbers for recycling;
9. clean-up of water in car wash systems;
10. precleaning of primary sewage sludge (1–2% solids);
11. removal of sand, scale and other organic or inorganic particles to protect heat exchangers, boilers, glands and seals;
12. recovery of salts from saline solutions;
13. cleaning of washwater for potato processing;
14. cleaning of washwater in coal or ore processing plants;
15. degritting of milk of lime;
16. clarification of seed oil;
17. removal of drill chips from drill mud;
18. primary treatment of wool scour effluent, or removal of any other fine particles from aqueous or non-aqueous suspensions.

6.9.2 Thickening

Thickening means concentrating the solids present in a suspension into a smaller amount of fluid; the goal is to produce high concentration of solids in the underflow and any loss of solids to the overflow is undesirable but of secondary importance. The feed solids concentrations in thickening applications are usually higher than in clarification duties.

If hydrocyclones are to be used to produce thick underflows (i.e. to dewater the solids), the total mass recovery of the feed solids must be sacrificed because throttling the underflow orifice inevitably leads to some loss of solids to the overflow. A hydrocyclone as a single unit cannot therefore be used for both clarification and thickening at the same time. The underflow concentrations that can be achieved with hydrocyclones may be as high as 50% by volume with some materials. Thus, for example, the AKW cyclone, 125 mm in diameter, can give 45% by volume with chalk. In this thickening performance hydrocyclones compare favourably with gravity thickeners and hydrocyclone systems are sometimes used to replace the much larger and more expensive gravity thickeners.

The limiting factor in how far the underflow orifice can be closed is the danger of blocking. In terms of design variables, there is no particular need for small cyclone diameters; smaller units are, in any case, more liable to block. Normally, narrow-angle cyclones are used. A variable underflow orifice is very useful here.

The following is a list of some reported applications of hydrocyclones in thickening of slurries:

1. thickening of the waste product from flue-gas desulphurization systems where cyclones are used to replace the more costly gravity thickeners;
2. dewatering of mine back-fill;
3. dewatering of silt in dredging operations;
4. densification of the recovered and cleaned medium in medium recovery plants in dense medium separation.

Apart from many existing applications in mineral processing, there are also actual and potential applications for hydrocyclones in the chemical industry, like recovery and concentration of ammonium chloride crystals or sodium bicarbonate.

A typical example of a hydrocyclone as a thickener is in prethickening of feed to vacuum filters, screens or dewatering centrifuges (*Figure 6.23*). Unless the feed is coarse, however, the overflow from the hydrocyclone used as a thickener must be clarified further (if it is required clean). The hydrocyclones are also used for preconcentration of slurries before decanting centrifuges as it improves overflow clarity. This may not affect the centrifuge capacity, however, because the limiting factor in dewatering duties is usually the mass flow rate of solids.

6.9.3 Classification of solids according to size

This application of hydrocyclones is for solid–solid separation by particle

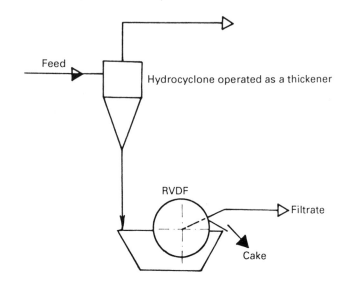

Figure 6.23. Prethickening of feed to a vacuum filter with a hydrocyclone

size. As the grade efficiency of a cyclone increases with particle size, it can be used to split the feed solids into fine and coarse fractions. This may be a process requirement, by which coarse and fine solids are separated to follow different routes in the plant, such as for example in closed-circuit wet grinding where the oversize particles are returned to the mill for further grinding (*Figure 6.24*). There are other possible grinding arrangements than that shown in *Figure 6.24*: open circuit, or open/closed (double function or two-stage), also with a post-desliming or post-refining stage. In this type of application, hydrocyclones are most frequently used either to remove coarse particles from the product (in a degritting or 'refining' operation) or to remove fine particles from the product (in a desliming or 'washing' operation). In other applications, the feed material may simply be split in half into two commercially useful products of narrow size distribution. Hydrocyclones are, therefore, used to engineer the size distribution of products or intermediates.

The requirement in classification is to be able to set the cut size to a predetermined value dictated by the application, and to achieve a good sharpness of cut (this requires a steep grade-efficiency curve). As pointed out previously, wide cone angles improve the sharpness of cut. There is a problem, however, in achieving low cut sizes with such cyclones. Consequently, when classifying at low or moderate cut sizes, say up to 50 μm, narrow-angle cyclones are still used to achieve those cuts. With increasing cut size, larger angles are used until at over 250 μm, such coarse cuts can in fact only be achieved with wide-angle or flat-bottomed cyclones. Liquid injection into the cone or an upside-down operation are also sometimes used to sharpen the cut.

Hydrocyclones are also used as classifiers simply to improve the perfor-

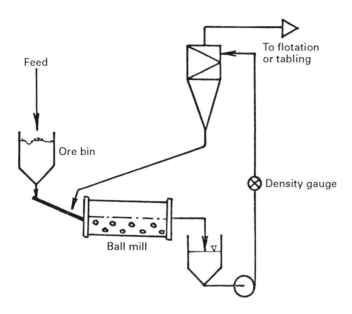

Figure 6.24. A hydrocyclone classifier in closed-circuit grinding

mance of other filtration or separation equipment. A good example is in applications where the cyclone separates the feed into coarse and fine particles, and the coarse material is fed onto the horizontal belt filter first, as a precoat, with fines to follow (see *Figure 6.25*). This will, if the cut point of the cyclone is set correctly, give good overflow clarity and extend the usefulness of vacuum filter to finer feeds without necessarily adversely affecting the moisture content of the cake.

Some further examples of hydrocyclones in a classification duty are:

1. desliming of minerals prior to flotation which leads to less reagent to be used and easier flotation;
2. desliming prior to leaching leading to higher extraction coefficients, desliming prior to filtration equipment to reduce the cake resistance to flow;
3. degritting of drilling muds;
4. separation of clay from barytes in drilling muds for control of specific gravity;
5. degritting of underground waters prior to pumps;
6. classification of 'raw' cement slurries before rotary kilns performed at high solids concentrations (to reduce thermal load on the kilns) as the high shear forces in the cyclone overcome the yield stress of the slurry which normally behaves as a Bingham plastic;
7. removal of impurities and unground particles from chalk;
8. desliming of phosphate rock;
9. classification of iron ore into coarse fraction (sinter feed) and fine fraction (pellet feed);

10. sizing of sand into fractions;
11. recovery of filter aid;
12. beneficiation of china clay (kaolinite).

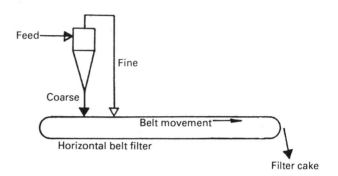

Figure 6.25. Precoating of a horizontal vacuum belt filter with hydrocyclone underflow

6.10 CONCLUSIONS

There are several recognizable trends and avenues in the future development and application of hydrocyclones; the recent international conferences on the subject have served as a good indicator in this respect. The following is a list of the author's main conclusions, and it necessarily represents a personal view.

Major changes in the conventional cyclone design for solid–liquid separation are unlikely, but many refinements will be made, based on the development of gradually more realistic flow models. Such models already give qualitative assistance in design decisions and provide the reasons why the previously empirically determined optimum geometries are better than others. In particular, the real value of the various pressure recovery devices will be established and other power-saving measures pursued (like the possibly beneficial effect of vorticity in the inlets of the second and further stages in multiple-stage systems). There will be, however, more changes in the geometry of cyclones for other phase separations where droplet or bubble break-up have to be minimized.

A better understanding of the performance characteristics of hydrocyclones, whether from fundamental or empirical work, will allow the best to be made of the advantages of hydrocyclones. In particular, combinations of hydrocyclones in series will continue to make it possible to overcome the inherent limits on the performance of a single hydrocyclone, and the design of such systems will be optimized.

Development is particularly expected to continue in liquid–liquid and liquid–gas separations, both in hardware and in the operating experience or performance.

It is the opinion of the author that hydrocyclones are undergoing a transformation from low to medium or high technology and that this process will continue well into the next century.

REFERENCES

1. Kelsall, D. F., 'A study of the motion of solid particles in a hydraulic cyclone', *Trans. Inst. Chem. Eng.*, **30**, 87–104 (1952)
2. Driessen, M. G., *Trans. Am. Inst. Min. (Metall.) Eng.*, **177**, 240 (1948)
3. Svarovsky, L., *Hydrocyclones*, Holt, Rinehart and Winston, London (1984)
4. Driessen, M. G., *Rev. Ind. Mining*, **Special Issue 4**, 449–461 (1951)
5. Criner, H. E., *'The Vortex Thickener'*, Int. Conf. on Coal Preparation, Paris (1950)
6. Bradley, D. *The Hydrocyclone*, Pergamon Press, London (1965)
7. Bradley, D. and Pulling, D. J., 'Flow patterns in the hydraulic cyclone and their interpretation in terms of performance', *Trans. Inst. Chem. Eng.*, **37**, 34–45 (1959)
8. Rietema, K., 'Performance and design of hydrocyclones, Parts I to IV', *Chem. Eng. Sci.* **15**, 298–325 (1961)
9. de Andrade Medronho, R., 'Scale-up of hydrocyclones at low concentrations', Ph.D. Thesis, University of Bradford, 1984.
10. Holland-Batt, A. B., 'A bulk model for separation in hydrocyclones', *Trans. Inst Min. Metall. (Sect. C: Min. Process Extr. Metall.)*, **91**, (1982)
11. Trawinski, H. F., *Filtration and Separation,* **6**, (1969)
12. Kutepov A. M., *et al.*, 'Calculation of separation efficiencies in hydrocyclones', *Izv. Vuzov, Chim. Chim. Technol.*, **XX**, 144–145 (1977)
13. Kutepov, A. M., 'Study and calculation of the separation efficiency of hydrocyclones', *Zh. Prikl. Khim.*, **51**, 614–619 (1978)
14. Rietema, K., 'The mechanism of the separation of finely dispersed solids in cyclones'. In K. Rietema and C. G. Verver (Eds), *'Cyclones in Industry'*, Chap. 4, Elsevier, Amsterdam (1961)
15. Bloor, M. I. G. and Ingham, D. B., *Trans. Ind. Chem. Eng.*, **53**, 1 (1975)
16. Duggins, R. K. and Frith P. C. W., 'Turbulence effects in hydrocyclones', *3rd International Conference on Hydrocyclones* (Oxford, 1987), Paper D1, Elsevier Applied Science Publishers, Barking (1987)
17. Schubert, H. and Neesse, T., A hydrocyclone separation model in consideration of the turbulent multi-phase flow', *Proc. Int. Conf. on Hydrocyclones* (Cambridge, 1980), Paper 3, pp. 23–36, BHRA Fluid Engineering, Cranfield (1980)
18. Neesse, T. and Schubert, H., 'Die Trennkorngrosse des Hydrozyklones bei Dunnstrom – und Dicht–stromtrennungen', *First European Symposium on Particle Classification in Gases and Liquids*, Nuremberg (1984)
19. Fahlstrom, P. H., 'Discussion', *Proc. Int. Min. Processing Congress 1960*, pp. 632–643, Inst. Mining and Metallurgy (1960)
20. Bloor, M. I. G., Ingham, D. B. and Laverack, S. D., 'An analysis of boundary layer effects in a hydrocyclone', *Proc. Int. Conf. on Hydrocyclones* (Cambridge, 1980), Paper 5, pp. 49–62, BHRA Fluid Engineering, Cranfield (1980)
21. Bloor, M. I. G. and Ingham, D. B., 'Theoretical analysis of the conical cyclone', *First European Conf. on Mixing and Centrifugal Separation* (Cambridge, 1974), Paper E6, BHRA Fluid Engineering, Cranfield (1974)
22. Laverack, S. D., *Trans. Ind. Chem. Eng.*, **58**, 33 (1980)
23. Lynch, A. J. and Rao, T. C., *Ind. J. Technol.*, **6**, 106–114 (1968)
24. Lynch, A. J., Rao, T. C. and Prisbrey, K. A., *Int. J. Min. Proc.*, **1**, 173–181 (1974)
25. Lynch, A. J. and Rao, T. C., 'Modelling and scale-up of hydrocyclone classifiers', *11th Int. Mineral Processing Congress* (Cagliari, 1975), Paper 9, pp. 9–25, Instituto di Arte Mineraria (1975)
26. Rao, T. C., Nageswararao, K. and Lynch, A. J., *Int. J. Min. Proc.*, **3**, 357–363 (1976)
27. Plitt, L. R., 'A mathematical model of the hydrocyclone classifier', *CIM Bull.*, **December**, 114–122 (1976)
28. Apling, A. C., Montaldo, D. and Young P. A., 'Hydrocyclone models in an ore-grinding context', *Int. Conf. Hydrocyclones* (Cambridge, 1980), Paper 9, pp. 113–125, BHRA Fluid Engineering, Cranfield (1980)
29. Gibson, K., 'Large scale tests on sedimenting centrifuges and hydrocyclones for mathematical modelling of efficiency', in *Proc. Symp. Solid–Liquid Separation Practice* (Leeds, 1979), pp. 1–10, Yorkshire Branch of the Institute of Chemical Engineers, Leeds (1979)
30. Medronho, R. A. and Svarovsky, L., 'Tests to verify hydrocyclone scale-up procedure', *2nd Int. Conf. on Hydrocyclones* (Bath, 1984), Paper A1, pp. 1–14 (1984)
31. Svarovsky, L., 'Selection of hydrocyclone design and operation using dimensionless

groups', *BHRA 3rd Int. Conf. on Hydrocyclones* (Oxford, 1987), Paper A1, Elsevier Applied Science Publishers, Barking (1987)

32. Gerrard, A. M. and Liddle, C. J., 'Numerical optimization of multiple hydrocyclone systems', *Chem. Eng.*, **February,** 107–109 (1978)

33. Bloor, M. I. G., 'On axially symmetric flow models for hydrocyclones', *BHRA 3rd Int. Conf. on Hydrocyclones* (Oxford, 1987), Paper D2, Elsevier Applied Science Publishers, Barking (1987)

34. Boyson, F., Ayers, W. H. and Swithenbank, J., 'A fundamental mathematical modelling approach to cyclone design', *Trans. Ind. Chem. Eng.*, **60**, 222–236 (1982)

35. Rhodes, N., Pericleous, K. A. and Drake, S. N., 'The prediction of hydrocyclone performance with a mathematical model', *BHRA 3rd Int. Conf. on Hydrocyclones* (Oxford, 1987), Paper B3, Elsevier Applied Science Publishers, Barking (1987)

36. Bloor, M. I. G., Ingham, D. B. and Ferguson, J. W. J., 'A viscous model for flow in the hydrocyclone', *Session I, Solid–Liquid Separation Practice III, 397th Event of the European Federation of Chemical Engineering* (Bradford, 1989)

37. Svarovsky, L., 'Evaluation of small diameter hydrocyclones', *Inst. Chem. Eng. Symp. on Solid/Liquids Separation Practice and the Influence of New Techniques* (Leeds, 1984), Paper 24, pp. 193–205, Institute of Chemical Engineers, Yorkshire Branch (1984)

38. *Hydrocyclone Design*, Fine Particle Software, Keighley, West Yorkshire.

7

Separation by Centrifugal Sedimentation

L. Svarovsky

Department of Chemical Engineering, University of Bradford

NOMENCLATURE

g	Gravity acceleration
G or $G(x)$	Grade efficiency curve
G_1	Grade efficiency curve of a single centrifuge
K	Constant
K_1	Constant
K_2	Constant
L	Length of the bowl
Q	Flow rate
r	Radius
r_1	Radius of the liquid surface
r_3	Inner radius of the bowl
t	Time
v	Liquid velocity
x	Particle size
x_{50}	Cut size of a centrifuge
X_{50}	Cut size of the whole separation plant
y	Variable
z	Position in the axial direction
$\Delta\rho$	Density difference between the solids and the liquid
μ	Liquid viscosity
Σ	Sigma factor
σ_g	Geometric standard deviation
ω	Angular speed

7.1 INTRODUCTION

Centrifugal sedimentation is based on a density difference between solids and liquids (or between two liquid phases); the particles are subjected to centrifugal forces which make them move radially through the liquid either outwards or inwards, depending on whether they are heavier or lighter than

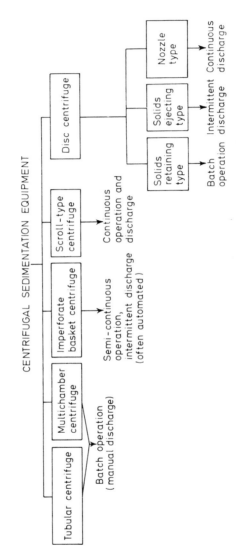

Figure 7.1. Classification of centrifugal sedimentation equipment

the liquid. Centrifugation may thus be regarded as an extension of gravity sedimentation to finer particle sizes and it can also separate emulsions which are normally stable in the gravity field.

A sedimenting centrifuge consists of an imperforate bowl into which a suspension is fed and rotated at high speed. The liquid is removed through a skimming tube or over a weir while the solids either remain in the bowl or are intermittently or continuously discharged from the bowl.

Five main types of industrial sedimenting centrifuges may be distinguished according to the design of the bowl and of the solids discharge mechanism. *Figure 7.1* shows a schematic classification of the equipment; the mode of discharge and operation is quoted for each type of equipment.

A general analysis, which forms a basis for predictions of performance, for a simple form of tubular centrifuge is given, followed by a detailed technical discussion of each available type of equipment and its particular design features, operation characteristics, performance and typical applications. Finally some notes and a guide to equipment selection are given.

7.2 THEORETICAL PERFORMANCE PREDICTIONS

If a particle of mass m is rotating with an angular velocity ω at a radius r from its centre of rotation it is acted upon by a centrifugal force $mr\omega^2$ in the radial direction. In sedimenting centrifuges the centrifugal acceleration $r\omega^2$ is very much larger than the acceleration due to gravity g (so gravity forces are neglected) and the ratio of $r\omega^2/g$ is used as a measure of the separating power of the machine. The centrifugal force is large enough to overcome the Brownian diffusion forces, which in gravity sedimentation hinder or prevent settling of very fine particles. As the separation efficiency is mainly affected by the behaviour of the smallest particles in the system, and as the fine particles moving in liquids have low Reynolds numbers (in the region of viscous resistance), it is common to assume, in describing particle motion in rotating liquids, that Stokes' law holds. It should be borne in mind therefore that the following analysis only applies to slowly moving fine particles with Re less than say 0.2 and that the resistance to motion of the larger particles would be in the transient or even Newton's law region (but that in any case all such particles would be separated with 100% efficiency anyway)—see chapter 1, section 1.2.2.

There are many other assumptions made in the following analysis (these are stated where they are introduced) but despite their often obvious oversimplicity the final results give a reasonable estimate of equipment performance. The analysis is made for a simple tubular centrifuge but the same approach, with slight modifications, may be used for other types of sedimenting centrifuges. An attempt is first made to derive the whole grade efficiency curve, which gives the only complete description of equipment separation efficiency (see chapter 3, 'Efficiency of Separation'), and the widely used Sigma concept, which gives an estimate of the cut size x_{50} only, is then derived as a special case.

7.2.1 Grade efficiency function for a simple tubular centrifuge

A simple tubular centrifuge consists of a vertical tube with a large length-to-diameter ratio which rotates at high speed about its vertical axis. The liquid is introduced at the bottom—see *Figure 7.2*—and the flow is essentially axial except in areas immediately adjacent to the inlet and the outlet. The radius of the liquid surface r_1 is determined by the radius of the outlet, which functions as an overflow dam or a weir.

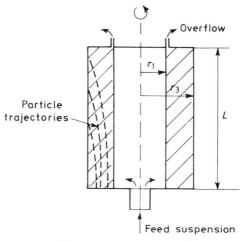

Figure 7.2. Schematic diagram of a simple tubular centrifuge

At the bottom of the tube, the particles that enter with the liquid are assumed to be homogeneously distributed in the annulus between the radius r_1 and the inner surface of the tube r_3. The suspension enters through a set of fixed quadrant plates which ensure that its angular velocity soon becomes identical with that of the cylinder. As the particles move upward with the liquid they are subjected to high centrifugal forces and follow trajectories (relative to the cylinder) similar to those shown in *Figure 7.2*.

The velocity of a particle (relative to the centrifugal bowl) at a point at a vertical distance z and radius r may be resolved into two components; one perpendicular to and the other parallel to the axis of rotation of the bowl (radial and axial velocity components).

If the input concentration of the solids is sufficiently low for any interaction between particles to be neglected, the magnitude of the radial velocity can be approximated by a modified version of Stokes' law for falling bodies (see chapter 1, section 1.2.2.2):

$$\frac{dr}{dt} = \frac{\Delta\rho x^2 r\omega^2}{18\mu} \tag{7.1}$$

where r is the radial position of a particle of size x, t is the time interval during which this particle is subjected to the centrifugal acceleration, $\Delta\rho$ is the solid–liquid density difference and μ is the viscosity of the liquid.

The assumption inherent in equation 7.1 is that as particles move from smaller radii to larger radii and are continuously accelerated, their instantaneous velocity on any point along their trajectory is equal to the terminal settling velocity corresponding to the centrifugal acceleration $r\omega^2$ at the appropriate radius r. In other words, equation 7.1 is derived from a simple balance between the drag force and the centrifugal force with the inertial force neglected.

The axial velocity may be approximated by the velocity of the liquid at the point (r, z) so that

$$\frac{dz}{dt} = v(r, z) \tag{7.2}$$

This of course assumes that there is no slip between the particles and the liquid flow in the axial direction.

If the end effects are neglected, the liquid velocity profile is only a function of the radius r. Schachman[1] derived an equation for the velocity profile $v(r)$ for the case of liquid flow in a concentric shell of fluid as follows

$$\frac{dz}{dt} = \frac{QK_1}{\pi(r_3^2 - r_1^2)}\left(r_1^2 \ln \frac{r}{2} + \frac{r_3^2 - r^2}{2}\right) \tag{7.3}$$

where K_1 is a bowl constant

$$K_1 = \frac{r_3^2 - r_1^2}{\frac{3}{4}r_1^4 + \frac{1}{4}r_3^4 - r_1^2 r_3^2 - r_1^4 \ln(r_1/r_3)}$$

and Q is the volumetric flow rate of the liquid.

Particle trajectories may be found using equations 7.1 and 7.3 thus leading to the theoretical limit of separation and the grade efficiency—see Bradley[2] and equation 7.21 in section 7.3.1.

The method of grade efficiency derivation will be demonstrated using a much simpler model in which the velocity profile in the liquid shell is assumed to be uniform, as in the so-called 'plug flow':

$$\frac{dz}{dt} = \frac{Q}{\pi(r_3^2 - r_1^2)} \tag{7.4}$$

The ratio of equations 7.1 and 7.4, when integrated between the limits $r=r_3$ at $z=L$ and $r=r_1$ at $z=0$, determines the smallest radius at which the sedimentation of a particle size x may begin in order that it may just reach the cylinder wall after covering the full length L of the separation zone (an approach commonly employed in the theory of settling tanks and settling chambers).

The radius r is then implicitly expressed as

$$\ln(r_3/r) = \frac{Kx^2 L\pi(r_3^2 - r_1^2)}{Q} \tag{7.5a}$$

where K is a sedimentation constant defined as

$$K = \frac{\Delta\rho\omega^2}{18\mu} \tag{7.5b}$$

The first parameter that can be derived from equation 7.5a is the limit of separation, i.e. the maximum particle size that would ever get a chance of escaping with the overflow; this size is obtained from the limiting particle trajectory which starts at radius r_1 at $z = 0$, i.e. $r = r_1$ in equation 7.5a thus giving

$$x_{max} = \left(\frac{Q \ln(r_3/r_1)}{K \pi L (r_3^2 - r_1^2)} \right)^{1/2} \tag{7.6}$$

The value of the grade efficiency for a particle size x is the fraction contained in an annulus between radii r and r_3 where r is, as previously stated, the smallest radius at which the sedimentation of a particle size x may start at $z = 0$ in order that it may just reach the wall of the cylinder at $z = L$. The grade efficiency is therefore

$$G(x) = \frac{r_3^2 - r^2}{r_3^2 - r_1^2} \tag{7.7}$$

where r is determined from equation 7.5a.

$$r = r_3 \exp \left(-K \frac{L\pi(r_3^2 - r_1^2)}{Q} x^2 \right) \tag{7.8}$$

Equations 7.6, 7.7 and 7.8 combined give the final formula for the grade efficiency function as

$$G(x) = \frac{r_3^2}{r_3^2 - r_1^2}[1 - \exp(-2K K_2 x^2)] \tag{7.9a}$$

or

$$G(x) = \frac{1 - \exp(-2K K_2 x^2)}{1 - \exp(-2K K_2 x_{max}^2)} \qquad \text{for } 0 \leqslant x \leqslant x_{max} \tag{7.9b}$$

and

$$G(x) = 1 \qquad \text{for } x > x_{max}$$

where K_2 is constant for a particular centrifuge geometry and volumetric flow rate and is equal to the residence time of the liquid because

$$K_2 = \frac{L\pi(r_3^2 - r_1^2)}{Q} = \frac{V}{Q} \tag{7.10}$$

where V is the effective volumetric capacity of the bowl.

Equation 7.9 gives the typical S-shaped grade efficiency curve (see *Figure 7.3*); the limit of separation x_{max} that appears in equation 7.9b may be determined from equation 7.6.

The method shown above is typical for the grade efficiency derivation for most of the following centrifugal equipment. It must however be emphasized that the theoretical efficiency functions are only as good as the assumptions made in their derivation.

An important parameter that can be derived from equation 7.7 (or 7.9) is the size corresponding to 50% on the grade efficiency curve, i.e. the equiprobable size or 'cut size' x_{50} (see chapter 3, 'Efficiency of Separation'). The corresponding radius r_{50} is the one that splits the annulus between r_1 and r_3 into equal areas hence

$$r_3^2 - r_{50}^2 = r_{50}^2 - r_1^2 \tag{7.11}$$

The cut size x_{50} can either be calculated from equation 7.9 by putting $G(x_{50}) = 0.5$ or by substitution of r_{50} from equation 7.11 into equation 7.5 thus

$$x_{50}^2 = \left(\frac{Q}{2\pi LK}\right) \left[\ln\left(\frac{2r_3^2}{r_3^2 + r_1^2}\right)\right] \left(\frac{1}{r_3^2 - r_1^2}\right) \tag{7.12a}$$

or, in terms of the liquid residence time K_2

$$x_{50}^2 = \frac{1}{2K K_2} \ln\left(\frac{2r_3^2}{r_3^2 + r_1^2}\right) \tag{7.12b}$$

An example of the comparison between the theoretical and practical grade efficiencies of a tubular centrifuge is given in section 7.3.1.

7.2.2 The Sigma concept

The so-called Sigma concept has been widely used in the field of centrifugal sedimentation ever since its first development by Ambler[3] in 1952. It is a simplified relation between the machine performance in terms of x_{50}, total volumetric flow rate Q and an index of the centrifuge size Σ. The cut size x_{50} is represented by its terminal settling velocity v_g in the given liquid under gravity so that from Stokes' law (using equation 7.5b for the definition of K)

$$v_g = \frac{x_{50}^2 \Delta \rho g}{18\mu} = x_{50}^2 K \frac{g}{\omega^2} \tag{7.13}$$

Equation 7.12a may thus be rewritten as

$$Q = 2v_g\left(\frac{\omega^2}{g}\right) \pi L \frac{r_3^2 - r_1^2}{\ln\dfrac{2r_3^2}{r_3^2 + r_1^2}} \tag{7.14}$$

or

$$Q = 2v_g \Sigma \tag{7.15}$$

where

$$\Sigma = \frac{\omega^2}{g} \pi L \frac{r_3^2 - r_1^2}{\ln\dfrac{2r_3^2}{r_3^2 + r_1^2}} \tag{7.16}$$

As equation 7.16 is rather cumbersome for routine calculations, an alternative expression may be used, based on an approximation of the logarithmic function

$$\ln y \simeq 2\frac{y-1}{y+1}$$

(taking the first term in a series) so that in equation 7.16

$$\ln \frac{\cdot 2r_3^2}{r_3^2 + r_1^2} \simeq \frac{r_3^2 - r_1^2}{\frac{3}{2}r_3^2 + \frac{1}{2}r_1^2}$$

and equation 7.16 itself becomes

$$\Sigma \simeq \frac{\omega^2}{g}\pi L(\tfrac{3}{2}r_3^2 + \tfrac{1}{2}r_1^2) \tag{7.17}$$

This is an approximation for which Ambler[4] claims a maximum error of 4%.

Equation 7.15 is the basic expression of the Sigma concept. It gives an estimate of the flow rate above which particles of size x_{50} will largely be unsedimented and below which they will mostly be separated. Σ is a constant containing factors pertaining only to the centrifuge; it is often called the theoretical capacity factor; it has the dimension of an area and it allows comparison between the performances of geometrically and hydrodynamically similar centrifuges operating on the same feed material. Theoretically, Σ represents the area of a settling tank capable of the same separational performance in the gravitational field; this is of course a false comparison because it ignores Brownian diffusion, convection currents and other effects which could mean that such a settling tank may hardly perform as well as the centrifuge, if at all. Equation 7.17 represents the Σ factor for a tubular centrifuge; similar expressions can be derived for other types of sedimenting centrifuges and these are given in the appropriate sections in the following. Extensive work on Σ comparisons in the past two decades, mostly made by equipment manufacturers, has shown that while experimental results differ from calculated values when different types of centrifuges are being considered, the scale up between centrifuges of the same type is fairly reliable; this is based on a simple application of equation 7.15

$$\frac{Q_1}{\Sigma_1} = \frac{Q_2}{\Sigma_2} \tag{7.18}$$

if x_{50} is to remain constant.

Despite some attempts[5] to extend the application of equation 7.18 to 'cross-type' scale up between different centrifuge configurations via the use of the efficiency factors μ_i using

$$\frac{Q_1}{\mu_1\Sigma_1} = \frac{Q_2}{\mu_2\Sigma_2} \tag{7.19}$$

where μ_i are relative efficiencies of different types, this has not found much response in practice. In fact it is difficult to find the actual values of these efficiencies in the literature except from the paper by Morris[5] (tubular bowl, 90%; imperforate basket, 75%; scroll type, 60% and disc type, 45%). Equipment manufacturers claim that this cross-type scale up is not reliable in practice unless many more factors than just the Sigma values are considered (see the discussion in Woolcock[6], p. 180).

Example 7.1 Use of the Sigma theory

A low-concentration suspension of clay (density 2640 kg m^{-3}) in water with a viscosity of 0.001 Ns m^{-2} and density 1000 kg m^{-3} is to be separated by centrifugal sedimentation. Pilot runs on a laboratory tubular bowl centrifuge operating at 20 000 rev. min^{-1} indicate that satisfactory overflow clarity is obtained at a throughput of 8×10^{-6} m^3 s^{-1}.

The centrifuge bowl is 0.2 m long, has an internal radius of $r_3 = 0.0220$ m and the radius of the liquid surface $r_1 = 0.0110$ m. If the separation is to be carried out in the plant using a tubular centrifuge 0.734 m long with an internal radius of 0.0521 m and $r_3-r_1 = 0.0295$ m, operating at 15000 rev. min^{-1} with the same overflow clarity, what production flow rate could be expected? Also determine the effective cut size.

Using equation 7.17

$$\Sigma_1 = \left(\frac{\pi\ 20000}{30}\right)^2 \frac{\pi\ 0.2}{9.81}\ [\tfrac{3}{2}(0.022)^2 + \tfrac{1}{2}(0.011)^2] = 221\ \text{m}^2$$

$$\Sigma_2 = \left(\frac{\pi\ 15000}{30}\right)^2 \frac{\pi\ 0.734}{9.81}\ [\tfrac{3}{2}(0.0521)^2 + \tfrac{1}{2}(0.0226)^2] = 2510\ \text{m}^2$$

hence from equation 7.18

$$Q_2 = \frac{\Sigma_2}{\Sigma_1}\ Q_1 = \left(\frac{2510}{221}\right)\ 8 \times 10^{-6} = 9.08 \times 10^{-5}\ \text{m}^3\,\text{s}^{-1}$$

The terminal settling velocity of the cut size under gravity is, from equation 7.15

$$v_g = \frac{Q_1}{2\Sigma_1}\left(= \frac{Q_2}{2\Sigma_2}\right) = \frac{8 \times 10^{-6}}{2 \times 221} = 1.81 \times 10^{-8}\ \text{m s}^{-1}$$

hence the cut size is, from equation 7.13

$$x_{50} = \left(\frac{18\,\mu v_g}{\Delta\rho g}\right)^{1/2} = \left(\frac{18 \times 0.001 \times 1.81 \times 10^{-8}}{1640 \times 9.81}\right)^{1/2} = 1.42 \times 10^{-7}\ \text{m}$$

$$= 0.142\ \mu\text{m}$$

There have been a few attempts to modify the Σ theory since its conception by Ambler. The most important improvements have been concerned with the introduction of more realistic flow patterns in different types of centrifuges, with account taken of end effects (Frampton[7] used $0.5L$ as the effective bowl length), with the introduction of particle shape factors etc.

Although experimental values of Σ can be obtained from the test data via the grade efficiency curve (from which x_{50} is determined—see chapter 3, 'Efficiency of Separation'), industrialists[8] prefer to measure the fraction of solids unsedimented $(1 - E_T)$ and plot this against the ratio of the measured flow rate and the calculated Σ value (which can be varied by changing the speed of rotation). This curve, which often comes out as a straight line on log–probability paper, is naturally a function of the size distribution of the feed but can be used to find the ratio of Q/Σ for acceptable efficiency with the given feed material. Extrapolation of the data over the linear parts of the graph can be made with caution.

There is however one major shortcoming of the Sigma concept: the cut size is insufficient as a criterion of separation efficiency because different total efficiencies can be obtained at a given cut size, if the size distribution of the feed particles differs. Murkes[9] recognized this but suggested a method which assumes a certain form of the feed distribution function, thus introducing an unnecessary limitation.

Zeitsch[10] recently pointed out and analysed the effect of the exchange of momentum between the feed entering a rotating bowl and the charge already in the bowl on the effective acceleration within the bowl. The feed slows down the rotation of the fluid in the bowl, and the lag between the bowl speed and that of the liquid surface in the bowl ranges in practice between 3 and 10%. Zeitsch concluded that the effect of the feed rate on separation efficiency is significant and derived a correction factor by which the nominal bowl speed has to be corrected to obtain the true effective centrifugal acceleration to which the particles in the flow are subjected.

The only way to describe fully the performance of a sedimenting centrifuge is by the grade efficiency curve; knowledge of this curve allows accurate and reliable (subject to the operation characteristics, the state of dispersion of solids and other variables remaining constant) predictions of total efficiencies with different feed solids. It appears that, rather than to keep modifying the Sigma concept to make it more flexible and complete, we should employ the grade efficiency concept; this of course requires many tests, together with deeper theoretical considerations. Research projects are in progress to fill this gap; some results of recently reported work[11] are included in the appropriate sections in the following account of equipment.

7.3 EQUIPMENT

7.3.1 Tubular centrifuge

The principle of operation of a tubular centrifuge was described in section 7.2.1 together with a derivation of the theoretical grade efficiency curve based on certain assumptions. One of these assumptions is that the amount of sedimented solids in the bowl is negligible throughout the operating cycle. In practice, however, these devices have to be stopped and cleaned (most often manually) when the solids content in the bowl reaches a certain level. During the operation period, a cake (or heel) of solids is gradually building up in the bowl, thus reducing the area available to flow. This in turn reduces the residence time of the liquid in the bowl and the efficiency of separation gradually drops. Reduction in efficiency reflects itself in increasing solids content in the overflow which can be monitored, for example by turbidity measurements. In order to avoid the necessity for excessively frequent cleaning, tubular centrifuges are usually used with suspensions which contain less than 1 % v/v of solids, i.e. for liquid clarification. For continuous operation two centrifuges are used alternately, one running while the other is being cleaned.

If no test data are available, use can be made of equation 7.9 to predict

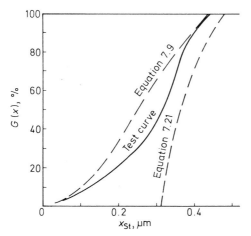

Figure 7.3. Example of an experimental grade efficiency curve for a tubular centrifuge in comparison with theoretical predictions

roughly the order of the cut size and the grade efficiency curve. Because of end effects, non-uniform velocity distributions etc., the predicted curve is always better than what can be expected in practice. *Figure 7.3* gives an example from a series of tests with the Sharples Supercentrifuge (standard 8RY separator: $r_3 = 2.223$ cm, $r_1 = 1.071$ cm, $L = 19.6$ cm) operated at 8000 rev. min^{-1} and a flow rate $Q = 7.57 \times 10^{-6}$ m^3 s^{-1} with a suspension of 0.5 % v/v of very fine TiO$_2$ ($\rho = 4000$ kg m^{-3}) in water at 23°C.

The experimental grade efficiency given in *Figure 7.3* was evaluated from size distributions of the feed and the overflow suspensions (measured by the Ladal X-ray Centrifugal Sedimentometer[12]), and from the total efficiency obtained by simple gravimetric concentration measurements (see chapter 3, 'Efficiency of Separation' for the method). *Figure 7.3* also gives the predicted curve from equation 7.9 ($K = 1.2 \times 10^{11}$ m^{-2} s^{-1}, $K_2 = 30.87$ s) which in this case becomes

$$G(x) = 1.302 \left[1 - \exp(-7.41 \times 10^{12} x^2) \right]$$

(in SI units).

Another theoretical curve may be obtained using the model by Bradley[2] who assumed the velocity distribution in equation 7.3 and used an approximation, specific to the Sharples Supercentrifuge 8RY, of the form

$$\frac{x^2 K}{Q} = 4.1 \times 10^{-3}\, r^{-1.2} \qquad (7.20)$$

Substitution of r from equation 7.20 into equation 7.7 gives

$$G(x) = \frac{r_3^2 - 1.05 \times 10^{-4}\,(Q/x^2 K)^{5/3}}{r_3^2 - r_1^2} \qquad (7.21)$$

which becomes, for our example

$$G(x) = 1.302\,(1 - 0.986 \times 10^{-15}\, x^{-10/3})$$

and this is also plotted in *Figure 7.3*.

As can be seen from the graph, Bradley's model gives conservative predictions of efficiency, mainly because he applied the approximation in equation 7.20, which was originally fitted to data obtained at radii between 1 and 2 cm, to radii up to $r_3 = 2.223$ cm where large discrepancies occur; this leads to underestimates of separation efficiency. Bradley's model is still useful because it gives a lower estimate of efficiency, the actual grade efficiency curves are usually found to lie between those predicted by equations 7.9 and 7.21.

Tubular centrifuges are available in both laboratory and industrial versions. The former can reach 50 000 rev. min^{-1} with flow rates between 0 and 0.1 m^3 h^{-1} (driven by an electric motor or by an air/steam turbine) whilst the latter reach 15 000 rev. min^{-1} and flow rates in the range of 0.4 to 4 m^3 h^{-1} (driven by an electric motor). Tubular centrifuges are the most efficient of all industrial sedimenting centrifuges because of their high speed and relatively thin settling zone; they are therefore used for the separation of very fine solids whose settling velocities are in the range 5×10^{-8} to $5 \times 10^{-7}\,\mathrm{m\,s^{-1}}$. The laboratory models, namely the widely used Sharples Supercentrifuge, have also been used for particle size measurement[2] and the classification of solids.

7.3.2 Imperforate basket centrifuge

This is an adaptation of the standard basket centrifuge used for centrifugal filtration made by replacing the bowl with an imperforate one.

The resulting arrangement is very similar to the tubular centrifuge in *Figure 7.2* but the length-to-diameter ratio is much smaller—usually about 0.6 compared to 4–8 for the tubular centrifuges. Subsequently the mathematical model used for the tubular centrifuge can be applied to the imperforate basket centrifuge as well, but the end effects become more significant here. The efficiency of separation is generally lower because of the relatively short clarification length of the bowl compared with the zone disturbed by feeding and it varies widely with different methods of feeding.

The basket centrifuges are usually operated with a vertical axis of rotation (see *Figure 7.4*) (except for the peeler centrifuges which usually have a horizontal axis), the feed is normally introduced near the bottom of the bowl, the solids separate at the bowl wall and the clarified liquid overflows the lip of the 'ring dam' at the top and is discharged continuously. At the end of an operating cycle, usually triggered by a cake thickness detection device,

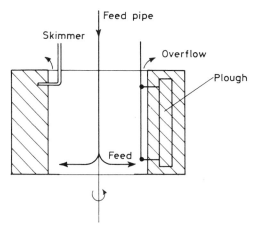

Figure 7.4. Schematic diagram of an imperforate basket bowl in cross-section

the supernatant liquid that remains on top of the cake may be first skimmed off to produce a drier cake on discharge. The mode of solids discharge depends on the type of solids handled; soft and plastic solids are skimmed at full speed and coarse and fibrous solids are removed at slower speeds by a ploughing knife; the cake drops through the open bottom of the bowl. Sometimes, for example in nuclear fuel processing applications, the cake is re-slurried by a liquid fed through a set of nozzles and flows out through the bottom of the bowl at a slow speed. The Donaldson Hydro-Clean Centrifuge uses a novel way of re-slurrying by the stirring action of the blades of a rotor; the rotor blades reach into the suspension inside the bowl and, during the normal operation, drive the freely suspended bowl via the resistance of the liquid. When the discharge operation is initiated, the bowl is simply stopped by a brake while the rotor blades continue to rotate. The cake is re-slurried in the residual supernatant liquid and is then discharged through the bottom of the bowl.

Several workers[13,14] have observed that, contrary to the common assumption that there is a regular stream of liquid through the whole cross-section available to flow within the bowl, the incoming liquid tends to flow within a thin layer near the surface of the liquid whilst the bulk of the liquid is essentially stagnant. As this is clearly detrimental to the separation efficiency, because it shortens the residence time of the liquid in the bowl, some manufacturers build in special baffles which are designed to stop the surface flow and make the liquid flow nearer to the bowl wall.

The solid content is usually low (3–5 % by weight) in basket centrifuges, again to prevent frequent cleaning. Speeds of rotation range from 450 to 3500 rev. min^{-1} with typical flow rates of between 6 and 10 m^3 h^{-1}. Typical applications include the dewatering of sludges[6] and the recovery of solids from a waste stream.

An interesting modification of the basket centrifuge is the triple bowl unit reported by Riesberg and Dudrey[13], and manufactured by Donaldson Co. Inc. as the Liqua Pac Centrifuge. As can be seen from *Figure 7.5* it consists of three bowls of different diameters mounted in one unit. The suspension comes in through an inlet manifold which divides it into three separate

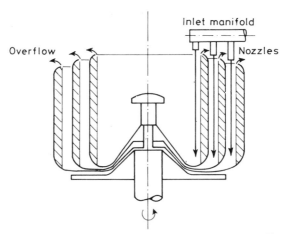

Figure 7.5. Schematic diagram of a triple bowl centrifuge[15]

streams and directs each stream into a different bowl. There are then three independent and parallel flows each discharging separately and bypassing the other two bowls. This makes a more efficient use of the space available within the bowl and, as it increases the residence time of the liquid in the bowl, a lower rotational speed (650 rev. min^{-1}) can be used for the same efficiency (this produces a number of advantages in design and operation). As in the multi-chamber centrifuge discussed in the next section, the solids holding capacity is also increased. The triple bowl centrifuge has been successfully applied to cleaning metal-working coolants[15].

7.3.3 Multi-chamber centrifuge

This type of centrifuge utilizes a closed bowl which is sub-divided into a number of concentric vertical cylindrical compartments through which the suspension flows in series—see *Figure 7.6*. The suspension is made to pass through zones of progressively higher acceleration and this results in a classification effect, with the coarsest fraction being deposited in the inner chamber and finest fraction in the outermost chamber. Another advantage is the large solids holding capacity; in a large machine; this may be up to as much as 751 ($= 7.5 \times 10^{-2}$ m^3).

Westfalia Separator AG make models with either a six-chamber bowl (five cylindrical inserts), or a two-chamber bowl (one cylindrical insert), which also incorporates a centripetal pump on the discharge side of the bowl (this converts the kinetic energy of the liquid on discharge into pressure). The Broadbent–Hopkinson centrifuge is a two-chamber type specially designed for reclaiming cutting-oil from grinding machines.

The efficiency of the multi-chamber centrifuges is high because of the long residence time of the liquor in the unit. A mathematical model similar to that used for the tubular centrifuge can be applied to the multi-chamber centrifuge provided that satisfactory assumptions and modifications are made.

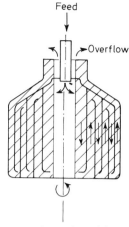

Figure 7.6. Schematic diagram of a multi-chamber bowl in cross-section

The cleaning of multi-chamber centrifuges is more difficult and takes longer than for the tubular type. The manual discharge generally limits this type of machine to suspensions containing less than 4–5% solids by volume.

Available speeds range between 4500 and 8500 rev. min^{-1}, flow rates from 2.5 to 10 m^3 h^{-1} and bowl diameters from 335 to 615 mm. Multi-chamber centrifuges, apart from the separation of grinding swarf from cutting oils already mentioned, are mainly used for the clarification of beer, wine, fruit juice and varnishes.

7.3.4 Scroll-type centrifuge

A characteristic feature of the scroll-type (often called 'decanter' type) continuous conveyor discharge centrifuge is the horizontal conical or cono-cylindrical bowl, with a length-to-diameter ratio of about 1.5 to 3.5, containing a screw conveyor that rotates in the same direction but at a slightly higher or lower speed (the difference being 5 to 100 rev. min^{-1} with respect to the bowl). The centrifugal fields are lower than in other centrifugal equipment; speeds range from 1600 to 6000 rev. min^{-1}. The operating principle is shown in *Figure 7.7*. The slurry enters through an axial tube at the centre of the rotor, passes through openings in the screw conveyor and is thrown to the rotor wall. Deposited solids are moved by a helical screw conveyor up a sloping 'beach' out of the liquid and discharged at a radius smaller than that of the liquid discharge. The liquid level is maintained by ports adjustable to the desired overflow radius.

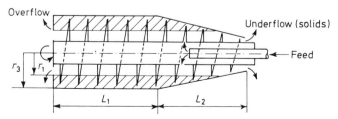

Figure 7.7. Schematic diagram of a scroll-type centrifuge bowl in cross-section (cono-cylindrical bowl)

For the conical conveyor type centrifuge the Sigma value can be calculated as[8] (see *Figure 7.7* for dimensions)

$$\Sigma = \frac{\omega^2}{g} \pi L \frac{r_3^2 + 3r_3 r_1 + 4r_1^2}{4} \tag{7.22}$$

where L is the length of the liquid layer or 'pond' measured at the liquid surface.

For the cono-cylindrical conveyor-type centrifuge the Sigma value is usually assumed to be the sum of the values for a true cylindrical section of length L_1 given by equation 7.17 and for a conical section of length L_2 thus (see *Figure 7.7*)

$$\Sigma = \frac{\omega^2}{g} \pi \left[L_1 (\tfrac{3}{2} r_3^2 + \tfrac{1}{2} r_1^2) + L_2 \left(\frac{r_3^2 + 3r_3 r_1 + 4r_1^2}{4} \right) \right] \tag{7.23}$$

This is of course a theoretical formula; two correction factors are usually applied in practice. Ambler[4] recommends that 6% should be deducted from this theoretical value of Sigma to allow for the finite volume of the conveyor flights in the suspension (this reduces the residence time). In addition to the volume of the conveyor flights, the sedimented solids which are being conveyed through the rotor also displace their own volume of liquid; Ambler[4] suggested a reduction factor of 0.62 for a conical conveyor type and 0.67 for a cono-cylindrical type to account for this effect.

Particles of settling velocities of the order of 1.5×10^{-6} to about 15×10^{-6} m s^{-1} are handled at medium to large flow rates (typical throughputs range from about 0.4 to 60 m^3 h^{-1}, depending on the machine and the application). When operating as a clarifier, this type of centrifuge recovers medium and coarse particles from feeds at high or low solids concentration 2–50% v/v. As a classifier, the flow rates are higher than for clarification. Particles smaller than 2 μm are normally not collected.

Gibson[11] has published some interesting results of tests with a scroll-type centrifuge, which show the cut size as a function of the feed rate and feed solids concentration—see *Figure 7.8*.

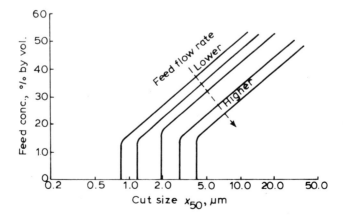

Figure 7.8. Effect of feed concentration and flow rate on cut size x_{50} for scroll-type centrifuge

Polyelectrolytes are widely used for the flocculation of slurries to be separated in the scroll-type centrifuges; the point of addition varies depending on the type of polyelectrolyte and the type of slurry. Anionic polyelectrolytes are usually dosed upstream of the centrifuge whilst cationics, which react faster, are usually added to the slurry within the centrifuge.

In design, considerable variations occur in the contour of the centrifuge shell, the flight angle and pitch, beach angle and length, conveyor speed and feed position. An alternative to the liquid overflow open outlet is a paring tube for skimming off the liquid. The most common position of the axis of rotation is horizontal but Pennwalt for example have introduced a large vertical machine (model P6000) specifically designed for higher temperatures and pressures.

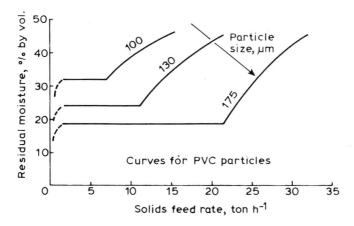

Figure 7.9. Scroll-type centrifuge dewatering performance declines at a critical feed rate

The scroll-type centrifuge has found a wide variety of industrial applications; as a clarifier for example, it is used to recover crystals and polymers. Most recently it has been used widely in the UK for the dewatering of both municipal and industrial sludges[16]. Stahl[17] has successfully applied the centrifuge to the dewatering of coarse solids, namely PVC particles with a median size between 100 and 190 μm. Some of his results are shown in *Figure 7.9*.

Efficient washing and dewatering may be achieved with the Broadbent Screen Bowl centrifuge, which is in fact an end-to-end combination of a solid bowl scroll-type centrifuge with a cylindrical scroll-screen-type centrifuge (the scroll is common to both), see *Figure 7.10* for a schematic diagram of the principle. A similar machine is available from Humboldt Wedag.

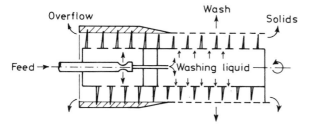

Figure 7.10. Schematic diagram of a solid/screen bowl scroll centrifuge

Another major area of application of the scroll-type centrifuge is in the classification of solids such as kaolin coating clay, TiO_2 etc.

7.3.5 Disc centrifuge

This type of centrifuge contains a stack of conical discs, as shown in *Figure 7.11*. The feed is introduced through the centre, passes underneath the disc stack and into the space between the stack and the wall of the bowl while both the stack and the bowl are rotating at a speed ω. The liquid then flows in thin layers between the discs, radially inwards, towards the outlet, which is an annulus at the top centre. Particles settle on the lower surface of the upper of the two discs which form each 'disc channel'. The settling motion of the particles is the first, usually decisive, stage of the separation process. The second stage is the downward–outward sliding motion of the particles on the disc surface towards the disc periphery and their subsequent impingement on the wall of the bowl.

The basic idea here of increasing the settling capacity by using a number of layers in parallel is the same as the Lamella principle[18] in gravity sedimentation.

Murray[19] considered possible improvements in the separation efficiency of disc centrifuges by reversing the liquid flow through the disc stack but found no advantage in using this method in practice.

The motion of a particle between two adjacent conical discs is shown schematically in *Figure 7.11*. The flow of the suspension is assumed to be divided equally between the spaces formed by the discs, so that the flow rate in each disc channel is equal to Q/n where Q is the throughput and n is the number of

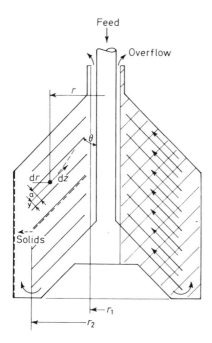

Figure 7.11. Schematic diagram of a disc centrifuge bowl in cross-section

channels. The flow is also assumed to be in a radial plane (i.e. having the same angular velocity as the stack) and directed parallel to the surface of the discs. Under these conditions, the velocity of a particle of size x at radius r from the axis of rotation can be resolved into two components: the radial velocity dr/dt, which can be approximated by Stokes' law in equation 7.1, and the velocity in the direction of flow z: dz/dt, which can be approximated by the velocity of the liquid at that point (if a plug flow is assumed)

$$\frac{dz}{dt} = \frac{Q}{2\pi nra} \tag{7.24}$$

where a is the perpendicular clearance between the adjacent discs.
   Since also

$$\frac{dy}{dt} = \frac{dr}{dt}\cos\theta \tag{7.25}$$

where θ is half the included angle of the discs and the y axis is perpendicular to the z axis, equations 7.1, 7.24 and 7.25 combined give (with K as defined in equation 7.5b)

$$\frac{dy}{dz} = Kx^2r^2\frac{2\pi na}{Q}\cos\theta \tag{7.26}$$

thus

$$\frac{dy}{dr} = -Kx^2r^2\frac{2\pi na}{Q}\cot\theta \tag{7.27}$$

since

$$\frac{dz}{dt} = -\frac{dr}{dt\sin\theta} \tag{7.28}$$

   Equation 7.27, if integrated between the limits $y = y_1$ at $r = r_2$ and $y = a$ at $r = r_1$ determines the distance y_1 from the bottom plate at which sedimentation of a particle size x may start in order that it may just reach the top disc after covering the full length of the space between the discs, i.e.

$$a - y_1 = Kx^2\frac{2\pi na}{3Q}\cot\theta(r_2^3 - r_1^3) \tag{7.29}$$

For the limit of separation this integration is done between $y = 0$ at $r = r_2$ and $y = a$ at $r = r_1$, leading to

$$a = Kx_{max}^2\frac{2\pi na}{3Q}\cot\theta(r_2^3 - r_1^3) \tag{7.30}$$

   The grade efficiency can now be determined if we assume a homogeneous concentration of particles at the entry to the separation zone at r_2 because

$$G(x) = \frac{a - y}{a} \tag{7.31}$$

which becomes (if equation 7.29 is substituted for $a - y$ and equation 7.30 for a)

$$G(x) = \frac{x^2}{x_{max}^2} \qquad \text{for } x \leqslant x_{max} \qquad (7.32)$$

and

$$G(x) = 1 \qquad \text{for } x > x_{max}$$

where x_{max} is, from equation 7.30

$$x_{max} = \frac{3Q}{2\pi n K \cot \theta (r_2^3 - r_1^3)} \qquad (7.33)$$

Equations 7.32 and 7.33 fully describe the theoretical grade efficiency curve which, as can be seen, is independent of the spacing between the discs but depends on the number of discs n. Equation 7.32 is a simple parabola and the cut size x_{50} can be determined from equation 7.32 since $G(x_{50}) = 0.5$ hence

$$x_{50} = x_{max}\sqrt{0.5} = 0.707 x_{max} \qquad (7.34)$$

The Sigma factor for a disc centrifuge becomes

$$\Sigma = \frac{\omega^2}{g} \tfrac{2}{3}\pi n(r_2^3 - r_1^3) \cot \theta \qquad (7.35)$$

(from equations 7.13, 7.15, 7.34 and 7.33).

Fitch[20] subjected this theoretical estimate of separation efficiency (equation 7.32) to a detailed investigation and tests, and found in practice considerably lower efficiencies than predicted. He found that this was not a result of the non-uniformity of the radial flow pattern between the discs because most disc centrifuges have radial spacing ribs which block tangential flow and cause the flow pattern to be fairly uniform except for variations close to the disc surfaces. Fitch found both theoretically[20] and more recently by taking photographs of the actual flow patterns in a disc centrifuge[21], that vortices exist in the disc sectors bounded by spacing ribs along the axis of rotation and these are the major cause of the deviations from theory.

The disc centrifuges are operated at speeds up to 12000 rev. min^{-1} depending on the bowl diameter, which is typically 150 mm to 1 m. Flow rates of up to 100 m^3 h^{-1} are obtained on easy separations; the range of theoretical settling velocities handled is 8×10^{-8} m s^{-1} at solids concentrations below 15%. Bowls normally have approximately equal height and diameter for optimum capacities, and the angle of the cones is sufficiently large for the deposited particles to slide on their surfaces—usually between 35 and 50 degrees. The primary process modifications to the disc centrifuge design concern the method of solids discharge. These modifications are discussed below.

7.3.5.1 Solids retaining type

The simplest disc centrifuge bowl (*Figure 7.11*) is designed with a non-perforate bowl wall parallel to the axis of rotation. In order to avoid frequent manual cleaning, this type is used with low solids concentrations (less than 1 % by volume). Cleaning is sometimes facilitated by using disposable paper liners. Typical dirt-holding capacity is 5 to 20 l. The most frequent application is the separation of cream from milk; applications are generally restricted to safe liquids because toxic materials may be a source of hazard during the manual cleaning.

7.3.5.2 Nozzle type

Continuous discharge of the solids as a slurry is possible with the nozzle-type disc centrifuges. The shape of the bowl is modified so that the sludge space has a conical section as shown in *Figure 7.12* thus providing a good storage volume and also giving a good flow profile for the ejected sludge. The walls of the bowl slope towards a peripheral zone containing a number of evenly spaced orifices. The number and size of these orifices, commonly called nozzles, are optimized in order to avoid build up of the cake or pluggage and to obtain a reasonably concentrated sludge. The number of nozzles ranges from 12 to 24, the size from 0.75 to 2 mm and the underflow-to-throughput ratio is normally between 5 and 50 %. The underflow flow rate is determined almost solely by the size and number of nozzles used. Similarly to hydrocyclones, the appreciable underflow flow rate in the nozzle-type disc centrifuges increases the proportion of the 'dead flux' of particles going in to the underflow and thus obscures the evaluation of the net separation effect. The concept of 'reduced' efficiency is applicable here, as described in chapter 3, 'Efficiency of Separation', section 3.4.

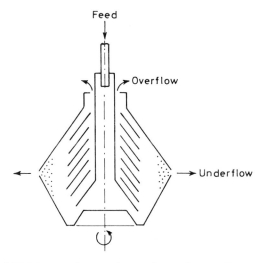

Figure 7.12. Schematic diagram of a nozzle-type disc centrifuge

A recently concluded series of large-scale tests[11] in the UK revealed a dependence of cut size on feed concentration and volumetric feed rate (*Figure 7.13*). The feed rate is bound by two limiting values: a minimum equal to the nozzle discharge rate (when no overflow occurs) and a maximum equal to the rate at which the centrifuge is flooded. The increase in cut size with increasing feed concentration is attributed to hindered settling.

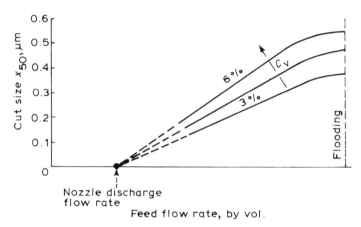

Figure 7.13. Dependence of cut size x_{50} on feed concentration and flow rate for nozzle-discharge disc centrifuge

There are several variations of the design such as the recirculation of part of the solids discharge, nozzles at reduced rotor diameter, a facility for washing before discharge, internal nozzles and a paring tube for pressurized solids discharge and others.

Important applications include kaolin clay dewatering, tar dehydration and clarification of wet process phosphoric acid.

7.3.5.3 Solids ejecting type

Intermittent solids ejection is achieved with the solids ejecting type of disc centrifuge (*Figure 7.14*). A number of peripheral ports, which are closed with valves, are provided and these are controlled either by a timer or by a self-triggering device that operates automatically depending on the cake depth. Sliding annular ring scales may also be used; these open wide slots around the periphery and handle considerably coarser and stiffer solids than the orifice ports could handle. The solids ejecting type is used where there is a medium concentration of solids, such as 2–6%, so that neither continuous discharge nor batch operation would be optimum. It is also useful for solids which break down or de-aggregate under the shear forces of nozzle discharge. A better concentration of solids which are slow to compact is also achieved. Depending on the length of time the rotor remains open, either total discharge (of the entire contents of the bowl) or partial discharge (solids only) may be obtained.

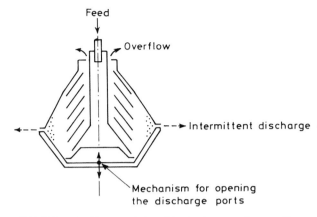

Figure 7.14. Schematic diagram of the solids ejecting type disc centrifuge

Applications for this type of centrifuge include the clarification of various juices and food extracts, and the purification of marine fuels.

7.4 FACTORS AFFECTING THE CHOICE OF CENTRIFUGAL EQUIPMENT

The engineer who is faced with the selection of centrifugal sedimentation equipment can first consult a guide similar to the one published by Lavanchy et al.[22], which is presented in *Figure 7.15* in a modified, metric version, and gives the limiting flow rates and settling capabilities of various types of

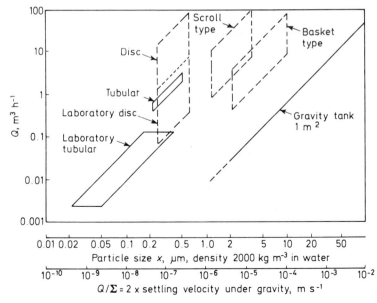

Figure 7.15. Performance of various centrifugal sedimentation equipment

equipment. The equipment can be located in *Figure 7.15* in its normal operating region (region of greatest utility) on the basis of nominal effluent flow rates (normal flows for good and economical clarification in standard applications) and the applicable equiprobable particle sizes or the actual Q/Σ values. Conversion to physical conditions other than those given ($\Delta\rho = 1\ \text{g cm}^{-3}$ and $\mu = 1\ \text{cP}$) can be carried out by converting the particle size using equation 3.12 in chapter 3. Note that a theoretical performance curve of a settling tank of unit area is also included in *Figure 7.15* for comparison.

The selection made using this guide may be narrowed by other material characteristics and process requirements. These include the solids content of the feed, the nature of the solids (sticky, fibrous or otherwise difficult), the nature of the liquid (corrosive or sensitive to contact with air), explosion hazards, and also the necessity of continuous, intermittent or batch operation. If all these factors are known, the information on each type of equipment selected has to be consulted and a compromise made by choosing the equipment which satisfies all demands to a tolerable degree.

7.4.1 Centrifuges in series or in parallel

When more than one sedimenting centrifuge is to be used in order to improve separation efficiency, the question arises whether it is better to use the units in parallel or in series. Both arrangements are beneficial to the overall recovery of the whole plant: the parallel arrangement increases the residence time which improves the separation, whilst the series arrangement (in the direction of the overflow) benefits the recovery by giving the particles that escaped the process in the first stage another chance to separate.

Let us take an example of two identical sedimenting centrifuges in a clarification duty, such as may be the case on board ships for cleaning fuel or lubricating oil. The question is whether the centrifuges would be best connected in parallel or in series. The general argument is made below for any type of sedimenting centrifuge, but when performance figures need be inserted a specific case of the disc-type centrifuge is used as an example and the conclusion then applies to that type only.

7.4.1.1 The series arrangement

The grade efficiency, G, of a series arrangement of two identical centrifuges, each having the same grade efficiency G_1, is according to chapter 3, equation 3.51

$$G = 2G_1 - G_1^2 \tag{7.36}$$

The series arrangement moves the grade-efficiency curve to the left, to a lower cut size and to higher values of $G(x)$.

In terms of the cut size, its value for the whole plant, X_{50}, is always less than that of the individual centrifuge, x_{50}, but how much less? This depends on the shape of the grade-efficiency curve: if G_1 is log–normal, for example, the improvement depends on the geometric standard deviation of the curve. *Table 7.1* gives three calculated examples. The lower the value of the multiplier in *Table 7.1*, the lower will be the cut size and the higher the resulting recovery of the plant; less steep grade-efficiency curves are clearly favoured in the series arrangements.

Table 7.1. THE IMPROVEMENT IN THE CUT SIZE IN SERIES CONNECTION

Geometric standard deviation of G_1 (X), σ_g	The cut size of the whole plant, X_{50}, μm
3	$0.53\ x_{50}$
2	$0.68\ x_{50}$
1.75	$0.73\ x_{50}$

7.4.1.2 The parallel arrangement

If the two centrifuges are arranged in parallel, each one takes one half of the total flow, Q, and the performance is compared with that of one centrifuge taking the whole of the flow. In terms of cut size, the Sigma theory (see section 7.2.2) can be used to write for the same machine (same $\sum$ value):

$$\frac{x_{50}}{\sqrt{Q}} = \frac{X_{50}}{\sqrt{Q/2}} \qquad (7.37)$$

because, for otherwise the same conditions, the cut size of a sedimenting centrifuge is proportional to the square root of the feed flow rate. Equation 7.37 leads to

$$X_{50} = 0.707\ x_{50} \qquad (7.38)$$

If this result is compared with the values in *Table 7.1*, the conclusion depends critically on the shape of G_1, the dividing line being at $\sigma_g = 1.85$ for a log–normal G_1 (the series arrangement being better for larger values of σ_g). As σ_g is very unlikely to be less than 1.85 for sedimenting centrifuges, it seems that the series arrangement is better.

Taking a specific example of disc centrifuges, some test results published in the literature are listed in *Table 7.2*. These data confirm the conclusion that for disc centrifuges the series arrangement gives better recovery than does the parallel one.

Table 7.2. MEASURED PERFORMANCE OF SOME DISC CENTRIFUGES

Centrifuge model	Ref.	σ_g
Alfa-Laval QX 412	11	
1.6 mm nozzles		2.35
2 mm nozzles		2.61
Alfa-Laval LAPX 202	23	
(solids-ejecting)		2.14

7.5 RECENT DEVELOPMENTS

In addition to the now more-or-less standard features and designs of the sedimenting centrifuges described in section 7.3, there have been many refinements and improvements added to the existing technology over the past decade or so. The most important ones are briefly reviewed here, with references given for further details and reading.

7.5.1 The scroll-type centrifuge

The most usual design of the scroll centrifuge is the one in which the solids and the liquid move in a countercurrent fashion in the bowl, with the feed point being well away from the liquid discharge point (refer to section 7.3.4). A relatively recent, co-current design offers certain advantages, mostly because it requires lower speeds for the same separation duty. The feed point is close to the liquid discharge end of the bowl but the slurry cannot bypass the process; it moves with the solids towards the beach until it reaches a point when it overflows into the scroll and then it moves back to the liquid discharge end, not interfering with the flow of the solids.

A co-current design is obviously a little more complex than a countercurrent one, because it requires a running seal to prevent the incoming feed from passing directly into the overflow, but it may well be worthwhile in applications where lower speeds are desirable, e.g. when there is a danger of the solids sliding down the beach at high speed. An alternative to the liquid overflow open outlet is a paring tube for skimming off the liquid.

The conventional, countercurrent decanters, if applied to very fine slurries, can suffer from problems of discharge of the solids. The separated sludge can flow back, countercurrent to the movement of the scroll conveyor; the flow may be through the clearance between the conveyor blades and the inner surface of the bowl and also along the helical canal formed by the conveyor blades and the cone. This highly undesirable phenomenon may cause variations in the solids discharge rate, with the accompanying fluctuations in the torque, poor centrate clarity and poor overall performance of the machine. Karolis and Stahl[24] studied the effect with the view of developing a model and proposing design and operating measures to counter the problems. The project has led to the development of grooved inner walls of decanters which reduce backflow problems[25].

Another important recent development is that of a slender decanter centrifuge[26]. Through a novel way of supporting the bowl, the centrifuge can be made much longer relative to its diameter than was previously possible. Both manufacturing costs and energy consumption are claimed to be reduced with the slender decanter as compared with the conventional short and thick unit of the same separating capacity. The G force for a given length of the decanter can also be increased. Further reductions in energy consumption can be achieved due to a design which allows a very small liquid surface radius for a given decanter diameter.

The industrial use of the scroll-type centrifuge has been growing, particularly in coal preparation[27] and in other mineral-processing applications where, in order to reduce abrasion, the internal walls are coated with a ceramic lining. Another major area of growing application of the scroll-type centrifuge is in the dewatering of municipal and industrial sludges[28].

7.5.2 The disc centrifuge

There are now several variations of the design of the nozzle-type bowl including the following features:

1. recirculation of a part of the solids discharge;
2. nozzles at reduced rotor diameter;
3. a facility for washing before discharge;
4. internal nozzles and a paring tube for pressurized solids discharge.

The solids recirculation system offered in the Merco Centrifuge, for example, takes part of the slurry discharged through the nozzles back into the bowl and reintroduces it close to the nozzles. This allows control of the underflow solids concentration without the need to change nozzles or use very small nozzles which are susceptible to blocking. Furthermore, washing liquid can be added to the recycled stream, thus allowing some co-current washing. The Merco Centrifuge is available with capacities of up to 230 m³ h⁻¹.

A relatively recent newcomer in this category of disc centrifuges is the Roto-Filter pump. This pump clarifies and pumps at the same time. The 400-mm diameter rotor has one flat disc in it and the suspension enters centrally, as in a centrifugal pump; it goes round the outside of the disc and back to the centre where it is collected by a stationary pipe. The solids discharge through nozzles situated on the periphery of the disc, as in the nozzle-type centrifuge described previously, except that the nozzles (1.5 mm diameter) face tangentially rather than radially. Only two or four nozzles are provided, making this centrifuge suitable for clarification only. The effective cut size is large by centrifuge standards, 10 μm or more for sand in water. This is because the residence time of the solids in the bowl is short, with some elutriation also taking place. Flow rates of 27 m³ h⁻¹ and delivery pressures of 6.7–27 bar can be achieved. The bowl speed is 3550 rev. min⁻¹. The use of this machine can only be justified if pumping is necessary and relatively coarse separation is acceptable. Otherwise, any conventional centrifuge will achieve much finer cuts.

7.5.3 Safety

Before leaving the subject of sedimenting centrifuges it is important to mention safety. As with other machines with parts rotating at high speeds, the necessary safety precautions must be observed. The enclosure must be designed to withstand possible disc failure and vibration switches are usually provided to give warning of excessive vibration due to bowl imbalance. When dealing with flammable solvents and vapours, extra precautions must be taken to prevent explosion: these include inert gas (nitrogen or carbon dioxide) blanketing, explosion suppression devices (which eject suppressant during explosion) and the elimination of possible sources of ignition.

Hermetically sealed decanters, for example, have been designed for this purpose[29].

REFERENCES

1. Schachman, H. K., *J. Colloid Phys. Chem.* **52**, 1034–1045 (1948)
2. Bradley, D., *Chem. & Proc. Eng.* November & December, 1–8 (1962)
3. Ambler, C. M., *Chem. Eng. Progress*, **48**, 150–158 (1952)
4. Ambler, C. M., *J. Biochem. & Microbiol. Tech. & Eng.*, **1**, 185–205 (1959)
5. Morris, B. G., *Brit. Chem. Eng.*, **11**, 347 (1966)
6. Woolcock, R. J., *Filtration and Separation*, **12**, 174–180 (1975)
7. Frampton, G. A., *Chem. & Proc. Eng.*, **44**, 402 (1963)
8. Purchas, D. B., *Industrial Filtration of Liquids*, 2nd edn, Leonard Hill Books, London (1971)
9. Murkes, J., *Brit. Chem. Eng.*, **14**, 636–641 (1969)
10. Zeitsch, K., 'Effect of the feed rate on the active acceleration of overflow centrifuges', *Trans. I. Chem. E.*, **56**, 281 (1978)
11. Gibson, K., 'Large scale tests on sedimenting centrifuges and hydrocyclones for mathematical modelling of efficiency', pp. 1–10. In *Proc. Symp. on Solid-Liquid Separation Practice*, Yorkshire Branch of the I. Chem. E., Leeds, 27–29 March (1979)
12. Allen, T. and Svarovsky, L., *Powder Technol.*, **10**, 23–28 (1974)
13. Bass, E., *Periodica Politechnica Eng. Machinen u. Burwesen, Budapest*, **4**, 1 (1960)
14. Reuter, H., *Chem. Ing. Tech.*, **39**, 311, 548 (1967)
15. Riesberg, K. D. and Dudrey, D. J., *Filtration Engineering*, May/June (1972)
16. Ford, J., 'The sludge dewatering centrifuge', *I. Chem. E. Symp. Series No. 41*, QI–Q14 (1975)
17. Stahl, W., *Personal Communication*, Krauss-Maffei A. G., München, West Germany (1975)
18. '*The Sala Lamella Thickener*, SALA Information SIM 374, Sala International, Sweden
19. Murray, K. R., *1st European Conf. on Mixing and Centrifugal Separation*, Paper F4, BHRA Fluid Engineering (1974)
20. Fitch, B., 'Separating power of disc centrifuges', *55th Int. Meeting of A.I. Chem. E.*, Preprint 11a (1965)
21. Willus, C. A. and Fitch, B., *Chem. Eng. Progress*, **69**, 73–74 (1973)
22. Lavanchy, A. C., Keith, F. W. and Beams, J. W., 'Centrifugal separation'. In Kirk, R. E. and Othmer, D. F., *Encyclopedia of Chemical Technology*, 2nd edn, vol **4**, Wiley Interscience, London, 710–758 (1964)
23. Allen, T. and Baudet, M. G., *Powder Technol.*, **18**, 131–138 (1977)
24. Karolis, A. and Stahl, W., 'Discharge of pasty material from decanter centrifuges', *Inst. Chem. Eng. Symp. on Solid/Liquids Separation Practice and the Influence of New Techniques* (Leeds, 1984), Paper 6.
25. Stahl, W. and Langeloh, T., 'Improvement of clarification in decanting centrifuges', *Ger. Chem. Eng.* **7**, 72–84 (1984)
26. Madsen, N. F., 'Slender decanter centrifuges', *Conference on Solid-Liquid Practice III, 397th Event of the EFCE* (Bradford, 1989), Inst. Chem. Eng. Symp. Ser. No. 113 (1989)
27. Osborne, D. G., 'Coal fines dewatering', *Trans. Soc. Mining Eng. AIME*, **276**, 1843–1849 (1982)
28. Spinosa, L. and Eikum, A., 'Dewatering of municipal sludges'. In L'Hermite, P. and Ott, H. (Eds), *Proc. 2nd European Symp. on the Characterization, Treatment and Use of Sewage Sludge* (Vienna, 1980), D. Reidel, Dordrecht (1980)
29. Jaeger, E. A., 'Hermetically sealed solid bowl decanter – a part of solvent chemistry', *Inst. Chem. Eng. Symp. on Solid-Liquid Separation Practice and the Influence of New Techniques* (Leeds, 1984), Paper 5

8

Screening

D. G. Osborne[*]

PT Kaltim Prima Coal, Jakarta, Indonesia

NOMENCLATURE

a	Clear opening or aperture	mm
d	Diameter of the screen wire	mm
F_{oa}	Open area	%
α	Slope angle of the screen	°
C	Feed rate of dry solids	t h^{-1}
X	Proportion of oversize (by weight)	%
A	Surface-area factor	—
ρ_s	Density of solids	t m^{-3}
d_m	Mean particle size	mm
M	Proportion of solids in size fraction	%
n	Number of grains	—
K	Shape factor	—

8.1 INTRODUCTION

Screening is the process of separating grains, fragments or lumps of a variety of sizes into groups in each of which are contained only fragments or grains in the size range between definite maximum and minimum size limits. The size of a non-uniform fragment cannot be readily defined, but it can be described in terms of a surface opening through which a fragment of that particular size will barely pass, or not pass at all. In other words, two openings, the smaller of which will retain all the fragments of the size group and the larger of which will pass all fragments, will define a size range.

In addition, the process of screening may also be used for dewatering. This occurs as a result of two mechanisms:

1. drainage of excess fluid from the solid fragments during the screening process;
2. distribution of surface moisture resulting from division into two or more size fractions, i.e. finer sized fractions contain more moisture than do coarser fractions due to their greater surface area.

*Helpful suggestions and other input to this chapter from R. J. Gochin (author of Chapter 19) are gratefully acknowledged.

Figure 8.1 shows the relationship between mean grain size and surface moisture content[1]. This may influence the selection of a flowsheet comprising screening and dewatering steps.

The uses for screening in process[2] engineering may be sub-divided into three groups, detailed below.

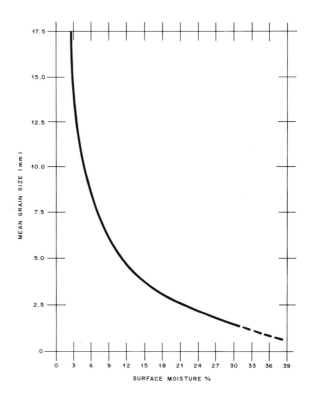

Figure 8.1. Graph to show the relationship between grain size and moisture; based on bituminous coal, 800 kg m^{-3}. (Courtesy of Allis-Chalmers)

8.1.1 Pre-screening (*Figure 8.2*)

This, as the name suggests, is an application prior to a particular sequence of preparatory processes. In mineral processing, pre-screening or primary screening is usually linked with the sequence of primary comminution (size reduction) events. Ore is reduced in size by stages in order to release valuable mineral from the ultimate waste or gangue. Combined with pre-screening could be a desliming stage involving a water wash to remove ultra-fine grains. Alternatively, the desliming stage could occur subsequent to crushing.

At this point it is perhaps worth noting that it is normal to distinguish two phases of sizing used commonly in process engineering. One is screening and the other is classification, and they are often confused. Screening relates only to the properties of size and shape while classification relates to size, shape and density. The process of classification involves the relative terminal velocities or settling rates of particles through a fluid, either moving or motionless. Such a fluid would most commonly be water or air.

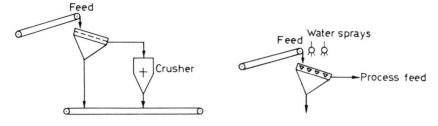

Figure 8.2. Pre-screening

8.1.2 Grading or sizing screening (*Figure 8.3*)

This is an application of screening machinery to divide a widely sized feed material into relatively closely grouped size ranges. (It is thus the application most closely related to the definition, yet strangely enough is not the most common.) Examples of grading commonly encountered are the sizing of coal and quarry stone. Obviously, the sizes graded are largely determined by market demand and the screen must incorporate a facility to allow changes to be made in products. Thus the screen plates are always easily replaceable.

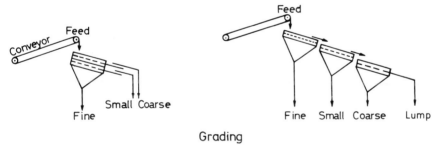

Grading

Figure 8.3. Grading screens

8.1.3 Dewatering (*Figure 8.4*)

This implies a solid–liquid separation function but can also be considered to include medium recovery screens used in dense-medium separation of ores and rocks.

Dewatering of coarser material occurs as drainage during screening operations. Applications of dewatering for finer materials are common and usually involve material with a size range from about 1 mm down to 0.1 mm. Dense-medium recovery screens, often referred to as drain and rinse screens (*Figure 8.4a*), are also common in the processing of the ores of tin, copper, lead and zinc as well as diamonds, coal and numerous other materials. In a widely used process a medium comprising finely ground magnetic iron oxide

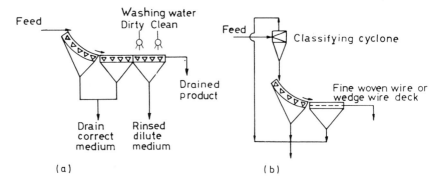

Figure 8.4. Dewatering screens. (a) Drain and rinse. (b) Dewatering

or ferro-silicon and water is used to form a fluid of intermediate density to the valuable and non-valuable fractions, thus allowing a float–sink separation to be achieved. This medium is too costly for more than minimal losses to be tolerated and it is recovered by drainage through a small aperture recovery screen over which the separation products are passed. To ensure maximum recovery a subsequent rinsing with water applied using flood boxes and sprays is carried out. This is normally conducted on the same recovery screen and the subsequent dilute medium, which is contaminated to some degree with fine grains of non-medium material, is passed through a magnetic separator which recovers the magnetic fraction.

The use of screening as a specific means of dewatering is most commonly applicable to material sized below 1 mm. This is distinguished by the size at which drainage occurs within a composite bed of material containing in its majority, fragments or grains smaller in size than the screen aperture. Above this size, dewatering can occur for a mono-layer of material through the screening surface. In order to distinguish between the two, we will refer to the latter as drainage screens and the composite bed type will be referred to as dewatering screens.

As shown in *Figure 8.4b*, dewatering 'screens' can, in fact, refer to a combination of size adjustment and dewatering steps. The featured example, commonly encountered in practice, comprises a classifying cyclone, to remove ultra-fine material (which may pass to another solid–liquid separation step), a sieve bend and a dewatering screen. The last component is in fact the only step where maximum solids retention is required, but all three are essential in order to optimize the performance of the dewatering process.

8.2 SCREEN DESIGN CONSIDERATIONS

8.2.1 General factors

Although a screen may appear to be a simple enough piece of equipment it is based on quite complex design features. These are too involved for detailed discussion in this chapter but the more important factors[3] in the design and construction of screens in general are as follows:

1. the main function of a sizing screen is to separate undersize material from a feed, whatever the range of material sizes in the feed, by letting undersize material pass through the apertures in the screen. This separation must be achieved as a stated efficiency and at a given feed rate, and the design and the construction must therefore comply with this basic requirement;
2. the construction must be sufficiently robust for the designed function and the flow design should be such as to minimize wear and tear. Mechanically, the machine must provide a suitable mode of transportation so as to allow separation to be achieved most effectively;
3. maintenance costs and power consumption, conveniently based upon tonnage handled, should be as low as possible;
4. from an operational standpoint the screen should function without producing excessive spillage or causing degradation of the material during transportation.

8.2.2 Dewatering screen design

The function of a dewatering screen is to separate solids from liquids with the highest possible recovery of solids and the highest possible reduction of moisture in the retained solids. For coarse solids, i.e. plus 0.5 mm, this does not present a problem as the apertures can be smaller than the smallest particle size and a 100% solids recovery may be achieved. In this easy dewatering process, there are numerous types of screens available in sizes of up to 5.5 m width. But the problem really becomes more acute in the minus 0.5 mm range of size. Consideration of the following parameters is therefore important for successful fines dewatering[4]:

1. particle shape;
2. particle size range and distribution;
3. screen deck and aperture;
4. open area (proportion occupied by holes);
5. specific gravity (SG) of material;
6. percentage liquids to solids in volume of feed to the screen;
7. drainage area required;
8. acceleration;
9. operating frequency, angle of throw and screen slope;
10. width-to-length relationship of screen.

How, then, do these parameters affect the dewatering process?

8.2.2.1 Particle shape

The particle shape determines with what ease the material can be dewatered. Spherical or cubic shapes leave enough interstitial space for drainage, while flat shapes do hinder the drainage and, depending on density, can settle out of suspension at a much slower rate.

8.2.2.2 Particle size range and distribution

The particle size range provides the top and bottom size of material fed into the screen. It has influence on the screen deck construction and the apertures.

The particle size distribution is perhaps the most important factor which influences dewatering and this is especially so in the case of fines dewatering.

In order that effective dewatering may be achieved, the selection of an appropriate aperture and open area is of paramount importance.

In dewatering, the effectiveness depends greatly on the formation of a filter bed created initially by coarser particles of larger than aperture size or by bridging of smaller particles to arrest percolation of solids with the drainage fluid. For a successful operation, a certain percentage of larger than average size particles is always preferable. The size distribution will best illustrate the most suitable aperture size and subsequently the best capacity for the screen. Generally speaking, 40% larger than mean size particle is required in a 50:50% volumetric solids-to-liquid ratio.

8.2.2.3 Screen deck and aperture

The type of screen deck selected usually determines the effectiveness of the dewatering process and the amount of fines that can be expected to be lost with the drainage fluid. The characteristics of the various deck materials currently available are widely known and may have varying attractions for different applications. However, the most widely applicable surface is the wedge wire cross-flow deck, which is perhaps now regarded as being an ideal surface because the particles are exposed to the aperture for only a relatively short duration while fluid drainage occurs almost continuously. Wedge wire decks are perfectly flat with no protrusions to hinder the flow of the self-generating filter bed and can also be turned periodically to maintain good sharpness of separation. Woven wire cloth surfaces present restriction to the filter bed and fine material is lost as a result.

Polyurethane decks can be used if panel construction is sufficiently rigid to ensure steady flow while other materials offering similar capabilities, of which rubber is the most common, are also used for dewatering screen applications.

8.2.2.4 Open area

The open area determines the drainage capacity of the screen. Thus, the larger the open area, the greater will be the drainage capacity of the screen. However, in the case of wedge wire decks, open area is also affected by wire bar profile and selection would relate to specific applications. Normally, Sb or isorod profiles are chosen for dewatering because they exhibit minimal blinding tendency. (See *Figure 8.18* later in this chapter.)

8.2.2.5 Specific gravity of material

This is usually only a factor if the specific gravity of the solid and liquid are close. In such instances, and they do exist, there is a danger that the mechanical action of a screen will agitate the suspension to such an extent that a filter bed cannot form. Obviously, if this occurs, solids losses with the drainage fluid will be high. The normal solution in such cases is to use a fixed screen with fairly low flow rate.

8.2.2.6 Liquid–solids in volume of feed to the screen

The dewatering process is really a reversed wet screening process in that liquid is removed while solids are retained. As such, the concentration of each component has a great influence on the efficiency and, ideally, a feed with a sufficiently low fluid volume to permit drainage largely by capillary action is desired. For minus 0.5 mm pulps this amounts to about 30–40% fines but often this must be achieved by a densification stage occurring prior to screening. Hydrocyclones or scroll classifiers are commonly used for this task—see *Figure 8.5*. For example, where cyclones are used a cut-point of between 100 and 75 μm is common in mineral process and coal preparation applications, with the resulting slimes being routed to other forms of dewatering. This aspect is discussed in more detail in section 8.4.

8.2.2.7 Acceleration

Acceleration components occurring at feed end and discharge end have almost contradictory effects. Ideally, a low acceleration should occur at the feed end to avoid overcoming capillary action and creating loss of solids through the deck. At the discharge end, where most of the liquid has drained away leaving a well-formed coherent filter cake, higher acceleration leads to further moisture reduction. Such a combination, although an objective of many designs is rarely accomplished in any real practical sense and, to provide the compromise, dewatering combinations of fixed followed by moving screens are often applied.

8.2.2.8 Operating frequency, angle of throw and screen slope

In the dewatering application, frequency selected coincides with optimized transportation and drainage effect and minimum disturbance of the drainage bed. Commonly, electromagnetic screens are operated at quite high frequencies of between 5000 and 7000 cycles per second and relatively short strokes, us ally less than 3 mm. High-speed vibrator screens driven by out-of-balance vibrators are operated at low frequencies of between 1450 and 1800 cycles per second.

The rate of travel is a resultant of the angle of throw and acceleration but can also be influenced by inclination of the screen. Experience has shown that a 40–45 degree angle of throw to the horizontal gives good results for the out-of-balance screens, transporting at rates[5] between 0.2 and 0.25 m s^{-1}.

Inclination of the deck against the direction of flow can increase retention time and this is a common practice for certain types of screen. *Figure 8.6*

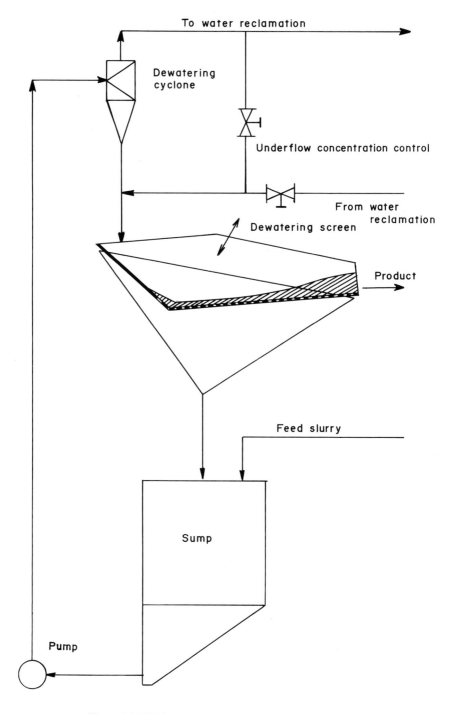

Figure 8.5. Thickening cyclone feed concentration system.

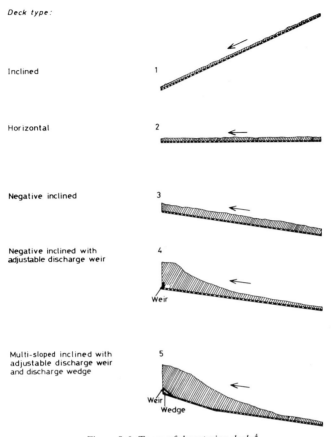

Deck type:

Inclined   1

Horizontal   2

Negative inclined   3

Negative inclined with   4
adjustable discharge weir

Weir

Multi-sloped inclined with   5
adjustable discharge weir
and discharge wedge

Weir
Wedge

Figure 8.6. Types of dewatering decks[4]

shows a variety of deck inclinations, all commonly applied in current
practice.

8.2.2.9 Width-to-length relationships

For most dewatering screen types the length is more or less fixed by the design
while the width determines capacity. Width is usually limited by mechanical
design and therefore unit capacity is usually small and necessitates numerous
units to satisfy total capacity requirements.

8.2.2.10 Drainage area required

The finer the feed the poorer the drainage as a result of the increase in surface
area, and it is difficult to quantify the extent to which material may be dried
without some form of empirical test to provide realistic data. Drainage
calculations are therefore usually based upon empirical formulae largely

developed for wedge wire decks with apertures below 0.5 mm and as such are greatly influenced by the relative proportions of undersize and oversize material in the feed pulp.

A typical formula is given below[6].

$$\text{Drainage area required (m}^2) = (0.25) \cdot C \cdot X \cdot (A/\rho_s)^{0.67}$$

where C is the solids feed rate (dry, t h^{-1}), X is the proportion of feed over mesh size, A is the surface-area factor, and ρ_s is the density of solids (t m^{-3}).

The surface factor (A) is calculated from the size analysis of the oversize solids. It is defined as

$$A = \Sigma (M_i/d_i)$$

where d is the mean particle size of fraction i and M is the proportion of solids in size fraction i.

Consider n grains of size d_i. Their total surface area is estimated as:

$$K_1 n d_i^2$$

where K is a shape factor. However,

$$n = \frac{\text{total mass of size } d_1 \text{ particles}}{\text{volume of one particle } \times \text{ density}}$$

$$= \frac{M_i}{K_2 \, \rho_s \, d_i^3}$$

Hence the total surface area of solids on a screen is given by:

$$\frac{K_1 M_i}{K_2 \rho_s d_i} = K \, (M_i/d_i) = KA$$

The open area of the screen will also influence the drainage capacity, particularly for certain types of screen surface and especially in the case of sieve bends. This phenomenon will be discussed more fully later when this type of screen is examined, but in the meantime the example given in *Table 8.1* will serve to demonstrate how the foregoing formulae are used.

8.3 SCREEN TYPES

Two basic categories of screen type exist. They are fixed or static screens and moving screens. Both are now described in more detail below.

8.3.1 Fixed screens

The two most common examples of this category lie, curiously enough, at the extreme ends of the size range of material treated in most processing circuits.

Table 8.1. DEWATERING COAL[6]: FEED = 120 t h^{-1} CONTAINING 73%
OVERSIZE ρ_s = 1.55 t m^{-3}

Screen fraction, mm	D, mm	% Weight	M, weight prop.	M/d
30 – 20	25.0	13.2	0.132	1.0053
20 – 10	15.0	19.9	0.199	0.0133
10 – 5	7.5	23.7	0.237	0.0316
5 – 3	4.0	14.1	0.141	0.0353
3 – 0.5	1.75	29.1	0.291	0.1663
3 – 1.5	2.25	13.5	0.135	0.0600
1.5 – 0.5	1.0	15.6	0.156	0.1560
				0.3015

Area$_1$ = (0.25) 120 (0.73) (0.2518/1.55)$^{0.67}$ = 6.48 m^2
Area$_2$ = (0.25) 120 (0.73) (0.3015/1.55)$^{0.67}$ = 7.31 m^2
Select the larger value to cater for the drainage requirements of the fines content.

8.3.1.1 The fixed grizzly

This screen consists of a set of parallel bars held apart by spacers at some predetermined opening, usually between 75 and 300 mm. The grizzly is commonly used before primary crushing in order to remove undersize material and in such cases it has an aperture of a little less than the setting of the crusher. It has no dewatering function.

8.3.1.2 Sieve bends

The first types of sieve bends were introduced by the Dutch State Mines (DSM) and are used both for dewatering suspensions and for producing an accurate classification from fine particle suspensions. In many instances sieve bends precede a vibrating screen. The conventional sieve bend comprises a curved wedge wire deck, mounted with wires at right angles to the direction of feed—see *Figure 8.7*. The feed to the screen, often a suspension of fine solids, is directed to the upper part of the curve and is distributed across the full width of the screen. Grains with a size approximately half of the spacing distance between each wedge pass through the slots into the underflow and separation is by classification with a centrifugal component assisting to distinguish fine from coarse.

The surface quality of the screen deck for such a separation is of prime importance. Impingement of the suspension on the edges of the wedge wire rounds the wires, and this results in the curious phenomenon of the size of grains screened out actually decreasing as the screen deck becomes worn. The situation may be restored by turning the deck around so that it is struck from the opposite direction. This allows a new edge to be presented while at the other end the blunt edge tends to become sharpened again. Depending on the material being treated, the screen may be turned round every few days or at longer intervals as deemed necessary.

The sieve bend comprises a wedge wire and curved screening surface with

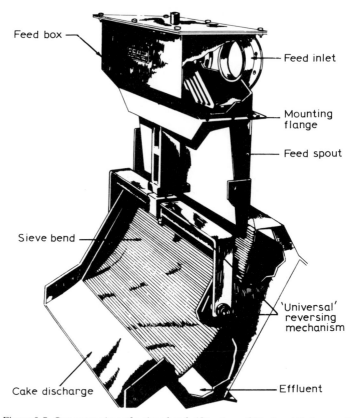

Figure 8.7. Cut-away view of a sieve bend. (Courtesy of Heyl and Patterson, Inc.)

an included angle which can range from 45° (as shown in *Figure 8.7*) to 300° for certain special applications. However, the range of 45–60° is most commonly used. Sieves in this range are widely applied as classification units but perhaps find greatest application as dewatering or drainage units where they can handle 8–10 m³ h⁻¹ of minus 0.5 mm coal fines slurry at 30–50% solids by weight for a 1-m wide unit. *Table 8.2* is a selection chart for different types and uses.

In many cases, vibrating screens are used in conjunction with sieve bends in drainage applications in order to cut down on the number of vibrating screens and to ensure effective dewatering prior to centrifugation. In other instances sieve bends operating in tandem provide extra drainage capacity at low operating cost.

The major factors affecting the sieve bend drainage capacity, as discussed earlier, are: the slurry density; deck inclination; aperture; and the open area of the screening surface. The density of the slurry affects drainage because of the effect which the fine particles in the slurry have upon viscosity. However, this is not usually of importance for a slurry unless a certain level of solids concentration is exceeded. For example, for coal slurries this value is 20% solids by weight; above this point the density markedly affects the drainage characteristics.

Table 8.2. SELECTION CHART FOR SIEVE BENDS

Angle, degrees	Capacity per metre of screen width, $l\,h^{-1}$	Size separation range, μm	Typical applications
45	5–100	2 000–200	Thickening mineral, organic and waste slurries
60	5–50	2 000–300	Sand, iron ore, copper ore, phosphates, steeped grain dewatering
120	15–60	150–40	Starch fibre washing, paper pulp fibre screening
270	30–60	300–40	Cement mill discharge at 65–70% solids
300	5–50	300–40	Raw sugar melt liquor, stillage liquors

The inclination affects the drainage by its influence upon the thickness of the slurry layer. If the screen surface is horizontal, the maximum attainable drainage capacity is achieved for a given opening size. This is due to the increase in static pressure which occurs as a result of increase in bed depth. As the surface becomes more and more inclined, the layer becomes thinner and the static pressure is reduced. Consequently, it is sometimes necessary to increase the static pressure by reducing the velocity of the slurry. This effect is achieved by the curvature of the sieve bend.

The drainage capacity is also affected by aperture size. Fine apertures tend to resist flow of liquids through them due to boundary conditions imposed by the opening. Friction at the edge of each opening comes into play more and more as the opening size is decreased. Normally this must be taken into account for openings smaller than 2.0 mm. For most sieve bends, openings larger than the separation size are used, the slot length being perpendicular to the direction of flow (i.e. cross-flow arrangement). The slot width may be up to twice the separation size on the steepest (first) section of the screen.

Another factor used in determining drainage capacity that is sometimes erroneously related to openings is that of open area. Similar types of screening materials (wedge profile wire, polyurethane etc.) are used for the conventional sieve bend/horizontal screen systems. The open area available for a given opening size is sometimes very different for the various screen surface materials. Drainage capacity is therefore not directly related to open area but is also influenced by the type of screen material selected. Empirical relationships are at present utilized in assessing this effect; some of these are given in *Table 8.3*.

A further development is the rapped sieve bend. Rapping the screen causes the profile wire to vibrate; the vibration dislodges particles that may be blinding the slot openings and so keeps the maximum open area available. The rapping mechanism is often air operated. Air is passed through a moisture trap and then to a relay valve activated by a solenoid, which in turn is governed by a variable timing unit. The valve periodically allows an air impulse to pass to a spring-loaded piston, which strikes a backing plate on the rear of the screen surface.

More recently, rapping devices have been replaced by high-frequency vibrators which perform with equal or better efficiency but which have a less harsh mechanical action and are thus less prone to mechanical failure and to causing screen damage.

Table 8.3. COMPARISON OF DESIGN CRITERIA FOR COAL SLURRY DESLIMING BY SIEVE BEND/VIBRATING SCREEN[7]

Design aspect	
Static sieve bend	
Water requirements	Up to 2.5 m³ h⁻¹ t⁻¹ of feed
Capacity	(a) 120 m³ h⁻¹ m⁻² for 34 Wb (i.e. for coarse
(750 mm radius bend 60° angle)	coal up to 50 mm)
	(b) 150 m³ slurry h⁻¹ m⁻² for 22 Wb (i.e. for
	small coal up to 19 mm)
Vibrating screen	
Water requirements	nil
Acceleration factor	3.5 – 5 g
Minimum length	3 m
Capacity	(a) 35 t h⁻¹ m⁻¹ width max.
	(i.e. for small coal up to 10 mm)
	(b) 40 t h⁻¹ m⁻¹ width max.
	(i.e. for coarse coal up to 50 mm)
Discharge moisture	18–20% for small coal up to 10 mm
Screen slope	nil

A recent modification to the DSM wedge wire bend is the use of a polyurethane surface instead of wedge wire. So far, application has been mainly confined to drainage screens in dense–medium circuits and, although the open area is less than for the wedge wire screen, medium losses have been lower. This is primarily due to the fact that polyurethane is less likely to blind and thus the percentage opening remains more or less unchanged in operation. The material also has excellent wear-resistant properties, which in many applications allow it to far out-last a stainless steel surface; it can thus be much more economical.

A conical version of the sieve bend was developed in the Polish coal industry and is now fairly widely used worldwide[8]. This screen, called the Vor-Siv, is shown in *Figure 8.8* and comprises a conical wedge wire screening surface. It has no mechanical parts and the feed is projected around the inner

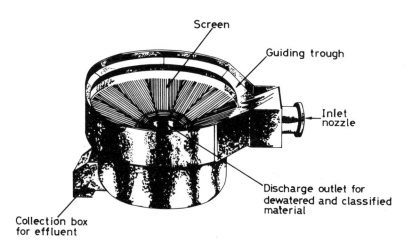

Figure 8.8. The Vor-Siv

periphery of the unit simply as a result of the pressure of the pumped feed. This channels the slurry into the circular raceway causing a stratification to commence. After the slurry has rotated around the raceway one to one and a half times, it loses enough of its energy so that it drops down on to the conical basket and the flow becomes a spiralling vortex. By the time the slurry reaches the bottom quarter of the basket, almost all of the fluid and ultra-fine material has filtered through the screen. The moist solids are discharged continuously throughout the central outlet while the effluent passes through a discharge outlet.

It is claimed that the 2 m diameter unit can handle 13–15 $m^3 h^{-1}$ of minus 0.5 mm coal fines, producing cakes with 35–50% moisture content, but the Vor-Siv is currently most commonly used as a classifier rather than a drainage screen.

8.3.2 Moving screens

8.3.2.1 Reciprocating or shaking screens

These may be either horizontal or inclined decks mounted on hangers or on rollers. They are normally operated by camshafts with eccentric bearings or by cranks and connecting rods. Such screens usually operate at relatively low speeds and have fairly long strokes, e.g.

slow: 150 rev. min^{-1} at 75–100 mm stroke;
fast: 200–300 rev. min^{-1} at 25–75 mm stroke.

Screens of this type are normally used as grading and/or drainage screens for relatively large-sized feeds. They find their greatest application in coal preparation.

8.3.2.2 Vibrating screens

Mechanically vibrated screens are mainly of two types:

high frequency: 500–2500 rev. min^{-1}; small stroke—less than 100 mm;
low frequency: 25–500 rev. min^{-1}; larger stroke—normally 15–30 mm.

High-frequency vibrating screens usually employ an out-of-balance weight arrangement to provide the vibration. Mechanically this can take several forms[9,10], of which the most common are:

1. Four-bearing type, employing an eccentric counterbalanced drive shaft. The stroke of these machines is predetermined by the amount of drive shaft eccentricity and cannot be changed except by substitution of the proper shaft for the desired stroke. In this application the counter-weights on each end of the shaft are used as balancers to provide smooth operation. Because of these balancing requirements no additions to the vibrated weight can be made without careful calculation of the transportation effect.
2. Two-bearing type. These are normally actuated by a concentric drive shaft which can be adjusted in operation, within the limits of bearing

capacity, by varying the amount of counterweight used, and by alteration of the speed. These vibrators produce a circular action and depend on gravity for their feed rate.

3. Horizontal screens actuated by off-centre shafts or weights geared together. These are also of the four-bearing type and the motion produced is virtually in a straight-line throw with a slight ellipse. The shafts or weights are geared together in a certain fixed relationship to produce a certain direction of action which can be varied between the vertical and the horizontal. The most commonly used throw is at 45° to the horizontal. Such screens are often employed where headroom is at a premium, or where sharp sizing, dewatering or feeding of a product is desired.

Positive stroke, four-bearing screens start and stop smoothly, while two-bearing units of any type of 'full-floating' screens have a characteristic stop-start bounce. However, from an application standpoint it may be said that, with only few exceptions, there is no job that a four-bearing type screen can do that a two-bearing unit, properly designed and applied, cannot do equally well if not better[11]. For the most part, the two types of unit have identical stroke characteristics, size for size, and the lump of material on the deck behaves more or less the same in each case.

Slowly vibrating screens are usually suspended on hangers or cables and are electrically driven through electric cams with little facility for adjustment. Such units are most commonly applied for fine solids or slurry dewatering.

Various methods of mounting are employed, such as massive springs or air cushions for floor mounting, or alternatively, suspension hangers, cables or springs. Examples of various commonly applied vibrating screens are shown in *Figure 8.9a–e*, and each figure is accompanied by a brief description of the different type of mechanism involved.

A typical example of a vibrating drainage screen is the elliptex dewaterer shown in *Figure 8.10*. The elliptical feeder action of the entire dewaterizer frame squeezes the material being screened against a series of 70° sided dams. The dams, of height 30 mm and about 1 m apart along the screen, serve to retard the flow of the lower strata of the bed, so allowing prolonged dewatering periods for the lower regions of the flowing bed. The conveying motion causes the material to travel uphill and over the dams allowing a dilation to occur which allows further release of water in a constant drainage process. An example of the effectiveness of the unit is reported for a coal feed ranging from 3 mm down to 0 being discharged with 12.5% surface moisture even after being flooded with washing sprays at the feed end. The deck is usually wedge wire. The normal feed range in dewaterizer applications is from 20 mm down to 0.5 mm but the concept of deep bed drainage (beds 30 mm thick or more) allows a larger proportion of plus 200 μm fine particles to be recovered in the bed in certain applications. Normally in such cases, the upper size would not exceed 10 mm.

These screens are most commonly used for drainage often preceded by a fixed screen or cross-flow sieve bend. Such screens would be fitted with apertures smaller than about 80% of the feed solids.

A 4.9 m long by 1.8 m wide screen dewatering > 0.5 mm coal fines would handle approximately 100 t h^{-1} of solids if preceded by a 1 m wide by 45°

sieve bend and assisted by spraying as shown in *Figure 8.10.* The capacity criteria given in *Table 8.2* would apply to this application.

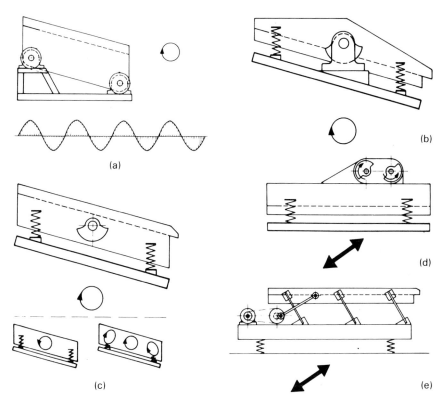

Figure 8.9. (a) Mechanically controlled, double-shaft screen. Screen movement: circular. The acceleration pattern has the form of a sine curve. (b) Mechanically controlled, single-shaft screen. Screen movement circular but may vary in the vertical plane at the edges of the screen. Acceleration pattern: sine curve. (c) Free-swinging screen with single-shaft vibrator. Screen movement: normally a circular or mainly circular movement in the direction of the feed but can also move against the direction of the feed. Alternatively, the movement can be circular at the shaft and elliptical at the ends of the screen. Acceleration pattern: sine curve. (d) Free-swinging screen with double-shaft vibrator. Screen movement: linear. Acceleration pattern: sine curve. (e) Counterweight screen. Screen movement: linear. Acceleration pattern: sine curve

8.3.2.3 Resonance screens (*Figure 8.11*)

In most other types of screens quite a lot of energy is used up in vibrating the screen, i.e. they require high-power motors. Because of continual changes in the direction of motion, much of the energy is wasted. Resonance screens, however, have been designed to save as much as possible on energy requirements and consequently lower power drives are used. The unit comprises a 'live' screen supported by flexible hangar strips and a balance frame three or four times heavier than the screen. Movement is imparted to the screen by an

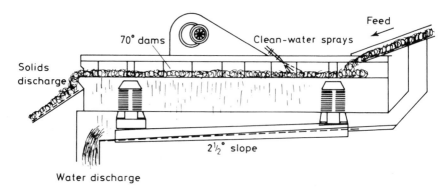

Figure 8.10. Elliptical dewaterizer. (Courtesy of Eimco)

eccentric drive and connecting rod. A rubber pad connects the frame to the screen.

Rubber trippers restrict the movement of the screen and serve to store up energy which is re-imparted to the 'live' frame. Hence, any movement given to the screen is transmitted to the balance frame which stands on rubber pads. The throw of the balance frame is less than that of the screen because its mass is so much greater.

Any motion thus given to the balance frame sets up vibrations which, instead of being wasted, are imparted back to the screening frame. Energy loss is therefore minimal.

In certain types of resonance screen, the deck movement is restricted by a series of stops which are attached to the deck and which operate between buffers attached to the frame. In addition to storing energy, the sharp return motion of the deck imparts a lively action to the deck and hence promotes good screening. This type of screen is frequently applied to dewatering a wide variety of materials ranging in size from about 50 mm down to 0.5 mm.

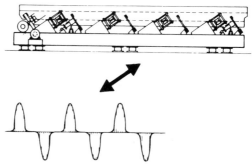

Figure 8.11. Resonance screen. Screen movement: linear. The acceleration pattern shows the characteristic peaks associated with resonance screens. These peaks are caused by rapid changes of acceleration in the end positions of the stroke

8.3.2.4 Electromechanical and high-frequency electrical screens

Electromechanically operated screens operate with a high-frequency motion of very small throw. The motion is usually created by a moving magnet which strikes a stop. There are other types in which an electromagnet attracts an armature (in this case impact is avoided because of the obvious likelihood of damage to the armature).

While the principle is currently widely applied to vibrating feeders, it is not often used for sizing screens and only applications to fine-particle screening and dewatering are likely to be encountered.

Two common examples are the Fordertechnik and Derrick types of dewatering system shown in *Figures 8.12* and *8.13*. They differ in both drive type and drainage surface employed. The Fordertechnik screen employs dual high-frequency electromagnetic vibrator mechanisms mounted above the screen frame which causes a cake bed to form and travel uphill to the discharge point. The screening medium is wedge-wire plate. The Derrick screen employs a high-frequency, out-of-balance rotor mechanism mounted above the screen which dilates the cake bed and causes it to travel downhill to the discharge point. The screening medium is slotted rubber. The Fordertechnik screen is shown in *Figure 8.14* in two circuits having a different type of pre-thickening stage before the screen. The screen surface is a number of wedge wire cross-flow panels which usually increase in aperture from the feed to discharge ends. Side panels at the feed end of the screen assist primary drainage.

The application of these dewatering screens is primarily in handling relatively coarse granular material, having an ideal minimum top size of 0.5 mm. It is preferable to have granular material of an even coarser size, say 2–3 mm in the feed. The bottom size does not present any great difficulty, providing that there is a reasonable distribution of coarse and fine material.

In practice, it is possible to go down to about 75 μm by preceding the screen with a classifying cyclone making a fine cut and then dewatering the thickened underflow product on the screen and recycling the screen underflow product on the screen and recycling the screen underflow product back to the cyclone feed. As an alternative for a slightly coarser feed, a spiral classifier may be used instead of the cyclone. The screening surface is usually stainless steel wedge wire or polyurethane panels.

In dewatering minus 0.5 mm coal fines this screen can produce cakes with moistures 20–30%. The negative slope of screen, which is approximately 5%, can be augmented by the addition of a discharge weir (as shown in *Figure 8.6*) to produce the lowest moisture contents. Generally, a discharge bed depth of 200–300 mm can be accepted with such a system. The Derrick Vacu-Deck screen is shown in *Figure 8.13*. The screening surface normally used is of slotted rubber. The combination of the Derrick high-frequency (3600 cycles per second) vibrating machine and this dewatering surface gives excellent dewatering capability. An intense patting motion helps to draw fluid rapidly down through the filter bed while, simultaneously, a horizontal flexing action of the rubber extracts draining fluid as the cake proceeds down the screen. The dewatering capability of the screen is further enhanced (up to an additional 20% of moisture removal) by means of the introduction of a negative pressure on the final dewatering section.

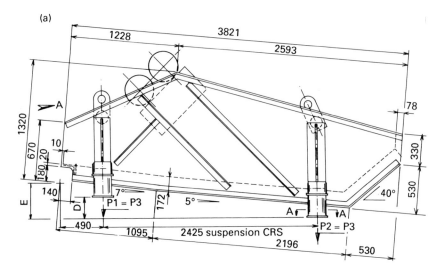

(a)

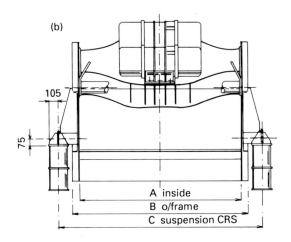

(b)

A inside
B o/frame
C suspension CRS

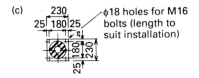

(c) 230
25 180 25
ϕ18 holes for M16
bolts (length to
suit installation)

Figure 8.12. Drawing of the Fordertechnik dewatering screen: (a) side elevation; (b) view on arrow A; (c) section A-A. (Courtesy of Tema-Siebechnik)

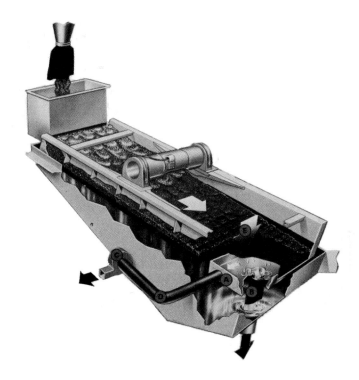

Figure 8.13. Derrick Vacu-Deck screen. A—underflow collection chamber. B—Rubber automatic underflow regulator valve. C—2.25-3.75 kW industrial blower. D—Suction chamber. E—Direction of air flow. (Courtesy of Derrick)

The flexing action of the 12.5mm thick rubber deck, besides providing for drainage, also resists pegging and blinding of trapped near-aperture size material or dry fines accumulating as a result of periodic shut down.

A Vacu-Deck type screen 3 m long and 1.2 m wide will handle about 30 t h^{-1} and produce a cake with a moisture content in the low 20s. This type of screen can also be effectively used in conjunction with either a thickening cyclone or classifying cyclone, forming part of a complete system similar to that shown in *Figure 8.15*.

The Linatex-Derrick screen shown in this circuit has a special sectioned deck with each section inclined at an increasing angle as shown in *Figure 8.16*. This multi-slope feature enhances flow potential. Initially, the relatively flat surface permits fluid to drain freely from the flowing slurry. As more and more moisture is removed the solid bed develops a greater resistance to flow. The increased slope therefore maintains velocity and bed thickness. At the same time, the finer material accumulates at the bottom of the bed and causes further resistance to flow which is compensated for by further increase in slope.

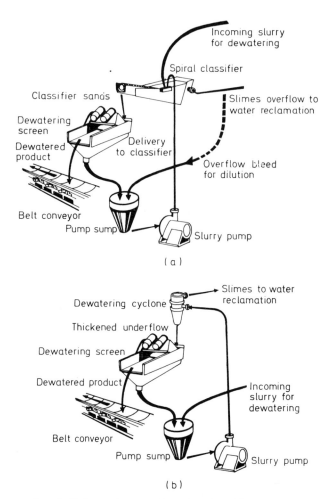

Figure 8.14. Fordertechnik dewatering screen in conjunction with (a) a spiral classifier, and (b) a desliming cyclone

8.3.2.5 Revolving screens

Commonly referred to as 'trommel-type' screens, these consist of perforated drums, often slightly conical in shape. It is usual for trommels of varying aperture to be arranged concentrically and rotated about an axis which is slightly inclined. Alternatively, they can be aligned in order to discharge undersize particles into successive trommels. Unfortunately the disadvantages of using trommels far outweigh the advantages. They are inefficient, create excessive breakage of soft materials, are costly and difficult to

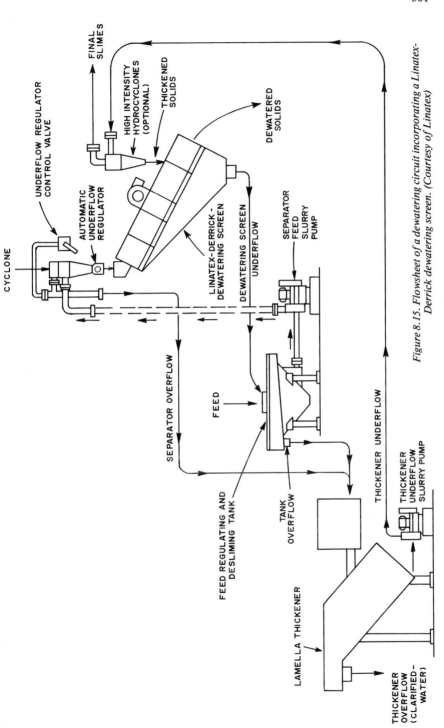

CYCLONE

UNDERFLOW REGULATOR
CONTROL VALVE

FINAL
SLIMES

HIGH INTENSITY
HYDROCYCLONES
(OPTIONAL)

THICKENED
SOLIDS

AUTOMATIC
UNDERFLOW
REGULATOR

DEWATERED
SOLIDS

LINATEX-DERRICK-
DEWATERING SCREEN

DEWATERING SCREEN
UNDERFLOW

SEPARATOR
FEED
SLURRY
PUMP

SEPARATOR OVERFLOW

FEED

FEED REGULATING AND
DESLIMING TANK

TANK
OVERFLOW

THICKENER UNDERFLOW

THICKENER
UNDERFLOW
SLURRY PUMP

LAMELLA THICKENER

THICKENER
OVERFLOW
(CLARIFIED-
WATER)

Figure 8.15. Flowsheet of a dewatering circuit incorporating a Linatex-Derrick dewatering screen. (Courtesy of Linatex)

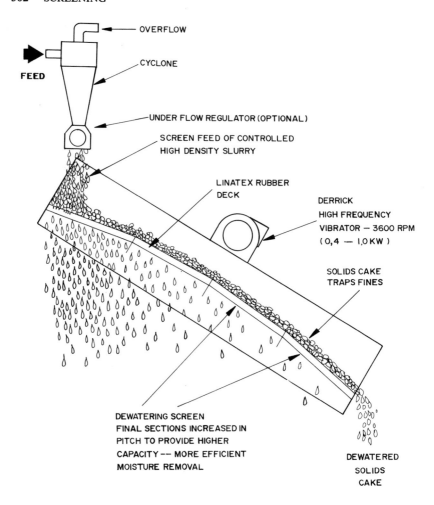

Figure 8.16. Diagram of a Linatex-Derrick dewatering screen

maintain, and are space-consuming and cumbersome. They have more or less fallen into obsolescence, although some attempts to utilize them as 'scrubbers' for removing fines and colloidal clay material have been reported recently.

8.4 SCREEN DECK MATERIALS

The apparent simplicity of choosing screening surfaces is often quite deceiving. It is true that the primary factors—the required aperture size and percentage of open area—are in most cases easily ascertained. Yet even here, decisions are not specified simply; the user's judgement and experience are always required. To this are added the frequent necessity of choosing the shape of opening, the most suitable weave or hole pattern and the construction material. Fortunately, there are standards that all manufacturers accept; the catalogues of most wire-cloth manufacturers exhibit their own particular products in a way that is easily understood in terms of the industry standard. Even the patented specialities developed by some makers are described in relation to those standards and this lends a degree of uniformity to specifications and ordering procedures that is lacking in the case of the screens into which they are to be fitted. Industrial screens may use woven wire (square or elongated holes) or perforated plates (round, square or slotted holes). There are many national standards which cover such screening media, e.g. BS 481 (parts 1 and 2), BS 1669, DIN 4192, 24041, 24042, 24043, ASTM E 454, to name but a few. Furthermore, textile screening cloths (bolting cloths), e.g. BS 1812, are used in the flour trade. Woven wire screen cloths, usually constructed from carbon and stainless steel, are by far the most commonly used screening surfaces, and they account for perhaps 80% of normal applications. About 15% of screen decks are constructed of closely spaced bars; these are usually wedge shaped although one special type of parallel bar deck uses rods. Steel rods are held loosely at each end in frames called 'combs', within which they are free to rotate as a result of the motion given to the whole screen deck. The motion of the rods prevents build-up of bridges of fine particles between the rods, hence more or less eliminating blinding. Applications are usually confined to the screening of damp fine particles and the aperture is normally of the order of 5–10 mm. A wedge wire screen deck consists of wedge-shaped stainless steel bars, which taper from about 2 mm wide to about 1 mm over a thickness of about 4 mm. Apertures range from about 1 mm to below 0.25 mm. The remaining 5% of screen decks are plates of metal or rubber perforated with square or round holes. One form of perforated surface, which is becoming increasingly used because of its exceptional resistance to wear, is polyurethane. Such screens are usually duroplasts of interlaced polyurethanes and thermoplasts of linear polyurethane. They often have embedded reinforcements of commercial mild steel or in some cases stainless or spring steel in order to facilitate tensioning and to provide the necessary intrinsic strength.

Another commonly used non-metallic surface is wear-resistant hard rubber. Such surfaces are applicable over a wide range of apertures from 300 mm down to below 20 mm. However, all non-metallic surfaces appear to have the common disadvantage of decreasing percentage open area with decreasing aperture.

Figure 8.17 gives the open area relationships for various types of screen surfaces.

As an example, the percentage open area for a woven-wire 10 mm square aperture screen constructed from 2.25 mm diameter carbon steel wire is (using the formula $F_{oa} = 100\,[a/(a+d)]^2$)

$$F_{oa} = \left(\frac{10}{10 + 2.25}\right)^2 \times 100$$
$$= 67$$

In practice, the open area can vary from 30 to 80% of the screen; such fine apertures, typical of dewatering application, have low open areas. *Figure*

Aperture	Formula
1. Rectangular opening 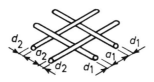	$$F_{oa} = \frac{a_1 a_2}{(a_1 + d_1)(a_2 + d_2)} \times 100$$ where F_{oa} is the open area, expressed in % d is the diameter of the wire (or the horizontal width of a bar (for plates)) a is clear opening or aperture.
2. Square opening (equivalent to a rectangular opening with $a_1 = a_2$ and $d_1 = d_2$)	$$F_{oa} = \left(\frac{a}{a + d}\right)^2 \times 100$$
3. Square opening expressed in terms of the mesh size m	$$F_{oa} = a^2 m^2 \times 100$$ since $m = \dfrac{1}{a + d}$
4. Wedge or parallel opening 	$$F_{oa} = \frac{a}{a + d} \times 100$$
5. Sloping screen For a sloping screen the effective aperture is the horizontal projection of the actual screen aperture.	$A_{horiz} = a_s \cos \alpha$ where a_s is the slope aperture and α is the slope angle.

Figure 8.17. Open area factors

8.18 shows a comparison of wedge profile wire types commonly used in dewatering screen applications.

8.5 SCREEN PERFORMANCE

The operation of a dewatering screen usually incorporates limited control of the actual screen functions. As can be seen from the screen-design notes included earlier, once a screen of a particular size and type has been installed, it is mainly constrained in its future performance by its physical size and its vibration mechanism. This is particularly evident in the case of dewatering screens, especially fixed-sieve type screens, which makes it all the more important to establish the design criteria properly *before* the screen is finally designed and installed.

In evaluating performance it is usually necessary to design and conduct a testing programme that will take into account the normal variability of day-to-day operation. To establish such a programme, it is advisable first to review operational data collected for the circuit(s) over a reasonable period of

Aperture size (slot width), mm

		0.315	0.4	0.5	0.63	0.8	1.0	1.25
Looped profile wire sieve	Open area: a = 1.8 mm, b = 3.1 mm, slot length = 70 mm	12.4	16.6	19.8	23.7	28.1	32.6	37.4
	Open area: a = 2.2 mm, b = 3.8 mm, slot length = 70 mm	10.4	13.8	16.6	20.0	24.0	28.1	32.6
	Open area: a = 2.5 mm, b = 4.5 mm, slot length = 70 mm	9.2	12.4	15.0	18.1	21.8	25.7	30.0
	Open area: a = 3.0 mm, b = 5.5 mm, slot length = 70 mm	—	—	—	15.3	18.6	22.0	36.9
Welded profile wire sieve	Open area: a = 1.5 mm, b = 4.0 mm, slot length = 40 mm	17.4	21.0	25.0	29.6	34.8	40.0	45.4
	Open area: a = 2.2 mm, b = 4.5 mm, slot length = 50 mm	12.5	15.4	18.5	22.3	26.7	31.2	36.2
	Open area: a = 2.8 mm, b = 5.0 mm, slot length = 50 mm	10.1	12.5	15.2	18.4	22.2	26.3	30.9
	Open area: a = 3.4 mm, b = 8.5 mm, slot length = 50 mm	8.5	10.5	12.8	15.6	19.0	22.7	26.9

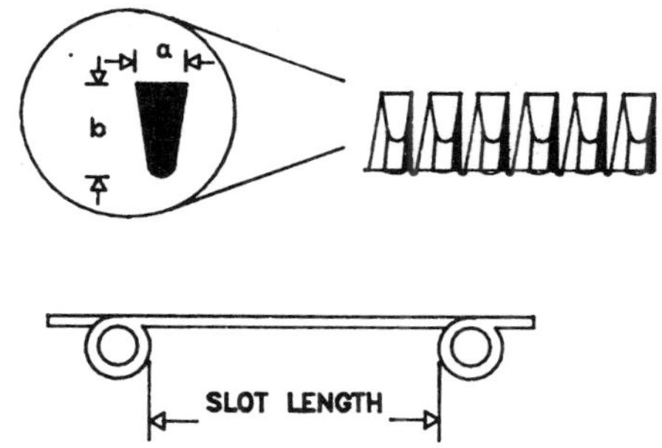

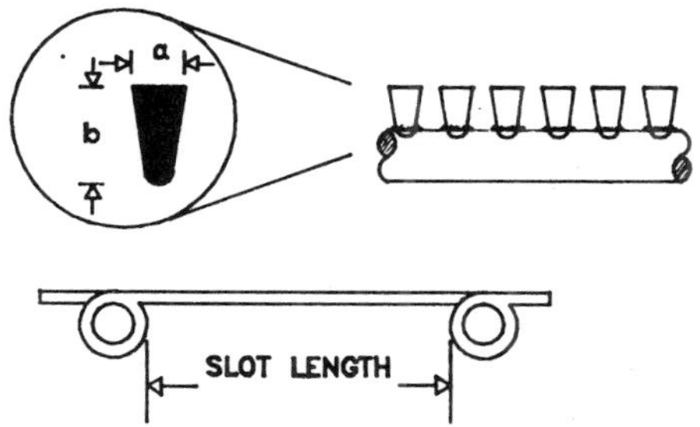

SLOT LENGTH

Figure 8.18 Comparison of wedge profile wire data[7]

time (2–3 months) for established installations. This will enable working ranges in operating conditions to be ascertained, e.g. feed solids content, mass flow rate, temperatures etc.

Once a performance evaluation programme has been decided upon, emphasis must then be placed upon obtaining samples which are truly representative of the feed and products. This may mean installation of sampling equipment or improving access to ensure correct sampling procedures. The data obtained are only as good as the sample material from which they originated. Sampling guidelines and specifications are available in numerous international standards for various granular materials. For example, ISO 1988 deals with coal sampling and includes details of sampling procedures, sampling equipment and analysis of results.

Specialized equipment for slurry sampling are available and some examples of wet-sample cutters are shown in *Figure 8.19*. Other types allow on-line sampling from a pipe. Most slurry samples take increments that are stored in a slurry which may later be re-agitated prior to sample reduction steps. There are also other types which include preparatory stages incorporating sample reduction and dewatering.

The following data require collection to enable performance evaluation of a dewatering screen:

1. Screen installation details:
   screen deck and aperture;
   open area;
   operating frequency, angle of throw, screen slope;
   acceleration;
   screen dimensions, width–length ratio etc.

2. Operating conditions:
   screen condition, pegging or blinding observed;
   flowsheet considerations (other process impacts?);
   ambient conditions (temperature, vibration etc.);
   reagents used.

3. Feed and product data:
   feed size range and distribution;
   product size range and distribution;
   solids content of the feed;
   density of the solids;
   other physical properties of the solids (type, shape etc.);
   density of the fluid;
   other properties of the fluid (type, evaporation etc.);
   flow rate of slurry.

4. Calculated results:
   solids flow rate;
   moisture content of the product;
   solids loss either per cycle (circulating mass) or for the total system;
   filtrate solids content.

5. Subsequent action:
   media change;
   operating condition change;

reagent dosage change;
pre-screen process change of classifier setting and feed solids concentration.

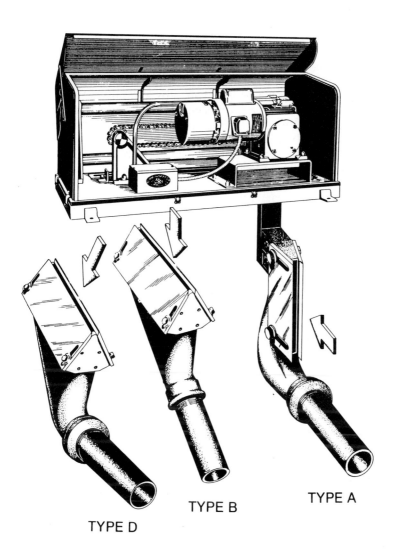

TYPE D

TYPE B

TYPE A

Figure 8.19. Slurry sampling equipment. Type A — with a vertical cutter opening for taking a
sample from the discharge of a horizontal stream; Type B — with horizontal cutter opening for
taking a sample from a free-falling vertical stream; Type D — with an inclined cutter opening for
taking a sample from the discharge of an inclined stream. (Courtesy of Ramsey)

8.6 COST OF SCREENING EQUIPMENT

The cost of vibrating or electromagnetic screening equipment is usually a function of three major components.

1. screen frame plus supports;
2. screen deck;
3. vibrating mechanism plus drive.

Each component is usually standardized to ensure competitive costs amongst the manufacturers of each specific type. Hence, it is usually possible for a manufacturer to estimate the cost of a new installation accurately once basic criteria have been determined. A typical questionnaire used for design and cost estimation by screen manufacturers is shown in *Figure 8.20*.

Frame and vibrator costs can be estimated approximately from the following cost formula:

$$\text{Capital Cost (US\$ 1985)} = 8000\ W^{0.8} I^{0.4}$$

where W is the screen width and I is the screen length (both in metres). Other similar forms of cost estimation have been published[1].

Accuracy in estimating screen-deck costs depends on detailed specification, in particular of open area and wire type (see *Figure 8.18*), but approximate costs can be obtained from the following:

stainless steel   US$ (1985) 110 per m^2
rubber   US$ (1985) 340 per m^2
polyurethane   US$ (1985) 600 per m^2

Equipment technical specification No. ....
Item No. ....: Slurry dewatering screen

1. *Duty*

   For dewatering slurry containing fine solids size below 0.5 mm and ranging in solids concentration from 35–45% solids by weight.

   The equipment shall be capable of continuous operation, 24 h per day, 7 days per week.

2. *Operating conditions*

   Feed (to each unit)

Received from	Water only cyclone Overflow after sizing in cyclones and sieve bend	
Size	0.5 mm × 0.15 mm	
Nominal feed rate:	Solids (t h^{-1}) (metric)	16
	Water (m^3 h^{-1})	15
Maximum feed rate:	Solids (t h^{-1}) (metric)	17.5
	Water (m^3 h^{-1})	16
Specific gravity of solids		1.45

   Discharge

Surface moisture content of product (max.)	27.8 %
Surface moisture	26.0% or less

   Consideration will be given to suppliers able to guarantee lower moisture than above specified.

3. *Performance requirements and equipment warranty*

The equipment shall be warranted to perform in accordance with the operating conditions of this specification.

4. *Equipment information summary*

The following information will be required for each screen:

Equipment No.: ....

Type of screen: ....

Proposed model: ....

Service: ....

Screen width: ....

Screen length: ....

Screen deck material: ....

Screen opening (size): ....

Screen opening expressed as per cent of total screen area: ....

Vibration frequency: ....

Vibration amplitude: ....

Vertical acceleration component: ....

Method of motor mount: ....

Screen rate of travel: ....

Type of vibrating mechanism: ....

Material of construction: Frame: ....

                                           Screen: ....

                                           Housing: ....

Drive motor type: ....

            motor speed: .... rpm, Power: .... kW

Drive motor starting torque: ....

Drive motor special features: .... kg

Weight of the total screen:

     Heaviest piece: ....

     Longest piece: ....

Dimension of the total screen: Length .... mm

                                            Width: .... mm

                                            Height: .... mm

The largest piece: Length: .... mm

                         Width: .... mm

                         Height .... mm

Exception to specification: ....

                                    ....

                                    ....

Alternatives: ....

                   ....

                   ....

Guaranteed screening efficiency: ....

Guaranteed surface moisture of product: ....

Existing operations using similar sized equipment:

(year of installation and operations contract)

....

....

Nearest major spare parts inventory: ....

                                   ....

Nearest service personnel location: ....

                                   ....

Technical assistance included: ....

                                   ....

Figure 8.20. Screen selection questionnaire

A further comparison of costs between stainless steel and polymer decks is given in Table 8.4. This includes operating costs of both types for which there is also a significant difference.

Table 8.4. EXAMPLE OF SCREEN-DECK COSTS[6]

	Cost, £ (sterling)	
	Wire	Polymer
Purchase price per deck	324	3348
Changes of deck per year	14	0.5
Total yearly cost for media including fitting costs ($3\,h \times 2\,men \times 10\,h^{-1}$)	4536	1574
Total yearly costs	5376	1634

However, it should be kept in mind that with decreasing aperture size the open area corresponding to each alternative becomes significantly larger in the case of steel wire which, in turn, can have an increasing impact on performance. Consequently, in fine-aperture applications (below 0.5 mm) metal media are still the ones most widely used.

REFERENCES

1. Mular, A. L., 'Mineral processing equipment costs', *Can. IMM.*, **18** (1978)
2. Osborne, D. G., 'Screening and classification'. In *Coal Preparation Technology*, Chap. 4, pp. 114–140, Graham and Trotman (1988)
3. Brereton, T. and Dymott, K. R., 'Some factors which influence screen performance', *Proc. 10th Int. Min. Cong. Trans.*' IMM, London (1976)
4. Norton, G. and Hambleton, G., 'Cleaning to zero seldom pays', *Coal Age*, **88**, 5055 (1983)
5. Ohl, E., Private communication regarding dewatering screen design (1980); 'New dimensions for vibrating screens', *Coal Mining Proc.*, **Sept.**, 80–82 (1979)
6. Gochin, R. J., Private communication regarding screen design (1988)
7. Leeder, W. R., Hogg, J. W., Jacobs, E. M., and Osborne, D. G., 'Application of high capacity multi-slope screens for coal desliming applications in heavy media plants', *Proc. 10th Int. Coal Prep. Congress* (Edmonton, Alberta, 1986), Paper 4.3, pp. 300–321
8. Anon, 'Screens — new ideas and new machines', *Mining Mag.*, **Oct.**, 229–319 (1980)
9. Brown, G. G., 'Screening'. In *Unit Operations*, pp. 9–24, Wiley, New York (1955)
10. Gluck, E. S., 'Vibrating screens: surface selection and capacity calculations', *Chem. Eng.*, 179–184 (1975)
11. Elder, J., 'Achieving top performance from vibrating screens', Part 1 — Installation, operation and testing; Part 2 — Lubrication, repair, trouble-shooting, *Coal Age*, **March** (1966)

9

Filtration Fundamentals

L. Svarovsky

Department of Chemical Engineering, University of Bradford

NOMENCLATURE

a	Constant
a_1	Constant
A	Face area of a filter
b	Constant
b_1	Constant
c	Solids concentration in the feed
f	Submergence or filtration time/cycle time
K	Permeability of a packed bed
K_o	Kozeny constant
L	Cake thickness
L_f	Final cake thickness
m	Mass of wet cake/mass of dry cake
n	Exponent
Q	Volumetric flow rate
Q_1	Flow rate at the start of a constant pressure period
R	Medium resistance
R_c	Cake resistance
S_o	Volume-specific surface of the bed
t	Time
t_c	Cycle time
t_s	Time at the start of a constant pressure period
v	Superficial velocity in a packed bed
V	Filtrate volume collected
V_f	Final volume of filtrate per frame
V_s	Filtrate volume at the start of a constant pressure period
w	Mass of cake deposited per unit area
α	Specific cake resistance
α_{av}	Average specific cake resistance
α_o	Cake resistance per unit pressure drop
Δp	Static pressure drop
Δp_c	Pressure drop of the cake

Δp_m Pressure drop of the medium
ϵ Voidage of a packed bed
μ Liquid viscosity
ρ Liquid density
ρ_s Solid (particle) density

9.1 INTRODUCTION

Filtration may be defined as the separation of solids from liquids by passing a suspension through a permeable medium which retains the particles. A filtration system can be shown schematically as in *Figure 9.1*.

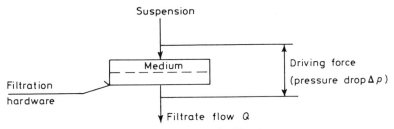

Figure 9.1. Schematic diagram of a filtration system

In order to obtain fluid flow through the filter medium, a pressure drop Δp has to be applied across the medium; it is immaterial from the fundamental point of view how this pressure drop is achieved but there are four types of driving force:

gravity,
vacuum,
pressure,
centrifugal.

Before introducing the basic filtration relationships it is worthwhile examining the actual process of particle removal.

There are basically two types of filtration used in practice: the so-called *surface filters* are used for *cake filtration* in which the solids are deposited in the form of a cake on the up-stream side of a relatively thin filter medium, while *depth filters* are used for *deep bed filtration* in which particle deposition takes place inside the medium and cake deposition on the surface is undesirable.

In a surface filter, the filter medium has a relatively low initial pressure drop and, as can be seen in *Figure 9.2*, particles of the same size as, or larger than, the openings wedge into the openings and create smaller passages which remove even smaller particles from the fluid. A filter cake is thus formed, which in turn functions as a medium for the filtration of subsequent input suspension. In order to prevent blinding of the medium, filter aids are used as a precoat which forms an initial layer on the medium. Some penetration of fine solids into the precoat or the medium itself is often inevitable.

Surface filters are usually used for suspensions with higher concentrations of solids, say above 1 % by volume, because of the blinding of the medium

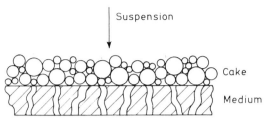

Figure 9.2. Mechanism of cake filtration

(or of the precoat) that occurs in the filtration of dilute suspensions. This can, however, sometimes be avoided by an artificial increase of the input concentration, in particular by adding a filter aid as a 'body feed'; as filter aids are very porous their presence in the cake improves permeability and often makes cake filtration of dilute and generally difficult slurries possible.

Note that the model described above and shown in *Figure 9.2* is that of conventional batch cake filtration where both the particles and the liquid approach the medium at an angle of 90° and no attempt is made to disturb the cake or prevent its formation. There is an alternative, which Tiller and Cheng[1] call 'delayed cake filtration', when the cake is prevented from forming or kept thin by hydraulic or mechanical means; the solids are thus continuously stirred back into the suspension, which gradually thickens. The effective particle motion is then parallel to the medium while the liquid approaches the medium at an angle. Continuous filter thickeners based on this principle are available; Tiller and Cheng[1] have demonstrated that appreciably higher filtration rates and lower cake porosities can be obtained by using mechanical agitators and have proposed a mathematical model for the delayed cake filtration process. A similar mechanism is used in the so-called by-pass centrifugal filtration where particles are removed from the medium by centrifugal forces while the liquid flows through it; this is further discussed in chapter 14 'Centrifugal Filtration'. Chapter 18 reviews the methods available for limiting cake growth.

In a depth filter—*Figure 9.3*—the particles are smaller than the medium openings and hence they proceed through relatively long and tortuous pores where they are collected by a number of mechanisms (gravity, diffusion and inertia) and attach to the medium by molecular and electrostatic forces.

The initial pressure drop across the depth filter is generally higher than that across a surface filter of comparable efficiency but the build-up of pressure drop as particles are collected is more gradual for a depth filter. Depth filters are commonly used for clarification, i.e. for the separation of fine particles from very dilute suspensions, say less than 0.1% by volume.

Although the above classification of filtration into cake and deep bed filtration is clear cut in most cases, in some instances, such as with some cartridge filters, it may be difficult to decide which of the two is the governing process.

Of the two types of filtration, cake filtration has the wider application, particularly in the chemical industry (because of the higher concentrations used) and the following discussion will, on the whole, be concerned with cake filtration and surface filters. Deep bed filtration is discussed further in chapter 11.

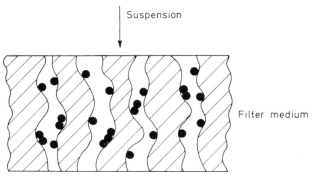

Figure 9.3. Mechanism of deep bed filtration

As with other separation equipment, the main characteristics of filters are the flow rate–pressure drop relationships and other performance characteristics such as the separation efficiency. In filtration however, these relationships are more complex as there are many variables and factors (cake thickness, mass of cake per unit area, specific cake resistance etc.) which greatly influence the process.

9.2 FLOW RATE–PRESSURE DROP RELATIONSHIPS

9.2.1 Clean medium

At the beginning of batch cake filtration, the whole pressure drop available (i.e. the driving force) is across the medium itself since as yet no cake is formed. As the pores in the medium are normally small and the rate of flow of filtrate is low, laminar flow conditions are almost invariably obtained.

Darcy's basic filtration equation relating the flow rate Q of a filtrate of viscosity μ through a bed of thickness L and face area A to the driving pressure Δp is

$$Q = K \frac{A \Delta p}{\mu L} \tag{9.1}$$

where K is a constant referred to as the permeability of the bed. Equation 9.1 is often written in the form

$$Q = \frac{A \Delta p}{\mu R} \tag{9.2}$$

where R is called the medium resistance (and is equal to L/K, the medium thickness divided by the permeability of the bed).

If the suspension were a clean liquid, all the parameters in equations 9.1 and 9.2 would be constant, resulting in a constant flow rate for a constant pressure drop and the cumulative filtrate volume would increase linearly with time, as shown in *Figure 9.4*.

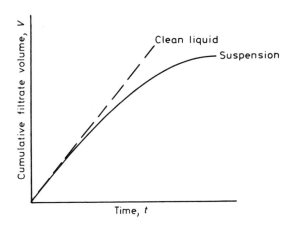

Figure 9.4. Plot of cumulative filtrate volume against time

In batch filtration, however, where the suspension contains particles, a cake starts to build up on the surface of the filter so that, gradually, a greater proportion of the available pressure drop is taken up by the cake itself. This results in an effective increase in the bed resistance thus leading to a gradual drop in the flow rate Q. The cumulative filtrate volume therefore slows down with time as shown in *Figure 9.4*.

9.2.2 Medium with a cake forming on its face

As explained in the previous section, the filtrate flow rate at constant driving pressure becomes a function of time because the liquid is presented with two resistance in series, one of which, the medium resistance R may be assumed constant and the other, the cake resistance R_c increases with time.

Equation 9.2 then becomes:

$$Q = \frac{A\Delta p}{\mu(R + R_c)} \tag{9.3}$$

In practice, however, the assumption made above that the medium resistance is constant is rarely true because some penetration and blocking of the medium inevitably occurs when particles impinge on the medium.

As the resistance of the cake may be assumed to be directly proportional to the amount of cake deposited (only true for incompressible cakes) it follows that

$$R_c = \alpha w \tag{9.4}$$

where w is the mass of cake deposited per unit area (in kg m^{-2} in SI units) and α is the specific cake resistance (in m kg^{-1} in SI).

Substitution of equation 9.4 for R_c in equation 9.3 gives

$$Q = \frac{\Delta p A}{\alpha \mu w + \mu R} \tag{9.5}$$

Equation 9.5 relates the flow rate Q to the pressure drop Δp, the mass of cake deposited w and other parameters, some of which can, in certain circumstances, be assumed to be constant. These parameters are discussed briefly.

9.2.2.1 Pressure drop

The pressure drop Δp may be constant or variable with time depending on the characteristics of the pump used or on the driving force applied. If it varies with time the function $\Delta p = f(t)$ is usually known.

9.2.2.2 Face area of the filter medium

The face area of the medium A is usually constant, but with a few exceptions such as in the case of equipment with an appreciable cake build-up on a tubular medium or a rotary drum.

9.2.2.3 Liquid viscosity

The liquid viscosity μ is constant provided that the temperature remains constant during the filtration cycle and that the liquid is Newtonian.

9.2.2.4 Specific cake resistance

The specific cake resistance α should be constant for incompressible cakes but it may change with time as a result of possible flow consolidation of the cake and also, in the case of variable rate filtration, because of variable approach velocity.

Most cakes, however, are compressible and their specific resistance changes with the pressure drop across the cake Δp_c. In such cases, an average specific cake resistance α_{av} should replace α in equation 9.5. α_{av} can be determined from

$$\frac{1}{\alpha_{av}} = \frac{1}{\Delta p_c} \int_0^{\Delta p_c} \frac{d(\Delta p_c)}{\alpha} \tag{9.6}$$

if the function $\alpha = f(\Delta p_c)$ is known from pilot filtration tests, bomb filter tests or from the use of a compressibility cell (which often provides useful data despite the fact that it replaces hydraulic pressure by mechanical pressure applied to the cake by a piston).

An experimental empirical relationship can sometimes be used over a limited pressure range[2]

$$\alpha = \alpha_0(\Delta p_c)^n \tag{9.7}$$

where α_0 is the resistance at unit applied pressure drop and n is a compressibility index (equal to zero for incompressible substances) obtained from experiments.

Using equation 9.7, the average cake resistance α_{av} can be shown to be (from equation 9.6)

$$\alpha_{av} = (1 - n)\alpha_0(\Delta p_c)^n \tag{9.8}$$

9.2.2.5 Mass of cake deposited per unit area

The mass of cake deposited per unit area w is a function of time in batch filtration processes. It can be related to the cumulative volume of filtrate V filtered in time t by

$$wA = cV \tag{9.9}$$

where c is the concentration of solids in the suspension (mass per unit volume of the filtrate, in $kg\ m^{-3}$). This makes no allowance for the solids that are in the liquid retained by the cake (see section 9.7) but in most cases this is negligible.

9.2.2.6 Medium resistance

The medium resistance R should normally be constant but it may vary with time as a result of some penetration of solids into the medium and sometimes it may also change with applied pressure because of the compression of fibres in the medium.

As the overall pressure drop across an installed filter includes losses not only in the medium but also in the associated piping and in the inlet and outlet ports, it is convenient in practice to include all these extra resistances in the value of the medium resistance R.

9.3 FILTRATION OPERATIONS—BASIC EQUATIONS, INCOMPRESSIBLE CAKES

The general filtration equation, equation 9.5, after substitution of $w(t)$ from equation 9.9, becomes

$$Q = \frac{\Delta p A}{\alpha \mu c(V/A) + \mu R} \tag{9.10}$$

As the total flow volume is an integral function of the flow rate

$$Q = \frac{dV}{dt} \tag{9.11}$$

Equation 9.10 can be rewritten in reciprocal form (thus giving time per unit flow), which is more convenient for further treatment

$$\frac{dt}{dV} = \alpha\mu c \frac{V}{A^2\Delta p} + \frac{\mu R}{A\Delta p} \tag{9.12}$$

It is useful for the mathematical simplicity of the final equations to define two constants a_1 and b_1:

$$a_1 = \alpha\mu c \tag{9.13}$$

this is a constant (if α, μ and c are constants) relating the properties of the feed suspension and of the suspended solids;

$$b_1 = \mu R \tag{9.14}$$

is a 'cloth–filtrate' constant.

Equation 9.12 then becomes

$$\frac{dt}{dV} = a_1 \frac{V}{A^2\Delta p} + b_1 \frac{1}{A\Delta p} \tag{9.15}$$

9.3.1 Constant pressure filtration

If Δp is constant, equation 9.15 can be integrated

$$\int_0^A dt = \frac{a_1}{A^2\Delta p} \int_0^V V\,dV + \frac{b_1}{A\Delta p} \int_0^V dV \tag{9.16}$$

giving

$$t = a_1 \frac{V^2}{2A^2\Delta p} + b_1 \frac{V}{A\Delta p} \tag{9.17}$$

Using equation 9.17 either t or V can be calculated from the value of the other variable provided all the constants are known.

For experimental determination of α and R, equation 9.17 is often put in the form

$$\frac{t}{V} = aV + b \tag{9.18}$$

where

$$a = \frac{a_1}{2A^2\Delta p} \quad \text{and} \quad b = \frac{b_1}{A\Delta p}$$

which gives a straight line if t/V is plotted against V—see *Figure 9.5*. The use of equation 9.18 and the plot in *Figure 9.5* are, of course, limited to those cases when it is possible to apply the given pressure drop right from the commencement of the filtration.

Often high initial flow rates through a clean medium have to be avoided in order to prevent the penetration of solids through the clean medium, which would cause contamination of the filtrate, and also to ensure an even deposition of the cake.

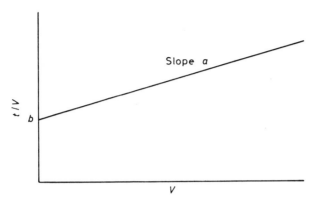

Figure 9.5. Plot of $t/V = f(V)$ for constant pressure filtration and incompressible cakes

It is therefore necessary in such cases to precede the constant pressure period with a period in which the applied pressure drop is gradually increased from a low value (this may be a nearly constant rate period).

The governing equation from which α and R may be evaluated, is then obtained by integrating equation 9.16 starting from a point t_s, V_s at the beginning of the truly constant pressure period; the resulting equation is

$$\frac{t - t_s}{V - V_s} = \frac{\alpha \mu c}{2 A^2 \Delta p} (V + V_s) + \frac{\mu R}{A \Delta p} \tag{9.19}$$

The actual procedure for the evaluation of results is best explained in an example.

Example 9.1 Evaluation of specific cake resistance α and medium resistance R from pilot scale tests

Filtration tests were carried out with a plate and frame filter press (for details of the construction of the press see chapter 12, section 12.2.1) under the following conditions:

solids: $\rho_s = 2710$ kg m^{-3}
liquid: water at 20°C, $\mu = 0.001$ Ns m^{-2}
suspension: concentration $c = 10$ kg m^{-3}
filter: plate and frame press, 1 frame, dimensions $430 \times 430 \times 30$ mm (the actual cake thickness can be larger by 5 mm because of a recess in the plates).

Table 9.1. DATA FROM FILTRATION EXPERIMENT, EXAMPLE 9.1

$10^{-5}\,\Delta p$, N m^{-2}	t, s	V, m^3	$\dfrac{t - t_s}{V - V_s}$ (calculated), s m^{-3}
0.4	447	0.04	12458
0.5	851	0.07	12326
0.7	1262	0.10	12120
0.8	1516	0.13	12765
1.1	1886	0.16	12857
1.3	2167	0.19	13809
1.3	2552	0.22	14175
1.3	2909	0.25	15540
1.5	3381	0.28	15250
1.5	3686	0.30	—
1.5	4043	0.32	17850
1.5	4398	0.34	17800
1.5	4793	0.36	18450
1.5	5190	0.38	18800
1.5	5652	0.40	19660
1.5	6117	0.42	20258
1.5	6610	0.44	20886
1.5	7100	0.46	21337
1.5	7608	0.48	21789
1.5	8136	0.50	22250
1.5	8680	0.52	22700
1.5	9256	0.54	23208

The frame was full of cake at $V = 0.56$ m^3.
The value of 0.30 m^3 corresponding to 3686 s was chosen as a starting point for constant pressure operation, i.e.

$$V_s = 0.3\ \text{m}^3, \qquad t_s = 3686\ \text{s}.$$

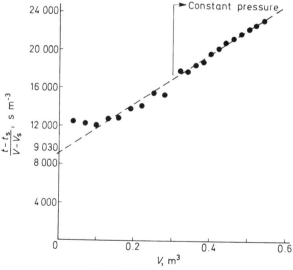

Figure 9.6. Plot of t/V for example 9.1

The data obtained during the filtration experiment are shown in *Table 9.1*; the initial stages of filtration were controlled manually before constant pressure filtration at $150000 \, \mathrm{N \, m^{-2}}$ was carried out. Determine the specific cake resistance α and the medium resistance R for this test.

Solution

Equation 9.19 can be used to evaluate the constants a and b:

$$\frac{t - t_s}{V - V_s} = a(V + V_s) + b$$

$(t - t_s)/(V - V_s)$ is plotted against V—as in *Figure 9.6*—and the slope (a) and the intercept on the vertical axis $(b + aV_s)$ of the best straight line drawn through the part of the graph that corresponds to constant pressure operation, i.e. for $V \geqslant V_s$ (i.e. $0.3 \, \mathrm{m^3}$) are measured.

The slope obtained is

$$a = 26219 \, \mathrm{s \, m^{-6}}$$

and the intercept is

$$b + aV_s = 9030 \, \mathrm{s \, m^{-3}}$$

from which

$$b = 9030 - 26219 \times 0.3 = 1164.3 \, \mathrm{s \, m^{-3}}$$

From the definition of the constants a and b (equations 9.13, 9.14 and 9.18)

$$a = \frac{\alpha \mu c}{2A^2 \Delta p} \qquad b = \frac{\mu R}{A \Delta p}$$

from which α and R can be calculated, using $A = 0.43 \times 0.43 \times 2 = 0.37 \, \mathrm{m^2}$ and $c = 10 \, \mathrm{kg}$ per cubic metre of suspension $= 10/(1 - 0.00369) = 10.037$ kg per cubic metre of filtrate (as the volume occupied by 10 kg of solids is $10/2710 = 0.00369 \, \mathrm{m^3}$)

$$\alpha = \frac{2A^2 \Delta p a}{\mu c} = \frac{2 \times 0.13675 \times 1.5 \times 10^5 \times 2.6219 \times 10^4}{0.001 \times 10.037}$$

$$= 1.069 \times 10^{11} \, \mathrm{m \, kg^{-1}}$$

and

$$R = \frac{A \Delta p b}{\mu} = \frac{0.37 \times 1.5 \times 10^5 \times 1164.3}{0.001}$$

$$= 6.4619 \times 10^{10} \, \mathrm{m^{-1}}$$

9.3.2 Constant rate filtration

If the flow rate Q is kept constant and the pressure Δp varied, equation 9.10 becomes

$$Q = \frac{\Delta p(t)A}{\alpha \mu c[V(t)/A] + \mu R} \qquad (9.20)$$

where V is simply

$$V = Qt \tag{9.21}$$

thus

$$\Delta p = \alpha\mu c \frac{Q^2}{A^2} t + \mu R \frac{Q}{A} \tag{9.22}$$

which, using the definitions of a_1 and b_1 from equations 9.13 and 9.14, becomes

$$\Delta p = a_1 v^2 t + b_1 v \tag{9.23}$$

where v is the approach velocity of the filtrate

$$v = \frac{Q}{A} \tag{9.24}$$

this is of course constant for constant rate filtration. A plot of Δp against t as in *Figure 9.7*, will, from equation 9.23, be a straight line.

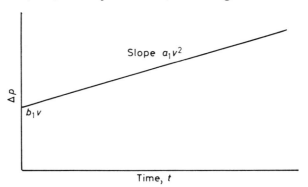

Figure 9.7. Plot of $\Delta p = f(t)$ for constant rate filtration, incompressible cake

9.3.3 Constant rate followed by constant pressure operation

In many cases, for example a plate and frame or a leaf press operated with the suspension supplied from a centrifugal pump, the early stages of filtration are conducted at a nearly constant rate. As the cake becomes thicker and offers more resistance to the flow, the pressure developed by the pump becomes a limiting factor and the filtration proceeds at a nearly constant pressure. For such a combined operation, the plot of Δp against time is as shown in *Figure 9.8*. The equations are (see equation 9.23)

$$\Delta p = a_1 v^2 t + b_1 v \qquad \text{for } t < t_s$$

and $\qquad\qquad\qquad\qquad\qquad\qquad\qquad\qquad\qquad\qquad\qquad$ (9.25)

$$\Delta p = \Delta p_s = \text{constant} \qquad \text{for } t \geqslant t_s$$

The plot of $t/V = f(V)$ will be as shown in *Figure 9.9*.

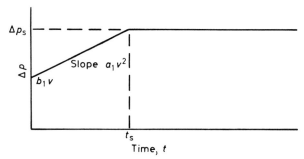

Figure 9.8. Plot of $\Delta p = f(t)$ *for constant rate followed by constant pressure operation (incompressible cakes)*

The equations are (see equation 9.21)

$$V = Q_1 t \qquad \text{for } V \leqslant V_s \qquad (9.26)$$

and

$$\frac{t - t_s}{V - V_s} = a(V + V_s) + b \qquad \text{for } V > V_s$$

(this is the same as equation 9.19 derived by integration of equation 9.15, and

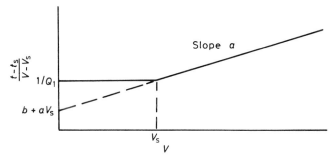

Figure 9.9. Plot of $t/V = f(V)$ *for constant rate followed by constant pressure operation*

is similar to equation 9.18) where Q_1 is the flow rate in the initial constant rate operation. V_s and t_s are related by

$$V_s = Q_1 t_s \qquad (9.27)$$

Example 9.2

A filtration experiment is carried out on a single cloth of area 0.02 m² to which a slurry is fed at a constant rate, yielding 4×10^{-5} m³ s⁻¹ of filtrate.

Readings taken during the test show that after 100 s the pressure is $4 \times 10^4 \, \mathrm{N\,m^{-2}}$ and after 500 s the pressure is $1.2 \times 10^5 \, \mathrm{N\,m^{-2}}$.

The same filter cloth material is to be used in a plate and frame filter press, each frame having dimensions $0.5 \, \mathrm{m} \times 0.5 \, \mathrm{m} \times 0.08 \, \mathrm{m}$, to filter the same slurry. The flow rate of slurry per unit area of cloth during the initial constant rate period is to be the same as that used in the preliminary experiment and the constant rate period is to be followed by constant pressure operation once the pressure reaches $8 \times 10^4 \, \mathrm{N\,m^{-2}}$. If the volume of cake formed per unit volume of filtrate, v is 0.02, calculate the time required to fill the frame.

1. Constant rate period
The filtrate approach velocity for the constant rate period is

$$v = \frac{Q_1}{A} = \frac{4 \times 10^{-5}}{0.02} = 2 \times 10^{-3} \, \mathrm{m\,s^{-1}}$$

which is the same for both the experimental and the filter press.

As two experimental values of Δp and t are known, the two constants a_1 and b_1 can be determined by substituting the values of Δp and v in equation 9.23:

$$4 \times 10^4 = a_1 \times 4 \times 10^{-6} \times 100 + b_1 \times 2 \times 10^{-3}$$

$$12 \times 10^4 = a_1 \times 4 \times 10^{-6} \times 500 + b_1 \times 2 \times 10^{-3}$$

giving

$$a_1 = 5 \times 10^7$$

and

$$b_1 = 10^7$$

so that $\Delta p = 200 \, t + 2 \times 10^4$ (this is applicable to both the experiment and the filter press) from which t_s can be determined because from the experimental data $\Delta p_s = 8 \times 10^4$, therefore

$$8 \times 10^4 = 200 \, t_s + 2 \times 10^4$$

thus

$$t_s = 300 \, \mathrm{s}.$$

A graph of $\Delta p (= f(t))$ against t can now be drawn (*Figure 9.10*). The cumulative volume V_s passed through one frame in 300 s can be calculated from equation 9.26

$$V_s = Q_1 t_s = vAt_s = 2 \times 10^{-3} \times 0.5 \times 300$$
$$= 0.3 \, \mathrm{m^3}$$

and the resulting thickness L_s of cake is

$$L_s = v \frac{V_s}{A} = 0.02 \times \frac{0.3}{0.5} = 0.012 \, \mathrm{m}$$

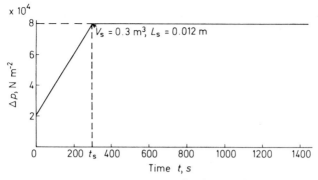

Figure 9.10. Plot of $\Delta p = f(t)$ for example 9.2

2. Constant pressure period

The final (total) volume of filtrate per frame V_f at the end of the whole combined filtration operation can be determined from the final cake thickness L_f, which is 0.04:

$$V_f = \frac{L_f A}{v} = \frac{0.04 \times 0.5}{0.02} = 1 \text{ m}^3$$

The time required to fill the frame (which is the total time of the combined operation) can be determined from equation 9.19

$$t_f = t_s + \frac{a_1}{2A^2 \Delta p_s}(V_f^2 - V_s^2) + \frac{b_1}{A \Delta p_s}(V_f - V_s)$$

$$= 300 + \frac{5 \times 10^7}{2 \times 0.25 \times 8 \times 10^4} \times (1^2 - 0.3^2) + \frac{10^7}{0.50 \times 8 \times 10^4}(1 - 0.3)$$

$$= 300 + 1137.5 + 175 = 1612.5 \text{ s}$$

Note that it can be shown that this result is independent of the filtration area.

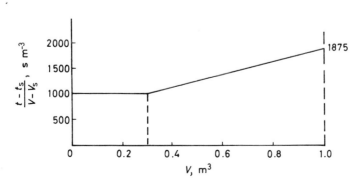

Figure 9.11. Plot of $(t - t_s)/(V - V_s)$ as a function of V for example 9.2

Compare this value of 1612.5 s with the time in which 1 m³ of filtrate would be filtered at all-constant rate operation

$$t = \frac{V}{Q} = \frac{V}{vA} = \frac{1}{2 \times 10^{-3} \times 0.5} = 1000 \text{ s}$$

The plot of t/v against V for this operation is shown in *Figure 9.11*.

9.3.4 Variable pressure–variable rate operation

If a centrifugal pump is used, its flow rate–pressure drop relationship may follow the curve shown in *Figure 9.12*.

Equation 9.10 can be written as

$$V = \frac{A}{\alpha \mu c}\left(\frac{\Delta p A}{Q} - \mu R\right) \tag{9.28}$$

where Δp and Q are related by the pump characteristics.

The time necessary for the filtration of a volume V of filtrate can be calculated by integration of the reciprocal rate $1/Q$ as a function of V since

$$dt = \frac{dV}{Q}$$

hence

$$t = \int_0^V \frac{dV}{Q} \tag{9.29}$$

The actual method of calculation is best explained by giving a worked example.

Example 9.3

Determine the time necessary for the filtration of 50 m³ of the same slurry as in example 9.1 in a plate-and-frame filter press with 25 frames of dimensions 1 × 1 × 0.035 m. Use the test data for cake resistance and medium resistance (the same cloth is used) obtained in example 9.1. The pump characteristic is shown in *Figure 9.12*.

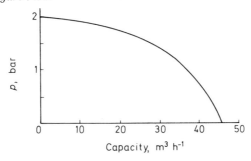

Figure 9.12. *Pump characteristics*

Data:

cake resistance, $\alpha = 1.069 \times 10^{11}\,\mathrm{m\,kg^{-1}}$
medium resistance, $R = 6.462 \times 10^{10}\,\mathrm{m^{-1}}$
viscosity, $\mu = 0.001\,\mathrm{Ns\,m^{-2}}$
concentration, $c = 10.037\,\mathrm{kg\,m^{-3}}$
filtration area, $A = 1 \times 1 \times 2 \times 25 = 50\,\mathrm{m^2}$

Solution

Equation 9.28 is used to find $V = f(Q)$ as follows:

$$V = \frac{50}{1.069 \times 10^{11} \times 10^{-3}\,10.037}\left(\frac{\Delta p}{Q}\,50 - 10^{-3} \times 6.462 \times 10^{10}\right)$$

$$V = 2.32 \times 10^{-6}\left(\frac{\Delta p}{Q} - 1.2924 \times 10^{6}\right)$$

Using the data from *Figure 9.12*, V can be calculated as a function of Q; a summary of the results is given in *Table 9.2*. To obtain the total filtration time, equation 9.29 must be integrated up to, in this case, $V = 50\,\mathrm{m^3}$; this can be done graphically from a plot of $1/Q$ against V (*Figure 9.13*). Graphical integration gives a value of 1.463 h (1 h 27 min 47 s).

Table 9.2

Q, m³ h⁻¹	$\Delta p \times 10^{-5}$, N m⁻²	V, m³	$1/Q$, s m⁻³
45	0.2	0.7	80
40	0.75	12.71	90
35	1.15	24.44	103
30	1.4	35.98	120
25	1.6	50.46	144
20	1.75	69.48	180
15	1.8	97.23	240

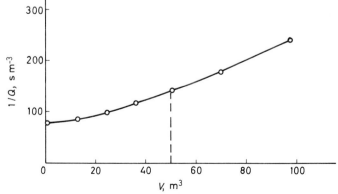

Figure 9.13. Plot of $1/Q = (V)$ for example 9.3

A check also has to be made as to whether there is enough cake-holding capacity available.

The filtration tests in example 9.1 showed the volume of cake formed per unit volume of suspension is

$$\frac{0.43 \times 0.43 \times 0.035}{0.56} = 0.01156 \text{ m}^3$$

The actual volume available in one frame is

$$1 \times 1 \times 0.035 = 0.035 \text{ m}^3$$

i.e. the maximum volume of suspension that can be filtered in one period is:

$$V_{max} = \frac{0.035 \times 25}{0.01156} = 75.69 \text{ m}^3$$

which is greater than the 50 m³ given in the example, and therefore the cake-holding capacity is sufficient.

9.4 FILTRATION OPERATIONS—BASIC EQUATIONS, COMPRESSIBLE CAKES

The compressibility of cakes, i.e. the increasing resistance of cakes with pressure, can be tested in various ways. As was briefly mentioned in section 9.2.2, one way of testing the relationship $\alpha = f(\Delta p_c)$ is in the compression–permeability cell. The solids compression is created by the mechanical action of the piston—an obvious assumption made is that hydraulic pressure can be simulated by mechanical compression.

Another method is to use pilot test data from constant pressure operations carried out at different pressures. This yields a family of lines on a plot of t/V against V (Figure 9.5) with varying slopes, from which the cake resistances α can be determined.

Jahreis[3] has suggested a procedure in which a single test run can be used for obtaining such data. The pressure is stepped up from one operating pressure to another, without interrupting the flow, and at each pressure step measurement of the cumulative volume of filtrate is started from zero again. Purchas[4] suggests an alternative plot of $\Delta p/Q = f(V)$ which makes the slope independent of pressure and yields a series of lines which diverge if the cake is incompressible.

An obvious way of dealing with compressible cakes is to use the concept of average cake resistance α_{av} as defined in equation 9.6. This is of course only possible if the final pressure is known, otherwise a solution can only be obtained by a series of iterative calculations.

If, however, an analytical relationship such as equation 9.7 is applicable, usually over a limited range of pressure drops, the function $\alpha = f(\Delta p)$ can be employed directly as shown in the following.

For an analytical solution of the filtration of compressible cakes which satisfy equation 9.7, pressure drops across the medium Δp_m and across the cake Δp_c have to be treated separately.

Let

$$\Delta p = \Delta p_c + \Delta p_m \tag{9.30}$$

$$\Delta p_m = \frac{\mu R Q}{A} \tag{9.31}$$

and

$$\Delta p_c = \frac{\alpha_{av} \mu c V Q}{A^2} \tag{9.32}$$

If equation 9.8 is substituted into equation 9.32

$$\Delta p_c = (1 - n)\alpha_0 \Delta p_c^n \frac{\mu c V Q}{A^2}$$

from which

$$\frac{\mu c V Q}{A^2} = \frac{(\Delta p_c)^{1-n}}{(1 - n)\alpha_0} \tag{9.33}$$

This is the basic equation from which the special cases can be derived.

9.4.1 Constant pressure filtration

This operation is of course unaffected by compressibility and, consequently, the governing relationships remain the same as in section 9.31 (where α is now the specific cake resistance corresponding to the given pressure).

9.4.2 Constant rate filtration

Substituting

$$V = Qt$$

from equation 9.20 into equation 9.33 gives

$$(\Delta p_c)^{1-n} = \alpha_0 (1 - n)\mu c \frac{Q^2}{A^2} t \tag{9.34}$$

A plot of log Δp_c against log t should give a straight line. The pressure drop across the medium is constant and given by equation 9.31.

9.4.3 Variable pressure—variable rate operation

This is a complicated case in which Δp_c, Δp_m, V, Q and t are all variable. Equation 9.33 can be written in the form (using equation 9.30):

$$V = \left(\frac{A^2}{(1 - n)\alpha_0 \mu c}\right)\left(\frac{(\Delta p - \Delta p_m)^{1-n}}{Q}\right) \tag{9.35}$$

where Δp and Q are related by pump characteristics (as in section 9.3.4) and Δp_m is given by equation 9.31.

Equation 9.35 can then be treated in a similar manner to equation 9.28 in section 9.3.4; the time of filtration can again be calculated from equation 9.29.

Example 9.4

As in example 9.3, determine the time necessary for the filtration of 50 m³ of the same slurry in the same filter but assuming a compressible filter cake with cake resistance α following the law

$$\alpha = 6.1094 \times 10^9 (\Delta p_c)^{0.24}$$

The medium resistance is constant and equal to (as in example 9.3) $R = 6.462 \times 10^{10}\ \text{m}^{-1}$.

Use the same pump characteristics as in example 9.3.

Solution

Equation 9.35 is used to find V as a function of Q:

$$V = \left(\frac{2500}{0.76 \times 6.1094 \times 10^9 \times 10^{-3} \times 10.037}\right)\left(\frac{(\Delta p - \Delta p_m)^{0.76}}{Q}\right),\quad \text{m}^3$$

A summary of the values for V, Q and Δp are given in *Table 9.3*. Note that

Table 9.3

Q, m³ h⁻¹	$\Delta p \times 10^{-5}$ N m⁻²	$\Delta p_m \times 10^{-5}$ N m⁻²	V, m³	$1/Q$, s m⁻³
45	0.2	0.1616	2.3	80
40	0.75	0.1436	20.8	90
35	1.15	0.1257	35.4	103
30	1.4	0.1077	49.2	120
25	1.6	0.0898	66.5	144

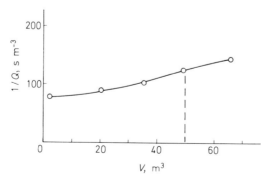

Figure 9.14. Plot of $1/Q = f(V)$ for example 9.4

pressure drops across the medium have been calculated from equation 9.31

$$\Delta p_m = \frac{10^{-3} \times 6.462 \times 10^{10} \times Q}{50}, \qquad N\,m^{-2}$$

A plot of $1/Q$ against V is drawn (*Figure 9.14*) and the time required for the filtration is obtained by integration of the area under the curve up to 50 m³ (from equation 9.29), giving

$$t = 4750\,s = 1.3194\,h$$
$$= 1\,h\,19\,min\,10\,s$$

9.5 RELATIONSHIP BETWEEN SPECIFIC CAKE RESISTANCE, POROSITY AND SPECIFIC SURFACE

Considering the flow of clean liquid through a packed bed, Kozeny[5] assumed that the pore space of a packed bed of powder could be regarded as equivalent to a bundle of parallel capillaries with a common equivalent radius, and with a cross-sectional shape representative of the average shape of the pore cross-section. Using Poiseuille's law for the flow of fluids through capillary tubes, Kozeny[5] and later Carman[6] were able to relate the permeability to the porosity, the specific surface and the density of the powder in the bed.

Using the Kozeny–Carman equation it can be shown that the specific cake resistance is equal to

$$\alpha = \left(\frac{K_0 S_0^2}{\rho_s}\right)\left(\frac{(1-\varepsilon)}{\varepsilon^3}\right) \tag{9.36}$$

where K_0 is the so-called Kozeny constant which approximates to a value of 5 in the lower porosity ranges; it generally depends on particle size, shape and the porosity. S_0 is the specific surface of the particles making up the bed,

$$S_0 = \frac{\text{surface area of solids}}{\text{volume of solids}}$$

ρ_s is the density of solids and ε is the porosity defined as

$$\varepsilon = \frac{\text{volume of voids}}{\text{volume of cake}}$$

Equation 9.36 combined with equation 9.5, indicates the high sensitivity of the pressure drop to the cake porosity and to the specific surface. Unfortunately, the importance of equation 9.36 in filtration is limited, despite its frequent quotations and derivations in textbooks. Although it has been found to work reasonably well for incompressible cakes over narrow porosity ranges, it cannot be used for compressible cakes and its most valuable use is for illustration of basic principles. For its derivation from the capillary model, the reader can consult either the original works by Kozeny[5] and Carman[6] or other texts, e.g. by Tiller[2].

9.6 CAKE MOISTURE CORRECTION—MASS BALANCE

In some cases, particularly in the filtration of highly concentrated suspensions, when working out the mass of dry cake from filtrate volume collected a cake moisture correction has to be applied, which accounts for the presence of moisture in the cake. The relationship that replaces equation 9.9 in such cases is obtained from solids concentration c (kg m^{-3}) in the original suspension and from the moisture content of the cake

$$m = \frac{\text{mass of wet cake}}{\text{mass of dry cake}}$$

(kg/kg) via the mass balance:

$$\text{mass of dry cake} = \text{volume of filtrate collected} \times \frac{\text{mass of solids}}{\text{volume of filtrate}}$$

$$+ \text{volume of filtrate in cake} \times \frac{\text{mass of solids}}{\text{volume of filtrate}}$$

hence

$$Aw = V\left(\frac{c\rho_s}{\rho_s - c}\right) + \left(\frac{m-1}{\rho}\right) wA \left(\frac{c\rho_s}{\rho_s - c}\right) \tag{9.37}$$

which gives

$$w = \frac{V}{A}\left(\frac{1}{c} - \frac{1}{\rho_s} - \frac{m-1}{\rho}\right)^{-1} \tag{9.38}$$

Comparison of equations 9.9 and 9.38 may be used to define a 'corrected' concentration $c_{\text{corrected}}$, i.e.

$$c_{\text{corrected}} = \left(\frac{1}{c} - \frac{1}{\rho_s} - \frac{m-1}{\rho}\right)^{-1} \tag{9.39}$$

which takes into account the presence of moisture in the cake. Clearly, for low values of c when $\rho_s \gg c$ and $\rho \gg c$:

$$c_{\text{corrected}} = c$$

Example 9.5

A suspension of incompressible solids at a concentration of 300 kg m^{-3} of slurry is to be filtered at constant pressure drop. Estimate the filtration area necessary to produce 50 kg h^{-1} of dry solids.

Data:

pressure drop,	$\Delta p = 10^5$ N m^{-2} (1 bar)
cake moisture content,	$m = 1.2$
cake specific resistance,	$\alpha = 10^{11}$ m kg^{-1}
medium resistance,	$R = 6.5 \times 10^{10}$ m^{-1}
liquid density,	$\rho_1 = 1000$ kg m^{-3}
solids density,	$\rho_s = 2600$ kg m^{-3}
liquid viscosity,	$\mu = 0.001$ N s m^{-2}

Solution

The corrected solids concentration (equation 9.39)

$$c_{\text{corrected}} = \left(\frac{1}{300} - \frac{1}{2600} - \frac{0.2}{1000} \right)^{-1}$$

$$= 363.8 \text{ kg m}^3$$

hence the volume of filtrate required to produce 50 kg of solids (equation 9.38) is

$$V = \frac{wA}{c_{\text{corrected}}} = \frac{50}{363.8} = 0.1374 \text{ m}^3$$

Substituting $\Delta p = 10^5$, $V = 0.1374$ and $t = 3600$ s in equation 9.17 gives:

$$A^2 - 0.0248A - 0.954 = 0$$

i.e.

$$A = 0.99 \text{ m}^2$$

9.7 FURTHER DEVELOPMENT OF FILTRATION THEORY

The filtration theory presented here is only very basic and the assumptions on which it is built are often too simplistic. This so-called classical filtration theory has been under considerable criticism from many research workers recently. A brief account of the main points is given in the following.

Starting with the medium resistance, R, which can be determined from the intercept in constant pressure filtration (section 9.3.1), this is subject to great uncertainty and in practice often comes out either unrealistically small or even negative. Tiller et al.[7] have pointed out that the plots of t/V against V have marked curvature at the start of the filtration experiment and this can easily be missed. They defined the intercept values as 'false medium resistance' and showed that a 'good' value of medium resistance is obtained only if the ordinate of the first experimental point $(1/Q)$ is close to the intercept value.

Another, even more troublesome variable is the specific cake resistance, α, and the concept of using an average value α_{av}. Even in a constant pressure experiment, the pressure varies through the depth of the cake and subsequently the porosity and cake resistance vary also. The average value of α_{av} defined by equation 9.6 is an average of the values of the effective cake resistances obtained from a series of constant-pressure experiments, which in turn are averages of the point values within the cake. The underlying assumption of the conventional theory is that the point and average values of α are functions of applied pressure only. In practice the porosity, and therefore the cake resistance also, depend on time (due to time consolidation of the cake), filtration velocity and solids concentration. Rushton et al.[8] have shown, for example, the effects of velocity and concentration on specific cake resistance of several inorganic materials. They found that the values of α

in a constant-pressure experiment go through a maximum with increasing concentration. For concentrations above that corresponding to the maximum, resistance decreases exponentially with concentration and this is attributed to shorter cake formation times at those concentrations and the resulting open cake structures. At concentrations below the maximum α, the lower cake resistance is attributed to higher initial fluid velocities which are also likely to produce open cake structures. Rushton et al.[8] also found that the maximum α itself increases exponentially with pressure.

It is now certain that the only truly rigorous approach to filtration theory is by using point values of α within the cake itself and those can only be obtained from studies of flow and local porosities in practical cakes. This is inevitably very complicated and beyond the scope of this book. The reader is referred to the already mentioned review by Tiller et al.[7] and to a theoretical paper by Wakeman[9].

The conventional filtration theory has been challenged by Willis et al.[10] (as an example of the many published) who have applied a two-phase theory to filtration and explained the deviations from parabolic behaviour in the initial stages of the filtration process. This new theory incorporates the medium as an integral part of the process and shows that it is the interaction of the cake particles with the medium which in fact controls the filterability. The authors define a cake–septum permeability which then appears in the slope of the conventional plots (Figures 9.5 and 9.7) instead of the cake resistance. This theory is not yet accepted by the engineering community, probably because it merely represents a new way of interpreting test data rather than a new method of sizing or scaling filters.

9.8 THE BENEFITS OF PRE-THICKENING

The feed solids concentration has a profound effect on the performance of any cake filtration equipment. It affects the capacity and the cake resistance, as well as the penetration of the solids into the cloth (which influences filtrate clarity and medium resistance). Thicker feeds lead to improved performance of most filters through higher capacity and lower cake resistance.

The effect on solids yield can be easily demonstrated using the following equation derived in appendix I of Svarovsky[11] (neglecting medium resistance):

$$Y = \left(\frac{2 \, \Delta p \, f \, c}{\alpha \, \mu \, t_c} \right)^{\frac{1}{2}} \tag{9.40}$$

where Δp is the pressure drop, c is the feed solids concentration, α is the specific cake resistance, μ is the liquid viscosity, Y is the solids yield (dry cake production in kg m^{-2} s^{-1}), f is the ratio of filtration to cycle time and t_c is the cycle time. For the same cycle time (i.e. the same speed), if the concentration is increased by a factor of four, production capacity is doubled. In other words, filtration area can be halved for the same capacity.

For given operating conditions and submergence, the dry cake production rate increases with the speed of rotation (see equation 9.40) and the limiting factor is usually the minimum cake thickness which can still be successfully discharged by the method used in the filter. Equation 9.41 (also derived by Svarovsky[11]) shows the dependence of the solids yield on cake thickness:

$$Y = \frac{2 \, \Delta p \, f \, c}{\alpha \, \mu \, L \, (1 \, - \, \epsilon) \, \rho_s} \tag{9.41}$$

where L is the cake thickness, ρ_s is the solids density and ϵ is the cake porosity. It can be seen that for constant cake thickness, doubling the feed concentration doubles the yield. This is the secret of the success of the so-called high duty vacuum drum filters which, by using a unique cake discharge method, allow very thin cakes to be discharged and can therefore be operated at very high speeds up to 25 rev. min^{-1}.

For those filters which allow variations in cake formation time within a fixed cycle time (such as the horizontal vacuum belt filter, for example), the advantage of pre-thickening is that for thicker feeds the cake formation time can be shortened, thus giving more time for dewatering, washing or other cake-processing operations. On the other hand, it can be shown (see Svarovsky[11] or section 12.2.1.3) that in an optimum cycle time, at constant-pressure operation and where medium resistance is low compared with cake resistance, the cake-formation time should be equal to the time the filter is 'out of service' for cake dewatering, washing or discharge. This result of course takes no account of any optimum operating conditions for minimum moisture content of the cake produced. In any case, this theoretical result may be outside the operational range governed by the design of a particular continuous filter: for example, design limitations on submergence in vacuum drum filters prevents the application of the above optimum condition. It can, however, be applied to any batch filters.

One additional benefit of pre-thickening not taken into account in the above analysis is the reduction in cake resistance which it causes. If the feed concentration is low, there is a general tendency of particles to pack together more tightly, thus leading to higher specific resistances. If, however, many particles approach the filter medium at the same time, they may bridge over the pores; this reduces penetration into the cloth or the cake underneath and more permeable cakes are thus formed. This effect of concentration is particularly pronounced with irregularly shaped particles. A possible explanation of the variation in the specific resistance is in terms of the time available for the particles to orientate themselves in the growing cake. At higher concentrations, but with the same approach velocities, less time (referred to as particle relaxation time) is available for a stable cake to form and a low resistance results.

An example of the concentration effect on the specific cake resistance is given by Svarovsky and Walker[12] who reported results of some experiments with a laboratory horizontal vacuum belt filter. In spite of the operational difficulties in keeping conditions constant, the effect of feed concentration on specific cake resistance was so strong that it swamped all other effects. *Figure 9.15* shows this in a log–log plot and the measured values correlate

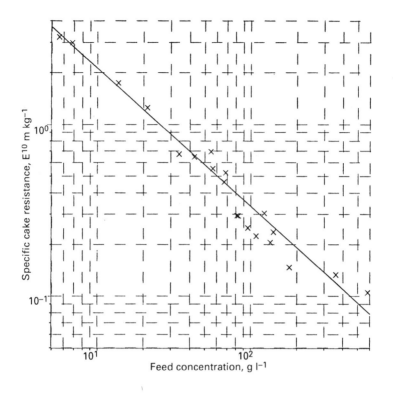

Figure 9.15. A plot of the specific cake resistance against the feed solids concentration for aluminium hydrate[12]

quite well in a straight line. The function fitted, for the aluminium hydroxide tested, was in the following form:

$$\alpha = 13.47 \times 10^{10} c^{-0.8031} \qquad (9.42)$$

where α is in m kg^{-1} and c is in g l^{-1} (kg m^{-3}) from 5 to 580 g l^{-1}.

The form of the above equation is in keeping with a similar study by Rushton *et al.*[8] who used a pressure filter to test a range of inorganic materials (calcium carbonate, silicate and sulphate, and two grades of magnesium carbonate) at constant pressure and obtained a range of values for the numerical constant in equation 9.42 from 9.14 to 37 800, and the exponent from -0.08 to -1.92, with feed concentrations ranging from 5 to 190 kg m^{-3} (g l^{-1}). Unlike Rushton *et al.*, however, we did not find a maximum on the curve of specific cake resistance versus feed concentration.

One current project at Bradford University is concerned with the effect of feed concentration (the present report includes the initial tests in this project) in which we hope to correlate the porosity of cakes at different feed concentrations with the relaxation effects for hard spheres in a model similar to that used in molecular studies[13]. Another project is concerned with precoating the belt filter with the underflow from a hydrocyclone and studying the properties of such highly stratified cakes[14].

The benefits of pre-thickening can be summarized as:

1. increase in dry cake production (see equation 9.40);
2. reduction in specific cake resistance (see *Figure 9.15*);
3. clearer filtrate;
4. less cloth blinding.

Further work is needed to build a physical model which allows prediction of the concentration effect from the primary properties of the slurry or from a limited amount of slurry testing.

Pre-thickening of filter feeds can be done with a variety of equipment such as gravity thickeners, hydrocyclones or sedimenting centrifuges. Even cake filters can be designed to limit or completely eliminate cake formation and, therefore, act as 'thickening filters' and be used in this thickening duty (see chapter 18).

REFERENCES

1. Tiller, F. M. and Cheng, K. S., 'Delayed cake filtration', *Filtration and Separation,* **14**(1), 13–18 (1977)
2. Tiller, F. M., *Filtration and Separation,* **12**, 386 (1975)
3. Jahreis, C. A., *Filtration and Separation,* **2**(4), 308 (1965)
4. Purchas, D. B., *Industrial Filtration of Liquids*, 2nd Edition, Leonard Hill Books, London (1971)
5. Kozeny, J., *Ber. Wien. Akad.,* **136a**, 271 (1927)
6. Carman, P. C., *Trans. Inst. Chem. Eng.,* **15**, 150 (1937)
7. Tiller, F. M., Crump, J. C. and Ville, F., 'Filtration theory in its historical perspective, a revised approach with surprises', *The Second World Filtration Congress 1979*, Filtration Society, London (1979)
8. Rushton, A., Hosseini, M. and Hassan, I., 'The effects of velocity and concentration on filter cake resistance', pp. 78–91, *Proc. Symp. on Solid–liquid Separation Practice*, Leeds, UK, Yorkshire Branch of the I. Chem. E., 27–29 March (1978)
9. Wakeman, R. J., 'A numerical integration of the differential equations describing the formation of and flow in compressible filter cakes', *Trans. I. Chem. E.,* **56**, 258–265 (1978)
10. Willis, M. S., Bridges, W. G. and Collins R. M., 'The initial stages of filtration and deviations from parabolic behaviour', *Filtech Conference 1981*, pp. 167–176, Filtration Society, London (1981)
11. Svarovsky, L., '*Solid–Liquid Separation Processes and Technology*,' Elsevier, Amsterdam (1985)
12. Svarovsky, L. and Walker, A. J., 'The effect of feed thickening on the performance of a horizontal vacuum belt filter', *1st World Congress Particle Technology, 331 Event of EFCE, Part IV*, NMA mbH, Nuernburg (1986)
13. Woodcock, L. V., 'Glass transition in the hard-sphere model and Kauzmann's paradox', *Ann. NY Acad. Sci.*
14. Zeitsch, K., Eine neue Theorie fuer die Permeabilitaet eines durch sedimentation enstandenen Filterkuchens, *Tech. Mitt. Krupp Forsch.-Ber.,* **42**, 71–76 (1984)

10

Filter Aids

G. R. S. Smith

W. L. Gore & Associates, Inc., Elkton, Maryland, USA

10.1 INTRODUCTION

In the introduction to chapter 9 on filtration fundamentals, filtration is divided into surface filters (cake filtration) and depth filters. The author goes on to say that surface filters operate in the area of above 1% solids and mentions the use of filter aids to alleviate the problems of pressure build-up and medium blinding. This chapter is about these filter aids.

10.2 AREAS OF USE

Filter aids serve to extend the areas of surface filtration economics, the ranges of particle sizes removed, concentration limits and the type of particles to be filtered.

Filter aids, then, operate in the area of cake filtration (see Figure 1.2 in reference 1) and result in a continuous surface filtration when used as precoat and body feed or as a precoat on a rotary vacuum precoat filter (RVPF).

Referring to *Figure 2.1* in chapter 2, filter aids operate in the particle size range $x < 5\,\mu$m (actually removing particles as fine as 0.1μm) to $x < 50\,\mu$m. In the $x < 5\,\mu$m size, filter aids are used where the particles cannot be settled or are compressible.

Thus, filter aids are used for removal of as little as 5 p.p.m. solids (reducing them to as low as 0.1 p.p.m.) from water where the initial solids amount is too small for flocculation and the solids are too fine for removal by fixed media. Or, the rotary drum precoat filter can be used for the removal of 6% or higher amounts of solids from fermentation broths where the median particle size is only 1–2 μm. In the larger particle size ranges, the RVPF is used to remove solids from clarifier underflows in alum treatment of water and in metal wastes.

Emulsions stabilized by the presence of solids can be broken by filtering the solids—usually on a rotary vacuum precoat filter. Petroleum refinery slop oil emulsions are an example.

10.3 FILTER AID CHARACTERISTICS

The term 'filter aid' is actually a misnomer in most instances, for it implies that filtration is merely aided. In actual practice, however, filter aids are the filter medium. That is, the separation of the solid material from the fluid being filtered is carried out by the filter aid. A filter aid, then, is a substance of discrete particles which can be suspended in a fluid and caused to form a layer, or filter medium, on a support material called a septum. Materials which make particles easier to filter, such as flocculants, may be thought of as preconditioning agents and are not, for the purposes of this discussion, filter aids. Filter aids are effective to the extent that they can provide a high degree of particle retention, they can 'hold' a large quantity of particles and provide high rates of flow. The principal properties that enable them to do this are:

permeability,
pore size (of filter aid cakes),
rigidity (incompressibility).

They should also be relatively inert to the liquid being filtered. In order that they will form a stable layer on the septum, and so that the septum may be relatively open for good permeability and cleaning, they usually have 3% or more (depending on the septum used) particles by weight on a 150 mesh screen. It will also be found that efficient filter aids have low cake densities in the order of 160–320 kg m^{-3} (10–20 lb ft^{-3}), depending upon the type of filter aid.

Table 10.1. RANGES OF KEY PROPERTIES FOR THE PRINCIPAL TYPES OF FILTER AIDS

Property	Filter aid		
	Diatomite	Perlite	Cellulose
Permeability, Darcies*	0.05–30	0.4–6	0.4–12
Median pore size, μm	1.1 –30	7–16	—
Wet density			
kg m^{-3}	260–320	150–270	60–320
lb ft^{-3}	16–20	9–17	7–20
Compressibility	Low	Medium	High

*The Darcy is defined in the *Dictionary of Scientific and Technical Terms*, McGraw-Hill (1974), as, "a unit of permeability equivalent to the passage of one cubic centimetre of fluid of 1 centipoise viscosity flowing in one second under a pressure of one atmosphere through a porous medium of one square centimetre area and one centimetre long". It can be expressed as $K^l = v\mu L/\Delta P$

where: v = filtrate volume, cm^3 s^{-1}
μ = viscosity, centipoise
L = cake thickness, cm
ΔP = differential pressure, atmosphere

Table 10.1 summarizes the ranges of key properties for the principal types of filter aid. Carbon and starch are also used on rare occasions as filter aids, but they are, for the most part, locally produced and the properties vary too widely to be included in this table. The different types of filter aids are discussed in more detail below.

Figure 10.1. Diatomite filter aid (× 200). (Courtesy of Manville Corp.)

10.4 TYPES OF FILTER AID

10.4.1 Diatomite

On a worldwide basis, diatomite filter aids (*Figure 10.1*) are the most commonly used. The reason for this can be seen in *Table 10.1* from the significantly wider permeability and pore size ranges for diatomite as compared to perlite and cellulose. Implied in these figures (as is actually the case) for a given permeability, the median pore size of diatomite filter aid will be smaller than that for perlite filter aid, indicating greater clarification efficiency. Diatomite filter aids are, therefore, used where a high degree of clarification is desired, or at the other end of the spectrum where flow rate requirements are high.

Diatomite is the prehistoric skeletal remains of aquatic plants and is processed by a selective mining, drying and air classification. To produce filter aids with higher permeabilities, sintering is accomplished by calcination with and without a flux. Diatomite being approximately 90% silica is practically insoluble in almost all liquids except hot caustic (although it is regularly used for the filtration of 50% caustic at ambient temperatures). Diatomite is used for filtering a wide range of liquids such as wine and beer where a very high degree of polish is required (especially if they are being filtered through membranes), through lube oils to antibiotic fermentation broths, algenates and mineral processing liquids.

Figure 10.2. Perlite filter aid (× 200). (Courtesy of Manville Corp.)

10.4.2 Perlite

Perlite filter aids are the next most commonly used filter aid and are made by expanding perlitic rock, then milling and classifying the expanded material. The flow rate, or permeability ranges for perlite filter aids fall in the middle of those for diatomite.

The median pore size of perlite filter aids, on the lower permeability end especially, is significantly larger than that for diatomite filter aids, and this is shown in their inability to remove micro-organisms and other fine particles to the same degree as that for diatomite. As can be seen from *Figure 10.2*, the particle shape of perlite is considerably different from that for diatomite, consisting of large flat impermeable surfaces. It is this shape that gives it its lower pore size-to-permeability ratio as compared to diatomite. The wet density of perlite filter aids is, however, significantly lower, especially in their mid-flow rate ranges, than that for diatomite filter aids resulting in thicker cakes for the same weight of material. The compressibility of perlite is slightly higher than that for diatomite. Perlite is approximately two-thirds aluminium silicate and one third silica (with small amounts of other minerals) so its solubility is similar to that for diatomite. Perlite filter aids are used primarily for the filtration of relatively large, compressible solids, such as would be found in the production of antibiotics and waste sludges.

10.4.3 Cellulose

Cellulose filter aids (*Figure 10.3*) are used to a much smaller extent than either diatomite or perlite due primarily to their high cost and low filtration efficiency. They find a place in the filtration of hot caustic solutions or in

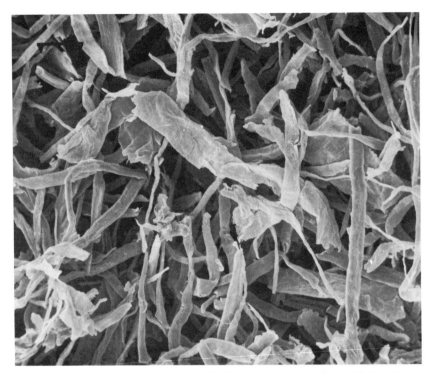

Figure 10.3. Cellulose filter aid, ball-milled (× 200). (Courtesy of Manville Corp.)

instances where it is desirable to incinerate the filter cake. Perhaps the widest usage of cellulose filter aids is in mixtures with diatomite and perlite filter aids to form a more stable precoat, or the cellulose filter aids may even be used as a pre-precoat to reduce filter aid bleed-through. They are also used in the filtration of boiler-feed make-up water (where the solubility of diatomite and perlite may be a problem) and various metal plating solutions. It is also possible to precoat a rotary drum precoat filter with cellulose filter aids, using special techniques. In general, the permeability ranges from 0.4 to 12 Darcies, but some of the faster flow rate grades go considerably higher in flow rate.

Cellulose filter aids may be made from bleached or unbleached wood pulp and, on occasion, wood flour is used. The finer grades of bleached and unbleached cellulose filter aids are made by ball milling cellulose pulp. As can be seen in *Figure 10.3*, the pore sizes are extremely large, so much so that in the removal of the fine particles from a washed, raw sugar solution, the finest grade of cellulose filter aid will not give the same clarification as the coarsest grade of diatomite or perlite.

10.4.4 Asbestos

Asbestos is one of the oldest types of filter aids, having been used for many years in combination with cellulose in pulp filters for the filtration of beer. It is also very effective (much more so than cellulose) in combination with diatomite or perlite used as a pre-precoat. Its clarification effectiveness is much higher than that for cellulose due to its extremely fine fibre diameter. By itself, however, it forms a compressible medium which greatly reduces flow rate. The adsorbent properties of asbestos are of advantage in the filtration or removal of bacteria whereas the high clarification obtained with asbestos in wine filtration is thought by some to be strictly mechanical.

In recent years the use of asbestos for filtration has been all but eliminated because of the industrial hygiene aspects of handling it.

10.4.5 Carbon

Non-adsorbent carbon filter aids have been used for many years in the filtration of very hot, caustic solutions (where even cellulose filter aids would soften or dissolve) and other liquids where the solubilities of other filter aids would be unacceptable or where it is desirable to incinerate the filter cake. Due to the impossibility of producing a highly irregular particle-shaped carbon, filtration efficiency is very low. One example of carbon's use is the removal of mercury from highly caustic liquors during the electrolysis of brine.

10.4.6 Mixtures

The advantages and disadvantages of the various types of filter aid have been discussed above.

The clarification advantage of diatomite was mentioned as compared to the density advantage of perlite. On occasion, users of filter aids will combine the two grades wherein they might use diatomite filter aid as a precoat followed by perlite filter aid as body feed. This combination is infrequently used in pressure filters, but is relatively common on the rotary vacuum precoat filter in the filtration of antibiotic broths. Here, diatomite is used for precoating while perlite is used for body feed. Laboratory studies show that mixtures of the two types of filter aid result in an average of the clarification properties of the individual filter aids.

By mixing cellulose or asbestos with diatomite, as mentioned above, the stabilizing effect of the fibrous filter aids can be obtained while their compressibility is reduced by the diatomite filter aid. Cellulose is commonly used at the rate of 15% of the mixture.

10.5 FILTER AID FILTRATION

There are two basic ways of using filter aids for filtration:

1. the precoat-body feed method, commonly known as pressure filtration;
2. the rotary drum precoat method which, in the vast majority of cases, operates under vacuum.

Filter aids can be used in a wide variety of filters such that a filter cake can be formed on the filter element and then removed at the end of the filter cycle. The internal hydraulics of the filter must be capable of keeping the filter aid in suspension.

All mention of flow rates in these sections refers to liquids with a viscosity of 10^{-3} N s m^{-2} (1 centipoise).

10.6 THE PRECOAT-BODYFEED (PRESSURE) FILTRATION SYSTEM

In this system a precoat is first formed by recirculating a filter aid slurry through the filter. Then, during filtration, filter aid is added to the unfiltered liquid before it enters the filter. When the filtration rate becomes too low, or the cake space in the filter is filled, or both, filtration stops and the filter cake is removed from the filter. Filter aid is also used, on occasion, only as precoat or body feed. The system for carrying out these operations is shown in *Figure 10.4* and consists of tanks for body feed and precoat slurries, filter feed and body feed pumps, lines and valves for connecting the different pieces of equipment.

10.6.1 Precoating

The purposes of the precoat are to protect the filter septum, to provide immediate filtrate clarity, and to facilitate cake removal at the end of the filter cycle. To do this, the precoat should be of sufficient fineness to prevent any significant penetration of unfiltered solids through the precoat where they can bind the precoat to the septum, should the solids be of a slimy nature. With the correct grade there is a very distinct line separating the precoat from the filter cake.

Precoating is carried out by the recirculation of a mixture of filter aid in a clean liquid. Ideally, the liquid is filtrate from the previous filter cycle but, should this be impossible, the liquid should be of similar viscosity and temperature. If dilution of the product with the precoat liquid is undesirable, it is possible, by proper valve manipulation, to blow the precoat liquid from the filter before filtering. The quantity of precoat varies from 250 to 700 g m^{-2} (5–15 lb per 100 square feet) of filter area depending primarily on the type of filter aid used and on the hydraulics in the filter. The more usual amount of precoat is 500–700 g m^{-2} (10–15 lb per 100 square feet).

The precoat tank should be sized to contain enough liquid to fill the filter and connecting piping, and to leave a small heel for recirculation of the precoat slurry. To this tank is added the precoat liquid and the desired amount of filter aid for the precoat. As the amount of precoat added is related to the filter area, this will result in a precoat slurry concentration of 0.3–0.6% depending primarily on the ratio of filter area to filter tank (and connecting piping) volume.

It should be possible to form a precoat within ten minutes. If the precoat slurry concentration is much less than 0.3%, precoating times will be prolonged and, depending on the openings in the filter septum, a precoat may be impossible to form.

Precoat rates must be high enough to keep the filter aid in suspension inside the filter (for water and upward flow of approximately $1\frac{1}{2}$ m min^{-1}—4 feet per minute—is required). Too high a precoat rate may prevent precoat formation

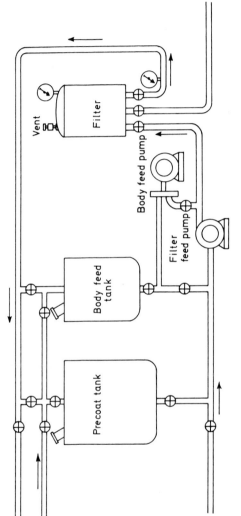

Figure 10.4. Typical diatomite filtration system. (Courtesy of Manville Corp.)

near the entrance of the precoat liquid to the filter should internal baffling be inadequate. The minimum precoat rate using water is in the order of 1.2 m³ m⁻² h⁻¹ (0.5 gallons per square foot per minute) although twice that rate is more desirable. Proportionately lower filtration rates may be used with liquids of higher viscosity and specific gravities. In general, the precoat rate is equal to, or slightly higher than, the average filtration rate. The precoat formed should be evenly dispersed over the entire area of the filter and is generally not more than 3 mm ($\frac{1}{8}$ inch) thick.

Poor precoating techniques and the formation of irregular precoats are major causes of poor filtering efficiency.

10.6.2 Filtration with body feed

Once the precoat is in place, filtration can begin. A body feed of filter aid completely mixed into the liquid is usually required. The purpose of body feed addition is to increase the permeability of the filter cake and thus lengthen the filter cycle.

This is generally done in one of two ways. The first, and simplest, is to add the proper quantity of filter aid into an agitated tank of the liquid to be filtered. This method is commonly employed when the processed liquid is batch-produced and filtered. The principal disadvantage of this method, where body feed concentrations are of the order of 0.1% of the weight of liquid or higher, is the degradation of the body feed as it passes through the filter feed pump, not to mention the long-term wear on the pump.

A second, and more common, method is to make up a slurry of body feed (see *Figure 10.4*) and then, by means of an injection pump, to add the slurry between the filter feed pump and the filter. The injection pump is usually of the plunger type, specially designed for abrasive slurries or the diaphragm (tube) type. Precoat slurry concentrations can be as high as 18% by weight but less trouble with plugging lines and pumps occurs if the concentration is kept near 10%. As most filter aids have some oversized particles in them, suction to the injection pump should be taken at a location off the bottom of the tank to prevent these lumps from getting into the body feed pump valve. A recirculation line from the filter back to the body feed tank for use when the filter is being cleaned will reduce maintenance on the body feed pump resulting from starting and stopping. The change from precoating to filtration must be done with minimum pressure fluctuation and with continuing flow through the filter to make sure the precoat stays in place.

10.6.2.1 Selection of body feed grade

In the selection of body feed, both the amount and the grade of body feed must be considered. The primary criterion for selection of both body feed and precoat, is to produce a filtrate of acceptable clarity to the filter user at acceptable filtration rates. Generally speaking, the most efficient grade of filter aid to use is the most permeable grade that will give the minimum desired clarity and (for the precoat) protect the septum.

While this is a general criterion for the selection of body feed, it does not always turn out that the fastest flow rate grade gives the most efficient filtration in terms of cubic metres of throughput per kilogram of filter aid. More recent work[2] has shown that that grade of filter aid which has a median pore size near that of the median particle size of the suspended solids being filtered will result in the highest filter aid efficiency. This is assuming that the particle size measurement method produces accurate or realistic results, and that the clarity with this grade of filter aid is acceptable.

It is frequently practical to use different grades of filter aid for precoat and body feed. That is, a finer grade of precoat filter aid may be used in order to protect the screen and give immediate clarity. On the other hand, it may be found that the optimum grade for body feed is coarser, or more permeable, than the precoat, or may even be a different type of filter aid (i.e., diatomite for precoat and perlite for body feed). In any case, the grade and type of filter aid selected for body feed must not be so coarse that it permits suspended solids to penetrate it and coat out on the precoat. There they can form an impermeable layer resulting in a reduction in cycle length. As a rule of thumb, the filter aids used for precoat and body feed should not be more than two grades apart in permeability.

10.6.2.2 Amount of body feed addition

Having selected the grade of filter aids for precoat and body feed, it is necessary to determine the amount of body feed addition. This will depend on the amount and type of suspended solids and the desired cycle length. If the suspended solids are incompressible, such as activated carbon and fine bleaching clays, 25% by weight of body feed on the weight of the solids will begin to be effective in lengthening the filter cycle, with the maximum effective amount of body feed usually being equal to two times the weight of suspended solids. For highly compressible solids, such as metal hydrates or yeast, the amount of body feed addition may be 1–8 times the weight of suspended solids. At the higher levels, it is frequently more practical to go to a rotary precoat filter.

The effects of body feed addition are summarized in *Figure 10.5*. Note that

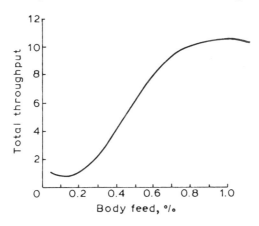

Figure 10.5. Effect of body feed amount on throughput. (Courtesy of Manville Corp.)

the curve has a high and a low point, and actually declines as it goes off the graph on the right-hand side. At the low point (between 0.1 and 0.2 on the graph), the amount of body feed has increased cake thickness but has not increased permeability and has therefore decreased filter cycles. At the upper end of the curve, the same thing occurs where body feed addition has greatly increased cake permeability up to that point, but additional body feed no longer increases permeability and so decreases cycle length. Thus, at the high end of the curve, the total throughput of 10 could be obtained at 0.8% body feed (on the weight of liquid) or, extending the curve off the right-hand side of the graph, at 1.1–1.2% body feed. It is important to run sufficient tests to know on which side of this maximum the data lie. At the point at which body feed is effective, between 0.2 and 0.7 in this example, the curve is quite straight so that data can be interpolated between any points on this curve.

10.6.3 Filtration rate

The next decision is to determine the rate at which filtration is to take place. If filter equipment is already in place, this rate is determined by production needs, so the amount and grade of filter aid must be selected accordingly. For a new installation, the selection of filter rate is a compromise between higher rates and filter aid consumption and lower capital costs, or the reverse.

Figure 10.6 shows the effect of filtration rate on cycle length. Theoretically, doubling the rate decreases the filter cycle by a factor of 4.

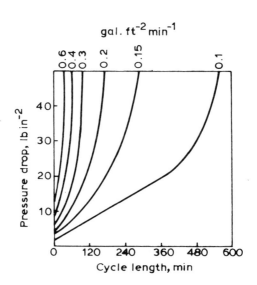

10.6. Effect of filtration rate on throughput. (Courtesy of Manville Corp.)

Actually the decrease is not quite as great in many instances, primarily as a result of a slight reduction in particle removal at the higher rates. In addition if the cake, even with body feed, is compressible, higher rates will increase compressibility and reduce permeability. It is for this reason that it is usually desirable to operate at constant flow rather than at constant pressure, especially if a high degree of clarity is required.

If the desired rate is much different from the rate at which the filter aid grade selection tests were run, it may be found that the filtrate clarity is also different. If the difference is significant, the grade selection tests should be reviewed.

Finally, as mentioned in the introduction, filter aid filtration is primarily a surface filtration in that it does not depend on depth for clarity. In practice, it is found that clarity increases with pressure and solids loading. This is unlike depth filtration. Even though relatively thick cakes can be built up, filtration takes place on the surface of these cakes so that filter aid filtration may be thought of as a continuous surface filter.

10.6.4 Filter cake removal

Poor filter cake removal and resultant plugged septums is probably the most common cause of short filter cycles. It is well worth making sure that all the cake is removed at the end of each cycle. One way to do this is to note precoating pressures. A rise in pressure indicates poor septum cleaning.

At the end of the filter cycle the cake can be removed by back-washing with an air bump (tubular or candle elements only), sluicing, dry cake discharge or combinations of the above. Back-washing is only effective where compressed air is used to 'bump' the cake off the candles.

Sluicing is one of the most effective cleaning methods but, like back-washing, it produces a slurry which may present disposal problems. The velocity of the sluice jet loses its effect within approximately 0.3 m after it hits the filter element. For cakes that are difficult to discharge, filters are made with either movable leaves or sluices in order to obtain the maximum sluicing over as large an area as possible. Sluice pressures should be at least 3.5 kg cm^{-2} (50 p.s.i.g.), but can be over 70 kg cm^{-2} (1000 p.s.i.g.) in special installations.

For dry cake discharge, the filter heel is blown from the filter either back to the filter feed tank or body feed tank (not precoat tank), and the cake removed by vibration. In some filters, the remaining amount of cake (usually only 1 or 2%) is removed from the leaves by sluicing.

10.6.5 Test equipment

Laboratory and pilot filters are available from most filter aid producers and filter manufacturers. One of the greatest causes of error in most test equipment is the substantially larger volume-to-area ratio that is encountered in full-scale filters. This can cause precoat and solids settling and distortion of filtration results because of the large amount of precoat liquid that must be displaced. One filter that overcomes this problem is the Walton[3] test filter (*Figure 10.7*), which has a variable volume chamber and which can also be used with the element in a vertical or horizontal position.

Figure 10.7. The Walton constant rate test filter with variable volume feed chamber, constant rate pump and interconnecting lines and instrumentation. (Courtesy of Manville Corp.)

10.6.6 Automation

Filter aid filtration systems can be completely automated from the precoating through to the cleaning steps.

10.7 ROTARY DRUM PRECOAT FILTER

Although the vast majority of these filters are operated as rotary vacuum precoat filters, there are some that operate as rotary pressure precoat filters where high temperatures and pressures are desirable, such as in the filtration of liquefied coal[4].

The primary advantage of the rotary precoat filter is that it can discharge very thin cakes of relatively large quantities of solids which are very fine or compressible, or both. Cake thickness may be as low as 0.4 mm (0.016 inches) and solids contents in the filter feed of 5% or more by weight or up to 100% by volume, as in the case of some fruit juice or wine materials.

The rotary vacuum precoat filter is essentially a rotary drum filter with vacuum applied over the entire surface of the drum and equipped with a knife, or doctor blade, running the entire length of the drum that can advance towards the drum at a slow rate. Before filtration, a layer (precoat) of filter aid up to 15 cm thick is first formed on the drum. During filtration, suspended solids remain on the surface of the precoat, where they may be washed if needed, and are then removed by the knife. Typical drum speeds range from 1/2 to 2 rev. min^{-1} with those on rotary pressure precoat filters projected as high as 10 rev. min^{-1}. When the precoat is cut to within 6–13 mm of the drum, filtration stops and the filter is cleaned of the remaining precoat heel (approximately 10 mm), and reprecoated.

Diatomite and perlite filter aids with permeabilities of 0.6 Darcies or greater are commonly used in aqueous systems. Permeability can be as low as 0.3 Darcies for light hydrocarbon systems.

The use of a precoat makes the drum filter operation very different from that of a drum or belt filter. For maximum efficiencies, special equipment considerations are necessary, and these are listed below.

Filter equipment considerations

1. Precoat thickness up to 15 cm.
2. Precoat uniformity: ±6 mm.
3. Knife advance rate: 0.0125 to 0.25 mm rev.$^{-1}$
4. Knife flexing: maximum 0.001 25 mm.
5. Drum speed: 2–3 rev. min^{-1} maximum.
6. Tub level control:
   10–15% of drum diameter for precoat;
   as required for filtering.
7. Vacuum: 1.2 m^3 per minute and m^2 area.
8. Internal piping: 'trailing edge'.
9. Septum: 80–200 Frazier permeability.
10. Septum wash sprays: 1379 kN m^{-2} (13.79 bar) flat sprays.
11. Agitator: two speeds:
    precoating—high;
    filtering—low.

10.7.1 Precoating

Precoating objectives are to apply a precoat of desirable thickness (usually $7\frac{1}{2}$–15 cm) for 20–90 minutes. The precoat should have maximum permeability, firmness, evenness and a minimum of cracking. By means of different precoat techniques, it is possible to apply a precoat that meets the above standards or one that takes 3–4 hours to apply with a maximum thickness of only $7\frac{1}{2}$–10 cm and a permeability so low as to make desired filtration rates unattainable.

The following are recommended for maximum precoat efficiency:

1. maximum vacuum (capacity should be 1.2 cubic metres per square metre of filter area; 4 cubic feet per square foot);

2. 1–2 rev. min^{-1} drum speed;
3. 7% or higher slurry concentration;
4. constant slurry concentration in the filter;
5. 15% max. drum submergence;
6. pH adjustment to that of the liquid being filtered;
7. flat sprays at 0.7–1.0 kg cm^{-2} (10–15 p.s.i.g.), if available.

10.7.2 Filtration

With the precoat in place, the precoat slurry that remains in the filter tub can be drained or unfiltered liquid added directly.

Filtration variables are:

1. grade (and resultant amount of penetration of unfiltered solids) of precoat;
2. precoating method;
3. cake filterability;
4. drum speed;
5. drum submergence;
6. knife advance rate;
7. filtrate viscosity;
8. cake removal efficiency.

10.7.3 Precoat grade

A grade of precoat should be selected such that the filtered solids do not penetrate into it more than 0.2–0.8 mm (0.001–0.003 inches), to minimize the rate at which the precoat must be removed by the knife to maintain flow rate. The amount of penetration will depend on the particle size of the solids being filtered and on the pore size of the precoat material. Because the solids being filtered are usually of a blinding nature, they must be completely removed on each revolution of the drum. The less the amount of penetration, the smaller the amount of precoat that must be removed with each drum revolution. Thus, the use of the more dense diatomite with its smaller sized pores (as compared to perlite) may result in enough lower knife advance rates to more than overcome its density disadvantage.

10.7.4 Cake filterability

The suspended solids in some cakes are sufficiently compressible that no filtration takes place after only 3–5 seconds of precoat submergence. Other types of solids may continue to form cake on the precoat for several minutes or more but, of course, flow rate drops as cake thickness increases. In addition, cutting efficiency may be poor due to the cohesive nature of the thicker cakes. The nature of cake filterability results in the manner in which the filter is optimized.

10.7.5 Drum speed

For cakes that seal off completely, there is little to be gained by operating at drum speeds and submergence times in excess of the cake sealing time. This may result in drum speeds of up to 3 rev. min^{-1} or more for vacuum filters, and as high as 10 rev. min^{-1} for pressure filters. In these situations the filtration rate may be directly proportional to drum speed with little or no increase in precoat consumption per unit of throughput, as drum speed is increased. For more filterable cakes, higher drum speeds will result in higher rates due to less cake resistance and also because cake removal efficiency may be increased with the thinner cakes.

10.7.6 Drum submergence

Drum speed is a primary factor affecting filtration rate. Drum submergence should be set at that level which will permit adequate cake drying at a given drum speed. As drum speed is increased, it will usually be necessary to decrease submergence to maintain cake dryness. If the cake is too wet, drum submergence would be decreased before reducing drum speed.

10.7.7 Knife advance rate

The knife advance rate required for a given throughput will depend on a combination of precoat permeability (solids penetration) and drum speed. *Figure 10.8* shows typical filtration rate versus drum speed for different knife

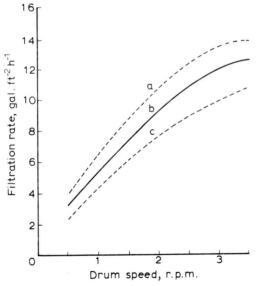

Figure 10.8. Filtration rate versus drum speed at different knife advance rates. Curve a—1.6mm per rev. knife advance. Curve b—0.8mm per rev. knife advance. Curve c—0.4mm per rev. knife advance. (Courtesy of Manville Corp.)

advance rates using a specific grade of precoat filter aid. For a finer grade of filter aid, increasing the knife advance rate from 2 ml rev.$^{-1}$ to 4 ml rev.$^{-1}$ may not result in any increase in rate as the cake is being completely removed with each cut. From the graph, equal rates can be obtained by different combinations of drum speed and knife advance. It is obviously desirable to select the combination which minimizes precoat usage. The data from *Figure 10.8* have been recalculated to show precoat usage versus drum speed at different knife advance rates (*Figure 10.9*). For instance, from *Figure 10.8*, a rate of 0.4 m h^{-1} (10 gallons per square foot per hour) can be obtained by a 0.13 mm (4 ml) per revolution knife advance and a drum speed of 1.75, or 0.025 mm (1 ml) per revolution knife advance and a 3 rev. min^{-1} drum speed.

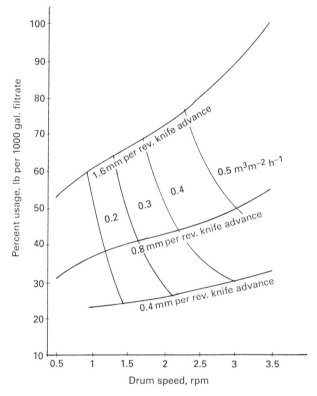

Figure 10.9. Precoat usage versus drum speed from Figure 10.8. Curve a—1.6mm per rev. knife advance. Curve b—0.8mm per rev. knife advance. Curve c—0.4mm per rev. knife advance. (Courtesy of Manville Corp. and Gordon Smith)

Figure 10.9 shows a precoat usage amount of almost 8.4 kg m^{-3} (70 pounds per thousand gallons) of filtrate for the 0.13 mm and 1.75 rev. min^{-1} drum speed combination as compared to 3.6 kg m^{-3} (30 pounds per thousand gallons) precoat usage for the 0.025 mm per revolution knife advance, 3 rev. min^{-1} combination. The lines crossing the precoat usage versus drum speed curves show the different combinations of drum speed and knife advance that can be used to obtain specific flow rates.

10.7.8 Cake removal efficiency

The above selection of high drum speed and low knife advance rate is only effective if the cake of filtered solids can be removed smoothly and evenly without causing gouging or chipping of the precoat. The thinner cakes are easier to remove evenly, which points to higher drum speeds and lower submergence up to a point. At that point the knife may no longer be able to maintain an accurate advance rate with a resultant decrease in cake-removal efficiency. Knife advance deviations from the desired rate of up to 0.13 mm (0.005 ml) which have been measured on commercial filters of 30 m^2 and larger areas with total knife movement twice this amount. The effect of this deviation on precoat consumption and filter rate can be seen from *Figures 10.8* and *10.9*. Knife flexing can be minimized by maintaining a sharp blade with a bevel facing up and by adjusting the knife angle so that the outward and downward forces against the knife, caused by the resistance of the precoat to the advancing knife and the drum rotation, are resolved along the plane of maximum knife strength.

10.7.9 Filter septa

Since filtration on a precoat filter takes place on the precoat and not the cloth, a more permeable cloth than is used on drum filters without precoat must be, and can be, used. There are, of course, many types of cloth, but general specifications for the rotary precoat filter are as follows:

1. Permeability: 80–200 Frazier (scf ft^{-2} at 0.5 in H$_2$O p.s.i.);
   (24–60m^3 × m^{-2} at 1.25 mm water pressure;
2. monofilament for minimum blinding;
3. plain or twilled weave with a typical thread count of 30 × 66, or
   44 × 108 (200–300gm^{-2});
4. a weight of approximately 8–10 oz yard^{-1}.

Polypropylene is most commonly used. In the metal cloths, the 24 × 110 Dutch weave gives a better combination of permeability and retention than almost any of the synthetic cloths found to date and will retain all grades of diatomite or perlite precoat materials that can be applied to a filter.

10.7.10 Testing

Test data can be obtained through use of a test leaf, such as the one in *Figure 10.10*, which is designed to simulate a 93 cm^2 area on a filter drum. In operation, a precoat up to 50 mm is formed in the leaf. The leaf is then immersed in the test liquid, raised for cake drying, the precoat (with cake) is advanced out of the end of the leaf, an amount equal to the desired knife cut by means of a screw at the back of the leaf, and the precoat and cake cut from the leaf. Auxiliary equipment includes a 'clutch' so the leaf can be left in position without hand holding, two graduated filtrate receivers and a cross-over valve arrangement.

10.7.11 Filter aid re-use and regeneration

With the exception of a few very large use locations (in the order of 3000 tonnes per year), it has not been found economical to regenerate filter aid. Work has been done on the recovery of filter aid by use of cyclones but, despite many tests, this is being carried out at only one location to the writer's knowledge.

Figure 10.10. Rotary vacuum precoat test leaf with auxiliary equipment

10.7.12 Filter aid packaging

Filter aid is most commonly packaged and shipped in 22 kilogram multi-wall paper bags. In large industrial areas, particularly in the USA, filter aid is shipped in pressurized railcars which are unloaded into silos for automatic handling at the use point.

REFERENCES

1. Svarovsky, L., *Solid–Liquid Separation,* 1st edn, Butterworth, London (1977)

2. Cain, C. W., 'Filter aid filtration—the interaction of filter aid pore size and turbidity pore size', presented at the *American Institute of Chemical Engineers Meeting,* Vancouver, Sept. (1973)
3. Walton, H. G., 'Laboratory procedure and filter for diatomite filtration tests', *Filtration and Separation,* Jan./Feb. (1978)
4. Electric Power Research Association, *Filtration Equipment Development for Coal Liquids,* EPRI Report No. AF 459-1, Phase 2, Palo Alto (1978)

BIBLIOGRAPHY

Bell, G. R. and Hutto, F. B., 'Analysis of rotary precoat filter operation—new concepts', *Chem. Eng. Progr.,* **54,** 69 (1958); **55,** (1959)

Cain, C. W., 'Filter aid, use in filtration', *Encyclopedia of Chemical Processing and Design*, p. 348, M. Dekker, New York (1984)

Mefford, M. H., Lowder, J. R. and Coors, J. K., 'D. E. filtration optimization and large scale filter systems start-up', *MBAA Tech. Quart.*, **20,** No. 3 (1983)

Smith, G. R. S., 'Improve your filter aid filtration', *Chem. Eng.,* Oct. 11, 187 (1965)

Smith, G. R. S., 'Filter aid regeneration and recovery', *Chem. Eng. Progr.,* **71,** 37 (1975)

Smith, G. R. S., 'How to use rotary vacuum precoat filters', *Chem. Eng.,* Feb. 16, 84 (1976)

Tiller, F. M., Crump, J. R., Chen, W. and Shen, Y. L., *'Cycle Optimization Involving the Use of Filter Aids',* Chem. Eng., University of Houston, Houston, TX.

Tiller, F. M. and Khatib, Z., *'Choosing the Correct Quantity of a Filter Aid Powder to Maximize Cycle Filtration Rates'*, AIChE, TX Filtration and Separation Society, Houston, TX (1987)

U.S. Patent No. 4,514,306, 30 April 1985. Method of and apparatus for controlling the quantity of filter aid fed to a sediment filter so as to maintain virtually constant a preselected optimum specific cake resistance

11

Deep Bed Filtration

K. J. Ives

Professor of Public Health Engineering, University College, London

11.1 INTRODUCTION

The principles of deep bed filtration are quite well-known. It is a clarification process using a deep bed of granular media, usually sand. As a sewage tertiary treatment process it can frequently produce filtrates containing only $5 \, \text{mg} \, \text{l}^{-1}$ or less of suspended matter. In conjunction with coagulation processes, either directly or with sedimentation, filtration may produce filtrates of exceptional clarity, with the suspended matter scarcely detectable by turbidimetric instruments.

The penalty paid for this clarification is that the filter becomes clogged, requiring more and more energy to sustain the required flow. When the energy required reaches the maximum available, the filter media have to be cleaned, usually by reverse flow backwashing, assisted by air scouring or auxiliary wash jets.

Rates of filtration are normally in the range of 5 to $15 \, \text{m} \, \text{h}^{-1}$ (100 to $300 \, \text{gall} \, \text{ft}^{-2} \, \text{h}^{-1}$), with a trend towards the higher rates. The media may range in size from about 0.4 to 2.5 mm, and be graded in various layers (dual media or multi-media) or be size-graded continuously, as in up-flow filters. The thicknesses of the layers may vary, but overall depths are normally about 0.6 to 1 m (2 to 3.3 ft), except for up-flow filters which are considerably deeper.

Open gravity filters usually operate to a head loss limit of 2.5 m (8 ft) water gauge, but pressure filters may use up to 2 or 3 times this value. Backwashing is normally employed at a rate sufficient to fluidize the media, and to ensure sufficient separation for multiple layer filters. Wash rates of up to 0.6 m min^{-1} (2 ft min^{-1}) can achieve this, although this depends on the media, and the water temperature. If air scour is used, typical rates are $0.4–0.6 \, \text{m}^3 \, \text{m}^{-2}$ min^{-1}. Normally, backwash water would represent about 1% of filtrate production; 3% would be high and quantities over 5% would be excessive.

Filter runs of about 24 h between washes are normal. Where the water to be filtered does not contain a high organic load, runs may be up to 100 h, but in sewage tertiary treatment filtration a daily washing is desirable to prevent anaerobic decomposition of the deposits retained in the filter pores. If filter runs fall to below 8 h it becomes difficult to maintain production, and washwater demands become excessive.

This brief review of existing filtration practice may inform the designer or operator whether his filter is within the accepted range. But it does not define for the enterprising designer the best selection of conditions for the filtrate quality criteria he is trying to satisfy and it does not inform the intelligent operator whether his filters are working in the best manner, or how he could increase their production.

11.2 THEORY

Filtration is a process which takes place within the pores of the granular material and should not be confused with straining, which takes place on the surface of a mesh or fabric. The consequence of this is that filters act in depth, with each pore having a certain probability of retaining particles from the suspension flowing through.

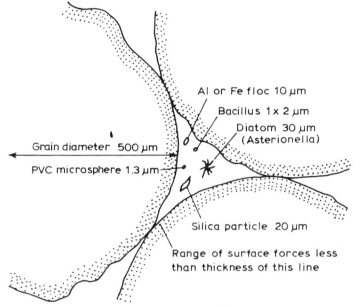

Figure 11.1. Typical small filter pore

This follows from the fact that the particles to be filtered are much smaller than the pore size (*Figure 11.1*). As the flow is laminar, that is streamlines do not mix, there must be forces which move particles across the streamlines to touch the filter grains. These actions which move particles up to the grain surfaces are called the 'transport mechanisms'[1].

11.2.1 Movement of particles

11.2.1.1 Transport mechanisms

The principal forces which act on a particle to move it across the streamlines are due to:

1. diffusion (Brownian motion), if the particle is very small, less than 1 μm;
2. gravity (Stokes settlement), significant if the particle is large enough, more than 5 μm, and appreciably denser than water;
3. hydrodynamic forces, which are due to the velocity distribution in the filter pores, together with the shape of the particle, which cause it to rotate and translate across the flow field (analogous to the curved flight of a spinning ball).

Another factor to be taken into account is the finite size of the particle so that when its centre approaches to one radius of grain surface it is intercepted.

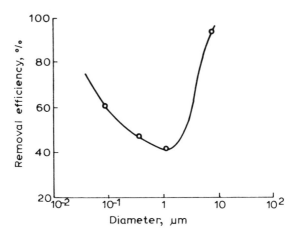

Figure 11.2. Filter efficiency depending on particle size (Yao)

The efficiency of removal of particles as a function of their size (at constant density) is shown in *Figure 11.2*. The minimum occurs at about 1 μm where the particles are too large for diffusion, too small for gravity and small for interception. It should be noted therefore that, somewhat surprisingly, very small sub-microscopic particles are more efficiently filtered than some which are larger but of microscopic size.

11.2.1.2 Attachment mechanisms[2]

When a particle has been transported to a grain surface, interface forces determine whether it will attach. Particles and grain surfaces in water carry small electric potentials (usually less than 20 mV), normally negative in sign. Depending on the dissolved ions in the water, an electrical repulsion may occur so that the particles cannot attach. This repulsion may be neutralized, or even reversed to attraction, by the presence of positive coatings on the grain, of aluminium or ferric hydroxide, or cationic polyelectrolytes.

Between the particles and grain surface, at very close approach (less than 0.1 μm) there exists the attractive London–van der Waals force, which will ensure attachment if the electrical repulsion force is not too large. The approach of the particle to the grain surface will be slowed by the viscous resistance of the water film between the two approaching surfaces, which has to be displaced radially for contact to occur.

In general, attachment mechanisms are operative unless the water is very low in dissolved salts (approaching distilled water) or the particles or grain surface has specifically adsorbed chemicals, thus causing a high surface electrical potential.

11.2.1.3 Detachment mechanisms

Particles, or aggregates of particles, may be detached from filter grain surfaces if their bonding is weak. The viscous drag of the water flowing through the pores creates a shear stress on the deposited particles. As the pore sizes diminish due to accumulated deposits, so the mean pore velocity increases, with an increased shear stress. There is evidence that this detaches deposits, especially if a sudden increase in flow rate occurs, or if it is created deliberately as in backwashing. Also, some particles which collide with existing deposits can create micro-avalanche effects detaching some particles. Such detachment mechanisms only appear to be significant when pores are considerably clogged with accumulated deposits.

11.2.1.4 Combined effects

The physics of particle behaviour in filter pores shows that:

1. filter efficiency depends on particle size (as shown in *Figure 11.2*), density and shape;
2. greater grain surface area, i.e. smaller grains, give higher efficiencies, and angular grains are better than spherical;
3. higher flow velocities give lower efficiencies;
4. filtration is more efficient at higher water temperatures, principally due to lower viscosity;
5. downflow is more efficient than upflow for gravity-dominated transport mechanisms, but there is no difference for diffusion- or hydrodynamic-dominated transport mechanisms.

These aspects of the performance of filters are discussed more fully in Ives[1].

11.2.2 Depth and time dependence

For a filter which is uniform at all depths, i.e. the same grain size throughout, consideration of the filter mechanisms shows that each layer should be equally efficient at removing particles of a given type. The removal efficiency can be defined as the fraction of particles removed per unit depth. Mathematically,

$$\lambda = -\frac{\delta C}{C}\frac{1}{\delta L} \tag{11.1}$$

where $-\delta C$ is the reduction of concentration of particles passing through a layer δL. Equation 11.1 can be rearranged into the form of the Iwasaki equation:

$$-\frac{\partial C}{\partial L} = \lambda C \qquad (11.2)$$

where λ is the filter coefficient. Integration of equation 11.2, at the commencement of filtration ($t = 0$), when the filter is clean and therefore uniform, gives

$$\ln \frac{C}{C_0} = -\lambda L \qquad (11.3)$$

where C_0 is the inlet concentration of particles being filtered.

The consequences of equations 11.1–11.3 are shown in *Figure 11.3*, where the filter has been arbitrarily divided into ten uniform layers, each 40%

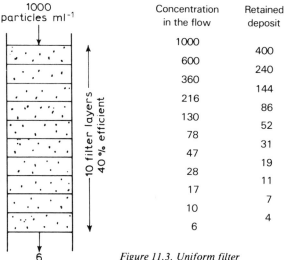

Figure 11.3. Uniform filter

efficient in removing particles from suspensions, with an inlet concentration of 1000 particles ml⁻¹. Note that the column showing the number of particles in suspension declines logarithmically with depth, in accordance with equation 11.3. However, it is important to note that the number of particles retained as deposit is not equal in all layers, but that the top layers contain most of the deposit and the lower layers contain very little. Therefore a uniform filter does not become uniformly clogged with deposit, but the upper layers carry the burden and the lower layers contribute little.

This situation is made worse if the filter grains are not uniform and are size-graded by backwashing, as is frequently the case in practice. As the filter coefficient, λ, is less for larger grains, the filter efficiency diminishes layer by layer down through the filter depth, as exemplified in *Figure 11.4*. This destroys the simple logarithmic decline of equation 11.3 but, more importantly, it exacerbates the loading of the upper layers with even less use of the lower layers. This causes rapid clogging of the top layers, leading to a shortened filter run, when the lower layers are greatly under-utilized.

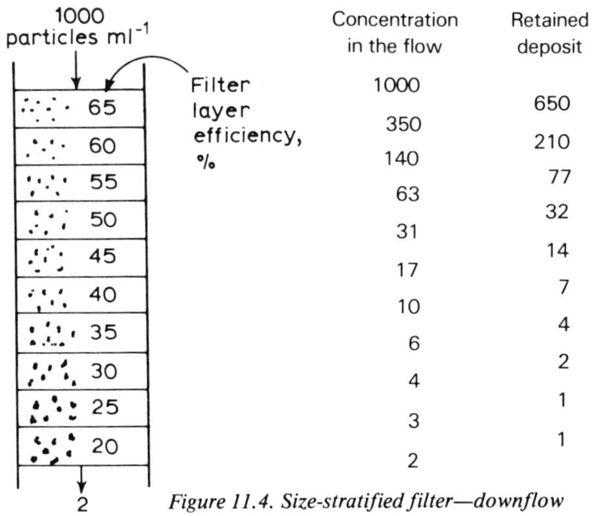

1000 particles ml^{-1}

	Filter layer efficiency, %	Concentration in the flow	Retained deposit
		1000	
65			650
		350	
60			210
		140	
55			77
		63	
50			32
		31	
45			14
		17	
40			7
		10	
35			4
		6	
30			2
		4	
25			1
		3	
20			1
		2	

2

Figure 11.4. Size-stratified filter—downflow

However, by using exactly the same filter, but in the upflow direction and assuming filter mechanisms which are not gravity dominated, the resultant pattern of removal and deposition is as shown in *Figure 11.5*. In this case the deposit load is spread much further into the filter, with better utilization of the filter material and consequent longer filter runs. Notice that the overall efficiency of the filters is high in each case (over 99% removal), although the internal distributions are significantly different.

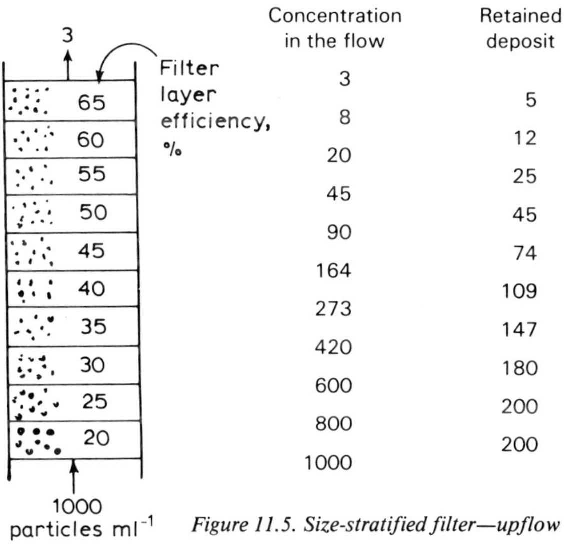

3

	Filter layer efficiency, %	Concentration in the flow	Retained deposit
		3	
65			5
		8	
60			12
		20	
55			25
		45	
50			45
		90	
45			74
		164	
40			109
		273	
35			147
		420	
30			180
		600	
25			200
		800	
20			200
		1000	

1000 particles ml^{-1}

Figure 11.5. Size-stratified filter—upflow

This illustrates one of the advantages of upflow filters, which can retain more deposit than downflow. However, for drinking-water treatment they have the disadvantages of requiring covers, to protect the filtrate, and of backwashing through the filtered water compartment, which is hygienically undesirable. Consequently, they are more applicable to tertiary filtration of sewage or industrial water treatment.

An alternative to upflow, which will present a coarse-through-fine grading sequence, is the dual or multilayer filter. An example of a three-layer filter of carbon, anthracite and sand is shown on *Figure 11.6*, similar to the filters

1000 particles ml^{-1}		Concentration in the flow	Retained deposit
20%	Activated carbon	1000	
		800	200
15		680	120
10		612	68
50	Anthracite	306	306
45		168	138
40		101	67
35		66	35
80	Sand	13	53
75		3	10
70		1	2

Filter layer efficiency, %

Figure 11.6. Three-layer filter

used at Siegburg (Bonn) in West Germany. Because uniform filter material cannot be obtained in practice, each layer has a local size-grading but the overall sequence has a progressive increase in filter efficiency with depth. As shown, this distributes the deposit uniformly allowing greater utilization of the deeper layers. Once again, the overall removal efficiency is high, over 99%.

11.2.2.1 Time dependence

The simple calculations in *Figures 11.3* to *11.6* are valid only when the filters are clean, with no deposit. The presence of accumulating deposits modifies

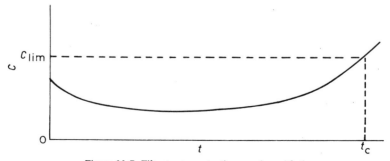

Figure 11.7. Filtrate concentration varying with time

the local filter coefficient in a complex manner, initially increasing it, then decreasing it ultimately to a complete non-retaining state, i.e. when the pores are saturated with deposit, filter efficiency in those pores is zero.

This means that the concentration of suspension emerging from any layer in the filter varies with time, as shown on *Figure 11.7*, first improving but later deteriorating. The latter part of this curve is frequently referred to as 'breakthrough'.

The accumulating deposits in a filter layer decrease its permeability. For a filter operated at constant rate, there is a consequent increase in the hydraulic gradient across the layer which is proportional to the quantity of deposit present:

$$\frac{\partial H}{\partial L} = \left(\frac{dH}{dL}\right)_0 + K\sigma \tag{11.4}$$

where $(dH/dL)_0$ is the initial hydraulic gradient across the clean filter material, and σ is the quantity of deposit per unit filter volume, called the 'specific deposit'.

The total head loss H across all the layers of the filter is obtained from integration of equation 11.4 with respect to depth L, at any time t. This integration is approximately linear with time, and can be given as

$$H = H_0 + KvC_0t \tag{11.5}$$

where H_0 is the initial, clean-filter head loss across all the layers, and v is the approach velocity equal to the volumetric flow rate per unit face area of the filter (Q/A).

Equation 11.5 is illustrated in *Figure 11.8*. Also shown is a line curving upwards, which may be observed in practice. This steeply mounting

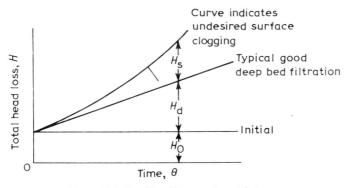

Figure 11.8. Total head loss varying with time

exponential curve of head loss is due to the formation of a surface mat of deposit on the inlet face of the filter material. This mat indicates that straining is taking place—the very antithesis to depth filtration. The consequence is a much shortened filter run with very little utilization of the filter layers below the inlet surface. If such steeply curving head loss lines are observed, the surface mat effect can be reduced by one or more of the following procedures:

1. reduce the inlet concentration by pretreatment (e.g. presettlement, possibly with flocculation);
2. remove the larger particles from suspension by coarse primary filtration or micro-straining;
3. replace the inlet layer of filter material with coarser grains;
4. increase the filtration velocity.

11.2.3 Optimization

It can be seen from *Figure 11.7* that the concentration of suspension in the filtrate will rise in the later part of a filter run. When this reaches the quality limit set for filtered water, e.g. 0.5 mg l^{-1}, the filter run must be terminated at time t_c. Similarly, it can be seen from *Figure 11.8* that the total head loss rises with time. When this reaches the limit of head loss set by the hydraulic conditions in the filter, the filter run must be terminated at time t_H.

If either of these times is reached first, the filter is being under-utilized with respect to the other performance criterion. So if $t_H < t_c$, the filter is under-utilized for its capacity to produce satisfactory filtrate, and vice versa if $t_c < t_H$. But if the two limit times are reached simultaneously, then $t_c = t_H$, and the filter operation is optimized. This is illustrated in *Figure 11.9*.

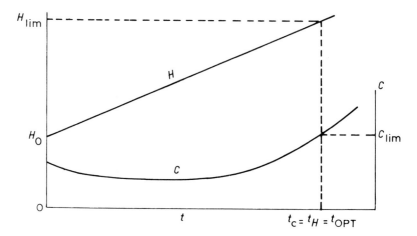

Figure 11.9. Operationally optimum filter performance

Formal procedures for optimizing uniform filters at the design stage are set out in Ives[1]; some of these are described later, in section 11.3.6. For size-graded or multilayer filters, optimization techniques are still being developed. For filters already operating, the most useful variable to adjust performance towards the optimum is the dose of polyelectrolyte to the filter inlet. Such procedures are described by Mints[3], but require continuous monitoring of filtrate quality and total head loss for each filter. Such instrumentation is becoming more and more normal, with several reliable continuous flow turbidimeters available on the market.

11.3 PROBLEMS OF DESIGN AND OPERATION

If the designer or operator faces a range of design variables, yet does not know the best combination of flow rate, grain size, depth, etc., for his particular problem he will have to consider the possibility of some pilot or experimental studies. Similarly, if he reviews filtration theory and finds that he cannot predict filter design and performance without empirical coefficients, he must again consider experimental tests.

Some methods of testing and their implications will be presented here, to act as a guide for those wanting design or operating information.

11.3.1 Design and operation factors

In design, the following factors have to be considered:

1. the quality of the water reaching the filter;
2. selection of media—type;
                   —grading;
3. depth of media;
4. rate of filtration;
5. backwash rate; provision for air scour;
6. underdrain system;
7. hydraulic equipment: pipes, valves, wash troughs, flow control, flow and head loss indicators;
8. filter structure.

In operation, some of the factors are similar:

9. the effect of chemical dosing (coagulants, polyelectrolytes), on filter performance;
10. can the filter run be made longer?
11. can the filtrate quality be improved?
12. can the rate of filtration be increased?

In general, operational problems are easier to solve than design problems because the pretreated water is available for testing changes in operation. In design the treatment works may not exist, and only raw samples may be available.

11.3.2 Water quality

Water samples can be analysed in the laboratory, by chemical and physical testing. These may reveal the biochemical oxygen demand (BOD) (or other parameter of oxidizable organic material), the total organic carbon (TOC), suspended solids, colour, pH etc., but they do not reveal how well the water will respond to chemical precipitation and coagulation, and filtration.

In chemical treatment the jar test may be used as a guide to the chemical conditions required to provide rapid settling precipitates and flocs. But a comparable test is required if, after settlement, the supernatant water can be successfully filtered.

Where there is no existing water treatment works, but the raw water is available, pilot units will be required to simulate the stages of treatment prior to filtration. If this is to be physico-chemical precipitation and coagulation, then jar testing may be a useful indicator. In this case the following procedure could be used.

1. Set up the jar test in the usual manner trying different chemicals and doses.
2. Note the conditions: formation of precipitates and flocs, rate of settling, for each test.
3. After the contents of a jar have settled for 10 min transfer the supernatant to a filtrability test apparatus, described later. This apparatus requires about 1 litre of sample; if the jars contain less than this, two or more jars will be needed, operating under identical conditions.
4. Measure the filtrability. The lowest value will be the best.
5. The best chemical dose can be chosen taking into account both the jar test and the filtrability. The best settling floc may give poor filtrability, or a poor settling floc may give a good filtrability, and a compromise may have to be made.

If this filtrability test is to be used as an operational guide, then water from the prior precipitation/coagulation/sedimentation stage is used in the filtrability apparatus.

11.3.3 Filtrability test

The filtrability test can be used in two ways:

1. to test changes in water quality against standard filter media at a standard rate;
2. to test changes in rate, or filter media, against a constant quality of inflow water.

In both cases the following considerations apply:

1. the head loss should not rise at too fast a rate;
2. the filtrate quality should not be too poor;
3. if the filter can operate at worse inlet qualities it is an advantage;
4. the higher the flow rate, the more economic will be filtration.

Bearing these in mind, a filtrability number F, is defined as

$$F = HC/C_0 vt$$

where

> H/t is the rate of rise of head loss
> C is the average filtrate quality during time t
> C_0 is the inlet quality
> v is the velocity of filtration.

In consistent units e.g. H, mm; v, mm s^{-1}; t, s; with C/C_0 a ratio, F is dimensionless.

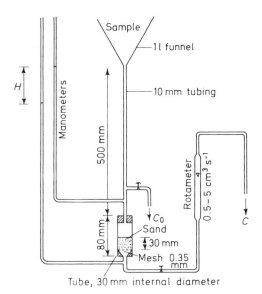

Figure 11.10. Filtrability apparatus

The apparatus is shown in *Figure 11.10*. Standard conditions for testing treated water could be 0.55 mm sand (25–30 sieves) at $v = 2.0$ mm s^{-1} (150 Imp gall ft^{-2} h^{-1}). For the sizes given on the diagram, 33 g of sand will be required. It is important always to put in the same weight of sand and bring it to the same depth mark in the tube. For the size shown, the test will run for about 11 min.

The apparatus should be filled with clean water to saturate the sand and remove all air bubbles. The water level should then be drawn down to the bottom of the funnel, and the sample put into the funnel. The timing clock should be started when the sample reaches the filter cell, which at the rate suggested above is 27 s after opening the valve to commence flow. (Increase or reduce this time proportionately if rate is below or above this value.)

The outlet valve may have to be adjusted during the test to maintain constant flow, as measured by the rotameter or stopwatch and measuring cylinder. Head loss can be measured during the test, the critical value is when the sample level reaches the bottom of the funnel. The quality of the influent can be measured by withdrawing a sample as indicated, or taking it from the funnel. The outlet quality can be sampled at intervals and averaged, or the entire filtrate collected and sampled, after the clean water has been displaced from the apparatus. Any suitable quality parameter can be used, e.g. suspended solids, turbidity, residual coagulant.

It must be stressed that this apparatus measures filtrability on a comparative basis. It is not a design tool for predicting media sizes, depths, rates, etc., although with experience it can be an aid to making such decisions. It can however, be a guide to operation control, and a preliminary step to more careful tests which are to follow.

The apparatus has restricted applicability; for example, nonsensical results would be obtained for the F value if $C = 0$, or if infinite values of C_0 or v were used. The apparatus must not be confused with the micro-straining filtrability test, which is not applicable to sand filters.

11.3.4 Selection of media

Selection of suitable media is principally concerned with size and size-grading. Some preliminary testing of uniform size samples can be made in the filtrability test apparatus. More detailed testing of size-graded media can be made in pilot units to be described later. However, certain other tests may be desirable, particularly if dual or multiple layer filters are contemplated. These are:

1. Microscopic test. Under low power (about $\times 20$) the media can be examined by transmitted and reflected light. Qualitative assessments of the shape, roughness, durability, staining can be made. Microscopic examination is not definitive, but with experience becomes a useful aid.
2. Solubility test. Loss of weight, and dissolution of staining material can be measured in 20% HCl. It should not exceed 2%, after 24 h immersion.
3. Durability test. Extended water washing, for 100 h, with a check for loss of weight, will assess the durability (resistance to attrition) for an equivalent of 3 years' working. Details of this test are given by Ives[4].
4. Density test. A standard density bottle test can be made. With anthracite material thorough soaking (24 h) is advised because of micropores which may contain air. Note that not all anthracites have the same density, e.g. Pennsylvania anthracite is 1600 kg m^{-3}, but Welsh anthracite is 1400 kg m^{-3}.
5. Settlement rate test. A uniform sieved size fraction is allowed to settle grain by grain through a 1 m deep column of water. The settling velocity of the upper layers in multiple layer beds must be lower than that of a layer below. *Figure 11.11* shows an example of a multiple layer media selection chart based on settlement rate.
6. Sphericity test. The extent to which a grain is non-spherical cannot be readily defined from geometric measurements so that it has meaning from a hydraulic point of view. A useful definition of sphericity is the ratio of the diameter (d) of a sphere of settling velocity equal to that of the grains, to the mean size of the passing and retaining sieves (d_s); for a sphere d/d_s is 1.0, for rounded sands about 0.85, for anthracite about 0.7. The value of d can be calculated from the settlement rate

measurement, using textbook procedures[5,6]. Note that most grains are too large to follow Stokes' law, which cannot, therefore, be used.

These tests on media require laboratory personnel, and it may be best to have a test service centrally located, at a government or university laboratory, with reference collections of media available for comparison.

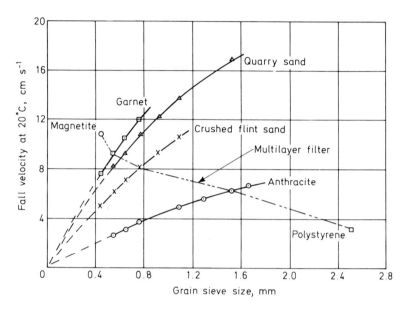

Figure 11.11. Settlement of media in multiple-layer filters

11.3.5 Pilot filters

Both from the point of view of design (see section 11.3.1, points 2, 3 and 4) and operation (section 11.3.1, points 9, 10, 11 and 12) a pilot filter can give valuable, or often definitive, information.

Some pilot filters for research purposes can be extremely complex and quite expensive units, built from perspex, stainless steel, etc., with automatic control and sophisticated instrumentation. Other pilot filters can be constructed as reinforced concrete reduced plan area units, similar to full scale filters. However, they are frequently inflexible, and require much labour to make any experimental changes.

Experience has shown that pilot filters, as small as 150 mm (6 in) in diameter give the same results as full scale units, providing they have the same vertical height. Consequently, simple pilot units can be erected at a water treatment works for experimental purposes.

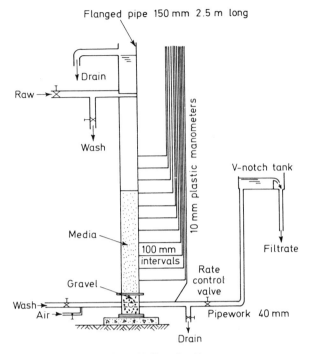

Figure 11.12. Pilot filters

Such a unit made from 150 mm (6 in) diameter flanged pipe is shown in *Figure 11.12*. It has the disadvantage that the media is not visible, as in perspex filters, but conversely biological growths in the media caused by sunlight will not occur. Flow control is manually controlled by a valve, and the flow is measured either by a rotameter, if available, or a weir tank. As shown, the weir tank will prevent the pressure in the filter falling below atmospheric; however, if the tank is set lower, more head is available to extend the filter runs.

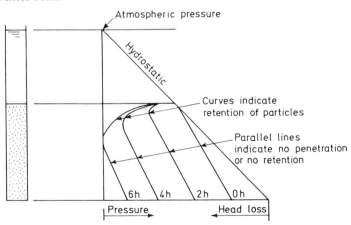

Figure 11.13. Pressure curves

With such an experimental filter, media can be replaced easily, so tests on different media at different flow rates can be tried. The manometer readings can be plotted on a graph as shown in *Figure 11.13* and so the clogging of each layer can be determined. This will show if the media is too fine or too coarse to spread the load through the bed.

Filter washing is an important aspect of the operation, and can also be tested on such a unit. The degree of expansion of the media can be measured during washing, by lowering a rod on which small cups are fastened at about 30 mm intervals. After washing has finished, the cups can be taken out—sand in the cups will indicate how far the media has expanded.

The pilot filter shown in *Figure 11.12* is a gravity down-flow filter. Similar units can be made for up-flow or bi-flow tests. In some cases, manufacturers have pilot units which can be hired or borrowed for on-site experiments.

11.3.6 Full-scale filters

The use of existing full-scale filters for experimental purposes is attractive because there are no problems of scale, and the investigator may feel a greater confidence in the results. If the objective is to test the effects of varying pretreatment, then this may have some justification, but it does not allow easy adjustment to the filter material, either in the thickness of the media, or the grain size.

It is possible to install filter probes for pressure measurement throughout the depth of the media, to give pressure profiles as with pilot filters. However, this requires holes to be cut through the filter walls (which are often reinforced concrete) for insertion of the probes, and it requires careful operation to maintain the manometer system because air may enter during the washing process.

There may also be practical problems in attempting to increase flow rates during experiments, because of the hydraulic limitations of the inlet and outlet piping, valves, etc. An interesting possibility is declining rate filtration, in which flow control is virtually discarded, and the filter output falls as the head loss due to clogging builds up. This will often overcome problems of filtrate deterioration ('breakthrough') towards the end of the filter run. The average filter output should be maintained by higher rates operating at the beginning of the run when the media are still fairly clean. Obviously, this requires several filters (probably a minimum of six), deliberately worked out of phase, to maintain the output of the treatment works[7].

Even if pressure probes cannot be fitted, a total head loss indicator, with recorder, can be interpreted to some advantage. If the total head loss remains approximately linear with time, the filter is performing as a deep bed, but if the plot of head loss against time curves upward with time, a surface layer has formed on the top of the media, which is undesirable. This may be a result of wrong pretreatment (too great a suspended solids load, or too much polyelectrolyte), or of media at the filter surface which are too fine, or of a filtration rate which is too slow.

A continuous monitor for the filtrate quality is also desirable; there are several continuous flow turbidimeters on the market which can do this. Interpretation of this record can indicate the onset of breakthrough, or can allow some simple optimization studies to be made[8].

Practical selection and design of underdrain systems, pipework, flow controllers, and backwash systems is outside the scope of this chapter, but reference can be made to the textbook by Fair *et al.*[5] and the review by Huisman[9].

Operational experience with filters for water treatment has been predominantly for final clarification following conventional flocculation and sedimentation processes. There is a growing literature, some containing details of experimental work either on pilot units, or on full-scale filters, describing such filtration processes, in the normal technical press dealing with water and waste water treatment.

11.4 CURRENT DEVELOPMENTS

With growing demands for water supply, stricter standards for drinking water, higher quality discharges from waste-water treatment plants for the protection of groundwater, rivers and lakes, and the increased interest in the renovation of waste water for re-use particularly for irrigation and industrial water recycling, filtration continues to play a principal role in treatment.

11.4.1 New filter media and filter aids

As sand is the preferred filter material for the majority of deep bed filters, there is an increasing exploitation of local sands from beaches, river beds, quarries and deserts. Most of these prove to be satisfactory as tested by the criteria of section 11.3.4, although many contain dust, organic matter, clays etc., which make them too dirty to use without careful previous washing.

In India, bituminous (soft) coal has been proposed as an alternative to anthracite. In some European countries pumice has been used, or as an alternative with about the same density, a prepared material known as expanded slate, or baked clay, in the form of pellets, is available. Literature from the USSR mentions ceramic grains, and plastic granular material. Some of these materials have been reported by Mörgeli and Ives[12].

The addition of filter aids, usually polyelectrolytes or similar flocculants to the inlet water of the filter, is receiving more attention as a means of saving expensive presettlement processes. Such direct filtration can provide for flocculation prior to the water entering the filter media, sometimes with flocculation stirrers, or can provide for flocculation within the filter media as described in section 4.10.4.1. Sometimes, to accelerate the initial improvement period of filter performance (shown as the first part of the filtrate concentration curve on *Figure 11.7*), polymer is added to the final stage of the washwater. This changes the surface potential on the filter grains, making them more receptive to the adhesion of suspension particles during the early stages of the filter run.

11.4.2 Semi-continuous filtration

Semi-continuous filters have been available for more than 40 years, the earlier forms being known as Hardinge Filters. They, and their recently updated versions, are moving-bridge filters (also known as 'automatic backwash filters'), all relying on the same principle of a semi-continuous mode of operation. The filter consists of a shallow layer of filter material (normally sand), about 300 mm deep, which is divided into compartments about 200 mm wide, running laterally across the filter width, typically 3–4 m. Consequently, in a filter unit 20 m long, there would be approximately 100 lateral strips with low vertical walls separating the sand and the underdrains of each compartment. Spanning across the width of the filter a travelling bridge carries a hood which fits over each strip compartment in turn. While the hood is over a compartment, a pump mounted on the bridge draws filtrate from adjacent underdrains up through the sand in the compartment which is covered by the hood. The resulting washwater is pumped into an adjacent drainage channel. This principle of operation is shown on *Figure 11.14*.

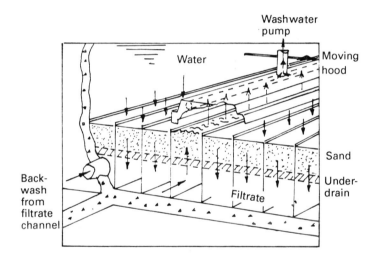

Figure 11.14. Principle of semi-continuous travelling-bridge filtration

The semi-continuous shallow filter bed contains quite fine sand, usually less than 1 mm grain size in tertiary sewage treatment. Consequently, the suspended solids are filtered principally by straining in and on the surface layer. Allowable head loss is small, not exceeding 0.3 m, and the filtrate overflow weir is usually above the sand level thus avoiding 'negative head' (pressures below atmospheric in the sand layer). Backwashing is initiated either by level probes as the water level rises due to head loss, or by a timing sequence. The travelling bridge washes each compartment in turn, usually taking 30–60 min to traverse the entire filter length. During the backwashing of a compartment, the rest of the filter continues to function so that only a small fraction of the full production is lost. For this reason, the filtration is frequently referred to as 'semi-continuous'.

Moving-bridge filters compare very favourably in cost with traditional designs, particularly as they do not require washwater storage, or large-capacity washwater pumps, and they operate at a lower head loss limit with a very shallow layer of filter medium. Their bridge mechanisms and control gear are sophisticated and probably operate better in covered buildings. However, there are some outside installations in warmer climates.

There are over 100 installations in the USA and a few in Europe, for industrial and municipal water supply, and waste-water treatment. Most installations are of small capacity, less than 5 Ml day^{-1}; the largest, for drinking water filtration, is 150 Ml day^{-1}; the largest and most recent for waste-water treatment (at Houston, Texas) is about 400 Ml day^{-1}. The most detailed unbiased account of such filters is that given by Lynam *et al.*[13], where comparisons are made with micro-strainers for tertiary sewage treatment. The semi-continuous filters produced filtrates below 7 mg l^{-1} suspended solids, at filtration rates of 7.5 m h^{-1}. There is further mention in the more recent book on advanced wastewater treatment by Culp[14].

11.4.3 Continuous filtration

The first practical continuous-flow filters employed radial flow from a perforated, axially vertical pipe, through a layer of sand in a drum-shaped compartment, through permeable walls to a peripheral filtrate collector. The sand moved slowly downwards whilst collecting a suspended particle as a deposit, and was then air-lifted through an external pipe, which separated the sand from the deposits. The clean sand was returned to the surface of the sand layer in the drum, thus ensuring steady-state operation, while dirty washwater was removed through a pipe above the sand surface.

This design was subsequently modified by removing the axial inlet pipe and allowing the influent to flow into the sand bed from above. By allowing the returned sand to form its natural angle of repose, the inlet face is an inverted cone-shaped incline. This modified filter is the Tenten (10:10 mg l^{-1} effluent standard for BOD: suspended solids), shown in *Figure 11.15*. Accounts of the use of the Tenten filter have been given by Lebref[15] (tertiary waste water, and industrial papermaking waste water), Critchard *et al.*[16] (tertiary treatment) and Barnhoorn and Tye[17] (iron removal).

An alternative design of continuous filter introduces the influent near the bottom of the sand bed, filtration is upwards (countercurrent to the descending sand) and filtrate is collected above the sand surface. A continuous airlift cleans and recycles the sand, and the washwater outlet is maintained below the level of the filtrate weir, ensuring that there is no contamination of the filtered water. This is the Dynasand filter, shown in *Figure 11.16*.

Some USA experience has been presented by Shimokubo[18] including filtration of secondary effluent from activated sludge (biological oxidation) treatment, and filtration of phosphate precipitates from sewage effluents. Flow rates varied from 7.5 to 21.5 m h^{-1} with inflow solids being highly variable from 1 to 100 mg l^{-1}. Washwater consumption was 2–3% of production, although other users reported up to 10% with continuous filters.

Head loss through both Tenten and Dynasand filters, was about 0.6 m in 1 m sand depth, which is a steady-state condition. This loss is higher than that

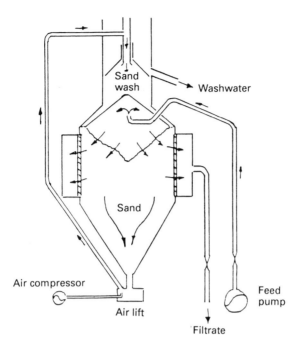

Figure 11.15. Tenten continuous filter; note that later versions have a central air-lift tube

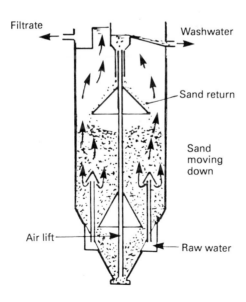

Figure 11.16. Dynasand continuous countercurrent filter

recorded in the fundamental study by Fourie and Ives[19] where 0.15 m was recorded through 0.6 m sand when filtering ferric hydroxide precipitates. Nevertheless, the values are significantly less than the 2–3 m of conventional filtration.

In Europe, principally in Sweden, more than 20 Dynasand units are used for waste-water filtration, industrial water treatment, and a few for drinking-water treatment. In the latter case, hygienic considerations require the filter to be covered to protect the filtrate from contamination.

REFERENCES

1. Ives, K. J., 'Capture mechanisms in filtration'. In Ives, K. J. (Ed.) *The Scientific Basis of Filtration*, Noordhoff, Leyden (1975)
2. Gregory, J., 'Interfacial phenomena'. In Ives, K. J. (Ed.), *The Scientific Basis of Filtration*, Noordhoff, Leyden (1975)
3. Mints, D. M., 'Preliminary treatment of water before filtration', *Proc. 8th Congr. Int. Water Supply Assoc. (Vienna)*, I.W.S.A., London (1969)
4. Ives, K. J., 'Specifications for granular filter media' *Effl. Wat. Trtmt. J.*, **15**, 296–305 (1975)
5. Fair, G. M., Geyer, J. C. and Okun, D. A. *Water and Wastewater Engineering, Vol. 2: Water Purification and Wastewater Treatment and Disposal*, Wiley, New York (1968)
6. Camp, T. R., *Water and Its Impurities*. Reinhold, New York (1962)
7. Cleasby, J. L., 'Filtration'. In Weber, W. J., (Ed.), *Physicochemical Processes for Water Quality Control*, Wiley–Interscience, New York (1972)
8. Baumann, E. R., 'Least cost design—optimization of deep bed filters'. In Ives, K. J., (Ed.), *The Scientific Basis of Filtration*, Noordhoff, Leyden (1975)
9. Huisman, L., 'Trends in the design, construction and operation of filtration plant', *Proc. 8th Congr. Int. Water Supply Assoc. (Vienna)*, I.W.S.A., London (1969)
10. Spielman, L. A., 'Flow through porous media and fluid-particle hydrodynamics'. In Ives, K. J. (Ed.), *The Scientific Basis of Filtration*, Noordhoff, Leyden (1975)
11. Ives, K. J., 'Mathematical models of deep bed filtration'. In Ives, K. J. (Ed.). *The Scientific Basis of Filtration*, Noordhoff, Leyden (1975)
12. Mörgeli, B., and Ives, K. J., 'New media for effluent filtration'. *Water Res.*, **13**, 1001–1007 (1979)
13. Lynam, B., Ettelt, G. and McAloon, T., 'Tertiary treatment at Metro Chicago by means of rapid sand filtration and microstrainers'. *J. Water Poll. Control Fed.*, **41**, 247–289 (1969)
14. Culp, R. L., Wesner, G. M. and Culp, G. L., *Handbook of Advanced Wastewater Treatment*, 2nd edition, pp. 148–149, Van Nostrand, New York (1978)
15. Lebref, J., 'Continuous filtration on a Tenten inclined bed filter' (in French), *Eau Indust.*, **22**, 73–76 (1978)
16. Critchard, D. J., Fox, T. M. and Green, R., 'A pilot plant comparison of the "Tenten" filter and three conventional static bed filters for tertiary treatment at the Aldershot sewage treatment works, *Water Poll. Control*, 382–388 (1979)
17. Barnhoorn, D. and Tye, D. C., 'The treatment of ferruginous groundwaters for river augmentation in the Waller's Haven, East Sussex'. *J. Inst. Water Eng. Sci.*, **38**, 217–230 (1984)
18. Shimokubo, R. M., 'Continuous sand filtration: an innovative approach to rapid sand filtration'. *Filtr. Sep.*, **20**, 376–380 (1983)
19. Fourie, J. and Ives, K. J., 'Continuous countercurrent filtration'. *Proc. Symp. Water Filtration* (Antwerp, 1982), pp. 3.7–3.18, KVIV, Antwerp (1982)

12

Pressure Filtration
Part I — Batch Pressure Filtration

L. Svarovsky

Department of Chemical Engineering, University of Bradford

12.1 INTRODUCTION

The driving force for filtration in pressure filters is usually the liquid pressure developed by pumping or by the force of gas pressure in the suspension feed vessel. Alternatively, or in addition, the liquid may be squeezed through and out of the cake by the mechanical action of an inflatable membrane, a piston or a porous medium pressed on top of the cake. Pressure filtration is, therefore, defined here as any means of surface filtration where the liquid is driven through the medium by either hydraulic or mechanical pressure, greater than atmospheric. The solids are deposited on top of the filter medium (as in all surface filters), with the possible exception of some cartridge filters which also use a certain amount of depth filtration. In this chapter, the suspension is assumed to approach the medium at 90° and this excludes the so-called dynamic filter/thickeners or cross-flow filters (also driven by pressure) which are dealt with in separate chapters (18 and 21).

Most conventional pressure filters are batch-wise in operation, i.e. the whole of the filter medium surface in each unit is going through a cyclic process of cake formation (possibly preceded by coating with filter aid—see chapter 10), followed by cake washing or dewatering and, finally, cake removal. This leads to discontinuous solids discharge from the filter, usually taking place during that part of the cycle when the filter is taken off pressure and opens to atmosphere (some do not have to open but the solids discharge is still occasional). The batch pressure filters are dealt with in part I of this chapter.

There have been some relatively recent developments of continuous pressure filters, of both the hydraulic and squeezing variety. The different parts of the filtration surface in these units at any point in time are undergoing different stages of the filtration and dewatering cycle, with one part continuously discharging the cake. The cake and filtrate production is continuous, therefore, with all the associated advantages of continuous or large-scale production processes. The continuous pressure filters are dealt with in part II of this chapter. Both parts of this chapter start with a brief fundamental introduction.

12.2 BATCH PRESSURE FILTRATION

12.2.1 The case for pressure filtration in general

High pressure drops have a two-fold effect: on capacity and on displacement dewatering which often follows. It is best to consider these two effects separately.

12.2.1.1 The filtration part of the cycle

The most important feature of the pressure filters which use hydraulic pressure to drive the process is that they can generate a pressure drop across the medium of more than 1 bar, which is the theoretical limit of vacuum filters. Whilst the use of a high pressure drop is often advantageous, leading to higher outputs, drier cakes or greater clarity of the overflow, this is not necessarily the case. For compressible cakes, an increase in pressure drop leads to a decrease in permeability of the cake and hence to a lower filtration rate relative to a given pressure drop. This reduction in permeability due to cake consolidation or collapse may be so large that it may nullify or even overtake the advantage of using high pressures in the first place and there is then no reason for using the generally more expensive pressure filtration hardware. Whilst a simple liquid pump may be cheaper than the vacuum pump needed with vacuum filters, if air-displacement dewatering is to follow filtration in pressure filters, an air compressor must be used and this is expensive.

The fundamental case for pressure filters may be made using the following equation for dry cake production capacity, Y (kg m^{-2} s^{-1}), derived from Darcy's law when the filter medium resistance is neglected (for the full derivation see appendix I in Svarovsky[1]):

$$ Y = \left[\frac{2 \, \Delta p \, f \, c}{\alpha \, \mu \, t_c} \right]^{1/2} \tag{12.1} $$

where Δp is the pressure drop, c is the feed solids concentration, α is the specific cake resistance, μ is the liquid viscosity, Y is the solids yield (dry cake production in kg m^{-2} s^{-1}), f is the ratio of filtration to cycle time, and t_c is the cycle time. For the same cycle time (i.e. same speed), if the pressure drop is increased by a factor of four, production capacity is doubled. In other words, filtration area can be halved for the same capacity but only if α is constant. If α increases with pressure drop and depending on how fast it increases, the increased pressure drop may not give much more capacity and, in some extreme cases, it may actually cause capacity reductions.

For most industrial inorganic solids such as minerals, etc., the increase in α with Δp ia not too large and thus, if the material to be filtered is too fine for vacuum filtration, pressure filtration may be advantageous and will give better rates.

Pressure filters can treat feed concentrations of up to and in excess of 10% solids by weight and having large proportions of difficult-to-handle fine particles. Typically[1], slurries in which 10% of the solid particles are larger than 10 μm may require pressure filtration, but increasing the proportion of particles larger than 10 μm may make vacuum filtration possible. The range of typical filtration velocities in pressure filters is 0.025–5 m h^{-1} and of dry solids rates is 25–250 kg m^{-2} h^{-1}. The use of pressure filters may also, in some cases (such as in filtration of coal flotation concentrates), eliminate the need for flocculation.

12.2.1.2 The dewatering part of the cycle

The key to the understanding of the dewatering process by air displacement is the capillary pressure diagram. *Figure 12.1* shows an example typical for a fine coal suspension; there is a minimum moisture content which cannot be removed by air displacement at any pressure, called irreducible saturation (about 12% in this case). There is also a threshold pressure (about 0.13 bar in this case) which must be exceeded in order that air may enter the filter cake.

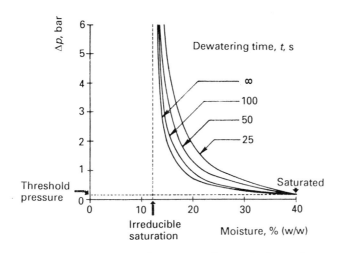

Figure 12.1. A capillary-pressure diagram for fine coal

The capillary retention forces in the pores of the filter cake are primarily affected by the size and size range of the particles forming the cake, and by the way the particles have been deposited when the cake was formed. There is no fundamental relation to allow the prediction of cake permeability but, for the sake of the order-of-magnitude estimates, the pore size in the cake may be taken loosely as though it were a cylinder which would just pass between three touching, monosized spheres. If d is the diameter of the spherical particles, the cylinder radius would be 0.0825 d. Using this simple argument and the surface tension of water at 20°C of 0.07275 N m^{-2}, the capillary pressure of 1 bar would correspond to $d = 17.6$ μm.

Bearing in mind that although most cakes consist of poly-disperse, non-spherical particle systems that are theoretically capable of producing more closely-packed deposits, the practical cakes usually have large voids and are much more loosely packed due to the lack of sufficient particle relaxation time available at the time of cake deposition (see a further discussion of this phenomenon in chapter 9) the above-derived value of 17.6 μm probably becomes nearer the often-quoted 10 μm limit when pressure air dewatering becomes necessary.

The lowest curve in *Figure 12.1* gives the moisture content at different pressure drops which can be achieved given infinite time of dewatering. The dewatering kinetics, are such, however, that many such curves can be measured and drawn, depending on the time available for dewatering. Although, strictly speaking, *Figure 12.1* or any such diagram only applies to one cake thickness, Bott *et al.*[2] have recently shown that, given a suitable filter cloth and a sufficiently high pressure differential, the filter speed (of a continuous pressure filter) can be increased nearly up to the speeds giving the minimum practicable cake thickness without any increases in moisture content. A capillary-pressure diagram, such as the one in *Figure 12.1*, obtained for a given cake thickness can, therefore, also be used for thinner cakes.

Figure 12.2 shows clearly that the same moisture content of the produced cake can be obtained in shorter dewatering times if higher pressures are used. If a path of constant dewatering time is taken, on the other hand, moisture content is reduced at higher pressures, with a parallel increase in cake production capacity. This is a clear and indisputable advantage of pressure filtration of reasonably incompressible solids like coal and other minerals: it has been substantiated by experiments by many workers, including Bott *et al.*[2] and workers in Bradford[3].

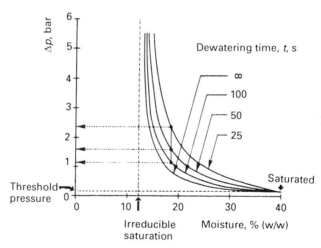

Figure 12.2. Diagram to show that higher pressure allows shorter dewatering times

12.2.1.3 Optimization of cycle times

In batch filters, one of the important decisions to be made is how much time is to be allocated to the different operations such as filtration, displacement dewatering, cake washing and cake discharge (which may involve opening the pressure vessel). All of this will have to happen within a cycle time, t_c, which itself is not fixed. However, some of the times involved, such as the cake discharge time, may be defined. The question we want to answer here is how much time is to be given to the filtration part of the cycle relative to the other operations. Are we to run the filtration for a long time, thus forming a thick cake, with the inevitable consequence of rapidly diminishing filtration rates towards the end, or should we stop the filtration earlier whilst the rate is still reasonably high?

If all the non-filtration operations are grouped together into a 'down-time', t_d, and this is assumed to be fixed and known, we can derive an optimum filtration time, t_{opt}, in relation to t_d by optimizing the average dry cake production obtained from the cycle. The full derivation may be found in appendix II of Svarovsky[1] and only the final result is given below.

For constant-pressure filtration and where the medium resistance, R, and the specific cake resistance, α, are constant, the following equation applies:

$$t_{opt} = t_d \left[1 + \sqrt{\frac{2 \mu R}{\alpha c \Delta p t_d}} \right] \qquad (12.2)$$

where Δp and c are the operating pressure drop and the feed solids concentration respectively.

When the medium resistance, R, is small compared with the specific cake resistance, α, the second term in the above equation becomes negligible and the optimum filtration time, t_{opt}, becomes equal to the down time, t_d. For any other situation t_{opt} is always greater than t_d.

It follows, therefore, that the filtration time should always be set to be at least equal to the sum of the other, non-filtration periods involved in the cycle.

12.2.1.4 Cake squeezing

The compressibility of many industrial filtration cakes, an unwanted phenomenon in filtration by hydraulic pressure, may be turned into an advantage by considering mechanical squeezing of the pre-formed cake as an alternative or additional dewatering method. The fact that a cake has a reasonable or high porosity does not necessarily mean that the liquid within it can be easily removed. Some of the liquid can, however, be removed simply by altering the cake structure by mechanical compression. In fact, the more compressible materials are often also less permeable. It is, therefore, the very cakes which consolidate under hydraulic pressure and render such filtration (and subsequent displacement dewatering) difficult that are the best candidates for mechanical squeezing, sometimes referred to as 'expression'. Compressibility of cakes can often be promoted by addition of flocculation agents and, indeed, some available hardware such as the belt presses almost entirely depend on chemical pre-treatment of slurries. There is very little that can be done using fundamental theory. The compressibility and permeability of practical filtration cakes, which affect the choice of equipment for dewatering, must be tested experimentally and this is best done in a compression filtration cell[4,5].

The specific volume of a cake, v, after compression is usually related to the applied pressure, p, in a way similar to that used in soil mechanics[6].

$$v = v_{o} - \lambda \ln p \qquad (12.3)$$

where v_{o} is the value of v corresponding to unit pressure p and λ is a characteristic of the material and may sometimes itself depend on p.

Carleton[7], who has studied the economics of using the relatively expensive compression filters, states that if the volume of the cake is not reduced by at least 2% when the applied pressure is doubled, then it is unlikely that a compression filter can be justified.

The values of v in the above equation are the equilibrium values after an infinite pressing time. There are kinetics involved here too in that the deformation of the cake due to a mechanical stress is not instantaneous but dampened by the necessary permeation of the liquid out of the cake. Baluais et al.[5] considered a rheological model for this in analogy with the action of a shock absorber in parallel with one spring and in series with another.

Those pressure filters which use the squeezing of the cake are, on the whole, expensive and their use can be justified only if the energy savings are large enough. The squeezing process itself may extend the down-time of the filtration cycle and a full economical analysis is needed[7] to optimize the choice of equipment and the operating conditions. It should be borne in mind, however, that compared with thermal drying, cake compression requires about 25 times less energy.

There are other positive features of cake squeezing in that it often improves cake release, it closes any cracks and tightens the cake, thus making any subsequent cake washing more effective.

Excluding variable-chamber presses, which rely on mechanical squeezing of the cake and which will be dealt with in a separate section, pressure filters may be grouped into two categories, plate-and-frame filter presses and pressure vessels containing filter elements. The latter group also includes the cartridge filters but these have their own section in the following.

12.2.2 Filter presses

In the conventional plate-and-frame press (*Figure 12.3*), a sequence of perforated, square or rectangular plates alternating with hollow frames is mounted on suitable supports and pressed together with hydraulic or screw-driven rams. The plates are covered with a filter cloth and the cloth also forms the sealing gasket. The slurry is pumped into the frames and the filtrate is drained from the plates.

The drainage surfaces are usually made in the form of raised cylinders, square-shaped pyramids or parallel grooves in materials such as stainless steel, cast iron, rubber or resin-coated metal, polypropylene, rubber or wood. Designs are available with every conceivable combination of inlet and outlet location: top feed, centre feed, bottom feed, corner feed and side feed, with a similar profusion or possible positions of discharge points, each combination having particular advantages, depending on whether washing is required and also on the application and the nature of the suspension.

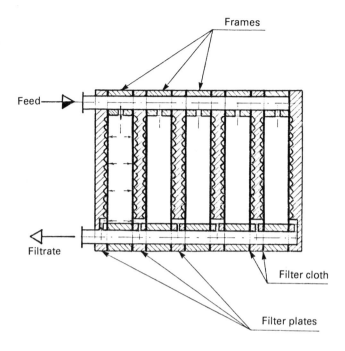

Figure 12.3. Scheme to show the principle of plate-and-frame presses

Sometimes, the discharge is through a separate cock on each plate rather than through a common filtrate port and manifold. This allows observation of the filtrate from each plate and sampling from it, thus enabling the operator to spot cloth failure and to isolate the plate, or segregate the cloudy filtrate. The cocks discharge into open channels or enclosed pipe systems (the latter being fitted with a sight glass).

Both flush plates and recessed plates can be specified, the latter obviating the need for frames but being a little tougher on filter cloths due to the strain around the edges. Recessed plate presses are more suitable for automation, however, because of the difficulty of the automatic removal of residual cake from the frames in a plate-and-frame press. Recessed plates with no frames limit the chamber width to less than about 32 mm (to limit the strain on the cloth), whilst plate-and-frame presses allow this width to be more than 40 mm if necessary. Plate sizes range from 150 mm^2 to 2 m^2, giving filtration areas up to 200 m^2. The number of chambers varies up to 100, and exceptionally up to 200.

Plate-and-frame filters are the most versatile filters because their effective area can be varied simply by blanking off some of the plates. Cake-holding capacity can be altered by changing the frame thickness or by grouping several frames together. These filters are available in a variety of semi-automated or fully automated versions that feature mechanical leaf-moving devices, cake removal by vibration or by pulling the cloth when the press is open, etc. An operator usually has to be present, however, because it is not certain that each and every chamber will discharge its cake unaided every time. Should manual intervention be necessary, the operator has to be protected from injury by the leaf-moving mechanism, by a suitable safety photo-electric device.

Some attempts have been made to re-slurry the filter cake without having to open the filter press. However, a number of problems tend to crop up, such as for example, bending of the plates due to uneven cake deposition or cavitation, uneven dewatering and washing within the frames or plugging of the inlet ports.

There are two ways in which washing can be performed in filter presses: simple washing and thorough washing. In simple washing, the wash liquid is introduced either through the main feed port or through a separate port into each chamber, and the washing is therefore in the same direction as the filtration process that formed the cake. In thorough washing, the wash liquid enters through a separate port, behind the filter cloth on every other plate, thus passing through the whole thickness of the cake in the chamber. Washing is less efficient with recessed chamber presses than with flush plate frame presses, probably because of poorer distribution of the wash liquid. In either case the amount of wash liquid necessary tends to be high.

Filter media for plate-and-frame presses include various cloths, mats and paper. In the case of paper, this usually has to be provided with a backing cloth for support.

The typical operating pressure of filter presses is around 6 or 7 bar, although some manufacturers offer presses for up to 20 bar or higher. As the pressure builds up during filtration, it tends to force the plates apart and this may be offset by a pressure compensation facility offered with some large mechanized presses.

Full mechanization of filter presses began in the late 1950s and this was closely followed by addition of the mechanical expression (i.e. cake squeezing) mechanism. Rubber or plastic membranes are sometimes fitted to compress the cake which is formed by conventional pressure filtration. The membranes normally rest on the plates and have grooves and openings in their surfaces for filtrate collection. The membranes are inflated at the end of the filtration cycle by air or, for pressures higher than 1000 kPa, by water or hydraulic fluid. The membranes should be designed to last up to or in excess of 10 000 pressings. The main advantage of using mechanical expression of cakes here is the additional dewatering usually achieved and also the ability to handle thin cakes. Another advantage is that the main filtration process can be done at lower pressures (so that a relatively cheap, centrifugal pump can be used) and the compression by the membrane then goes to higher pressures.

The automation of filter presses brought about several other advantages and developments. Plate-shifting mechanisms have been developed, allowing the cloths to be vibrated, filter-cloth washing (on both sides) has been incorporated to counteract clogging from the expression, and, most importantly, the down-times have been reduced with automation, thus increasing capacities.

One of the recent developments in this filter category is the vertical recessed plate automatic press shown schematically in *Figure 12.4*. While the conventional filter press usually has the plates hanging down and linked in a

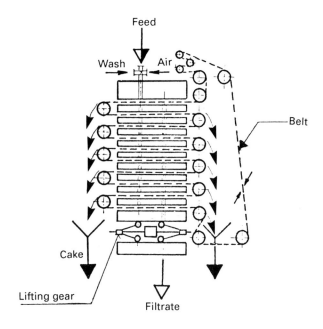

Figure 12.4. Scheme to show the vertical automatic filter press

horizontal direction, this type has the plates in a horizontal plane and placed one upon another. This design offers semi-continuous operation, savings in floor space and easy cleaning of the cloth, but it allows only the lower face of each chamber to be used for filtration. The filter usually has an endless cloth, travelling intermittently between the plates via rollers to peel off cakes. The disadvantage of this is, however, that if an endless cloth is damaged anywhere, the whole cloth has to be replaced, which is a difficult process. As the filter cloth zig-zags through the filter, the filtering direction is reversed each time and this tends to keep the cloth clean. Most of these filters incorporate membranes for mechanical expression, and cakes sometimes stick to the membranes and remain in the chamber after discharge. Some vertical filters are available with a separate cloth for each frame. The cloths may be disposable and such filters are designed to operate with or without filter aids.

As the height of the vertical press makes maintenance difficult, the number of the chambers is restricted, usually to 20, with a maximum of 40, with filter areas of up to 32 m^2.

The application of filter presses spans virtually all areas of the processing industries due to their versatility. Examples of use include the clarification of beer and juices or waste-water and activated sludge filtration in breweries, paper mills and petrochemical plants, the dewatering of fine minerals and lime mud separation and washing in the sugar industry. The filtration velocities are usually 0.025 to 1 m s^{-1} and the dry solids handling capacities are less than 1000 kg m^{-2} h^{-1}, the higher values being more usual with the automatic presses due to their shorter down-times.

12.2.3 Pressure vessel filters

There are several designs of pressure vessel filters. They all consist of pressure vessels, housing a multitude of leaves or other elements which form the filtration surface and which are mounted either horizontally or vertically. With the horizontal leaves there is no danger of the cake falling off the cloth and they are most suitable where thorough washing is required, whereas with the vertical elements, a pressure drop must be maintained across the element to retain the cake. The disadvantage of the horizontal-leaf types is, however, that half the filtration area is lost because the underside of the leaf is not normally used for filtration (because of the danger of the cake falling off). Discharge of the cake may also be more difficult in this case.

The elements or leaves normally consist of a coarse stainless-steel mesh over which a fine (often metal) gauze or filter cloth is stretched and sealed at the edges. The leaves are in parallel, each connected to a header and, almost without exception, the filtration is from the outside inwards through the gauze. These filters are all essentially batch operated and most require the use of a filter aid for pre-coating, because cloudy filtrates and blinding would otherwise occur.

One operational probelm with these filters is concerned with the disposal of the heel of unfiltered slurry which is still in the vessel at the end of the filtration period. This is particularly troublesome with the vertical-leaf filters because compressed air cannot be used to complete the filtration of this heel (as the air would preferentially escape through the tops of the leaves as soon as they emerged from the suspension). It is common practice to install a separate filtering element at the bottom of the vessel as a 'scavenger filter' for the filtration of the residual slurry. During the scavenge filtration the main leaves are usually isolated so that compressed air is not lost through them.

12.2.3.1 Cylindrical element filters (*Figure 12.5*)

Cylindrical element filters, often referred to as 'candle filters', have cylindrical elements or sleeves mounted vertically and suspended from a header sheet, which divides the filter vessel into two separate compartments. The filtration takes place on the outside of the sleeves. The inlet is usually in the bottom section of the vessel and the filtrate outlet in the top section above the header sheet. A less usual design is to locate the filtrate outlet at the bottom of the elements and thus allow the top chamber to be opened for easy inspection of the elements during operation.

The tubes are generally 25–75 mm in diameter, and up to 2 m in length, made from metal or cloth-covered metal, and provide filtration areas of up to 100 m^2. Alternatively, the tubes can be made of stoneware, plastics, sintered metal or ceramics. The elements may be made deliberately flexible, sometimes filled with loose packing. Tank diameters of up to 1.5 m are available. Cake removal is performed by scraping with hydraulically operated scraper rings, by vibration or by turbulent flow 'bumping'. The mechanical strength of the tubular element makes it ideal for cleaning by the sudden application of reverse pressure. Physical expansion or 'flexing' of the tubular elements on application of the reverse flow aids cake discharge.

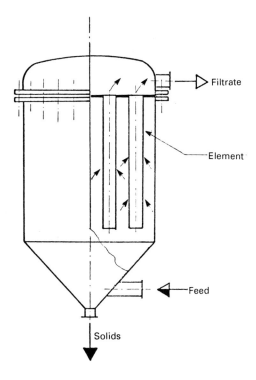

Figure 12.5. A cylindrical element (candle) pressure filter

Cylindrical element filters find wider use where cake washing is not required. Their inherent advantage is that, as the cake grows on the tubular elements, the filtration area increases and the thickness of a given volume of cake is therefore less than it would be on a flat element. This is of course only of importance where a thick cake is being formed and the rate of increase in the pressure drop is less with tubular elements in such cases.

The pressure filter with tubular elements has also been used as a thickener, when the cake, backwashed by intermittent reverse flow, is redispersed by an agitator at the bottom of the vessel and discharged continuously as a slurry. In some cases the filter cake will build up to a critical thickness and then fall away without blow-back.

12.2.3.2 Vertical-tank, vertical-leaf filters (*Figure 12.6*)

These are the cheapest of the pressure leaf filters and have the lowest volume-to-area ratio. Their filtration areas are limited to less than 80 m². Large bottom outlets, fitted with rapid-opening doors, are used for dry cake discharge and smaller openings are used for slurry discharge. Wet discharge may be promoted by spray pipes, vibrators, reverse flow, bubble rings, scrapers etc., while dry discharge is usually caused by vibration. As all vertical leaf filters, these are not suited for cake washing.

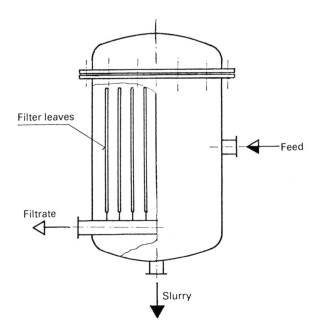

Figure 12.6. A vertical-tank, vertical-leaf filter

In one somewhat different design, known as the Scheibler filter, the filter leaves take the form of bags, each suspended in a rectangular pressure vessel from a horizontal tube which acts as the filtrate outlet. The sides of the bag are prevented from meeting by looped chains which are attached at the tops of their loops to the horizontal tube and hang downwards. With this method of separating the cloth surfaces by chains the bags can be wide enough to hang in pleats and the filtering area can thus be as much as three times the area of the frames which are inside the bags. Filtration areas of up to 250 m^2 are available, applications being mostly in the chemical industry.

Special mention should also be made of the pressure version of the Moore filter developed in France for the sugar industry. The vertical cylindrical vessel, which houses a set of radially arranged leaves, is twice the height of the leaves, allowing the leaves to be raised, rotated and lowered into the different compartments in the bottom half of the vessel. Positive air pressure must be maintained throughout the operation to prevent cake fall-off and the cake is blown off the leaves by air blow-back.

12.2.3.3 Horizontal-vessel, vertical-leaf filters (*Figure 12.7*)

In a cylindrical vessel with its axis horizontal, the vertical leaves can be arranged either laterally or longitudinally. The latter arrangement is less common; such filters may be designed as vertical-vessel, vertical-leaf filters but mounted horizontally. Those designs are suitable for smaller duties and

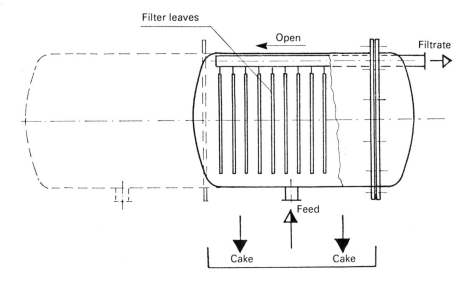

Figure 12.7. A horizontal-vessel, vertical-leaf filter

the leaves can be withdrawn individually through the opening end of the vessel.

Filtration areas of up to 120 m² are available with the second longitudinal arrangement, the Kelly leaf filter. This was probably the earliest pressure leaf filter and it has been used for the filtration of very viscous liquids such as glycerin and concentrated sugar solutions. The leaves, the height of which varies according to the space available at their location in the vessel, are attached to the removable circular front cover, each leaf having an outlet connection through this cover. The leaves together with the cover and outlet pipes are attached to a carriage which can be run into and out of the shell to facilitate cake discharge outside the vessel by air blow-back or rapping. This arrangement requires a considerable amount of floor space and to minimize this drawback the presses are often constructed in pairs on a single runway, with their opening ends facing each other. Thus, each filter can be opened in turn into the common space between them.

By far the greatest use of the horizontal-vessel filters is made of those types which have vertical leaves arranged laterally (i.e. in a plane perpendicular to the axis of symmetry of the vessel) because this design provides easy access to the leaves. Most of these filters open in a way similar to the Kelly filter, sometimes by moving the shell rather than the leaf assembly, so that the filtrate pipe can remain permanently connected; sometimes withdrawal of one leaf or a bundle of leaves at a time is possible. The leaves may be rectangular, circular or of some other shape, and may be designed to rotate during cake discharge. Sluicing by sprays is sometimes used for wet discharge, with or without rotation of the leaves. If the leaves are designed for rotation they are invariably circular and mounted on a central hollow shaft which serves as the filtrate outlet. Dry cake discharge may be carried out with rotating leaves by application of a scraper blade. If this is to be done without opening the vessel, then the bottom of the vessel must be shaped like a hopper, with a screw conveyor if necessary.

The original Vallez filter developed in the USA for the sugar industry, and the more modern developments of it, also rotate the leaves at about 1 rev. min^{-1} during the filtration operation, in order to keep the solids in suspension and achieve a more uniform cake.

One significant departure from the standard end-opening design is known as the Sweetland filter, which has the cylindrical shell split in a horizontal plane into two parts, where the bottom half can be swung open for cake discharge. The upper half is rigidly supported and both the feed and the filtrate piping are fixed to it. The lower part is hinged to the upper along one side and is counterbalanced for easy opening. Cake discharge is either by sluicing or by dropping, assisted by some scraping. If much scraping has to be done, there is not much advantage in using this type of filter. As the leaves are stationary, the cake deposited on them may be uneven, with the greater mass of the cake being at the bottom of the leaves.

In general, the horizontal-vessel, vertical-leaf filters with the leaves arranged laterally can be designed up to filtration areas of 300 m^2. Cake washing is possible but it must be carried out with caution because there is a danger of the cake falling off.

Horizontal-vessel filters with vertical rotating elements have been under rapid development recently with the aim of making truly continuous pressure filters, particularly for the filtration of fine coal—see part II of this chapter.

12.2.3.4 Vertical-vessel, horizontal-leaf filters (*Figure 12.8*)

These filters, like all horizontal-leaf filters, are advantageous where the flow is intermittent or where thorough cake washing is required. Filtration areas are limited to about 45 m^2. Special mention must be made of the pressure version of the Nutsche filter which, strictly speaking, falls into this category. These are either simple pressurized filter boxes or more sophisticated agitated Nutsches, much the same in design as the enclosed agitated vacuum filters described in another section. These are extremely versatile, batch-operated filters, used in many industries, for example in agrochemistry, pharmaceuticals or dyestuff production.

An obvious method of increasing the filtration area in the vessel is to stack several plates on top of each other; the plates are operated in parallel. One such design, known as the plate filter, uses circular plates and the stack can be removed as one assembly. This allows the stack to be replaced after the filtration period with another, clean stack, so that the filter can be put back into operation quickly. The filter consists of dimpled plates supporting perforated plates on which filter cloth or paper is placed. The space between the dimpled plates and the cloth is connected to the filtrate outlet, which is either into the hollow shaft or into the vessel, the other being used for the feed. When the feed is into the vessel, a scavenger plate may have to be fitted because the vessel will be full of unfiltered slurry at the end of the filtration period. This type of filter is available with filtration areas of up to 25 m^2 and gives cakes of up to 50 mm thick.

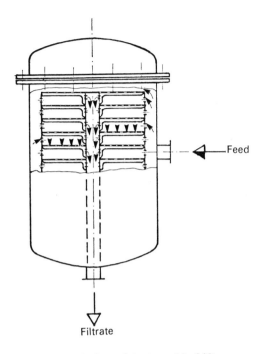

Figure 12.8. A vertical-vessel, horizontal-leaf filter

Centrifugal discharge filters form another group in this category. As the name suggests, the cake discharge is accomplished by rotating the stack of plates around the hollow shaft. The cake slides off the plates due to the centrifugal action; sometimes it is necessary to supplement this process by sluicing with a suitable liquid, in which case the discharge is wet. The filtrate leaves through the hollow shaft. These filters lend themselves to automation and, as opposed to manually operated leaf filters, they can be operated with short cycle times and very short down-times, which is very economical. Many different makes are available, with various ways of driving the shaft and locations of the electric motor as well as with other varying constructional details. The sizes available vary up to 65 m^2.

Another available design allows discharge of the cake by vibration of the circular plates, which are then slightly conical, sloping downwards towards the outside of the plates. This design allows higher pressures to be used as no rotating seals are necessary.

12.2.3.5 Horizontal-vessel, horizontal-leaf filters (*Figure 12.9*)

These filters consist of a horizontal cyclindrical vessel with an opening at one end. A stack of rectangular horizontal trays is mounted inside the vessel; the trays can usually be withdrawn for cake discharge, either individually or as the whole assembly. The latter case requires a suitable carriage to be supplied. One alternative design allows the tray assembly to be rotated through 90° so

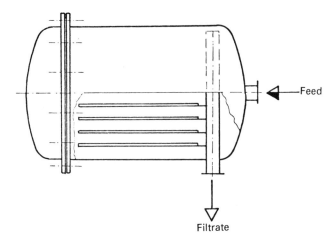

Figure 12.9. A horizontal-vessel, horizontal-leaf filter

that the cake can fall off into the bottom part, designed in the shape of a hopper and fitted with a screw conveyor.

The leaves are sometimes fitted with rims, thus forming trays and this is particularly useful for flooding the trays in washing operations. Scavenger leaves are often used. Filtration areas of up to 50 m^2 are available. Like all horizontal-leaf filters, they are particularly suitable when thorough washing is needed.

12.2.4 Cartridge filters (*Figure 12.10*)

This short review of pressure filters would not be complete without a mention of cartridge filters. These use easily replaceable, tubular cartridges made of paper, sintered metal, woven cloth, needle felts, activated carbon or various membranes of pore size down to 0.2 μm. Filtration normally takes place in the direction radially inwards, through the outer face of the element into the hollow core. Cartridge filtration is limited to liquid polishing or clarification, i.e. removing very small amounts of solids, in order to keep the frequency of cartridge replacements down. Typically, suspensions of less than 0.01% volume concentration of solids can be treated with cartridge filters and such filters are favoured in small-scale manufacturing applications.

Cartridge filters can be divided into depth and surface types, according to where most of the solids separate, although the precise demarcation line is sometimes difficult to draw. Probably the most common type of depth cartridge is the yarn-wound type which has a yarn wound around a central core in such a way that the openings closest to the core are smaller than those on the outside. The aim is to achieve depth filtration, which increases the solids-holding capacity of the cartridge. The yarn may be made of any fibrous material, ranging from cotton or glass fibre to the many man-made fibres like polyester, nylon or teflon. The spun staple fibres are brushed to

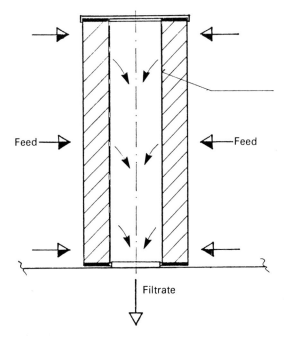

Figure 12.10. Scheme to show the principle of cartridge filters

raise the nap and this makes the filter medium. The cores are made from polypropylene, phenolic resin, stainless steel or other metals or alloys. The nominal rating of this type of cartridge varies from 0.5 to 100 μm. Cartridges intended for higher viscosity liquids are often made of long, loose fibres, again either natural or synthetic, impregnated with phenolic resin. Such bonded cartridges are usually formed into the shape of a thick tube by a filtration technique and do not normally require a core because they are self-supporting. The porosity of the medium can be graded during the formation process, again to increase the solids-holding capacity. Bonded cartridges are available in somewhat coarser ratings of 10–75 μm.

Depth-type cartridges cannot be cleaned but have high solids-holding capacity, and are cheap and robust. Considerable standardization of the cartridge size throughout industry (approximately 25 cm long, 6.3–7.0 cm overall diameter and 2.5 cm internal diameter) allows testing of different cartridge makes and types in the same holder.

In the category of surface cartridges is another very common type, the pleated-paper construction, which allows large filtration areas to be packed into a small space. Oil filters in the automobile industry are of this type. The paper is impregnated, for strength, with epoxy or polyurethane resin. Any other medium in sheet form, similar to cellulose paper, such as wool, polypropylene or glass may be used. The nominal rating varies between 0.5 and 50 μm. Pleating in the radial direction is most usual, but some cartridges have axial pleating, or have hollow discs of lenticular shape, in an effort to pack as much filtration surface into a given space as possible; up to 3 m^2 of surface in one cartridge is possible. The solids-holding capacity is low but some applications allow the prolonged build-up of solids on the surface, until the pleats are completely filled up.

Another type of cartridge is the edge filter, which contains a number of thin discs mounted on a central core and compressed together. The discs are usually made of metal, although paper or plastics are sometimes used. Filtration takes place on the surface of the cylinder, with the particles unable to pass between the discs. The main advantage of the edge-type filter is that it is cleanable, and this is done by reverse flow, ultrasonics or by scraping the outside surface of the cylinder with a mechanical scraper. Edge filters made from paper discs have been known to retain particles as fine as 1 μm but the metal variety retains solids larger than 50 μm or so.

Other designs of cartridges use active carbon, Fuller's earth, sintered metal or other specialized media.

The most important characteristics of cartridge filters are the filter rating, i.e. the largest spherical particle which will pass through the filter (usually, the 98% retention cut size is used), the relationship between the pressure drop and solids-holding capacity, and the maximum allowable pressure drop beyond which the cartridge will fail structurally. Both the retention and the solids-holding capacity depend on filtration velocity and this must be borne in mind when testing cartridges. Thermal or shock stresses sometimes lead to cracking of cartridges, with the subsequent loss of overflow clarity.

The housings of cartridge filters are simple pressure vessels designed for one cartridge or a number of cartridges in parallel, in multi-element filters. Some housings are designed to withstand pressures of up to 300 bar. Proper sealing of the elements is a necessary prerequisite of their efficient use. Frequent replacement of cartridges should be facilitated by quick-opening clamping fittings.

Cartridge filters are used to clean power fluids, lubrication oils, wines, fruit juices or pharmaceutical liquids. They are also used to protect other equipment, e.g. in reflux control systems or automatic valves. Low capital and installation costs, low maintenance costs, simplicity and compactness are the main advantages of cartridge filters. The running costs are high, particularly when disposable cartridges are used. It is most important, therefore, that a full economic analysis, based on reliable cartridge replacement frequency, is carried out before adopting a cartridge filtration system; the low cost of the basic hardware may be deceptive.

12.2.5 Compression filters

In conventional cake filtration the liquid is expelled from the slurry by fluid pressure in a fixed-volume filtration chamber, while in mechanical compression this is achieved by reducing the volume of the retaining chamber. This compression of either a slurry or a cake (which might have been formed by conventional filtration) offers advantages to a wide variety of industries handling a variety of different materials. Such materials include highly compressible, sponge-like solids, very fine particles such as clays, fibrous pulps, gelatinous mixtures like starch residues or some pharmaceuticals, and flocculated waste-water sludges.

The compressibility of filter cakes is a nuisance from the point of view of filtration theory, but in practice it means that with increasing pressure cakes become more compact and therefore drier. The resistance to flow increases due to reduced porosities, however, and, with some materials such as paper-mill effluents higher pressures do not necessarily give increased flow rates. In cakes undergoing conventional pressure filtration, only the bottom layers closest to the medium are subjected to the highest compression forces, whereas the top layers are only subjected to light hydraulic forces and are not compacted so tightly. If a mechanical force is applied to the top of the filter cake, the distribution of pressure through the cake is more uniform, and drier cakes can thus be achieved than by using high pumping pressures of the feed suspension.

In the past decade or so, a number of new filters have appeared on the market, utilizing some form of mechanical compression of the filter cake, either after a conventional pressure filtration process or as a substitute for it. In most designs the compression is achieved by inflating a diaphragm which presses the slurry or the freshly formed filter cake towards the medium, thus squeezing an additional amount of liquid out of the cake—see *Figure 12.11*.

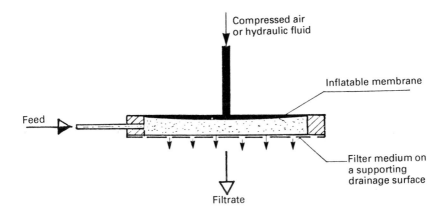

Figure 12.11. Scheme to show the variable-chamber principle

Other designs squeeze the cake between two permeable belts or between a screw conveyor of diminishing diameter (or pitch) and its permeable enclosure. The available filters which use mechanical compression can be classified into four principal categories:

1. membrane plate presses,
2. tube presses,
3. belt presses,
4. screw presses.

Only the first two are dealt with here as the others are continuous and are, therefore, included in part II of this chapter (section 12.3.5).

The advantages of using mechanical compression with compressible cakes are in the increased solid contents of the cakes (leading to reduced energy

requirements if thermal drying has to follow, or to better handling properties), improved washing efficiencies, increased filtration rates and easier or
automatic cake discharge. Invariably, however, the capital cost of such filters
is higher than for conventional pressure filters.

12.2.5.1 Membrane plate presses

Membrane presses are closely related to the conventional plate-and-frame
presses described in section 12.2.2. They consist of a recessed plate press in
which the plates are covered with an inflatable diaphragm which has a
drainage pattern moulded into its outside surface. The filter cloth is placed
over the diaphragm as shown in *Figure 12.12*. During the first stage of filling
the filter chambers with the slurry and conventional pressure filtration, the
membrane is pushed against the plate body. When the chambers are full of
cake, the feed is terminated and the membranes are inflated by pumping
compressed air or hydraulic fluid in between the membranes and the plates,
so that the cake in the chamber is compacted as the two membranes move
towards one another. Washing of the cake may follow and can be carried out
more effectively than in a normal press, because the cake is compacted to a
more uniform density by squeezing. The resulting advantage is in the
reduction of the washing time and washwater requirements. The squeeze
pressures vary from 6 to 15 bar. Additional reductions of up to 25% in the
moisture content over that obtained with conventional filter presses can be
achieved. The membranes are usually made of a solvent-resistant rubber
compound.

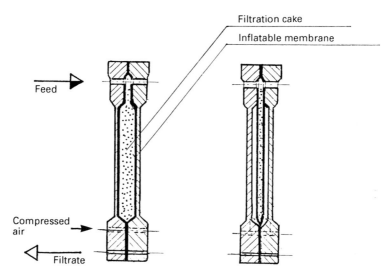

Figure 12.12. A recessed plate in a variable-chamber press

Another major advantage of the membrane plate is in its flexibility to cake
thickness: thinner cakes can be handled easily without loss of dryness. Cake-

release characteristics are also improved by deflating the membrane prior to cake discharge. Alternating arrangements have also been used in which the membrane plates and the normal recessed plates alternate, and this reduces the cost.

In the plate press filter the pneumatically operated membranes are replaced by flexible seals and by compression by a hydraulically powered ram. *Figure 12.13* depicts the compression stage after the cake has been formed in the chambers between the hollow circular frames carrying the filter medium and the flat rectangular discharge plates. The frames are sealed against the discharge plates by rubber rim seals which form the filtration chambers. As the rim seals are compressed by the hydraulic ram, the cake inside is squeezed. The filter is then opened, the cake adheres to the discharge plates and, as the plates are lowered and raised again, the cake is removed by scrapers. The process is fully automated and filtration areas of up to 20 m^2 are available. No washing of the cake is possible.

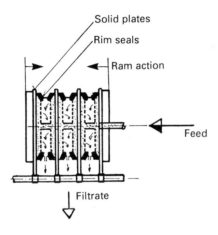

Figure 12.13. A plate-press filter

A similar principle is used in the OMD leaf filter from Stella Meta Filters. This is a vertical-leaf filter with a rubber diaphragm suspended between the leaves. The cake which forms on the leaves eventually reaches the diaphragm at which point pump pressure is used to inflate the diaphragm and compress the cake. Cake discharge is by vibration.

Another variation of the same principle is the DDS vacuum pressure filter[8] which has a number of small discs mounted on a shaft which rotates discontinuously. The cake is formed on both sides of the discs when they are at the bottom position, i.e. dipped in the slurry. When the discs come out of the slurry and reach the top position, hydraulically driven pistons squeeze the cake and the extra liquid then drains from both sides of the cake. The cake is removed by blow-back with compressed air.

Cake compression by flexible membranes is also used in the new auto-mated vertical presses that use one or two endless cloth belts, indexing between plates (as described in section 12.2.2). Filtration and compression take place with the press closed and the belt stationary, the press is then opened to allow movement of the belt for cake discharge over a discharge roller of a small diameter. This allows washing of the belt on both sides (see *Figure 12.4*). Cycle times are short (typically between 10 and 30 min) and the operation is fully automated. Sizes of up to 32 m² are available and the maximum cake thickness is 35 mm.

Washing and dewatering (by air displacement) of cakes are possible. Applications are in the treatment of minerals, in the sugar industry, and in the treatment of municipal sewage sludge and fillers like talc, clay, whiting etc.

12.2.5.2 Cylindrical presses

Another group of filters which utilize the variable-chamber principle are those with a cylindrical filter surface. There are two designs in this category, both of which originate from Britain.

The VC filter shown in *Figure 12.14* consists of two concentric hollow cylinders mounted horizontally on a central shaft. The inner cylinder is perforated and carries the filter cloth, the outer cylinder is lined on the inside with an inflatable diaphragm. The slurry enters the annulus between the cylinders and conventional pressure filtration takes place, with the cake forming on the outer surface of the inner cylinder. The filtration can be stopped at any cake thickness or resistance, as required by the economics of the process, and hydraulic pressure is then applied to the diaphragm which compresses the cake.

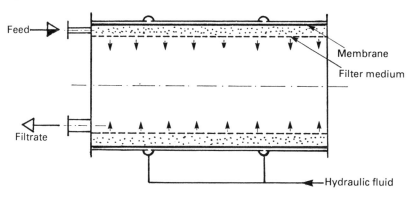

Figure 12.14. Scheme to show the principle of the VC filter

As with other filters of this type, washing can be carried out by deflation of the diaphragm and introduction of washwater into the annulus. Re-inflation of the membrane then forces the wash liquor through the cake, thus displacing the mother liquor. At the end of the process, the inner cylinder is withdrawn from the outer shell and the cake is either discharged manually or blown off with compressed air. Sizes available range from small, mobile test units of 0.4 m² filtration area to large, fully automated machines of 6.1 m² filtration area. A choice of two alternative core sizes is offered, giving annuli of 6 or 2.8 cm available for the cake. The hydraulic pressure for operating the membrane goes up to 14 bar. Although originally developed for filtration of dyestuffs, the VC filter has been used successfully for the filtration of gypsum, china clay, cement, industrial effluents, metal oxides, coal washings, nuclear waste and other slurries.

The second filter in this category is the ECLP Tube Press. This is smaller in diameter and, unlike the VC filter, it is operated in a vertical position. It uses compression only, both for filtration and for squeezing the cake. The space between the cylindrical rubber membrane and the cloth tube is first filled with slurry and the hydraulically operated membrane is used to drive the liquid through the cloth. It follows, therefore, that this filter is suitable for higher solids concentrations, usually in excess of 10% by weight, in order to obtain the minimum cake thickness necessary for efficient cake discharge of about 4 mm. At the end of the process, the central core is lowered by about 300 mm and the cake is removed by a blast of compressed air from the inside. The hydraulic operating pressures are higher than those of the VC filter at about 100 bar but the single tube area is only about 1.3 m^2. Multiple tube assemblies have to be used to treat larger flows. Cake washing is possible but with some solids there is a danger of the cake falling off the inner core whilst the annulus is being filled with water.

The ECLP Tube Press was originally developed for the filtration of china clay[9] but it has been used since with many other slurries such as those in mining, TiO_2, cement, sewage sludge, etc. The usual cycle time is about 4 min or more.

In conclusion, the method of mechanical compression of the cake is very attractive and can lead to very dry cakes. The compressibility of cakes can be very easily tested and most manufacturers provide such tests. Whenever a rubber membrane is used for the compression, however, this inevitably increases the capital cost and thus variable-chamber filters tend to be more expensive than conventional filters of the same category.

Pressure Filtration
Part II — Continuous Filtration

12.3 CONTINUOUS PRESSURE FILTRATION

12.3.1 The case for continuous pressure filtration

The following notes should be read in conjunction with section 12.2.1 where the general aspects of pressure filtration are discussed. This section only adds some notes specific to continuous pressure filtration.

The first advantage of any continuous filter is the absence of or reduction in down-times which means gains in capacity. Bearing in mind the effects of compressibility of cakes discussed in section 12.2.1.1, the effect of higher driving force on capacity can be seen in equation 12.1 which is derived for constant-pressure filtration from the well-known law of Darcy. The effect of resistance of the filter medium is neglected here and the cake is assumed to be incompressible.

If everything else remains constant, the equation can be rewritten in the form of equation 12.4, whereby the dry cake production capacity, Y, is proportional to the square root of the applied pressure drop, Δp, and to the square root of the speed of rotation, n:

$$Y = \text{constant} \cdot \sqrt{\Delta p\, n} \qquad (12.4)$$

Note that the above equation is only valid for incompressible cakes, negligible medium resistance and constant submergence, f.

The increased pressure drop in pressure filtration clearly affects the capacity in two ways: it increases capacity directly through higher driving force, and it allows the speed of rotation, n, to be increased without increasing moisture contents (because higher pressures lead to better dewatering, as shown in section 12.2.1.2). Higher speeds (at constant feed concentrations) lead to thinner cakes, and the limiting condition is the minimum cake thickness that can still be successfully removed from the filter medium and discharged. Another practical limit is the air consumption in the dewatering period that follows filtration: thin cakes offer little resistance to air flow and air rates (and energy consumption) become excessive.

It seems, therefore, that continuous pressure filters should be built to run at high pressures and high speeds (see equation 12.4). There are, however, two additional considerations: the higher the design pressure, the more expensive is the pressure vessel with its associated hardware, and thin cakes and high pressures lead to excessive air rates (and energy consumption). There must be compromise, therefore, between cake thickness and pressure: if high pressures are used, the cakes have to be thicker than for lower pressures. Different companies tackle this problem differently: KHD for example[10] limit themselves to moderate pressures (2.5 bar) and run their disc filter at high speed (up to 3 rev. min^{-1}). Amafilter, on the other hand[11], use a gauge pressure of 6 bar but run their disc filter more slowly (about 1 rev. min^{-1}) at higher submergence, which results in thicker cakes. As the equivalent speeds (taking into account different lengths of time provided for dewatering in their cycle) are in about the same ratio as the pressure drops used in the two machines, the cake production capacity with the same solid would probably be similar but the KDF filter is likely to have the edge on moisture content because it operates higher on the capillary pressure curve (*Figure 12.1*). Having a sophisticated computer-controlled system, it can also optimize the operating variables for maximum capacity, minimum moisture content, lowest energy consumption or any combination of these as required.

A continuous pressure filter may be defined as follows: it is a filter which operates at pressure drops greater than 1 bar and its operation does need to be interrupted to discharge the cake. The cake discharge itself, however, does not have to be continuous. As there is little or no down-time involved, the dry solids rates can sometimes be as high as 1750 kg m^{-2} h^{-1} with continuous pressure filters.

Most of the continuous pressure filters available today[18] have their roots in vacuum-filtration technology. It is obvious that a rotary drum or rotary-disc vacuum filter can be adapted to pressure by enclosing it in a pressure cover but the disadvantages of this measure are equally obvious. The enclosure is a pressure vessel which is heavy and expensive, the progress of filtration cannot be watched and the removal of the cake from the vessel is the most difficult problem of the whole design. Other complications of this method are caused by the necessity of arranging for two or more differential pressures between the inside and outside of the filter, which requires a troublesome system of pressure-regulating valves. Despite these disadvantages, the advantages of high throughputs and low moisture contents in the filtration cakes have justified the recent and vigorous development of continuous pressure filters, and the trend is certain to continue in the future.

Horizontal- or vertical-vessel filters with vertical rotating elements in particular have been under rapid development recently with the aim of making truly continuous pressure filters, particularly, but not exclusively, for the filtration of fine coal. There are basically three categories of continuous pressure filters presently available: disc filters, drum filters and belt filters, the last category including both the hydraulic and compression varieties. For the purpose of this review, we shall also briefly consider some semi-continuous, indexing cloth filters and other types with intermittent movement of elements. Only those filters which discharge the solids in the form of a semi-dry cake are considered here and the so-called filter-thickeners (which discharge the solids in a thick slurry) are, therefore, excluded; these are reviewed in chapter 18.

The advantages of continuous pressure filtration are clear and indisputable, particularly with slow-settling slurries and fairly incompressible cakes. Such filters are expensive, however, both to install and to run, and the most likely applications are either in large-scale processing of products that require thermal drying after the filtration stage (fine coal or cement slurries in the dry process) or in small-scale processing of high-value products, like in the pharmaceutical industry.

There are, obviously, many technical problems to be solved when developing a new commercial and viable filter; all are solvable and some of the recently developed and proven systems show this. However, the filtration hardware in itself is not enough: as the control of a continuous pressure filter is much more difficult than that of its equivalents in vacuum filtration, the necessary development may also include an automatic, computerized control system. This moves pressure filtration from low to medium or even high technology.

12.3.2 Disc filters

12.3.2.1 The McGaskell and Gaudfrin disc filters

One of the earliest machines in this category is the McGaskell rotary pressure filter[12] which is essentially a disc-type filter enclosed in a pressure vessel. Just like the vacuum rotary-disc filters, the rotating discs in this filter are each composed of several wedge-shaped elements connected to a rotary filter valve at the end of the shaft. This filter was originally designed for the filtration of waxes in the oil industry and was equipped for a gradual increase in pressure with cake build-up. This is said to have produced high filtrate clarity. Pressures of up to 7 bar have been used.

The slurry reservoir is divided into pockets or 'crenellations', which have spring-loaded scrapers. The scrapers press against the discs and direct the cake into the spaces between the pockets around the discs which lead to a chute connected to an inner casing in which is placed a worm gear. The worm conveys and compresses the cake, thus squeezing more liquid from it, through a filter cloth surround. The compressed cake forms a plug round a spring-loaded, tapered discharge valve, and the plug prevents leakage of gas. The cake can be washed but probably not very effectively.

This filter is reported in filtration literature[12] but whether or not it is available on the market at present is not known to the author; it is, however, included in this review for completeness and also to show that the idea of a pressure disc filter is by no means new.

Another disc-type pressure filter based on a similar principle is the Gaudfrin Disc Filter, originally designed for the sugar industry and available in France since 1959. It is also similar in design to a vacuum disc filter but it is enclosed in a pressure vessel with a removable lid. The discs are 2.6 m in diameter and composed of 16 sectors. The cake discharge is by air blow-back, assisted by scrapers if necessary, into a chute where it may either be reslurried and pumped out of the vessel or, for pasty materials, it can be pumped away with a monopump without reslurrying. The Gaudfrin Disc Filter is designed for only relatively low pressures of 1 bar on average and it provides for cake washing which can be in two stages, in two separate compartments within the same vessel.

12.3.2.2 The KDF filter

A recently developed KDF filter (*Figure 12.15*) from Amafilter[11] is based on a similar principle to that of a disc filter. It was developed mainly for the treatment of mineral raw materials like coal flotation concentrates or cement slurries. It can produce a filter cake of low moisture content, at very high capacities of up to 1750 kg m^{-2} h^{-1}. The pressure gradient is produced by pressurized air above the slurry level which, besides providing the necessary driving force for the filtration, is also used for displacement dewatering of the cake.

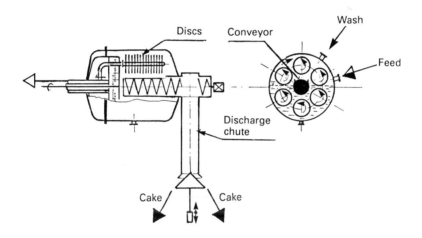

Figure 12.15. The KDF continuous-pressure filter

Assemblies of small discs are rotated in a planetary movement around a central screw conveyor. The discs are mounted on six hollow axles and the axles revolve on overhanging bearings from the gearbox at one end of the vessel where they are driven, via a drive shaft, by an electric motor. The filtrate is collected from the discs via the hollow shafts and a filter valve into a large collecting pipe. The hollow shafts also collect the water and air from the dewatering process, in another part of the rotational cycle. The number of discs mounted on the shafts can be adjusted for different materials, depending on the required capacity and the cake thickness to be used. As the vessel is only about half-filled with slurry, the discs become coated with the cake when immersed, the cake is then dewatered when they emerge from the slurry and scraped or blown off (by reverse blow) into the central conveyor, which takes the cake to one end of the vessel. The planetary action and the slow movement of the discs through the feed slurry ensure exceptionally good homogeneity of the cake which is critically important for good dewatering characteristics; the typical speed of rotation of the planetary system of shafts is 0.8–1.0 rev. min^{-1}.

A screw conveyor was used for conveying the cake at first but this has been replaced with a chain-type conveyor in the filter marketed and sold recently. The first prototype used a tapered screw to form a plug before discharge into atmospheric pressure but this has now been replaced by compaction in a vertical pipe.

The cake discharge is initiated and stopped by two level indicators inside the vertical pipe. The cake is actually discharged using the pressure inside the vessel; a specially designed, hydraulically operated discharge valve momentarily opens and the cake 'shoots' out. The air pressure used for driving the slurry through the filter is 6 bar and filtration areas are available up to 120 m^2. Cake washing is possible but it has not been reported as actually being performed.

The KDF filter was first tested in prototype on a coal mine in northern Germany. It was installed in parallel with existing vacuum filters and it produced filter cakes consistently lower in moisture content by 5–7% than the vacuum filters. Two production models have now been installed and operated at a coal mine in Belgium. The filter is controlled by a specially developed computer system; this consists of two computers, one monitoring the function of the filter and all the detection devices installed, and the other controlling the filtration process. The system allows optimization of the performance, automatic start-up or shut-down and can be integrated into the control system of the whole coal-washing plant.

12.3.2.3 The KHD pressure filter

Another development of the disc filter has been reported more recently from KHD Humboldt Wedag AG[10]. A somewhat different system, probably a predecessor, was also patented by the same company[13]. The patented system[13] has stationary discs mounted inside a pressure vessel (horizontal vessel, vertical discs) which is mounted on rollers and can rotate slowly about its axis. A screw conveyor is mounted in the centre of rotation but this is stationary; it conveys the cake, which is blown-off the leaves when they pass above the screw, to one end of the vessel where it falls into a vertical chute. The cake-discharge system involves two linear slide valves which slide the cake through compartments which gradually depressurize it and move it out of the vessel without any significant loss of pressure. The system relies entirely on the cake falling freely from one compartment to another, as the valves move across, and this may well be a rather unrealistic assumption, particularly with sticky cakes. This, combined with lots of sliding contact surfaces which are prone to abrasion and jamming, makes one wonder if the idea is practicable at all. There is another major disadvantage involved here in that there are two large running seals involved in the main body of the filter as the vessel rotates around a stationary central arrangement; this seal is another potential source of trouble. All in all, this version has, in my view, very little chance of commercial success and it seems to have been shelved in favour of a more conventional system of a stationary vessel as reported recently[10]

The newer version, actually built and tested with coal slurries in a pilot plant facility and also, most recently in production, with a 90-m^2 version, has the rotating discs and all the driving elements inside a stationary vessel. The discs, according to the manufacturer's literature, range from 1300 to 3000 mm in diameter, with up to 10 being housed in one vessel, giving filtration areas of up to 480 m^2. The cake discharge is through a rotary lock discharger which has cylindrical compartments rotating around a vertical axis. Once again, sliding surfaces are involved and the cake is assumed to be non-sticky so that it will fall out when the compartment opens to the atmosphere. The filtration area can be varied by changing the size and number of discs; no overall vessel dimensions or further details of the filter construction, other than those quoted above, are given.

The test results reported show the advantages of pressure filtration quite clearly: the dry-cake production capacity obtained with the test solids (coal suspensions) was raised by 60–70% by increasing the pressure drop from 0.6 bar to 2 bar. At the same time, the final moisture content of the cake reduced by as much as 5–7%. There is, obviously, a law of diminishing returns at play here, in that further increases in the operating pressure bring about less and less return in terms of capacity and moisture content.

The authors claim that, if the pressure ". . . is increased beyond 2 bar, the residual moisture content cannot be further reduced remarkably", but their own data does not seem to support this. The reduction continues at pressures beyond 2 bar, albeit at a lower rate, and, surely, statements on whether or not it is feasible to go to higher pressures should be supported by an economical analysis in which the cost of thermal drying is considered. The tests also show fairly predictable effects of particle size and cake thickness on dry-cake production capacity.

Another reason for keeping the operating pressure low is the air consumption: in order to obtain high capacities, the manufacturers designed the discs to spin fairly fast (up to 2 or 3 rev. min^{-1}) and this leads to thin cakes. Thin cakes, however, give little resistance to air flow and, consequently, the only way to keep this flow within economical limits is to reduce the air pressure in the vessel—see the previous discussion of this in section 12.3.1.

12.3.3 Drum filters

Another idea borrowed from vacuum filtration to produce a continuous pressure filter is the rotary-drum filter. This filter has the disadvantage that it makes relatively poor use of the space available in the pressure vessel and the filtration areas and capacities of such filters cannot possibly match those of the disc pressure filters. In spite of this disadvantage, however, the pressure drum filter has recently been extensively developed.

The drum is sometimes mounted in a vertical rather than a horizontal vessel. Once again the pressure is created by pumping compressed gas into the vessel. Sometimes, the intake of the compressor is connected to the filtrate side of the filter and the gas goes round in a closed circuit. The method used to discharge the cake continuously from the high pressure inside the vessel into the atmosphere is, once again, the real heart of the design. Variable pitch screws, star valves, alternating or serial decompression chambers, mono-pumps (for pasty and thixotropic cakes) and other similar devices have been tried with a varying degree of success, obviously depending very much on the properties of the cake. Another, rather obvious, alternative is the use of two storage vessels into which the cake is alternately discharged at the same pressure as in the filter, the pressure is later released and the cake is discharged from the vessel under atmospheric pressure.

12.3.3.1 The development at 'TU Karlsruhe'

Professor Stahl and his co-workers[2] at the University of Karlsruhe have worked in continuous filtration for many years and have developed a test unit of a small drum filter (total filter area of 0.7 m^2 with 30% submergence) housed in a large horizontal pressure vessel and, with it as a model, developed and tested several interesting concepts.

One such concept is the so-called hyperbar vacuum filtration. This is a combination of vacuum and pressure filtration in a pull–push arrangement, whereby a vacuum pump of a fan generates a vacuum downstream of the filter medium, whilst a compressor maintains higher-than-atmospheric pressure upstream. If, for example, the vacuum produced is 0.8 bar (i.e. an absolute pressure of 0.2 bar) and the absolute pressure before the filter is 1.5 bar, a total pressure drop of 1.3 bar is created across the filter medium. This is clearly a new idea in principle, but in practice it requires three primary movers: a liquid pump to pump in the suspension, a vacuum pump to produce the vacuum and a compressor to supply the compressed air. The cost of having to provide, install and maintain one. additional primary mover is probably the main reason why the idea of hyperbar vacuum filtration has not yet taken off: only Andritz in Austria offer a system commercially.

Apart from the hyperbar vacuum filtration, the work at Karlsruhe also includes fundamental investigations into cake filtration generally, pressure filtration of coal and ore suspensions, and studies of dewatering of cakes. As an alternative to the small drum filter, they have apparently also used a small belt filter in the same pressure vessel. Some commercial developments, like the KHD pressure filter, are reported to have originated from the work at Karlsruhe.

12.3.3.2 The TDF drum filter

The TDF drum filter was developed at Krauss Maffei and the product is a fairly conventional drum filter housed in a vertical pressure vessel. Some test data can be found in the company literature[14], obtained with the smallest model of only 0.75 m^2 filtration area; larger models have also been announced, ranging up to a filtration area of 46 m^2 but at this limit the vessels become very large. The operating pressures quoted are moderate (2.5–3.5 bar) and the drum speeds are fairly conventional (0.3–1.5 rev. min^{-1}). The range of dry-cake production quoted is 250 to 650 kg m^{-2} h^{-1} for fine coal.

The cake is scraped-off with a conventional knife arrangement and is then conveyed in a screw conveyor to one end of the vessel where it enters the discharge system; there are apparently four design alternatives depending on the cake to be processed. A tapered rotary valve (with a horizontal axis), a pump (presumably a mono-pump), a rotary valve (vertical axis) with a blow-through, similar to the one used in pneumatic conveying, and a vertical pipe compactor similar in design to the Fuller-Kinyon pump in pneumatic conveying. It is not known to what extent these systems have been developed or whether they merely represent possibilities yet to be proven.

A plate-type filter has also been mentioned in the Krauss Maffei literature; the PDF filter which, instead of a drum, uses a paddle wheel (like in a paddle steamer) with radial, longitudinal plates covered with filter cloth and manifolded to the filter valve at one end of the vessel. This filter apparently used a horizontal pressure vessel and was built to have only 0.75 or 1.5 m^2 a ea and operated at 2.5 bar. A central screw conveyor collected the cake blown-off the plates and conveyed it to the discharge end of the vessel.

12.3.3.3 The BHS-Fest Filter

A different approach to the use of a drum for pressure filtration is made in the BHS-Fest filter[15] (*Figure 12.16*). This permits a separate treatment of each filter section, in which the pressure may vary from vacuum to a positive pressure of several bar; pressure regulation is much less difficult than in the conventional enclosed drum-type pressure filter.

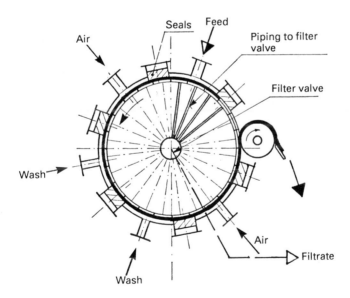

Figure 12.16. The BHS-Fest pressure filter

The BHS-Fest pressure filter has a rotating drum, also divided into sections but the separating strips project above the filter cloth and thus form cells. The drum is almost completely surrounded by an outer shell and the space between the shell and the drum is divided into a number of compartments.

The compartments are separated by seals under adjustable pressure. As the drum rotates, each cell on the drum passes successively through the series of compartments, thereby undergoing different processes such as cake formation, dewatering, cake washing or cake drying, and these can be carried out in several stages under different pressures or even under vacuum.

Cake discharge occurs at atmospheric pressure by the action of a roll or a scraper, assisted by blow-back. The cloth may be washed by a spray before the cycle starts again. Filtering areas range up to 8 m^2 and drum diameters up to 2 m. The necessity for large seals limits the operating pressure to less than 3 bar, typically. The cake thickness can be from 2 to 150 mm; depending on machine sizes and the speed of drum rotation is up to 2 rev. min^{-1} (usually 0.3–1 rev. min^{-1}). Applications occur in the manufacture of pharmaceuticals, dyestuffs, edible oils and various chemicals and minerals.

12.3.4 Horizontal belt pressure filters

As much as the vacuum horizontal belt filters have made a considerable impact in vacuum filtration, one would expect that the same principle may be used in pressure filtration, by enclosing the whole or part of the belt in a pressure vessel. This idea has to be seen as entirely separate from the so-called belt presses (section 12.3.5.1) which use the squeezing of cake by a second belt and do not use the pressure of the liquid or air to drive the filtrate through and out of the cake.

Horizontal belt filters have a great advantage in cake washing application due to their horizontal filtration surface. In the context of pressure filtration and the requirements of good dewatering, however, they have a major disadvantage because the cake is not very homogeneous; gravity settling on the belt and the inevitable problems of distribution of the feed suspension over the belt width give rise to particle stratification and non-homogeneous cakes.

It has already been mentioned in the previous section that a horizontal belt filter has been used in place of the small drum filter in the filtration studies at Karlsruhe[2]. In this case, the whole of the filter was placed in a large pressure vessel and there were, therefore, no moving parts passing through the filter shell. To the best of my knowledge, there is no commercial filter based on this principle; the utilization of the space inside the pressure vessel would be rather poor and the filtration areas very limited indeed.

Another possibility is to enclose only the working, top part of the horizontal belt in the pressure vessel and to pass the belt through the sides of the vessel. Inevitably, the operation has to be intermittent because the belt cannot be dragged over the support surface with the pressure on and, also, the entrance and exit ports for the belt must be sealed during operation to prevent excessive losses of air. The movement of the belt is intermittent, therefore, and is synchronized with decompression in the vessel. This means that the whole of the vessel volume must be depressurized in every cycle and this is wasteful; there is also the inevitable down-time involved but there are no problems with discharging the cake because this is done at atmospheric pressure. Strictly speaking, such filters do not fall in the scope of this review because the movement is not continuous, and they are only briefly reviewed in the following.

The idea is not new: the flat-bed pressure filter by Hydromation Engineering Co. Ltd[16], for example, is based on this principle. The pressure compartment consists of two halves, top and bottom. The bottom half is stationary whilst the top half can be raised to allow the belt and the cake pass out of the compartment and to be lowered onto the belt during the filtration and dewatering stages. The filter can be considered as a horizontal filter press with an indexing cloth; in comparison with a conventional filter press, however, this filter has a disadvantage in that it allows only the lower face of the chamber to be used for filtration.

A recent patent from KHD AG[17] describes the same idea except that the top half of the pressure compartment is not lifted but opens and closes little gates for the belt to pass through. This is another filter proposed for dewatering of fine coal in particular.

One relatively recent development in this filter category is the vertical recessed plate automatic press (shown schematically in *Figure 12.4*), also described in section 12.2.2. This filter has many filtration chambers placed vertically one upon another. It has an endless cloth belt, travelling intermittently between the plates via rollers to peel off the cake. As the filter cloth zig-zags through the filter, the filtering direction is reversed each time and this tends to keep the cloth clean. As it goes round, the filter belt can be washed on both sides as shown in *Figure 12.4*. The height of the vertical press makes maintenance difficult and the number of the chambers is, therefore, restricted, usually to 20, maximum 40, with filter areas of up to 32 m². Cycle times are short, typically between 10 and 30 mins, and the operation is fully automated. The maximum obtainable cake thickness is about 35 mm, washing and dewatering (by air displacement) of cakes is possible. Applications include the treatment of mineral slurries, sugar, sewage sludge, and fillers like talc, clay, whiting and similar. The filtration velocities are usually from 0.025 to 1 m h^{-1} and dry solids handling capacities are up to 1000 kg m^{-2} h^{-1} due to the short down-times involved.

12.3.5 Continuous compression filters

The variable-chamber principle applied to batch filtration, as described in section 12.2.5, can also be used continuously, in belt presses and screw presses.

12.3.5.1 Belt presses

The next category of continuous pressure filters is the belt filter press which usually combines gravity drainage with mechanical squeezing of the cake between two running belts. The first example is the Manor Tower Press, a Swiss invention, which is also available in the UK. A schematic diagram of the principle of this press is shown in *Figure 12.17*. Essentially it consists of two acutely angled vertically converging filter belts running together downwards. The shallow funnel formed between the belts (and the ends sealed by a special edge-sealing belt) is filled from the top with the slurry to be filtered and the slurry moves together with the driven belts down the vertical

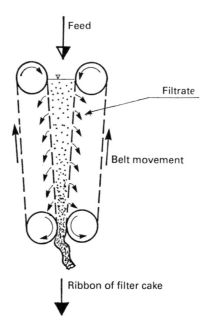

Figure 12.17. Scheme to show the principle of the Manor Tower press

narrowing gap where filtration takes place. The hydrostatic pressure causes the solids to be deposited on the faces of the belts until the two cakes combine to form a continuous ribbon of filter cake which is squeezed in the closing gap at a pressure of 2.5 bar (gauge pressure) and discharged at the bottom. In order to protect the press against overload, the final gap at the bottom is automatically controlled so that the above-mentioned pressure is not exceeded. Although originally designed for the continuous filtration of conditioned sewage sludges (as are most of the filter belt presses available), the Manor Tower Press is increasingly used for the treatment of paper mill sludge, coal or flocculated clay slurries.

There is a bewildering variety of horizontal filter presses, with probably more than 20 designs on the market; most of these were developed in the past decade. They owe their existence to the availability of cationic polyelectrolytes which promote the release of water from organic sludges which is the area where belt-filter presses are most used. Some are therefore called sewage-sludge concentrators. As an example, the Unimat Belt Filter Press, which has most of the features characteristic of many belt presses, is shown in *Figure 12.18*. The flocculated feed is first introduced onto the horizontal drainage section where the free water is removed by gravity. Sometimes a system of ploughs may be employed to turn over the forming cake and allow any free water on the cake surface to drain through the belt mesh. The sludge is then sandwiched between the carrying belt and the cover belt and compression dewatering takes place; liberated water passes through the belts. The third zone ('shear zone') is designed to produce yet drier cake by shearing the cake by flexing it in opposite directions during passage through a train of rollers in a meander arrangement. In order to prevent the released water

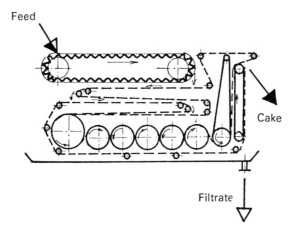

Figure 12.18. The Unimat belt filter press

being absorbed back into the cake on the cake release, scrapers or wipers are sometimes installed to remove the water from the outside of the belts. Belt-filter presses are made up to a width of 2.5 m and produce a final solids concentration of the discharge sludge in the range 35–60%.

It should be added here that, apart from the specially designed belt presses described above, the compression principle is also utilized in some conventional vacuum drum or belt filters by the addition of compression rollers or belts. The benefit in terms of further dewatering in such cases is small, if indeed there is any advantage, because the compression time is short and the excluded liquid might be sponged back into the cake before it can be removed. Such devices probably have greater value in tackling cake-cracking problems.

12.3.5.2 Screw presses

Another way of achieving compression of the cake is by squeezing in a screw press—see *Figure 12.19*. This process is only suitable for the dewatering of rough organic materials, pastes, sludges or similar materials, because it does not include a filtration stage. The material is conveyed by a screw inside a perforated cage, and the available volume continuously diminishes. This can be done either by reducing the pitch of the screw in a cylindrical cage, or by reducing the diameter of the screw in which case the cage is conical, as shown in *Figure 12.19*. The cage is either perforated or constructed from longitudinal bars in a split casing. The solids discharge is controlled by a suitable throttling device which controls the operating pressure. Washing or dilution liquid can be injected at points along the length of the cage. The power requirements are high.

The Stord Twin Screw Press uses two counter-rotating, intermeshing screws in a perforated cage. The gradual reduction in the space between the

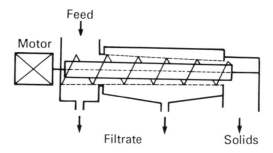

Figure 12.19. Scheme to show the principle of a screw press

screw flights and the strainer plates is achieved by a combination of reduction in the cage (and screw) diameter and of increase in the diameter of the hollow trunk of the screw body. Although more than one type of material can be handled in the same press, the press usually has to be adapted to different raw materials, according to their dewatering characteristics. Typical applications include dewatering of sugar-beet pulp, fish meal, distillers and brewers spent grains, starch residues, fruit, potato starch by-products, grass, maize, leaves and similar materials.

REFERENCES

1. Svarovsky, L., *Solid–Liquid Separation Processes and Technology, Vol. 5, Handbook of Powder Technology*, Elsevier, Amsterdam (1985)
2. Bott, R., Anlauf, H. and Stahl, W., 'Continuous pressure filtration of very fine coal concentrates', *Aufbereitungstechnik*, **5**, 245–258 (1984)
3. Svarovsky, L., 'Pressure filtration of coal flotation concentrates', *Conference on Solid–Liquid Practice III, 397th Event of the EFCE* (Bradford, 29–31 March 1989), Institute of Chemical Engineers Symposium Series, London (1989)
4. Wakeman, R. J., 'The application of expression in variable volume filters', *J. Separ. Proc. Technol.*, **2**, 1–8 (1981)
5. Baluais, G., Rebouillat, S., Laclerc, D. and Dodds, J. A., 'Modelisation of compression dewatering and the application to filter design, I', *Chem. Eng. Symp. on Solid/Liquids Separation Practice and the Influence of New Techniques* (Leeds, 3–5 April 1984), Paper 3, pp. 32–41, Institute of Chemical Engineers, Yorkshire Branch (1984)
6. Carleton, A. J., and Moir, D. N., 'Optimisation of compression filters, I', *Chem. Eng. Symp. on Solid/Liquids Separation Practice and the Influence of New Techniques* (Leeds, 3–5 April 1984), Paper 2, pp. 19–31, Institute of Chemical Engineers, Yorkshire Branch (1984)
7. Carleton, A. J., 'Choosing a compression filter', *Chem. Eng.*, **April**, 20–23 (1985)
8. Johnsen, A. F., Madsen, R. F. and Nielsen, W. K., 'DDS-vacuum pressure filter', *2nd World Filtration Congress 1979*, pp. 163–171, Filtration Society, London (1979)
9. Gwilliam, R. D., 'The EEC tube filter press', *Filtr. Sep.*, **March/April**, 1–9 (1971)
10. Blankmeister, D. and Triebert, Th., 'Dewatering ultrafine coal slurries by means of pressure filtration', *Aufbereitungstechnik*, **1**, 1–5 (1986)
11. Dosoudil, M., 'The application of the continuous pressure filter to flotation concentrates', *Aufbereitungs-Technik, Jahrgang,* **25**, 259–265 (1984)
12. Mills, F. D., *Report on Filtration*, Imperial Chemical Industries (1957)
13. Heintges, S., German Patent DE 3316561 A1 (1984)
14. Krauss Maffei, Information, Trommeldruckfilter TDF
15. BHS Trenntechnik, Sonthofen, BHS-Fest-Filter
16. Purchas, D. B., *Solid/Liquid Separation Technology*, Uplands Press, Croydon (1981)
17. Heintges, S., US Patent 4,477,358 (1984)
18. Svarovsky, L., 'Recent developments in continuous pressure filtration', *Aufbereitungs-Technik,* **5**, 242–250 (1986)

13

Vacuum Filtration — Part I

D. G. Osborne

PT Kaltim Prima Coal, Jakarta, Indonesia

NOMENCLATURE

A_c	Total area of a continuous filter	m^2
B	Drum or plate width	m
E	Feed slurry rate	$m^3\,s^{-1}$
F	Filtrate rate	$m^3\,s^{-1}$
h	Cake thickness	m
K	Proportionality constant	—
L	Length of filter belt	m
n	Drum or plate speed	s^{-1}
ΔP	Filtration pressure drop (overall)	$N\,m^{-2}$
Q_f	Form volumetric filtration rate	$m^3\,m^{-2}\,h^{-1}$
R_m	Filter media resistance	m^{-1}
r	Radius of the drum	m
s	Coefficient of cake compressibility	—
U	Radial velocity	$m\,s^{-1}$
w	Solid cake mass per unit volume of filtrate	$kg\,m^{-3}$
Z_c	Filter cake yield per cycle	$kg\,m^{-2}\,cycle^{-1}$
Z_f	Form solids filtration rate	$kg\,m^{-2}\,h^{-1}$
α	Cake-specific resistance	$m\,kg^{-1}$
α_0	Specific resistance at zero pressure	$m\,kg^{-1}$
μ	Filtrate viscosity	$N\,s\,m^{-2}$
Φ_d	Dewatering portion of the filter or belt	$\%$
Φ_f	Filtering portion of the filter or belt	$\%$
θ_d	Cake dewatering time per cycle	s^{-1}
θ_f	Cake formation time per cycle	s^{-1}

13.1 INTRODUCTION

All vacuum filtration techniques, like all other methods of cake filtration, employ a porous filter medium to support the filter deposit. Unlike other filter forms, however, vacuum filtration employs a low driving force and it is

415

this which allows it to be used in its own special areas of application. There are obvious advantages and disadvantages for such a system. The disadvantages stem mainly from the obvious limitation in driving force, which may lead to low filtration rates and higher than acceptable cake moistures. However, there are also some very substantial advantages.

For compressible cakes, higher pressures may lead to high cake and medium resistances, a factor often encountered in the use of filter presses. The design problems encountered for both batch and continuous vacuum filters tend to be less difficult to overcome than for filters which operate at high pressures or at high speeds of rotation, for example centrifugal separators. Because vacuum filters can be simply designed, it is generally possible to employ a wide variety of materials in their construction in order to achieve an almost complete suitability for a wide range of applications.

Perhaps the most significant advantage afforded by the vacuum system is that of continuous operation under relatively simple mechanical conditions, and although there is a range of batch vacuum filters currently in use, by far the greatest application of vacuum filters lies in continuous operations. Continuous vacuum filter machines, and particularly those of the rotary type, have long since proved themselves to be capable of extremely satisfactory commercial operation in a very much wider field of application than any other type of continuous filter. The notes which follow are not intended to provide a great insight into the more academic aspects of cake filtration; this is dealt with in chapter 9. Only brief references, where pertinent to rotary vacuum filter (RVF) units, are included and the operating formulae are given in section 13.4 in a more or less summarized form, although it should be stressed that filter design is still based upon a combination of laboratory and pilot-scale testwork and empirically derived data derived from operational experience. This chapter includes reference to the whole range of vacuum filters currently available and in use, together with comments about their cost, filter construction and the various applications of each type of machine. No attempt has been made to include much information about filter media or filter aids, as these topics are discussed in chapters 10 and 17.

13.2 VACUUM FILTRATION EQUIPMENT

Before beginning to discuss the various operational characteristics of vacuum filters, it is necessary to categorize and describe the types of machines currently in use. In this context, some indication of their merits and demerits, together with the normal applications of each unit, will also be included. The classification shown in *Figure 13.1* provides a useful reference to each type of machine and its relation to the others.

13.2.1 Batch vacuum filters

The simplest type of vacuum filter is illustrated by the more or less standardized laboratory Buchner funnel, which can be used for both fast and slow draining slurries and may in some cases be relied upon to provide filtration data for commercial application. Commercial batch filters are not common

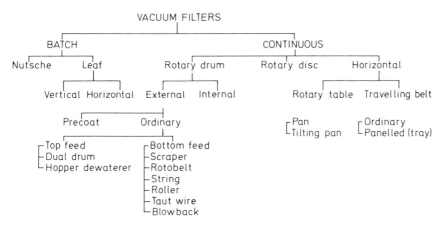

Figure 13.1. Classification of vacuum filtration equipment

and are normally only used in the few cases when continuous machines cannot be employed.

There are two types in usage (see *Figure 13.2*). One is the vacuum Nutsche, which is used for quite large industrial separations and can be arranged so that the cake can be discharged by ploughs or rakes. The other type is typified by the Moore filter, which consists of a battery of filter leaves that can be dipped into the slurry to be filtered and then arranged for back-flow discharge or for removal of the leaves for cake discharge. An improved vacuum leaf filter has been developed[1] for use with china clay slurries. There are also horizontal, single-pan filters which have been developed primarily for handling free settling slurries. They are used for small-scale batch production where they have all the advantages of a single Nutsche filter, but with the additional advantages that the solids are removed simply by means of either a manual or a pneumatic/hydraulic system, and excellent cake washing and rewashing can be incorporated into normal operations. Filters of this type can also be used for small-scale pilot plant work, particularly in establishing parameters for horizontal filter operation[2].

13.2.2 Continuous vacuum filters

The three main types of continuous vacuum filters are: rotary drum, rotary disc and rotary horizontal. Of these, the rotary drum and disc types, in that order, are by far the most widely used.

Rotary drum types of filter range in size from 0.05 m² to well over 100 m² and can be sub-divided into three categories, of which the first is the most common; they are: bottom feed, top feed (including ordinary dual drum and hopper dewaterers) and precoat, which is usually an extension of the bottom feed type.

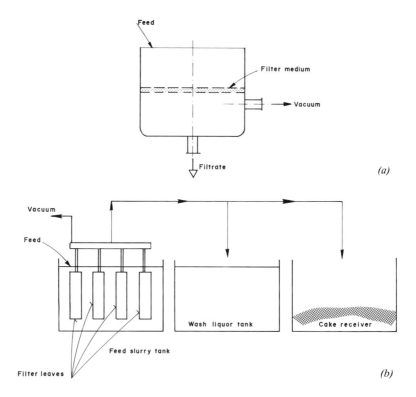

Figure 13.2. Two examples of simple batch vacuum filters: (a) Nusche vacuum filter; (b) Moore vacuum leaf filter

13.2.2.1 Rotary drum filters

13.2.2.1.1 Orthodox drum filters

These consist of a drum, with connections from the periphery to an automatic filter valve. The drum rotates with its circumference dipping into a tank of slurry from which the filtrate is drawn and the cake is deposited on the drum surface. As the drum rotates, the cake becomes drier and is eventually discharged. As the cake forms on a filter cloth, which is supported on some drainage medium, the filtrate passes through the drainage sections and thence through channels to the valve. A cross-section of the drum is shown in *Figure 13.3*. The manufacturers employ different methods of providing the drainage system and securing the filter cloth. Generally, however, most modern drum filters are of multi-panel design with the cloth secured by caulking or wire winding. The drum deck is usually sub-divided into sections with division strips incorporating a dovetail groove, and the ends of the drum are closed with end rings. The drum deck is formed by replaceable drainage grids held in position by the division strips. These grids support the filter cloth and are of such a design as to provide maximum free area and hence maximized potential filter capacity. Nowadays, these grids are made from

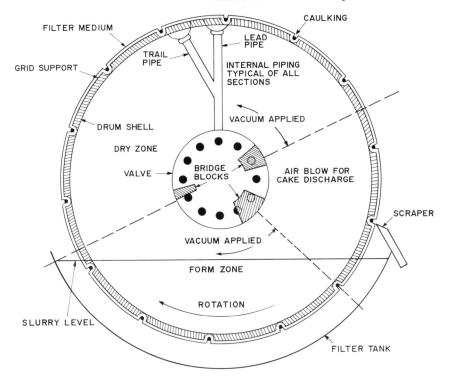

Figure 13.3. Cross-section of a drum filter

polypropylene and other thermoplastics in order to minimize adhesion and scale build-up, and to ensure long life.

The method of discharging the cake is of great importance in maintaining a clean cloth. Over the past few years, developments have taken place whereby the filter cloth is not secured to the drum but is wrapped around it, travels with it, and is arranged to come away from the drum at the discharge point, to pass around a discharge roll, and thence into a wash box for cleaning before returning to the drum for cake pick-up. This technique is very useful for slimy cakes and blinding materials, but for normal operations the provision of this facility greatly increases the price and may make such an installation much more expensive than a conventional filter. *Figure 13.4* shows an example of the latest type of drum filter incorporating a belt-type of discharge system.

Figure 13.5 shows the most commonly applied forms of cake discharge mechanisms for drum filters.

For cakes which do not blind and on which good filtrate clarity is required, a layer of cake can be built up on the drum to the limits of the position of a fixed knife, which scrapes off the excess cake. The scraper blade is mounted at a tangent to the filter drum but not in contact with either the cloth or the wire winding. Scrapers are usually fitted with hard plastic tips. The wire winding is usually wound on 5 cm centres. Scraper-type discharge can be used for almost all filtration applications where a cake thickness of more than 3 mm is obtained. For thin cakes, when the filtration rate is lower and where the cake itself tends to blind, a knife can be arranged to be near the surface of

420

Figure 13.4. Drum filter. (Courtesy of M. Vernay)

the filter cloth, and for the cake to be discharged on to it by means of blow-back applied, as each panel reaches the discharge point, via the rotary valve at the discharge point (*Figure 13.5a*).

Another technique used when the cake has some strength is to discharge it by means of endless strings (and in some cases fine chains), which are wrapped around the drum but which lift away at the discharge point passing over a series of rollers and through an aligning comb before returning to the drum. In suitable applications a complete sheet of filter cake is removed and the lifting away of the cake has a self-cleaning effect on the cloth. Variations of this system, using wire, chains and coil springs, have also been developed (*Figure 13.5b*). A comb ensures uniform spacing of the strings or chains.

The continuous type of belt discharge system shown in *Figure 13.4* probably represents the main avenue of development in recent years. This form of discharge was designed with the specific purpose of dealing with difficulties which the other forms could not satisfactorily overcome, i.e. handling thin, sticky, filter cakes without blinding of the filter medium[3].

The endless filter cloth passes over a series of rollers at the end of each filtration cycle, a shown in *Figure 13.5c*. Whilst away from the drum surface, the cake is discharged and the filter cloth is washed, usually on both sides. At the discharge roll, the direction of travel of the filter cloth is caused to change suddenly. The cake usually breaks cleanly from the belt and the flexing of the medium assists in dislodging adhering cake from the interstices of the woven cloth. Because the cake, in effect, self discharges it is not necessary to use mechanical devices or compressed air to assist in its removal and the risk of blinding is eliminated. Following cake discharge, the filter medium passes through a washing zone and high-pressure sprays remove remaining adhering solids.

Due to the fact that the cloth is washed during the period of separation from the drum periphery enabling, if required, wash liquid to be collected in a separate trough, it is possible to wash the cloth with a specific cleaning agent once every revolution. This can further ensure that a high degree of cleaning is maintained. After cleaning, the belt returns to the filter drum after passage over an adjustable return or tension roll.

Early difficulties encountered with this form of discharge were mainly due to misalignment of the belt. Misalignment is usually caused by four phenomena; edge misalignment, lateral misalignment, uneven stretching (sometimes called mooning) and wrinkling. Edge misalignment occurs when one edge of the belt becomes displaced longitudinally in relation to the other, and may be caused by poor belt fabrication or drum and roller misalignment. Lateral misalignment occurs when filter cake loading is uneven and is often caused by poor cake discharge or by roller misalignment. When edge misalignment occurs, integrated spraying of the belt may prove adequate to correct the problem.

Mooning is typified by a crescent shape occurring in the belt, hence the name. Typical causes are high cake load in the centre of the belt and uneven or excessive washwater which individually result in leading and trailing crescent formations.

Wrinkling, which is self-explanatory, may be caused by a combination of any of the malfunctions discussed. In general, all misalignment problems originate from some form of cloth fault being enhanced by poor filter

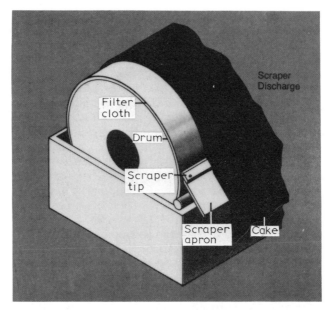

Figure 13.5. (a) Scraper discharge

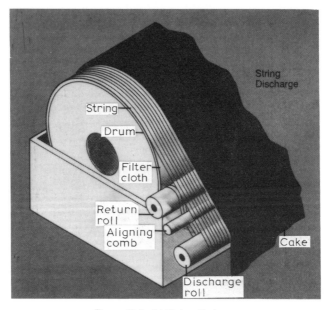

Figure 13.5. (b) String discharge

operation which ensues. Therefore, the more symmetrical the travelling web of the belt the less will be the difficulties encountered in tracking it. A single-width cloth having no longitudinal joints will give the fewest operational problems. This, combined with the use of a simple and effective tracking mechanism should result in minimal operating stresses in the belt and hence a

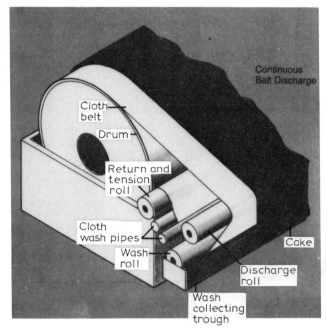

Figure 13.5. (c) Continuous belt discharge

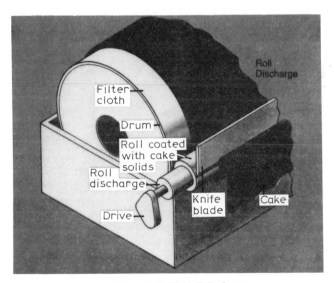

Figure 13.5. (d) Roll discharge

long belt operational life.

A number of tracking devices have been developed and two commonly employed systems, i.e. the edge track and the split roll, are shown in *Figure 13.6*. With roll-type cake discharge (*Figure 13.5d*), the cake is picked up from the drum by a roll which is driven from the drum drive and rotates in the

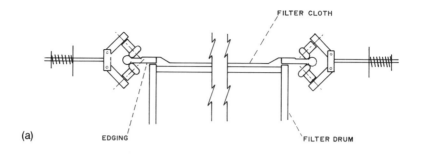

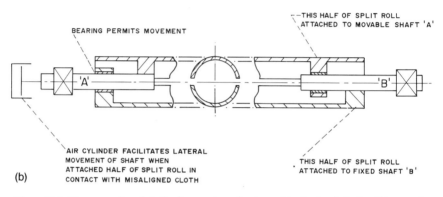

Figure 13.6. Two examples of cloth-alignment mechanisms: (a) edge track device; (b) operation of a split roll

opposite direction to the drum. The peripheral speed of the roll is a little higher than that of the drum. The cake is removed by scraper or comb, a residual layer of the cake being left on the roll to ensure a better adhesion to the filter cake, which is therefore easier to pick up. Filters with roll-type discharge do not require wire winding. They are used in all cases where very finely divided solids, which form very thin cake thicknesses of only 0.5–3.0 mm, are to be filtered.

The simplest type of drum filter operates with approximately 30–40% of the drum submerged in the tank. The geometry of this arrangement is such that glands at the trough ends to take the drum shaft and valve are not required. For some operations, where the cake formation stage is the limiting factor and precoat filters are used for clarification duties, high submergence (60–75% of the drum surface) is used. Such units require the use of stuffing boxes and glands and are therefore more expensive than the normal low submergence type.

The slurry in the filter trough must be kept in suspension. This is achieved by the installation of agitator gear which is normally in the form of a reciprocating rake, either rotating about the drum shaft or pivoted above the gear box and valve as shown in *Figure 13.7*. One design aspect which is of importance is to ensure that the agitator gives adequate suspension without

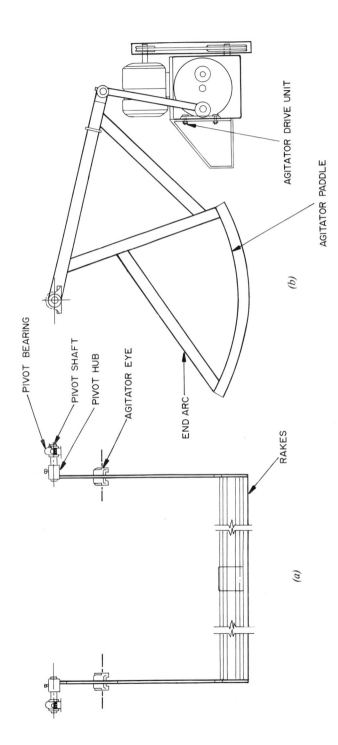

PIVOT BEARING

PIVOT SHAFT

PIVOT HUB

AGITATOR EYE

END ARC

RAKES

AGITATOR DRIVE UNIT

AGITATOR PADDLE

(a)

(b)

Figure 13.7. Diagram of a typical rake agitator used with drum filters: (a) front view; (b) end view

disturbance of the filter cake and for this reason the agitator is often fitted with a variable speed drive.

Cake washing is achieved by placement of either one or several washwater pipes located radially, close to the cake periphery in the drying zone. These are used in cases where maximum mother liquor recovery is sought. In such a wash system, cake compression can be obtained by means of a wash blanket, which avoids cake disturbance or leakage of liquor and air through cracks. They allow wash liquor to be applied much nearer to the point where cake emerges from the slurry. Compression rolls incorporated with such blankets provide an element of control of cake compression.

The performance of drum filters is controlled by adjustments in three main variables: drum speed, vacuum and submergence. Changes in any of these affect all aspects of cake formation, drying, throughput and discharge, simultaneously. Drum filters are normally supplied with a variable speed drive and, for maximum throughput, would be run at the fastest possible drum speed consistent with obtaining the required dryness and a dischargeable cake. In addition, it may be possible to achieve enhanced dryness by means of either a blanket or a roll press, or by using a steam hood. The submergence for maximum throughput would also be the highest that could be used except for fairly free-filtering materials where it may prove necessary to reduce submergence or to throttle the vacuum.

When considering the possibility of reducing submergence, care must be taken to ensure that the rotary valve is suitable for such operations. The advantages and disadvantages associated with orthodox drum RVFs are as follows.

Advantages:
1. continuous and automatic in operation and hence low operating labour is required;
2. design and operational variations (method of discharge, drum speed, vacuum, cloth, submergence, cycle, etc.) cater for a wide range of suspensions;
3. clean in operation;
4. low maintenance cost;
5. effective washing and dewatering;
6. wash filtrate separation is possible.

Disadvantages:
1. high capital cost;
2. relatively large floor area occupied per square metre of filtering area;
3. limitations imposed by vapour pressure of hot or volatile liquids;
4. incapable of handling products that form explosive or inflammable gases under vacuum;
5. unsuitable for quick settling slurries;
6. tendency of cloth blinding due to thin cakes and short cycles (which may to some extent be overcome by the application of belt or string discharge);
7. wetter cakes if blow-back is used, and shorter filter cloth life if blow-back is used in conjunction with a scraper knife.

Table 13.1. OPERATING DATA FOR SOME VACUUM FILTER APPLICATIONS[a]

Application	Type of vacuum drum filter frequency used[b]	Type of discharge	Solids content of feed, wt/wt	Average solids handling rate, kg dry solids m^{-2} h^{-1} filter surface[c]	Moisture content of cake, wt/wt	Average air flow, m^{-3} m^{-3} h^{-1} filter surface[d]	Operating vacuum, kPa
Chemicals							
Alumina hydrate	Top feed drum	Blow	40	650	15	90	16.7
Barium nitrate	Top feed drum	Blow	80	1250	5	450	33.3
Barium sulphate	Drum	Scraper/blow	40	50	30	18	66.7
Bicarbonate of soda	Drum	Scraper/blow	50	1750	12	540	40.0
Calcium carbonate	Drum	Scraper/blow	50	125	22	36	66.7
Calcium carbonate (precipitated)	Drum	Scraper/blow	30	150	40	36	73.3
Caustic lime mud	Drum	Scraper	30	750	50	108	50.0
Sodium hypochlorite	Belt	Belt	12	150	30	54	66.7
Titanium dioxide	Drum	Scraper	30	125	40	36	66.7
Zinc stearate	Drum	Scraper/blow	5	25	65	54	66.7
Minerals							
Frothed coal (coarse)	Top feed drum	Blow	30	750	18	72	40.0
Frothed coal (fine)	Drum	Scraper/blow	35	400	22	54	50.0
Coal tailings	Drum	Scraper/blow	40	200	30	36	73.3
Copper concentrates	Drum	Blow	50	300	10	36	70.0
Lead concentrates	Drum	Blow	70	1000	12	54	73.3
Zinc concentrates	Drum	Blow	70	750	10	36	66.7
Flue dust (blast furnace)	Drum	Scraper/blow	40	150	20	54	66.7
Fluorspar	Drum	Scraper	50	1000	12	90	50.0

Notes:

[a] The information given should only be used as a general guide because slight differences in the nature, size range and concentration of solids, and in the nature and temperature of liquor in which they are suspended, can significantly affect the performance of any filter.

[b] It should not be assumed that the type of filter stated is the only suitable unit for each application. Other types may be suitable, and the ultimate selection will normally be a compromise based on consideration of many factors regarding the process and the design features of the filter.

[c] The handling rate (in kg m^{-2} h^{-1}) generally refers to dry solids except where specifically referred to as filtrate.

[d] The air volumes stated are measured at the operating vacuum (i.e. they refer to attenuated air).

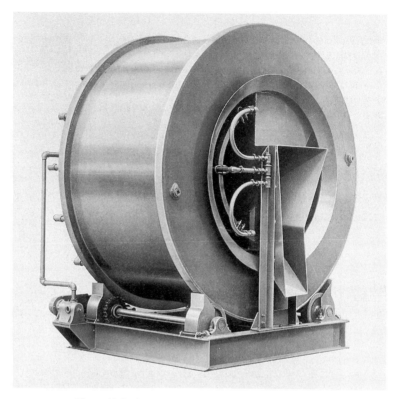

Figure 13.8. An internal drum filter. (Courtesy of Dorrco)

The applications of rotary drum vacuum filters are very widespread and include clays, fine coal concentrates, flocculated raw coal slurries, ore concentrates and tailings, pulp and paper, chemicals, petrochemicals, plastics, sewage and waste materials. Table 13.1 gives examples and operating data for some of these applications.

13.2.2.1.2 Internal drums

These filters are, in effect, of the multiple compartment type which have their filtering surface on the internal instead of the external drum periphery as shown in *Figure 13.8*. Filters of this type are particularly suited to handling quick settling materials and have a distinct advantage over external drums in that:

1. the initial cost is much lower because there is no slurry tank or agitator required;
2. variations in the feed consistency do not cause much difficulty;
3. the filter may be conveniently insulated if elevated temperatures are required.

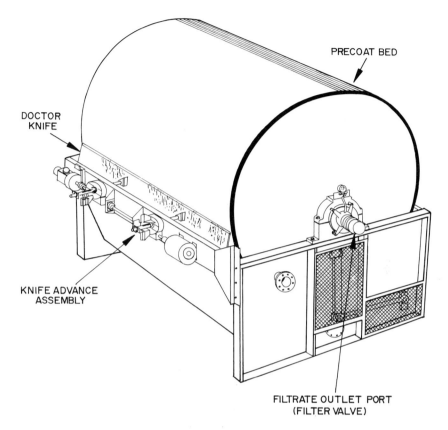

PRECOAT BED

DOCTOR
KNIFE

KNIFE ADVANCE
ASSEMBLY

FILTRATE OUTLET PORT
(FILTER VALVE)

Figure 13.9. The precoat filter scraper blade arrangement

The disadvantages are:

1. only limited areas of the drum can be utilized;
2. there is limited washing time because of the short cycle and it needs to be directed against gravity;
3. coherent cakes are necessary, otherwise the cake falls off before the discharge point, so creating loss of vacuum and inconsistencies in the feed slurry;
4. chute discharge is limited to friable cakes;
5. re-clothing is slower and more difficult.

Because the disadvantages tend to outweigh the advantages for most potential applications of this system, it has been largely superseded by rotary filters of the horizontal type.

13.2.2.1.3 Dewaterers and top feed units

For fast settling solids, which dewater easily, slurry will be fed to the top of the drum instead of being picked up from a trough. For very coarse free-filtering cakes, which will not adhere to the drum surface, it is sometimes necessary to have the surface of the drum in the form of a number of buckets, into which the slurry is fed at top dead centre, and the material is dewatered and discharged. These filters are called 'hopper dewaterers'.

With free-draining crystalline materials it is also possible to enclose the filter and to pass hot air or steam through the cake. Units have been operated at up to 400°C, to give bone-dry filter cakes on such materials as sodium sulphate. Units like this are often cheaper to install and operate than a combination of filter and dryer, or centrifuge and dryer. The details of design are different from those of normal drum filters in that, since liquor and gas rates are often high, great care must be taken in ensuring low pressure drops. On filter dryers, the differential pressure across the cake can be as low as 20 kN m^{-2}. Dual drum units consist of two orthodox multiple compartment units, synchronized and rotated in opposite directions with retaining dams to provide an overdrum feed pond.

13.2.2.1.4 Precoat drums

Technically speaking, these filters are not continuous since the precoat must be renewed periodically. Therefore the filter operation is only continuous for the life of the precoat, which may vary from six to eight hours up to as much as two weeks depending upon the character of the slurry. The conventional drum filter unit is fitted with a special cake discharge blade as shown in *Figure 13.9*. By means of a knife advance system, an extremely thin layer or coating of cake is continuously removed from the surface of the precoat. In order to achieve this, the knife advance system may move less than 0.025 mm per revolution of the drum, although the normal working range is regarded as 0.1–0.5 mm per revolution. A new precoat layer is applied when the lowest admissible layer thickness is reached. At this point advance is stopped automatically so as to ensure efficient filtration and maintenance of acceptable filtrate clarity. A new precoat is then filtered on to the drum. After reaching the required thickness, which may be 100 mm, slurry feed is re-initiated. The precoat material is held in a separate precoat slurry preparation tank from which a separate slurry pump provides the coating cycle feed. Once the coat is complete, the system reverts to the normal slurry feed circuit. The filter is able to produce highly clarified filtrates and handle sticky, slimy solids otherwise unsuited to vacuum filtration. On the other hand, the filter has the disadvantage of being limited to low viscosity slurries and the cake is always contaminated with the precoat material. Initial installation costs are generally higher than for the conventional type and operation is more expensive because of the cost of precoating materials, which are most commonly either diatomaceous earths or expanded perlite.

13.2.2.2 Rotary disc filters

These consist of a number of flat filter elements, mounted on a central shaft and connected to a normal vacuum filter valve (*Figure 13.10*). As the panels

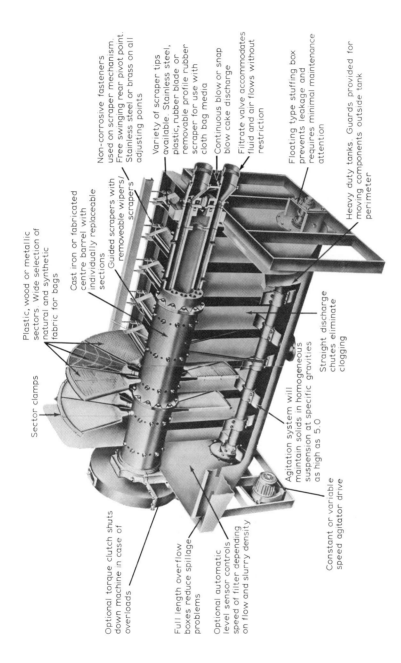

Non-corrosive fasteners used on scraper mechanism. Free swinging rear pivot point. Stainless steel or brass on all adjusting points

Variety of scraper tips available. Stainless steel, plastic, rubber blade or removable profile rubber scraper for use with cloth bag media

Continuous blow or snap blow cake discharge

Filtrate valve accommodates fluid and air flows without restriction

Floating type stuffing box prevents leakage and requires minimal maintenance attention

Heavy duty tanks. Guards provided for moving components outside tank perimeter

Cast iron or fabricated centre barrel with individually replaceable sections

Guided scrapers with removeable wipers/ scrapers

Plastic, wood or metallic sectors. Wide selection of natural and synthetic fabric for bags

Sector clamps

Straight discharge chutes eliminate clogging

Agitation system will maintain solids in homogeneous suspension at specific gravities as high as 5.0

Optional torque clutch shuts down machine in case of overloads

Full length overflow boxes reduce spillage problems

Optional automatic level sensor controls speed of filter depending on flow and slurry density

Constant or variable speed agitator drive

Figure 13.10. Cut-away view of a typical rotary vacuum disc filter. (Courtesy of Eimco)

Figure 13.11. Dorr–Oliver disc filter, Taconite concentrate

rotate they go through similar pick-up and dewatering operations to those carried out on drum filters; and at the discharge point the cake removal is assisted by means of blades or knives located on either side of the disc. To aid more positive cake discharge, the blow-back system described earlier for the drum filter is usually utilized in conjunction with the blades. The dislodged cake falls into a steep-sided discharge chute. There are further developments of this simple system, one of which is the patented Simonacco electropneumatic pulser valve. This works in conjunction with a multi-layering technique which permits resubmergence of undischarged cake creating in some cases substantially improved cake yields. Wetzel[5] describes an exact approach to disc filter design but generally the main difference in the various designs is in the detail of the filter support surface, the number of segments and methods of cloth fixing. In operation, disc RVFs are available with areas from 0.05 m^2 to approximately 300 m^2 achieved by using as many as 16 discs with a single unit. *Figure 13.11* shows a typical mineral concentrate dewatering installation, and *Figure 13.12* shows an exceptionally large disc filter manufactured by Krauss Maffei, which has a single disc filtering area of 37.5 m^2 (disc diameter 5.3 m). Several filter units of this type, fitted with three discs (112.5 m^2 filtering area) are in operation. Each disc in this unit is composed of 30 segments compared to the 8–12 of the more conventional range of disc filters.

Figure 13.12. Krauss Maffei large-diameter disc filter installation under final assembly

The advantages and disadvantages associated with disc filters in general are as follows.

Advantages:
1. low capital cost per unit (less expensive than drum filters);
2. large filter areas on minimum floor space requirements;
3. rapid medium replacement;
4. it is possible to handle different slurries on one unit simultaneously by partitioning the filter tank and using one or separate automatic valves;
5. generally used for handling large volumes of relatively free-filtering solids (50–200 mesh, i.e. 0.3–0.075 mm).

Disadvantages:
1. good washing is difficult on vertical cake surfaces and also because of limited cake drying time;
2. wetter cakes than with the drum type;
3. excessive filtrate blow-back on some designs;
4. the discharge of thin cakes is more difficult;
5. inflexible in operation, since submergence variation is limited;
6. rate of medium wear inclined to be high with scraper discharge;
7. unsuitable for non-coherent cakes;
8. no means of separating different filtrates if the unit is used to filter more than one slurry simultaneously.

Disc filters are being successfully operated in: cement, starch, sugar, waste

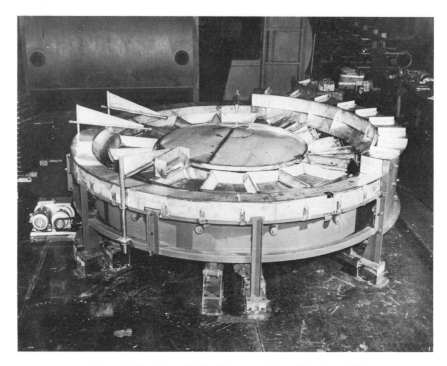

Figure 13.13. Figure 13.13. Tilting pan filter. (Courtesy of Eimco)

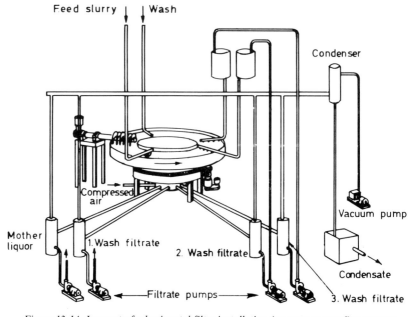

Figure 13.14. Lay-out of a horizontal filter installation (countercurrent flow system)

sludges, fine coal and mineral slurries, and flotation concentrates, flue dusts, paper and pulp, etc.

13.2.2.3 Rotary horizontal filters

This class of continuous vacuum filters is characterized by a horizontal filter surface in the form of a table, a belt or multiple pans in a circular arrangement. The principle of this filter type is the same as the rotary drum or disc. The difference is the fact that the filter medium is horizontal and, because of this, these filters have relatively high capacity and lend themselves to the treatment of granular fast-filtering materials and concentrates with high specific gravity.

Continuous horizontal filters can be divided into two groups:

1. belt-type filters, which include the continuous rubber belt and the tray belt-types of machine;
2. horizontal disc filters, which include single disc units with either scroll or paddle cake discharge and the more common tilting pan filter.

The following generalized advantages and disadvantages apply to each group.

Advantages:
1. excellent washing, and sharp wash liquid and filtrate separations. In this respect, it is better than the drum filter;
2. flexible operation;
3. filtration is assisted by settling of the solids, thus it is suitable for handling fast settling slurries;
4. large tonnages per unit with rapid dewatering slurries (it allows cakes 10–200 mm to be formed and washed);
5. no tank or agitator is needed.

Disadvantages:
1. large floor space requirements;
2. in the case of the belt, only about 45% of the area is effective;
3. higher initial cost per unit area than a drum unit, but this disadvantage is offset by the higher capacity per unit area because it handles thicker cakes at higher speeds;
4. in general, it is limited to free filtering materials, which build up about a 20 mm porous cake, in the case of the straight disc type as the scroll has a clearance of 3–4 mm;
5. in the case of the straight disc there is no adequate way of cleaning the medium, consequently filtrate clarity tends to be poor since an open medium is necessary;
6. the discharge is more expensive than dewaterers, and is sometimes more expensive than internal drum filters.

13.2.2.3.1 Tilting pan filter (Figure 13.13)

In this unit, the rotating annular table is divided up into sectors, each one of which is a physically independent unit surrounded with sides, and a pan connected by a radial arm to a central vacuum valve. These pans are carried in a horizontal framework which rotates around a vertical axis. At the point of cake discharge a mechanism inverts the pan, which can be tipped by either mechanical or hydraulic devices. The specific advantages of a tilting pan unit are complete wash containment, good cake discharge, filter medium washing and feasibility of construction in very large sizes, while disadvantages are high cost, particularly in the smaller sizes, and their mechanical complexities.

13.2.2.3.2 Horizontal table filter (Figure 13.14)

The horizontal table (shown in *Figure 13.14*) is similar to the tilting pan except that the sectors are not divided by pans and some short circuiting of the feed may occur during distribution around the table. The slurry is pumped on to the table at one point and the cake is removed a few degrees countercurrent to this point by a scroll conveyor which elevates it over the side of the filter. A residual cake, or heel, must be left on the filter medium to protect it and this is a disadvantage peculiar to the horizontal table unit. By using a special valve with multiple connections with vacuum receivers, it is possible to make additional separations of filtrates.

13.2.2.3.3 Horizontal belt filter (Figure 13.15)

The horizontal belt type filter (*Figure 13.15*) is a relatively new development. It comprises a slotted or perforated endless belt supporting a filter fabric which is also an endless belt. The supporting belt usually acts as drainage facility in addition to being the means of transportation of the filter cloth. The two belts combined together pass over a suction box. The feed slurry is pumped onto the filter at one end, wash liquor as needed is applied at one or more points along the path of belt travel and the cake is discharged at the other end. At this point, the support belt and the filter medium are parted to be directed along separate lines of pulleys beneath the filter. The medium is washed on its return journey to the head end of the filter where it rejoins the drainage belt. An advantage of this type of filter is the particularly effective filter medium washing and complete cake removal. Its peculiar disadvantage is the fact that half of the potential filtering surface cannot be utilized.

Enlarging on the list of credits, the rotary table and pan units permit an independent choice of cake thickness, washing time and drying cycle. Filter cakes of 10–13 cm in thickness are often handled. Due to its characteristic wide, open drainage, high hydraulic and pneumatic capacities are possible. Often up to 12 m^3 min^{-1} m^{-2} vacuum capacity is feasible for minimum moisture objectives. Sharp separations between countercurrent wash liquors are obtainable due to the good drainage.

The horizontal belt type has high capacity per square metre of area under vacuum; similar to the horizontal rotaries, it can be an excellent tool for a countercurrent filtration (CCF) circuit (see chapter 15). These filters are very effective in filtering heavy, dense solids, and allow flooding of the cake with wash solvent so that the cake may actually be steeped in the wash liquid. This

437

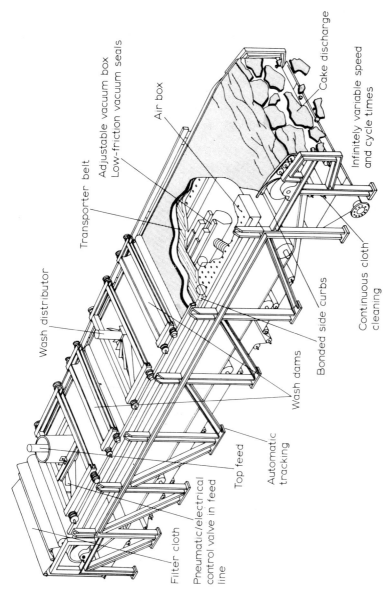

Wash distributor

Transporter belt

Adjustable vacuum box
Low-friction vacuum seals

Air box

Cake discharge

Infinitely variable speed
and cycle times

Continuous cloth
cleaning

Bonded side curbs

Wash dams

Automatic
tracking

Top feed

Pneumatic/electrical
control valve in feed
line

Filter cloth

Figure 13.15. Horizontal belt filter

is the reason why they are very adaptable to the countercurrent leaching or washing.

Countercurrent decantation (CCD) involves the use of conventional thickeners. If the filtration characteristics of the leach slurries are favourable, horizontal belt countercurrent filters (CCF) may be used for the recovery of soluble values resulting in a low moisture content disposable filter cake. In addition to the obvious capital cost savings, the wash ratio, i.e. volume of washwater divided by the volume of cake liquor prior to washing, is much lower. This results in a higher concentration and lower volume of liquor to be treated in each subsequent operation, hence creating reduced reagent cost. Also, since a number of washing cycles can be carried out on a single filter, multiple units may not be required in order to ensure high efficiency. *Table 13.2* shows a comparison of CCD and CCF as obtained from treating uranium slurries[6]. *Figure 13.16* shows the type of CCF installation used for this comparison.

Table 13.2. COMPARISON OF CCD AND CCF USING HORIZONTAL BELT VACUUM FILTERS[6]

	CCD	*CCF*
Wash ratio	3	1
Discharge slurry, % solids by mass	55	80
Unit loading, m^{-2} day^{-1}	43	64
Number of units	7	2
Size of unit, m	19.8 diam.	3.5 wide × 23 long
Power, kW	112	260
Cost, US$ 1985	1.7 million	2.2 million

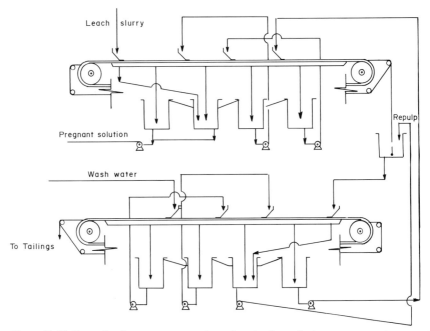

Figure 13.16. Example of a countercurrent decantion circuit employing two or more horizontal belt filters

Figure 13.17. A horizontal tray

The original horizontal tray or pan type of belt filter, although suitable for fairly granular or fast settling solids, suffered the disadvantage of being sectionalized and therefore vulnerable to mechanical problems, in particular, leakage. This was especially evident when filters of this type were applied to slurries having more difficult filterability. A typical example is shown in *Figure 13.17*.

A recent new horizontal belt filter development is the Dorr–Oliver 'rigid belt' horizontal filter[7] as shown in *Figure 13.18*. This unit is claimed to be the only completely continuous section-type unit commercially available. It consists of a series of in-line sections which provide the filtration area. A free-flowing feed slurry is equally distributed on to the filter cloth by means of a feed distribution box. The filter cake which forms is gradually drained of moisture and/or washed along the length of the cloth. The cake discharges when the vacuum is cut off as the sections pass around the drive drum. The cake drops off by gravity and the cloth is then washed before it approaches the tensioning drum to commence another cycle. Polypropylene grids in the sections form the filter deck, which is covered by a separate fabric filter cloth. Sections and cloth move continuously. The bottom of each section has a filtrate outlet connected to the vacuum chamber, which also serves as a collection box for the filtrate.

Washing rates, vacuum, cake thickness, cycle time and wash/filtering zones can be adjusted to achieve optimum filtering conditions and a fully automatic version is also available. Filter cloth tensioning can be controlled precisely by means of an adjustable water-filled counter-weight device. No compressed air is required for cake discharge and cloth tracking is accomplished by a single tracking device.

This type of filter has many varied applications and can be used in single or multi-stage dewatering and countercurrent washing systems.

Typical filter areas vary from 0.5 to 120 m^2 (i.e. machines with tray widths of 1–4.25 m and lengths of 9.5–33.5 m). Small, 0.5-m^2 test units are available for testing purposes. Applications include the following.

1. Chemical: catalysts, dye stuffs, enzymes, pigments, fertilizers, salts etc.

Figure 13.18. A rigid belt filter

2. Pharmaceuticals: antibiotics, fermentation products, vitamins etc.
3. Food products: sweeteners, oils, proteins.
4. Minerals processing: alumina, leaching residues, gypsum from flue gases in power plants, tailings, phosphates etc.

13.2.2.4 Further dewatering improvements

Various other forms of energy have been utilized to improve either the filtration or dewatering or both components of a vacuum filter cycle[8]. These include:

1. use of infrared radiation;
2. use of steam or other heat source for preheating of the feed slurry;
3. use of steam to improve cake dewatering;
4. use of ultrasound to aid filtration and dewatering;
5. use of electrical current (electro-osmotic) to aid dewatering;
6. addition of oil to cause agglomeration of hydrophobic solids (e.g. coal) to improve dewatering.

In general, the addition of heat reduces the viscosity of the fluid, thereby improving the filtration rate. Slurry warming practice can only usually be justified if waste heat is available, but moisture improvements of 1–2% have been obtained.

Infrared radiation is provided by a source located adjacent to the emerging formed cake. Due to the capacity of most substances to absorb infrared radiation, heating is mainly confined to the outer layers of the cake. Infrared radiation panels fitted onto drum filters maintained at 850°C at a distance of 70 mm can reduce moisture by 1–3%, but exposure time is usually short (1 min) to avoid overheating the cake surface.

Steam-aided dewatering has been attempted for several years with some reported successes. Drum, disc and horizontal filters can be fitted with specially designed hoods for directing superheated steam to the surface of the filter cake. Penetration of the steam and subsequent condensation in the pores of the cake causes some evaporation of moisture content but also creates a reduction in viscosity of the migrating filtrate. Moisture reductions of between 0.5 and 1.5% are common, but economic considerations may limit the application for such a small moisture reduction.

Ultrasonic dewatering has been used to aid the filtration and dewatering of coal slurries. With a sound pressure of 160 dB the cake moisture was reduced from 24.5 to 21% for coal sized between 50 and 75 μm and from 21 to 17% for the size range 75–100 μm, at a pressure drop of 68 kN m^{-2}.

Electro-osmotic dewatering involves the application of a d.c. potential across a low permeability cake which causes electro-osmotic flow to occur. This assists with drainage, particularly in the case of tightly packed cakes composed of particles of very fine size.

Oil agglomeration using low- to medium-density oils occurring in the filter feed bowl can result in improved dewatering of fine coal and other hydrophobic solids. The oil addition further enhances the hydrophobicity of the solid improving drainage rate, solids recovery and moisture content.

Coal slurries filtered after addition of 3% kerosene/fuel oil (following emulsification of the reagent) have resulted in moisture reductions of up to 17%, i.e. from 27 to 10%.

13.2.3 Filter systems

Figure 13.19 shows a standard vacuum filter installation and *Figure 13.20* shows a photograph of a similar arrangement.

The valve head forms the connection between filter cells and filtrate

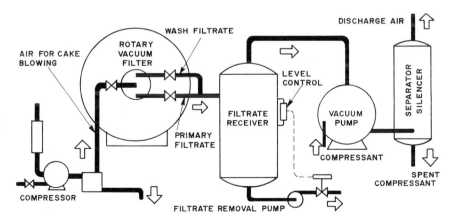

Figure 13.19. Standard layout, Paxman filter installation

Figure 13.20. A standard vacuum drum filter installation of the type shown in Figure 13.19.

receiver. This is sub-divided into various zones for each individual filtration process and although there are a number of varied designs of valve head a fairly typical one is clearly illustrated in *Figure 13.21*, which shows an exploded view of a face valve head end of a rotary drum filter.

The lower pipe connection from the filter valve transmits liquid pulled through the filtrates pipe during the formation zone while the upper valve connection carries liquid and air pulled through the cake in the drying zone. Finally, the third valve transmits low-pressure air to discharge the cake.

Liquid and air enter the side connection of the filtrate receiver where the liquid drops down to the filtrate pump and the air is pulled through the top connection of the receiver to a moisture drop or condenser, usually at a velocity of below 1 m s^{-1}. A filtrate pump is necessary to overcome the suction head created by vacuum. A check valve is placed on the filtrate pump discharge to ensure that no air is sucked back into the system. Should this occur, an inoperable system would develop. For efficient pump operation the discharge conditions must be based on correctly selected values of total dynamic head, flow capacity and net positive suction head. The discharge head available is the sum of the total dynamic head minus suction head and friction head. The practical application of net positive suction head is to allow the static head from the bottom of the receiver to the eye of the pump impeller to be equal to or greater than the required pump net positive suction head. To ensure further protection against loss of prime in the pump, a balance line should be connected from the top of the receiver to the eye of the pump impeller.

The air drawn through the receiver usually enters a tangential connection close to the base of the condenser. The spiralling air upon entering, expands giving up head and causing moisture in the form of water vapour and entrained droplets from the receiver to fall to a drainage sump. If a piston-

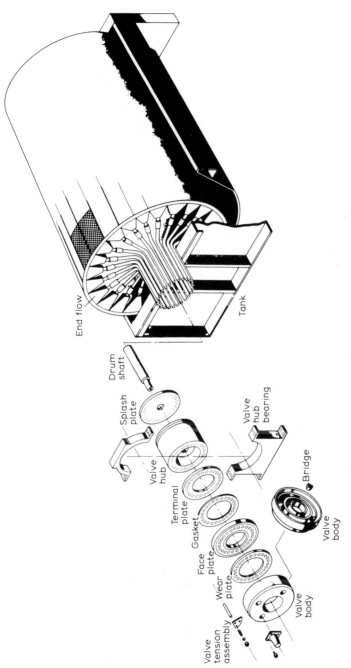

End flow

Tank

Drum shaft

Splash plate

Valve hub bearing

Valve hub

Terminal plate

Bridge

Gasket

Face plate

Valve body

Wear plate

Valve body

Valve tension assembly

Figure 13.21. Exploded view of the valve end of a rotary vacuum drum filter

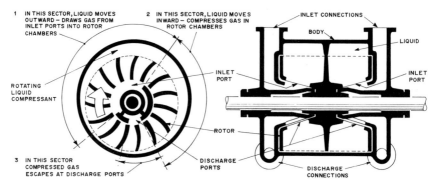

Figure 13.22. Components of the Nash vacuum pump. (Courtesy of Nash)

type of vacuum pump is used this liquid drainage offers protection to the pump in the event that the filtrate pump should fail. Normally, however, the required pressure differential for vacuum filtration is economically produced by a water ring-type vacuum pump, as shown in *Figure 13.19*. Fresh water must be used for seal water and a discharge silencer should be placed on the pump discharge to reduce the noise.

Many liquid-sealed blower-type vacuum pumps can handle small amounts of moisture and some solids carryover. A reciprocating piston pump may become difficult to maintain in such service because of lubricant dilution and valve fouling. Rotary compressor types of pump such as the Nash pump[9] shown in *Figure 13.22* have only one moving part, a balanced rotor which runs without internal lubrication. As shown in the diagram, liquid compressant almost fills, then partly empties each rotor chamber during a single revolution. This creates a pulsing effect similar to that of a piston. Stationary cones inside the rotor have closed sections located between the ported openings which separate gas inlet and discharge flows. A portion of the liquid compressant passes out with discharged gas. Make-up is provided by means of orifices and also by manual valve adjustment. Because of the liquid compressant concept, steady percolation of filtrate liquid causes no harm to the pump operation. Slugs of liquid can, however, disrupt operation and must be catered for by the filtrate handling system. This is achieved either by means of a barometric leg into a seal tank or by a filtrate removal pump which operates at vacuum. Both these arrangements are commonly used and the simple diagrams in *Figure 13.23* show their operation.

The piping from the filter valve should ideally be placed in a horizontal plane or dropped vertically to the receiver side connection. All equipment should be placed as close together as possible and unnecessary pipe bends and turns should be avoided to minimize friction head losses.

Filtrates from form and drying zones can be collected separately in filtrate receivers and discharged either by the use of suitable pumps, as described earlier, or by the barometric leg. This latter type of system is beneficial because filtrate pump problems are thus avoided. However, discharging filtrate by barometric leg is not always possible because of plant elevation limitations, since the leg requires a vertical column height of more than 9 m dependent upon local atmospheric pressure.

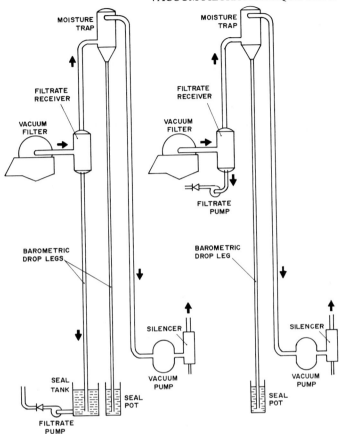

Figure 13.23. Two alternative arrangements of a vacuum filter barometric leg

Feeding filters is extremely important and every effort must be made to provide the filter installation with a consistent slurry. Pre-concentration must be properly controlled and proper feed piping to the filter to ensure uniform concentration of solids and even-size consistency is essential for good operating conditions. The feed point location will vary with the type of filter selected but side feeding via an overflow box to the filter tank is most common for rotary vacuum filters. The introduction of various controlling and safety transducers to filter installations have made significant improvements in operational efficiency and availability. Among these are density gauges for the feed slurry, water-sensing electrodes in air lines from the filtrate receivers, level devices in the filter tank, pump and drive failure electronic interlock systems, etc.

13.2.4 Materials of construction

Development of the mechanical design and materials of construction of continuous vacuum filters has meant a widening of the fields of application. Nowadays, most manufacturers will offer numerous sizes of drum filters available in three main standard materials of construction, that is, mild steel, stainless steel and acid-resistant covering (plus an extra range of epoxy resin filter bodies), with five alternative cake discharge mechanisms and two degrees of drum submergence. This so far represents a multiplicity of permutations from a completely pre-engineered range. Add to this the more specialized facilities available, for example totally enclosed filters, cloth discharge filters, filters with press rolls, press bands, washing equipment, variable agitation, etc. and there is an incredibly large selection potential which clearly indicates why drum RVF units are employed in such diverse user-industries as food processing, mineral processing, heavy and fine chemicals, pharmaceuticals, cement and clay, sewerage, sugar, lubrication products and effluent treatment.

Rubber-coated filter components may cost up to 20% more than stainless steel for drum or disc troughs, agitators, valve heads and other items dependent further upon filter type[2]. In extreme cases very costly materials such as titanium may be used even though it costs 150% more than stainless steel. Titanium is used in special applications such as when the ambient atmosphere is a severely oxidizing one. Titanium has a remarkable ability to withstand many corrosive substances in widely varying concentrations and temperatures but has the disadvantage of being difficult to fabricate.

Under more normal circumstances, plastic components made in polypropylene, polyvinylchloride (PVC), polyuretyhane epoxy resins and other similar materials have become fairly commonly used, due to their low cost, ease of fabrication, or manufacture and general resistance to corrosive and erosive wear. If properly designed, plastic components can have high strength and great durability in many filter applications. Many plastics now have good thermal, as well as good chemical resistance and are, therefore, applicable to filter components for units treating high-temperature slurries. Plastics are probably most widely used for components such as discharge rolls and bearings, chutes, blades and rotary valve inserts for which polyethylene, polyurethane and polytetrafluoroethylene are often used and drainage grids, enclosing hoods and canopies for which polypropylene, PVC and polyester-glass are most common.

13.2.5 Cost of filter equipment

Costs of filtration equipment are not discussed openly very often but a useful capital cost manual has been produced for costs in Canadian dollars of process equipment used in the minerals industry. The graph shown in *Figure 13.24* is used to obtain an order of magnitude cost for either a rotary drum or a disc installation. The example quoted in the text is as follows:

Example 13.1

A laboratory leaf filter test showed that for a safety factor of 0.65 a filter that handles 34 lb of dry solids/ft^2/h is required (0.046 kg m^{-2} s^{-1}). What size of disc filter is required to treat 100 tons (90.7 tonnes) of dry solids per day and

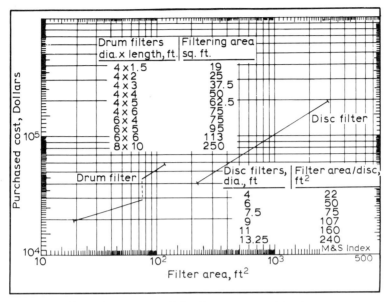

Figure 13.24. Drum and disc filter costs (imperial units)

what is the cost of the unit? What size drum would suffice and at what cost?

From $100 \times (2000/24) = 8333$ lb/h (3780 kg h^{-1}), it is found that the total surface area of filter needed is $8333/34 = 245$ ft^2 (22.76 m^2). From a table of standard disc filter sizes (see *Figure 13.24*), a 6 ft (1.83 m) diameter disc filter with five discs has a nominal surface area of 250 ft^2 (23.23 m^2). A four-disc filter of the same diameter has a nominal surface area of 200 sq. ft (18.58 m^2). Thus one possible size of interest is a filter of diameter 6 ft (1.83 m) with five discs.

From *Figure 13.24*, it is apparent that the cost of this disc filter is approximately \$41 000.

An equivalent area drum filter is 8 ft (2.44 m) in diameter and 10 ft (3.05 m) in length. This would cost (on extrapolation of *Figure 13.24*) \$96 000 (1980 terms).

13.3 FILTER SELECTION

The foregoing sections have provided descriptions of the various types of vacuum filter together with an indication of the advantages and disadvantages normally associated with each type. In every case, the mode of filtration incorporates the separate operations of cake formation, filtrate removal and cake discharge, and it is the relative requirements specified for each of these operations which largely determine the selection of filter type ultimately made. It should, however, be kept in mind that in many cases more than one type of filter could be selected and perform the task effectively. In some cases, technical objectives may be sacrificed to some extent because of cost considerations and reliability may be the major requirement at some sacrifice of cost in others.

In specifying requirements and arranging for the necessary test work, a technical specification similar to that shown in *Figure 13.25* should be compiled. Without doubt the most important factors to be considered when selecting a particular type of vacuum filter are the size distribution and density of the solids to be filtered and the solids concentration of the slurry.

A. *General data*
1. Material to be filtered
    (a) Solids ....
    (b) Liquid ....
2. Feed rate of slurry m^3 h^{-1} ....
3. Density of slurry   S.G. ....
    % solids by weight ....
    % solids by volume ....
4. Slurry temperature ....   pH ....
5. Source and method of feed ....
6. Product of value: Filtrate ......   Solids ......   Both?
7. Flocculant/coagulant usage ......   Type ....
8. Filter-aid usage: Type ....   Limitations ....
    ....
9. Other considerations   Volatility ....
                           Toxicity ....
                           Stability ....
                           Inflammable ....
                           Explosive ....
                           Corrosive ....
                           Fibrous ....
                           Other, please specify ....
                           ....

B. *Solids data*
1. Specify gravity ....   Bulk density ....
2. Size distribution (sieve analysis or other ....)
    ....
    ....
3. Moisture content (filter cake) ideal .... %
4. Solubility (state conditions) maximum .... %
        in mother liquor ....
        in washwater ....
5. Filter cake description
        amorphous .... crystalline ....
        colloidal .... granular ....
        thixotropic .... fibrous ....
        other comments ....
6. Can cake solids be compressed during dewatering?
        Yes ....   No ....
7. Will cake be thermally dried?
        Yes ....   No ....
8. How will cake be handled? ....

C. *Filtrate data*
1. Specific gravity ....
2. Admissible solids content: ideal .... %
                                 maximum .... %
3. Can wash liquor be used   Yes ....   No ....

4. Wash liquor separate from mother liquor? ....
5. Description of wash liquor (if known) ....
   ....
6. Are there temperature limitations? ....
7. What are the possible effects of wash liquor on
   (a) solids (dissolution) ....
   (b) combined filtrate ....

D. *Supplementary data*
1. Existing filtration equipment used ....
   ....
2. Specific materials of construction requirements
   ....
3. Electrical characteristics and ambient limiting conditions for motor drives ....
   ....
4. Operational details: h day^{-1} ....
                   day week^{-1} ....
5. Local conditions (and climate)
   ambient temperature range °C ....
   ambient barometric pressure ....
6. Flowsheet: if known, or prepared, a flowsheet incorporating the filter unit(s) should be submitted.
7. Layout drawings: if prepared, layout drawings showing filter unit(s) location should be submitted. Note: headroom or space limitations; lifting capability or access.
8. Painting/surface treatment requirements.

Figure 13.25. The technical specification for a vacuum filter

The solids content of a slurry is an important basic parameter, particularly in respect of filter capacity. As a criterion, it can be assumed that suspensions with a solids concentration of greater than 20 g l^{-1} or suspensions which, with a suction period of five minutes, do not form a cake thickness of at least 3 mm, are, except under special conditions, unsuitable for continuous vacuum filtration. For cake thicknesses below 3 mm, special methods for cake discharge are available as can be seen in *Table 13.3*

Table 13.3. MINIMUM CAKE THICKNESS FOR VACUUM FILTER DISCHARGE[6]

Type of unit	Range of design cake thickness, mm	Practical minimum value, mm
Disc	10 – 15	8
Drum belt	3 – 5	3
Roll discharge	1 – 2	1
Scraper	5 – 10	5
Pre-coat	0 – 3 max.	0
Horizontal belt	3 – 5	2
Tilting pan	20 – 25	20

The presence of coarser material is often associated with a higher feed solids concentration since it involves a straightforward densification step employing some form of classifying device, either cyclones, settling cones or inclined trough classifiers, to control the density required.

A major difficulty commonly encountered in vacuum filtration is inconsistency in stratification of the filter cake resulting from fluctuation in the size

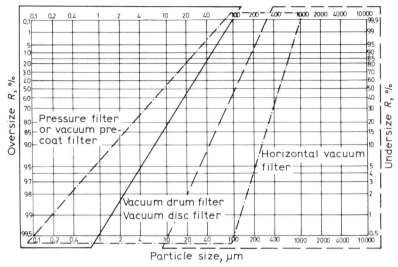

Figure 13.26. Diagram for filter selection according to size distribution. (Courtesy of Dorr-Oliver)

distribution of the solids in the slurry. Some filter types are more seriously affected than others by such fluctuations. *Figure 13.26* is a chart which allows preliminary filter selection to be made according to size distribution.

Having made the preliminary selection of the filter type, the most effective and economical design must be established for the overlapping ranges, utilizing the data obtained from filtration testing and empirical sources. In this respect, factors such as capacity, cake moisture, cake washing and cake discharge must be carefully considered. To assist with these factors, empirical data of the type given in *Tables 13.4* to *13.8* are used in conjunction with test data[4].

13.4 FILTRATION THEORY FOR CONTINUOUS FILTERS

The basic filtration theory applicable to vacuum filters has been dealt with in chapter 9. Application of filtration equations to practical filter installations, especially for design purposes, is usually dependent upon empirical data obtained from previous applications used in conjunction with operational requirements and test data, compiled in similar format to the filter specification given earlier in *Figure 13.25*. Continuous vacuum filters combine three functions, each of which can be determined by calculation; these are:

1. rate of cake formation;
2. rate of cake dewatering;
3. rate of cake washing.

The latter aspect is discussed in detail in chapter 15.

Table 13.4. EXAMPLES OF TYPICAL VACUUM FILTER PERFORMANCE[4]

Filter type	Application	Discharge	Solids rate, kg m⁻² h⁻¹	Solids concentration, %	Cake moisture, %	Vacuum, kPa	Air flows, m³ m⁻² min⁻¹
Disc	Frothed coal	Scraper/blow	200 – 400	15 – 30	16 – 30	50 – 70	15 – 20
Disc	Coal tailings	Scraper/blow	40 – 160	25 – 30	25 – 35	50 – 70	10 – 15
Disc	Copper concentrates	Scraper/blow	250 – 425	55 – 70	9 – 14	50 – 80	9 – 15
Disc	Lead concentrates	Blow	325 – 650	70 – 85	6 – 8	50 – 70	9 – 15
Drum	Frothed coal	Blow	250 – 450	25 – 30	15 – 30	50 – 70	15 – 20
Drum	Yellow cake (U_3O_8)	Belt	60 – 150	20 – 30	30 – 40	50 – 70	9 – 15
Drum	Gold cyanide leach	Scraper	150 – 300	50 – 60	20 – 30	60 – 80	5 – 10
Drum	Magnesium hydroxide	Belt	25 – 125	15 – 30	40 – 50	50 – 70	4 – 9
Horizontal belt	Cement	Belt	200 – 800	30 – 65	12 – 35	50 – 70	9 – 15
Horizontal belt	Gold cyanide leach	Belt	120 – 350	45 – 60	25 – 35	50 – 75	5 – 10
Tilting pan	Phosphoric acid	Scraper	200 – 350	25 – 40	20 – 25	50 – 70	15 – 18

Table 13.5. FACTORS AFFECTING SELECTION OF TYPE OF FILTER AND CHARACTER OF PULPS HANDLED

Typical materials	Character	Vacuum, kPa	Approx. filter capacity, kg m^{-3} day^{-1}
Cyanide slime	Finely ground quartz ores	60–85	1 900– 10 000
Flotation concentrates	Minerals, finely ground	60–85	1 900– 9 000
Gravity concentrates and sand	Metallic and non-metallic minerals almost free from slime	7–20	50 000–350 000
Cement slurry	Finely ground limestone and shale, or clay, etc.	60–85	1 900– 10 000
Pulp and paper	Free-filtering fibres	20–70	1 000– 6 000 and 0.6–0.8 m^3 m^{-2} min^{-1}
Crystals, salt, etc.	Granular, crystalline	7–20	15 000– 59 000
Cane-sugar-liquor clarification, beverages, etc.	Syrups and solution with small percentage of solids with filter aid	36–45	1.5–57 m^3 m^{-2} day^{-1}
Pigments	Smeary, sticky, finely divided, non-crystalline	67–90	1 000– 2 500
Sewage sludge	Colloidal and slimy	36–46	Batch operations 100– 1 200
Varnish	Cloudy viscous liquid, filter aid used for clarification. Filter hot	25–80 65–80	0.2 m^3 m^{-2} h^{-1}
Mineral oils, with or without wax	Removal of bleaching clay from petroleum products. 1–20% clay used	36–46	0.1–1.0 m^3 m^{-2} h^{-1}* 1–3 m^3 m^{-2} h^{-1}**
Cane mud	Vegetable fibre and cane juice		

* Lubricating oils. ** Gasoline.

Table 13.6. FILTER SELECTION: SLURRY CHARACTERISTICS

	Fast filtering	Medium filtering	Slow filtering	Dilute	Very dilute
Slurry characteristics					
Cake-formation rate	cm s^{-1}	cm min^{-1}	0.13–0.64 cm min^{-1}	0.13 cm min^{-1}	No cake
Usual solids concentration, %	20	10–20	1–10	5	0.1
Settling rate	Very rapid	Rapid	Slow	Slow	—
Leaf test rate, kg h^{-1} m^{-2}	2500	250–2500	25–250	25	—
Filtrate rate, m^3 h^{-1} m^{-2}	12.2	0.05–12.2	0.02–0.05	0.02–4.9	0.02–4.9
Filters					
Continuous vacuum filters:					
Multi-compartment drum	x	x	x	—	—
Single-compartment drum	x	—	—	—	—
Dorrco	x	—	—	—	—
Top feed	x	—	—	—	—
Horizontal table	x	x	x	—	—
Disc	—	x	—	—	—
Precoat	—	—	x	x	x
Batch vacuum leaf	—	x	x	x	x

Table 13.7. CHEMICALS FOR IMPROVED MINERAL FILTRATION

Type of slurry being filtered	Type of chemical as filter aid	g t^{-1}	Expected results
Carbonate of Li, Ba, Ca, etc.	Anionic and non-ionic flocculants	22.7–453.6	Production increase 20–50% Reduced moisture 5–10%
Uranium 'yellow cake' U_3O_8	Anionic flocculants	22.7–453.6	Production increase 15–30% Reduced moisture 10–20%
Heavy metal oxide and sulphide forms of Cu, Mo, Zn, etc.	Surface-active drying aids	2.3–226.8	Reduced moisture
Coal	Anionic flocculants	9.1–31.8	Increased production 15–30% Reduced moisture 10–20%
Clay	Polyamine coagulants	22.7–226.8	Production increase 15–30%
Hydroxide precipitates of Ca, Mg, etc.	Polyamine coagulants	22.7–453.6	Production increase 15–30% Reduced moisture 5–15%

Table 13.8 VALVE SETTINGS (°) FOR VACUUM FILTERS[10]

	Feeding	Washing	Discharge*
Disc — with agitation / without agitation	5 – 40	NR	5 – 45
Drum — scraper / roll / string / belt / precoat	5 – 50	0 – 30	0 – 60
Top feed drum	5 – 15	0 – 20	25 – 70
Horizontal belt	5 – 90	0 – 90	5 – 90
Horizontal table	5 – 70	0 – 70	5 – 75

*Discharge is only the 'live' portion of the cycle.

The rate of cake formation can be determined from an equation based upon Poiseuille's law[10]

$$Z_f = \left[\frac{K \, w \, \Delta P}{\mu \, \alpha \, \theta_f} \right]^{1/2} \tag{13.1}$$

Although this equation considers ideal filtering conditions it can provide a useful estimation of filter performance. Z_f is the form filtration rate expressed as weight of dry solids per unit area per unit time of cake formation. It is normally given as kilograms of dry (solids) per square metre per hour of form time, i.e. kg m^{-2} h^{-1} (or kg h^{-1} m^{-2}). The rate of cake formation does not take into account the remaining part of the total filter cycle (i.e. washing time, dewatering time and dead zones catering for submergence or cake discharge etc.).

The form filtration rate expressed as a volume per unit area per unit of time during the cake formation period is obtained by dividing both sides of equation 13.1 by the weight of dry cake solids per unit volume of filtrate, w, i.e.

$$Q_f = \left[\frac{K \, \Delta P}{\mu \, \alpha \, w \, \theta_f} \right]^{1/2} \tag{13.2}$$

The influence of viscosity and specific resistance of the filtering media on cake formation has been discussed in the chapters that deal specifically with cake formation, but in continuous vacuum filters their effects are perhaps less significant than with pressure-types of filter. Nevertheless, they are factors to be considered together with the properties of the solids forming the filter cake, i.e. size distribution, type of solids (including influence of shape) and effects of coagulants or flocculant. These will have a significant effect on the specific resistance of the filter cake, which often incorporates the media resistance, because it has a relatively small value in continuous filter rate equations.

Other, more specific forms of filtration rate equation, taking into account filter media resistance and cake compressibility, have been developed[11] to obtain greater accuracy for drum, disc and horizontal-belt vacuum filters. Drum and disc filter performance is given by the following relationship:

$$\frac{\Phi_f \, A_c}{F} = \frac{\mu \, \alpha_o \, w}{2 \, \Delta \, P^{(1-s)}} \left[\frac{F}{n \, A^c} \right] + \frac{\mu R_m}{\Delta P} \tag{13.3}$$

Vacuum filters of the horizontal-belt type are described by the equation:

$$\frac{L \, B}{F} = \frac{\mu \, \alpha_o \, w}{2 \, \Delta \, P^{(1-s)}} \left[\frac{F}{\mu \, B} \right] + \frac{\mu \, R_m}{\Delta \, P} \tag{13.4}$$

The above equations relate the basic suspension characteristics to the equipment parameters. It is assumed that during the filtration process the filter media resistance remains constant; this is important in filter selection. The filter design must be such as to ensure very thorough and consistent filter-cake removal.

The above equations permit the calculation of the filter area starting with known input data. The most meaningful type of calculation involves assumption of the desired filter-cake thickness, h. In the case of a rotary vacuum drum, the wet-filter-cake volume is calculated from: $Z = (E - F)$, based upon a particular filter-cake requirement. For an RV drum this may be rewritten as: $Z = B \cdot n \cdot U$; $U = 2\pi r \cdot n$, and $A_c = 2\pi r \cdot B$. Therefore:

$$\frac{E - F}{h} = A_c \cdot n \tag{13.5}$$

which, substituted in equation 13.3 gives:

$$A_c = \frac{F}{\phi_f} \frac{\mu \alpha w}{2 \Delta P^{(1-s)}} \left[\frac{F h}{E - F} \right] + \frac{R_m \mu}{\Delta P}$$ (13.6)

Finally, it should be confirmed whether the value obtained falls within the practical limits of a drum speed which can be verified using:

$$n = (E - F)/A_c \cdot h$$

Cake dewatering rate is a function of the change which occurs in the moisture content of the cake and the size distribution of the solids[10]; i.e.

$$M_c = \text{function } [F_a \cdot F_d \cdot M_e]$$ (13.7)

where M_c is the discharge cake moisture, M_e is the equilibrium moisture occurring when 100% saturated gas is drawn through the cake at a pressure drop of ΔP, F_a is known as the approach factor, and F_d is a particle size factor. Therefore, under normal conditions of vacuum filtration the three parameters M_e, F_a and F_d most influence the eventual moisture content of the discharged cake and, in turn, F_a is related to the major variables in the dewatering period; i.e.

$$F_a = \text{function } \left[\frac{\Delta P}{Z} \cdot \theta_d \cdot Q_d \right]$$ (13.8)

where θ_d is the cake dewatering time per cycle and Q_d is the quantity of gas drawn through the cake during the dewatering time of the cycle. This latter value, when determined for a value of ΔP, will give the requirements for the vacuum pump or compressor for the filter. Also, by increasing θ_d or decreasing Q_d, the productivity per unit area of filtration will be reduced and hence capital and operating costs are influenced in optimizing the moisture content which, in turn, may influence the type of discharge system selected.

13.5 VACUUM FILTER PERFORMANCE AND PREDICTION

Empirical formulae are useful in predicting filtering and dewatering performance of commercial filters from test data obtained from representative samples. The term 'representative' is the all-important one as usually such samples are not easy to obtain, especially for new installations. Nevertheless, it is important to be able to predict performance and determine operational requirements.

A great deal of work has been carried out to achieve reliable prediction of continuous filter performance directed mainly towards either dry cake yield or filtrate flow and dewatering rates. Some of the empirical relationships which have resulted from this work were given in the previous section. Many contributory factors have been studied ranging from the direct operational variables such as filter time, operating vacuum, filter media, submergence,

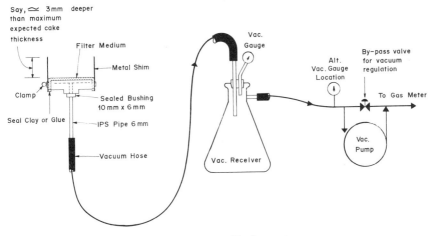

Figure 13.27. A typical leaf-test set-up

agitation, washing, etc. to the effects of the slurry, i.e. concentration, particle size, chemical or surface chemical effects and the use of filter aids, and to side-effects such as cake cracking, cake drop-off and agitator stripping effects, viscosity and temperature variation effects. Throughout all such considerations lies the common need to relate small-scale laboratory tests to commercial scale operation. Dahlström[12,13] in particular, has contributed significantly with practical comments relating to scale-up. Each rate contributor to the total cycle time must be carefully considered in attempts to allow reliable scale-up and leaf test data may not be reliable and should be supported if possible by data from small scale models of the proposed filter.

The majority of laboratory generated filtration data are usually obtained using a 0.01-m² filter leaf with an apparatus similar to that shown in *Figure 13.27*. The need to adopt a standardized method for laboratory testing is demonstrated by recent efforts by the ISO to produce a standard test method for comparison of flocculants for use on vacuum filters in coal-preparation plants. The apparatus specified in the proposed standard is similar to that shown in *Figure 13.27*. A more versatile apparatus suitable for simulating all three modes (i.e. top, bottom and side feed) is shown in *Figure 13.28*.

The method of calculating filter rates is quite simple. The cake is weighed wet, placed in an oven, dried and then reweighed. The solids rate is then calculated from:

$$\frac{\text{Dry weight} \times \text{cycles h}^{-1}}{\text{Test leaf area}}, \text{ in kg m}^{-2}\text{ h}^{-1}$$

and filtrate rate during the form period is obtained from:

$$\frac{\text{Volume collected} \times \text{cycles h}^{-1}}{\text{Test leaf area}} \text{ in m}^{3}\text{ m}^{-2}\text{ h}^{-1}$$

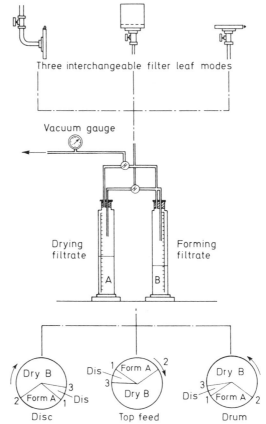

Figure 13.28. Interchangeable filter head assembly for simulator leaf testing.
(After Osborne, 1975)

When filtration leaf tests are carried out, in addition to obtaining all the rate functions and selecting the optimum filter type, various other items can be readily measured or tested. These include the following[14]:

1. feed solids concentration,
2. solids size distribution,
3. vacuum level or pressure drop,
4. cake weight (wet) and thickness,
5. cake moisture content,
6. volume of filtrate,
7. suspended solids content of the filtrate,
8. slurry agitation (if applicable),
9. slurry chemical treatment (if applicable),
10. cake formation time,
11. cake washing time (if applicable)
12. cake drying time (dewatering),
13. volume of wash liquor,

14. soluble solids content of the cake and mother liquor,
15. rate and volume of gas drawn through the cake,
16. temperature of slurry (and ambient temperature),
17. cake discharge mode.

All the rate functions can be determined from the above data and, providing that every effort has been made to simulate operating conditions of commercial scale filtration, valuable performance prediction is usually possible. However, supportive test work, using a pilot-scale filter of the type shown in *Figure 23.29* is always advisable in order to establish design parameters more firmly. Work carried out using leaf apparatus[15] (see *Figure 13.28*) in simulated filter conditions, has yielded reliable data which are further substantiated by means of pilot scale RVF tests (*Figure 13.30*). The results of a careful study into the differences between upward and downward filtration mechanisms, described by Rushton[16], further emphasize the need to incorporate simulation into leaf-scale filter tests if the true filtration characteristics of a slurry with a particular type of machine are to be obtained. Much of the published research work relating leaf test data with either pilot scale or full-scale operation appears to suggest that an overestimation is obtained using the leaf test. This appears to be borne out by early work reported by Dobie[17], who reported leaf tests as giving an overestimation of capacity actually obtained on full-scale units. However, in comparing the predictions of leaf and Buchner funnel tests for drum RVF filter prediction the leaf test, although high, proved more reliable than the Buchner funnel which grossly overestimated the cake yield value. He also pointed out that if the cake is

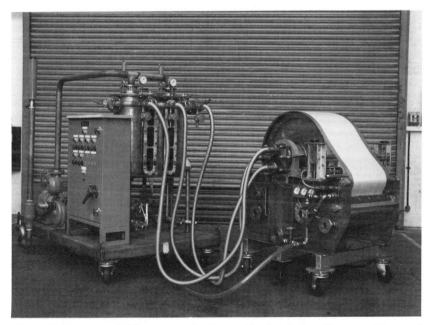

Figure 13.29. Paxman 10 ft² transportable pilot test filter installation. (Courtesy of Paxman)

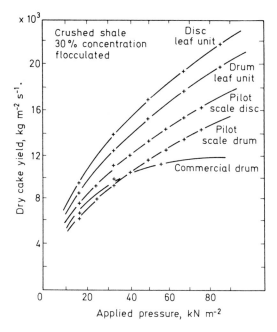

Figure 13.30. Comparison curves for leaf pilot scale and commercial scale RVFs

expected to crack during filtration then it is best to perform leaf tests in order to predict the vacuum at which the cracking might occur.

Pilot tests serve to provide an indication of the effects of factors which occur in commercial operations but which do not significantly affect leaf tests and such factors as cake drop-off during form and drain behaviour due to filter aid usage can be carefully examined by pilot scale tests as well as the more obvious factors such as variation of the major operating parameters and slurry conditions. *Figures 13.31* and *13.32* demonstrate the variations observed[14] in filtration of coal concentrates by means of leaf, pilot scale and commercial vacuum drum and disc filters, and serve to indicate the variations likely to occur in filtering a fairly free filtering material under widely differing filter conditions.

When the dry cake yield and cake formation time values, shown graphically in *Figure 13.32*, are plotted logarithmically, a straight line should be obtained. Different linear relationships will be obtained with variation in feed solids concentration, applied vacuum and slurry temperature. This relationship is evident from the rate equation 13.1 given in the previous section and demonstrated in *Figure 13.33* for a coal slurry. According to Dahlström[10], the line should have a slope of 0.5 unless some migration of cake solids occurs after deposition, in which case the value may range from −0.5 to −1.0. This is applicable to all types of vacuum filters except for precoat filters, where the precoat resistance may exceed that of the deposited cake and, therefore, the slope could be between zero and −0.5. For non-precoat filters, a shorter full-scale filter cycle is indicated by increasingly negative slope, providing that adequate cake thickness is achieved for effective cyclical discharge.

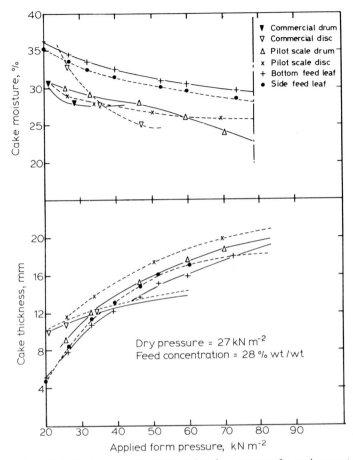

Figure 13.31. Cake thickness and moisture against form pressure for coal concentrate slurry

To obtain a log–log plot similar to that given in *Figure 13.33*, Dahlström[10] recommends about five tests be run at different form times spread over a 5:1 to 10:1 cake formation time range. The validity of the results obtained can be checked by back-calculation of the feed solids concentration for comparison with the measured value. This back-calculation may result in a slightly lower value due to settling of solids in the feed slurry which may require compensation by agitation or reduced flocculation during the test programme.

The solids rate obtained from the leaf test is usually multiplied by a scale-up factor of about 0.8 to obtain the predicted full-scale rate, but may vary to account for fluctuations in feed consistency, filter media effects and other operational factors. Cake thickness, governed by the type of filter, the nature of the solids and filter operating conditions would be specified as a optimum value in order to ensure that higher production could be achieved when required.

Hence, from *Figure 13.33* select a cake thickness of 12.5 mm corresponding to 500 kg m^{-2} h^{-1} formation time and a cake formation time of 1 min.

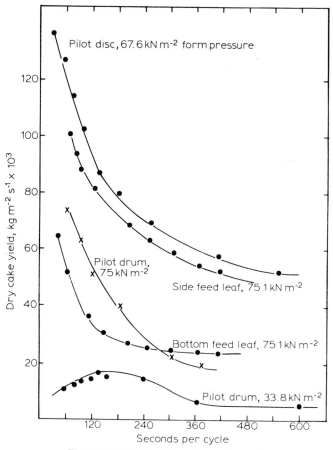

Figure 13.32. Cake yield against cycle speed

This corresponds to 35% effective submergence in a drum filter and hence, a full-scale design rate would be determined as follows:

$$500 \times \frac{35}{100} \times 0.8 \ = \ 140 \text{ kg dry solids m}^{-2} \text{ h}^{-1}$$

and $Z_f = B \cdot h \cdot U$ (from section 13.4), where $U = 2\,rn$, $Z_f = B \cdot h \cdot 2\,rn$, $B \cdot r \cdot n = 1782.5$, from which area = 37.3 m².

If the average speed is 3 min per revolution, the optimum sized filter based on *Table 13.9* would be a 2.435 m diameter by 6.550 m wide rotary vacuum drum. For a cake thickness of 12.5 mm a scraper blade with air blow discharge would be required.

The cake dewatering rate can be obtained from a graph similar to that shown in *Figure 13.34* which gives moisture content as a function of θ_d/Z_c from equation 13.5 in section 13.4.

Dahlström[10] suggests that a sufficient spread of the correlating factor

Table 13.9. TYPICAL STANDARD SIZES OF COMMERCIAL ROTARY VACUUM DRUM FILTERS (FOR USE AS A GUIDE ONLY)*

Filter area, m²	7.5	12.5	20	32.5	37.5	45	55	75	95	115
Drum diameter, mm	1875	1875	1875	2435	2435	3040	3040	4304	4304	4304
Overall length, mm	2986	3836	5112	6000	6550	6950	7495	7680	9150	10630
Overall width, mm	2715	2715	2715	3880	3880	4800	4800	6080	6080	6080
Overall height, mm	2214	2214	2214	2817	2817	3475	3475	4752	4752	4750
Approx. dry weight, t	5.5	6.0	7.0	10.5	11.5	18	20	26	33	38
Typical working weight, t	7.5	8.5	10.0	15.0	16.5	28	32	40	48	54

*From Paxman.

should be tested to determine the inflection point in the curve and extend a further 100% beyond it. This normally involves a range in θ_d/Z_c of about 10:1. In order to convert this to a production rate, the required moisture content must be used to determine an appropriate value of θ_d/Z_c. The value obtained should then be multiplied by a scale-up factor of about 1.2. The cake weight per unit area per cycle required to obtain a dischargeable cake at the desired moisture can then be used to determine the required dewatering

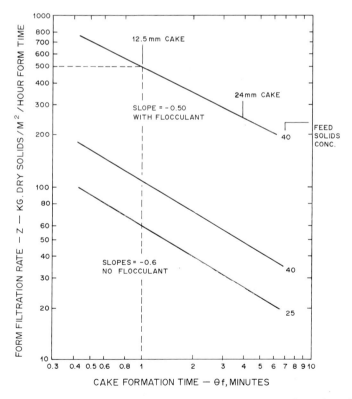

Figure 13.33. Graph to show the typical form filtration rate versus form time. (Source: Dalhström[10])

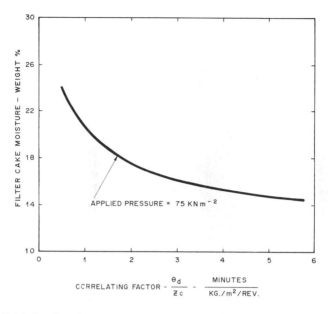

Figure 13.34. Graph to show typical moisture content versus correlating factor curve. (Source: Dahlström[10])

time. This can be related to the filter speed, in minutes per revolution (or cycle) by dividing dewatering time by the portion of the cycle to be devoted to cake dewatering, i.e. ϕ_d:

$$1.2 \ \frac{\theta_d}{\phi_d} \ \text{minutes per cycle}$$

Referring to *Figure 13.34*, select 18% moisture and 75 kPa applied pressure for a drum filter devoting 60% dewatering period. Hence:

$$\text{filtration rate} = \frac{60 \cdot 0.6}{1.2 \cdot 0.18} = 166.7 \text{ kg h}^{-1} \text{ m}^{-2}$$

for which the filter cycle time to form a 12.5-mm cake (where cake density is 1610 kg m^{-3}) would be

$$\theta_d = \frac{0.18 \cdot 1.2 \cdot 1610 \cdot 12.5}{1000}$$

$$= \ 4.35 \text{ min}$$

or

$$\frac{4.35}{0.6} = 7.25 \text{ min rev.}^{-1}$$

Using the same approach, the cake moisture content can be estimated as a function of the filtration rate by selecting various values of the correlation coefficient θ_d/Z_c from a similar graph to that shown in *Figure 13.32*.

To determine gas flow requirements, a measurement of gas flow rate through the cake as a function of dewatering time must be made during the filter tests. This must be carried out under correctly simulated conditions and at as close to the required cake thickness as possible. The maximum measured rate is usually employed and a scale-up factor of 1.1–1.2 is normally used for the internal drainage network inside the filter.

Vacuum pump quantities are usually based on the total area; an equation of the following type is applicable:

$$\text{Vacuum pump capacity} = \frac{1.2 \cdot Q_d \cdot \phi_d}{60 \cdot 100} \text{ m}^3 \text{ m}^{-2} \text{ h}^{-1}$$

REFERENCES

1. Brociner, R. E., 'An improved vacuum leaf thickener for china clay slurries,' *Filtration and Separation*, **9**, 562–565 (1972)
2. Bosley, R., 'Vacuum filtration equipment innovations', *Filtration and Separation*, **11**, 136–150 (1974)
3. Kelegan, W., 'An introduction to the Eimcobelt Filter', Paper presented to the Minerals Engineering Society (February 1981)
4. Moos, S. N. and Klepper, R. P., 'Selection and sizing of non-sedimentation equipment'. In *Design and Installation of Concentration and Dewatering Circuits* (A. L. Mular and M. A. Anderson, Eds), Chap. 10, pp. 148–170, Society of Mining Engineers, London (1986)
5. Wetzel, B., 'Disc filter performance improved by equipment re-design', *Filtration and Separation*, **11**, 270–274 (1974)
6. Mular, A. L., 'The estimation of preliminary capital costs'. In *Mineral Processing Plant Design* (A. L. Mular and R. B. Bhappu, Eds), Chap. 3, pp. 52–70, Society of Mining Engineers, London (1980)
7. Brabhu, P. S. and Giberti, R. A., Personal communication, Dorr-Oliver (1988)
8. Wakeman, R. J., Mehrotra, V. P. and Sastry, K. V. S. 'Mechanical dewatering of fine coal and refuse slurries', *Bulk Solids Handling*, **1**, 251–293 (1981)
9. Anon, 'Nash vacuum pump systems', *Commercial Publication No. 472G*, November (1983)
10. Dahlström, D. A., 'How to select and size filters', Chapter 28, pp. 578–600
11. Machej, J., 'Filtration study practical application of test data', *Int. Chem. Eng.*, **14**, 27–32 (1974)
12. Dahlström, D. A. and Puchas, D. B., 'Proceedings of a Symposium on scaling-up of plant and processes', *Trans. Inst. Chem. Eng.*, S120–S131 (1957)
13. Dahlström, D. A., 'Scale-up methods for continuous filtration equipment', *Int. Symp. on Chemical Processing*, Conference Proceedings, London (1988)
14. Osborne, D. G., 'Rotary vacuum filtration of coal flotation concentrates', *Int. J. Mineral Proc.*, **3**, 175–191 (1976)
15. Osborne, D. G., 'Scale up of rotary vacuum filter capacity', *Trans. Inst. Min. Metall.*, **84**, C158–C166 (1975)
16. Rushton, A. and Rushton, A., 'Sedimentation effects in filtration', *Filtration and Separation*, 254–267 (1973)
17. Dobie, W. B. *Trans. Inst. Chem. Eng.*, **43**, T225 (1965)

Vacuum Filtration
Part II—Horizontal Vacuum Belt
Filters

H. G. Pierson

Pierson and Company (Manchester) Ltd., Wellingborough, Northants

13.6 INTRODUCTION

More and more process industries are going over from batch filtration to continuous filtration. In many industries batch filtration has still continued because of the necessity to obtain well-washed filter cakes and/or because of the necessity to be able to handle widely varying quality of material.

In many cases, rotary drum filters in all their configurations have been tried, but not always entirely successfully. During the last 8–10 years, the process industry in the Western world has realized that horizontal vacuum belt filters offer advantages not available in rotary filters, yet enabling continuous production.

Various types of horizontal belt filters and their essential differences are being discussed, compared largely with rotary drum filters.

To make the text of this chapter slightly more readable I will use the words 'belt filter' only, but it must be understood that this does mean the horizontal

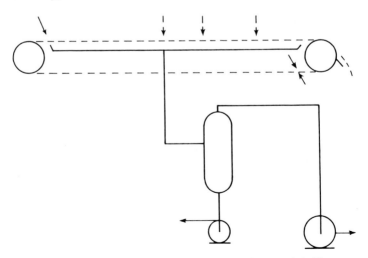

Figure 13.35. Schematic diagram of a horizontal vaccum belt filter

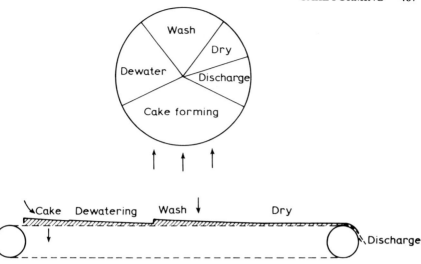

Figure 13.36. Comparison between a rotary drum filter and a horizontal belt filter

vacuum band filter and not the rotary belt filter, or the gravity belt filter, for all of which at times are used the words 'belt filter' only.

As the name indicates, a horizontal vacuum belt filter (*Figure 13.35*) is substantially a belt conveyor-type construction with a filter cloth supported on a vacuum grid, which enables the filtrate to be carried from the suspension, i.e. cake, to the normal receivers in a similar manner to that used on rotary vacuum filters.

The essential difference between a rotary vacuum filter is, therefore, in its configuration. Whereas the rotary drum filter provides a perfect circle, i.e. drum sub-divided into segments for cake formation/cake drying/optional washing/cake discharge, the belt filter provides these sections along its full length. At any one time, therefore, the suspension or cake travels in a horizontal manner as opposed to the situation as it exists on a rotary drum filter where, at one stage, the cake may be upside down, the next minute vertical, then horizontal and finally vertical again—*Figure 13.36*

13.7 CAKE FORMING

Whereas the rotary drum filter sucks the solids and mother liquor from a trough holding the solids against a screen by vacuum and gradually allows drainage to take place, the horizontal filter deposits its suspension on to a horizontal surface, and normally allows gravity draining to take place initially and then applies vacuum merely to remove mother liquor, not to hold the cake in place.

It is immediately clear, therefore, that purely from an energy point of view the horizontal filter has a theoretical advantage over the rotary drum. Most slurries will dewater by gravity to some extent and the horizontal belt filter uses this to the maximum. Unlike the rotary drum filter it does not use vacuum to hold the solids against the cloth but merely to dewater the residual slurry. This also means that a relatively lower vacuum can be applied at times since there is no need to keep the cake in place.

13.8 ADVANTAGES OF BELT FILTER OVER ROTARY DRUM FILTER

Although the belt filter is by no means a sole competitor of the rotary drum filter it is, like the rotary drum filter, a continuous vacuum filter and, as such, most comparisons must be related to the rotary drum filter rather than to other units.

The specific advantages over the rotary drum filter are as follows.

(a) The suspension is deposited on a horizontal surface. It is therefore not necessary to maintain agitation in a trough, nor is it necessary to maintain a constant level, which in practice means overflowing from the trough and repumping. This means, of course, that fragile material which does not stand up to mechanical handling has a far better chance of good filtration on a horizontal filter than on a rotary. Equally, materials which can only be filtered with the addition of flocculation agents like, for instance, activated sludge, perform far better on a horizontal belt than on a rotary drum filter. Differences in output per square metre of as much as 50% have been measured.

Heavy suspensions, namely suspensions with a fraction which tends to settle quickly, are obviously very difficult to filter on a rotary drum since they require extensive agitation. Clearly, on a horizontal surface this does not apply since the material would settle readily and rapidly on to the surface of the filter medium and allow, to some extent, natural drainage to take place before vacuum is applied.

(b) Most slurries contain, apart from some medium-sized material, a fraction of fines. On the rotary drum these fines, no matter how efficient the agitation, will always be present in a disproportionate percentage in the top layer in the trough. As a result the fine material is sucked preferentially against the filter cloth of the rotary drum filter and so blinds, or partially blinds, the filter cloth before the larger particles have even had a chance to be sucked against the cloth also. This, of course, does not happen on a horizontal belt filter since the reverse takes place, namely the largest particles, i.e. the heaviest particles, fall towards the filter cloth first, to be followed by the finer ones on top of the larger ones, so forming an ideal stratification, using spontaneously the heavier fraction as a type of precoat (*Figures 13.37* and *13.38*).

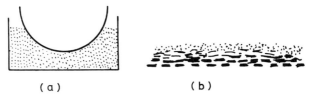

(a)   (b)

Figure 13.37. The difference in stratification of particles between (a) the rotary vacuum filter, and (b) the horizontal belt filter

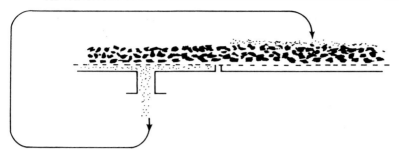

Figure 13.38. Recycling of 'fines' on a belt filter

(c) Many cakes, after dewatering, shrink and crack, thereby allowing air to pass freely through the cake, effectively reducing the vacuum and thereby the dewatering effect during the last stages. On the rotary drum filter this is a serious and rather difficult situation, which can only be remedied with a measure of success by allowing closing blankets or rakes or squeeze rollers to be applied. On the belt filter a simple sheet of impervious material (*Figure 13.39*), as for instance polythene sheet trailed over the cake, is adequate to maintain the maximum vacuum.

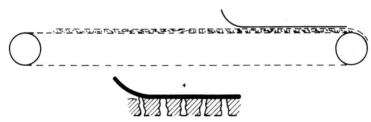

Figure 13.39. Maximum vacuum can be maintained by a trailing polythene sheet if cake cracking occurs

(d) In many continuous filtration applications cake washing is of primary importance and, indeed, many rotary filters are used for this purpose. The very configuration of a rotary filter is, however, dubious for cake washing efficiency. It is a prerequisite that the size of filter is thus chosen, that the cake in its fully dewatered form arrives at the topmost portion of the filter drum prior to cake washing being applied. Even at that stage, however, part of the

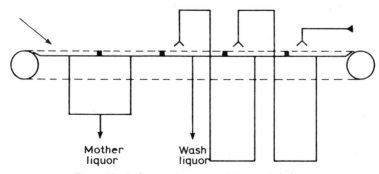

Mother
liquor

Wash
liquor

Figure 13.40. Countercurrent washing on a belt filter

Table 13.10. GENERAL COMPARISONS BETWEEN ROTARY DRUM (RVD) AND BELT FILTERS

	RVD filter	*Belt filter*
Floor space	Good	Poor
Filter area usage	Good–poor	Good
Energy usage	Good	Good
Maintenance	Good	Good–poor
Reliability	Good	Good–poor
Capital cost	Good–average	Good–poor
Control of process	Average	Good
Flexibility	Average–poor	Good
Materials of construction	Average	Good
Choice of cloth	Good	Good
Cloth life	Good	Average–good

washing water is likely to run back into the unwashed cake, effectively diluting the mother liquor, or to run into the so-called washed cake, effectively upsetting the washing balance. In addition, it is obvious that countercurrent washing can never be applied successfully on one rotary filter. None of these disadvantages applies to the horizontal filter where the cake arrives as a ribbon of unwashed cake to be gradually washed to the required purity, either cocurrently or countercurrently (*Figure 13.40*). The very fact that the cake is horizontal clearly allows for displacement as well as solution washing to the client's requirement. *Table 13.10* gives general comparisons between rotary drum and belt filters.

13.9 IS THE BELT FILTER UNIVERSAL?

From the foregoing it might appear that the belt filter is a universal tool with no competitors. The very fact that not every industry has gone over to belt filter usage rather proves that this is clearly not the case. First of all we have only compared the belt filter with the rotary filter and, although this is possibly the most logical way of viewing the matter, it must be remembered that belt filters should be compared equally with filter presses, centrifuges, sheet filters and a host of other cake filters.

Despite this, however, it is true to say that purely technically the belt filter is more universal and more generally acceptable than the rotary or, indeed, any other filter. If vacuum filtration will give the required results then there is no doubt that theoretically, at least, the belt filter will always have the edge over a rotary filter. From many hundreds of tests carried out it has been established that it is virtually impossible to find a slurry which will filter better on a rotary drum filter than on a belt filter. The reverse, however, namely the situation where the material will filter better on a belt filter than a rotary, has been found many times. This does not mean that commercially or practically the belt filter should always be the first choice. Its main disadvantages are its size, which almost inevitably means a greater floor area, sometimes the price and sometimes the fact that its inefficiency factor for a given slurry and a given capacity could be greater than on a rotary drum filter.

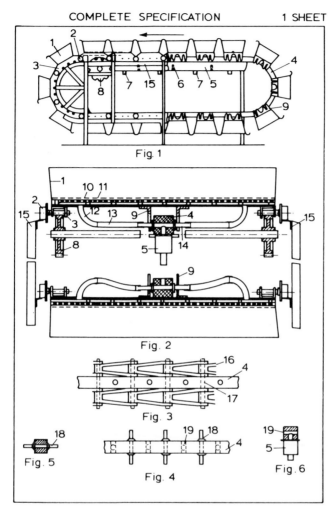

Figure 13.41. Continuously travelling pan filter. (Reproduced from the original on a reduced scale)

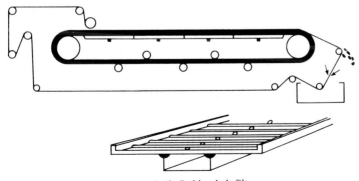

Figure 13.42. Rubber belt filter

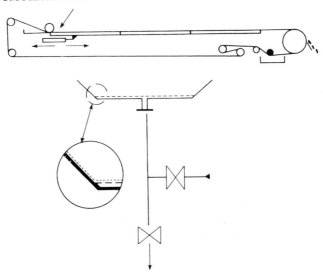

Figure 13.43. Reciprocating pan filter

It is important, therefore, to view the various types available. Unlike the rotary drum filters which, small manufacturing details apart, are substantially similar regardless of their make, the belt filters can be divided into four groups:

1. continuously travelling pan (*Figure 13.41*);
2. rubber drainage type (*Figure 13.42*);
3. reciprocating pan (*Figure 13.43*);
4. intermittent belt moving (*Figure 13.44*).

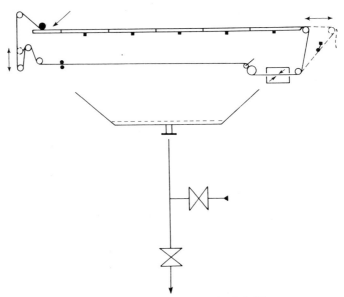

Figure 13.44. Intermittently moving belt filter

13.9.1 The continuously moving pan system

The continuously moving pan system has a series of vacuum pans which travel in a Jacob's-ladder fashion under a continuously moving filter cloth. On the top strand, vacuum is applied to the continuously moving pans and filtrate is carried away through one single or, more often, a number of valves, while on the bottom strand vacuum is being cut off and cloth washing takes place. This system appears to have fallen into disuse, possibly because of its mechanical sensitivity.

13.9.2 The rubber drainage belt type

Possibly the most popular and well-known system, this relies on a hollow rubber drainage/carrier belt. This belt is quite literally a double hollow belt with perforations on either side. The belt itself provides the strength required to apply vacuum to filter cloth, which in itself would not have this strength. The belt is dragged over stationary vacuum boxes and the filter cloth lies inside a belt. The major disadvantages of this system are the cost of the belt, which requires periodical replacement, the extreme difficulty in fitting and tensioning a belt of this kind, the tracking problems of the belt and, above all, the material limitations which a pliable rubber belt puts on industrial applications. Typically, of course, all solvents are not applicable to this type of filter. Although some manufacturers will claim to be able to operate at full vacuum it is my belief that in most cases it is safe to operate these filters only at vacuum not in excess of 500 mmHg, since the drag on the belt causes a vicious circle. If the vacuum is increased, the belt strength has to be increased which inevitably means a thicker, heavier, stiffer and stronger belt which, in turn, means a heavier drive which, in turn, means a stronger belt which, in turn, means a heavier drive *ad infinitum*. For practical purposes I believe, therefore, that the efficiency of this type of filter is limited by the extent to which it can apply vacuum.

13.9.3 The reciprocating tray filter

The reciprocating tray filter employs ordinary filter cloth as an endless belt without support belt. This can be achieved by allowing the vacuum trays to move partially forward with the filter cloth and then break the vacuum, allowing the trays to return to the starting position and to pick up the belt again under full vacuum. This system has the overriding merit of being relatively low on maintenance cost and being able to maintain, during filtration, a vacuum as high as the application deserves. The disadvantage of the system is that an inefficiency factor comes in during the return stroke when the trays are pulled back without vacuum. Since the filter belt continues to travel forward, the return stroke must be regarded as dead time.

Normally the return stroke is approximately 2 s on these filters, which in certain cases can, of course, give a very high inefficiency factor if, for instance, the forward stroke is only 6 or 8 s. The efficiency of this filter is, therefore, entirely dependent on the capacity, i.e. volume and filtration rate.

A further limitation of this system lies in the material of construction which tends to be primarily stainless steel, titanium or other steel alloy, but does not really allow for rubber-coated systems or homogeneous plastic constructions. Equally, it is my opinion that there is a limit to the width of the filter, which I would not consider to be practical beyond 3 m.

13.9.4 Intermittently moving belt filter

This filter tries to combine the advantages of the above systems. It operates with stationary vacuum trays similar to the rubber drainage system, but still relies on ordinary filter cloth for the belt without support belt. The belt is moved forwards in increments of approximately 600 cm (2 ft) during periods when the vacuum is switched off. In this respect it is similar to the intermittent tray movement, except that the light belt only moves instead of the trays. As can be appreciated, this is mechanically a far simpler construction. Because of this it is also possible to ensure a faster movement of the belt during the period that vacuum is not applied, which means that the dead time is reduced. In addition, the makers have been able to use a geometry which enables only a small amount of physical movement to effect a large amount of belt movement in a short space of time, so that the dead time as compared with the intermittent tray movement of the previous filter is approximately 50%. Since the filter relies on stationary vacuum trays there is no practical limit to the total width of the filter and, therefore, the filter area,

Table 13.11.

	Rubber belt	Reciprocating pan	Intermittently moving belt
Chemical resistance	6	7	10
Very fast filtration with low solids	10	8	8
Very fast filtration with high solids	8	8	8
Slow filtration with low solids	8	10	10
Slow filtration with high solids	5	10	10
Cakes less than 2 mm	6–7	5	8
Down-time risk	6	8	8
Foundation requirements	8	8	10
Cloth life	10	8	9
Maintenance	7	7	10
Size limits	10	6	8
Capital cost	7–8	7	10
Vacuum levels	7	10	10
Cloth wash liquor consumption	8	8	10
Vacuum loss/efficiency	7–8	10	10

Table 13.12. FACTORS WHICH INFLUENCE EFFICIENCY LIMITS

Rubber belt filters	Limit on vacuum
	Limit on slurry depth/cake thickness
Reciprocating pan filters and intermittently moving belt filters	Limit on belt speed/ displacement frequency

nor is there any limit to the materials of construction, which could be steel, rubberized steel, plastics, etc. according to the client's requirements.

Comparisons of different types of belt filters are shown in *Tables 13.11 and 13.12.*

13.10 FLOOR AREAS

All belt filters suffer substantially from a floor area problem in the sense that they require more space than a rotary drum filter. In this respect, however, the intermittent tray and intermittent belt movement filter suffer less from this problem than the other two since they tend to be of a lighter construction and do not require overhead cranes etc. to allow for rubber belts to be fitted or taken off. It is therefore possible to build these lighter filters on top of one another on a relatively light construction. Intermittent tray filters have a slight disadvantage here since the impact of the returning trays can be quite high and may still require a relatively strong superstructure (a ton of steel travelling at a speed of 0.3 m s^{-1} (1 ft s^{-1}) does require some cushioning). Clearly, the intermittent belt movement does not have the same problems.

14

Centrifugal Filtration

K. Zeitsch

Consultant, Köln, West Germany

14.1 INTRODUCTION

If filtration is understood as a process aimed at obtaining a clear liquid, then 'centrifugal filtration' is a misnomer as filter centrifuges are normally used to produce well-drained solids rather than a clear filtrate. Nevertheless, the given heading has been maintained for historical reasons.

14.2 FLOW THROUGH THE CAKE OF A FILTER CENTRIFUGE

The common feature of all filter centrifuges is a rotating basket equipped with a filter medium. Although this is a rather simple characteristic, its actual application has led to an astounding variety of different designs. Treating all of them would be far beyond the scope of this chapter, and the theoretical treatment of the filtration process will therefore be limited to cylindrical baskets. Conical and other possible shapes can be dealt with by the same approach.

To further simplify the theoretical treatment, the filtration resistance of the filter medium will be considered to be negligible as compared to the filtration resistance of the cake. For most practical applications of filter centrifuges, this assumption is well justified. In addition, the introductory treatment to follow will assume the filter cake to be homogeneous. The effect of stratification due to large particles settling more rapidly than small ones will be considered later.

In most practical cases, centrifugal sedimentation proceeds at a much faster rate than centrifugal filtration so that after filling the basket and some transient settling phenomena the process can be characterized by a filter cake submerged in an annulus of clear liquid. This situation is shown schematically in *Figure 14.1*.

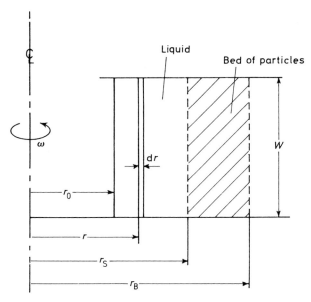

Figure 14.1. Geometry of centrifugal filtration

14.2.1 Incompleteness of Darcy's Law

The fundamental filtration law as formulated by Darcy[1] states that the flow of a liquid through a porous bed of particles requires a pressure difference across the bed; the original form of the equation for the linear case is:

$$Q = \frac{\kappa \Delta P A}{L} \qquad (14.1)$$

where

   Q is the flow rate of liquid passing through the cake
   κ is a proportionality constant
   ΔP is the pressure drop across the cake in the direction of the flow
   A is the filter area
   L is the thickness of the cake.

This equation deviates from equation 9.1 by using

$$\kappa = \frac{K}{\mu} \qquad (14.2)$$

where

   K is the permeability of the bed
   μ is the dynamic viscosity of the liquid.

κ is used not only because of its historical foundation but also for its practical value, as otherwise two measurements would have to be made to obtain a proportionality constant instead of one. κ is called the 'intrinsic permeability' of the system and is measured in the centrifugal field which will be used. The test procedure is discussed in section 14.4.

When Darcy's law is applied to centrifugal filtration however, it leads to erroneous conclusions[2]. With reference to *Figure 14.1*, let us consider the moment when the inner surface of the liquid annulus has just reached the cake, $r_0 = r_S$. At this instant, the pressure at r_S and r_B will be equal to the ambient gas pressure and equation 14.1 might be interpreted as indicating that, with ΔP equal to zero, no further flow will occur. Of course, this is contrary to common experience, and thus warrants a more detailed discussion of the phenomenon.

As will be shown, Darcy's law is incomplete as it does not take into account the effect of mass forces.

14.2.2 Pressure distribution in the liquid layer

According to the general laws of hydrostatics[3], the pressure in an ideal liquid exposed to a force field F can be computed from the differential equation

$$\text{grad } P = F \tag{14.3}$$

where F is the force per unit volume of liquid so that

$$F = \rho b \tag{14.4}$$

where
  ρ is the density of the liquid
  b is the acting acceleration

For cylindrical coordinates and a radial acceleration, integration of equation 14.3 yields

$$P = \rho \int b \, dr + \text{constant} \tag{14.5}$$

Thus, in an ideal liquid exposed to a centrifugal field caused by rotation at an angular velocity ω, the pressure at a radius r is found to be

$$P(r) = \rho \int_0^r r\omega^2 \, dr \tag{14.6}$$

which gives

$$P(r) = \tfrac{1}{2}\rho\omega^2(r^2 - r_0^2) \tag{14.7}$$

where r_0 is the radius at the liquid surface and where the pressure at the liquid surface is taken to be zero so that consideration is limited to pressures above the ambient pressure.

With reference to *Figure 14.1*, the pressure at the surface of the cake has the value

$$P_S = \tfrac{1}{2}\rho\omega^2(r_S^2 - r_0^2) \tag{14.8}$$

The acceleration due to gravity can be disregarded, as its effect in modern filter centrifuges with high centrifugal accelerations is too small to justify the mathematical complications that its inclusion would incur, and in any case the effect is only noteworthy for machines with a vertical axis.

14.2.3 Pressure distribution in the cake

If there were no centrifugal effect for $r > r_S$, application of Darcy's law would be a simple matter. In cylindrical coordinates, equation 14.1 has the differential form

$$\frac{dP}{dr} = -\left(\frac{Q}{2\pi\kappa W}\right)\left(\frac{1}{r}\right) \tag{14.9}$$

With the previous restriction to pressures above the ambient pressure, the boundary conditions are
  (a) $P = 0$ at $r = r_B$
and
  (b) $P = P_S$ at $r = r_S$

and lead to

$$P(r) = \frac{Q}{2\pi\kappa W} \ln\frac{r_B}{r} \tag{14.10}$$

and

$$Q = \frac{2\pi\kappa W P_S}{\ln(r_B/r_S)} \tag{14.11}$$

so that

$$P(r) = P_S \frac{\ln(r_B/r)}{\ln(r_B/r_S)} \tag{14.12}$$

Thus, for the given unrealistic assumption that there is no centrifugal effect for $r > r_S$, the pressure in the bed of particles would decrease with increasing radius according to a simple logarithmic function. In reality, the centrifugal effect does not cease to exist for $r > r_S$ so the true picture must be different.

To obtain the desired relationship in an elementary manner, it is convenient to imagine the filter cake to have straight conical pores in radial direction providing a constant porosity ε.

The flow of liquid through the pores of the cake can be caused by two driving forces, one resulting from any difference in pressure at the boundaries, and the other from the centrifugal action. For the linear case, the force resulting from a difference in pressure at the boundaries is

$$F_P = \Delta P \varepsilon A \tag{14.13}$$

so that

$$\Delta P A = \frac{F_P}{\varepsilon} \tag{14.14}$$

Substituting this in equation 14.1, we find

$$Q_P = \frac{\kappa F_P}{\varepsilon L} \tag{14.15}$$

If in addition to F_P a centrifugal force F_C is acting on the liquid of the pores as well, the general formulation will be

$$Q = \frac{\kappa(F_P + F_C)}{\varepsilon L} \tag{14.16}$$

For a ring element of thickness dr and height W at a radius r, and with porosity ε, the force on the liquid in the pores from the pressure drop dP across the element would be

$$dF_P = - 2\pi r W \varepsilon \frac{dP}{dr} dr \tag{14.17}$$

and the centrifugal force on the liquid in the pores from the rotation at the angular velocity ω becomes

$$dF_C = 2\pi r W \, dr \varepsilon \rho \omega^2 \tag{14.18}$$

so that the differential form of equation 14.16 for a rotating ring element is

$$Q = \frac{\kappa}{\varepsilon \, dr}(- 2\pi r W \varepsilon \, dP + 2\pi r^2 W \varepsilon \rho \omega^2 \, dr) \tag{14.19}$$

or

$$Q = 2\pi r \kappa W \left(\rho \omega^2 - \frac{dP}{dr} \right) \tag{14.20}$$

leading to the differential equation

$$dP = \left(r\rho \omega^2 - \frac{Q}{2\pi \kappa W} \frac{1}{r} \right) dr \tag{14.21}$$

With the boundary conditions (a) and (b), it is found that

$$Q = \frac{2\pi \kappa W}{\ln (r_B/r_S)} \left[P_S + \frac{\rho \omega^2 r_B^2}{2} \left(1 - \frac{r_S^2}{r_B^2} \right) \right] \tag{14.22}$$

and

$$P(r) = \frac{\ln (r_B/r)}{\ln (r_B/r_S)} P_S + \frac{\rho \omega^2 r_B^2}{2} \left[\frac{\ln (r_B/r)}{\ln (r_B/r_S)} \left(1 - \frac{r_S^2}{r_B^2} \right) - \left(1 - \frac{r^2}{r_B^2} \right) \right] \tag{14.23}$$

Hence, in the filter cake of a centrifuge the dependence of the pressure on the radius coordinate turns out to be a rather complex function, its first term being identical with equation 14.12, and its second term depending on the angular velocity. Note that the function contains neither the permeability nor the porosity.

The dimensionless form of equation 14.23 is

$$\frac{P(r)}{P_S} = \frac{\ln(1/\lambda)}{\ln(1/\alpha)} + \varsigma \left[\frac{\ln(1/\lambda)}{\ln(1/\alpha)}(1 - \alpha^2) - (1 - \lambda^2) \right] \tag{14.24}$$

where

$$\lambda = r/r_B \tag{14.25}$$

$$\alpha = r_s/r_B \tag{14.26}$$

and

$$\zeta = \frac{\rho\omega^2 r_B^2}{2P_S} \tag{14.27}$$

The expression in brackets in equation 14.24 is termed ξ:

$$\xi = \frac{\ln(1/\lambda)}{\ln(1/\alpha)}(1 - \alpha^2) - (1 - \lambda^2) \tag{14.28}$$

As can be seen from equation 14.23, when multiplied by

$$\tfrac{1}{2}\rho\omega^2 r_B^2$$

ξ will determine the pressure distribution in the cake for the case of $P_S = 0$, i.e. for the final state of the filtration process when there is no more liquid

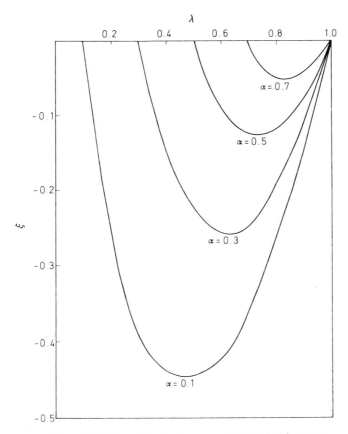

Figure 14.2. The residual cake pressure distribution function

'above' the cake. For this reason, ξ is called the 'residual cake pressure distribution function'. A plot of ξ against λ is shown in *Figure 14.2*. The negative values of ξ indicate that when the cake of a filter centrifuge is not covered by a liquid layer, the liquid in the cake will be submitted to pressures below the ambient pressure.

The relationship ceases to hold when the boiling point is reached as then the volume of the fluid in the cake will increase, thus violating the continuity law incorporated in the derivation.

14.2.4 Flow for different boundary conditions

If the boundary conditions (a) and (b) are reversed so that the pressure at r_S is zero while an elevated pressure P_B is applied at r_B, the analogue of equation 14.22 will be

$$Q = \frac{2\pi\kappa W}{\ln(r_B/r_S)}\left[\frac{\rho\omega^2 r_B^2}{2}\left(1 - \frac{r_S^2}{r_B^2}\right) - P_B\right] \tag{14.29}$$

The relative magnitudes of the terms in the square bracket will determine whether the liquid flows in a centrifugal or a centripetal direction. When the first term is greater than the second, the liquid will flow from a low to a high pressure although Darcy's law in the formulation of equation 14.1 would give the opposite direction.

14.2.5 Simplified treatment

Equation 14.22 has the same form as equation 14.11 which represents the normal formulation of Darcy's law for a ring if the interior pressure at r_S is P_S and the exterior pressure at r_B is zero. Thus, the flow through the cake of a filter centrifuge can be determined from the normal formulation of Darcy's law, but without the centrifugal term, if the pressure on the inner surface of the cake is supplemented by

$$\frac{\rho\omega^2 r_B^2}{2}\left(1 - \frac{r_S^2}{r_B^2}\right)$$

In other words, the term in the square bracket in equation 14.22 can be considered to be an 'apparent pressure difference' ΔP^* for the application of Darcy's law.

Consideration of equation 14.8 leads to

$$\Delta P^* = \tfrac{1}{2}\rho\omega^2(r_B^2 - r_0^2) = \tfrac{1}{2}\rho\omega^2 r_B^2\left(1 - \frac{r_0^2}{r_B^2}\right) \tag{14.30}$$

Introducing

$$h = r_B - r_0 \tag{14.31}$$

yields

$$\frac{r_0^2}{r_B^2} = 1 - 2\frac{h}{r_B} + \frac{h^2}{r_B^2} \tag{14.32}$$

where the last term on the right-hand side can be neglected as it is rather small in comparison with the other terms, hence equation (14.30) can be written in the approximate form

$$\Delta P^* \cong \rho \omega^2 r_B h \tag{14.33}$$

or

$$\Delta P^* \cong \rho g C_B h \tag{14.34}$$

where g is the acceleration due to gravity

and

$$C_B = \frac{r_B \omega^2}{g} \tag{14.35}$$

is the dimensionless 'centrifugal number of the centrifuge' referred to the radius r_B of the basket. Substituting equation 14.34 in equation 14.22 gives the approximate flow rate

$$Q \cong \frac{2\pi \kappa W}{\ln(r_B/r_S)} \rho g C_B h \tag{14.36}$$

14.2.6 False concept resulting from the simplified treatment

Equation 14.36 offers the simple concept of employing the normal form of Darcy's law, in cylindrical coordinates, and assuming that the pressure at the inner surface of the cake is equal to the total head of liquid 'above' the basket wall when this liquid is submitted to a uniform centrifugal acceleration corresponding to the value at the radius of the basket wall.

Although the practical merit of equation 14.36 is unquestioned, it should be kept in mind that the concept associated with it is fundamentally false, its unrealistic features being the following:

1. The pressure actually prevailing at the surface of the cake is by no means equal to equation 14.30 or equation 14.34. Its correct value is given by equation 14.8.
2. The height of liquid 'above' the cake is not equal to h but $(r_0 - r_S)$ which is a smaller value.
3. The liquid 'above' the cake is by no means submitted to the centrifugal acceleration prevailing at the wall of the basket but to smaller accelerations.
4. There is no such thing as a uniform centrifugal acceleration, i.e. a centrifugal acceleration which does not depend on the radius. As the liquid 'above' the cake has a finite height, it is submitted to a range of centrifugal accelerations commensurate with the range of radii involved.

14.2.7 Numerical example

With reference to *Figure 14.1*, a filter centrifuge characterized by $r_S = 0.60$ m and $r_B = 0.80$ m rotates at 500 rev. min^{-1} in an ambient gas atmosphere of 10^5 N m^{-2}. The density of the liquid phase is 1000 kg m^{-3}. The vapour pressure of the liquid phase is 0.1×10^5 N m^{-2}.

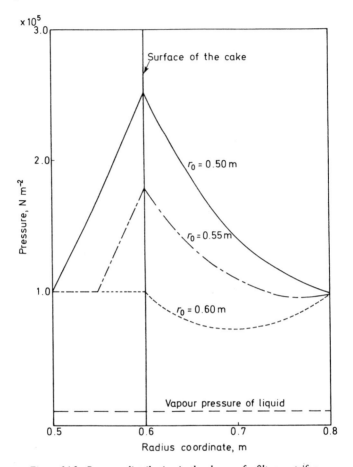

Figure 14.3. Pressure distribution in the charge of a filter centrifuge

Question: What is the pressure distribution in the charge for $r_0 = 0.50$ m, $r_0 = 0.55$ m, and $r_0 = r_S = 0.60$ m?

Solution: Adding the ambient pressure to equation 14.7 and equation 14.23 yields the curves shown in *Figure 14.3*.

14.3 THE FILTRATION PERIOD IN A CENTRIFUGAL FIELD

With reference to *Figure 14.1*, the 'filtration period' is defined as the time required for the level of liquid to be lowered from r_0 to r_S.

The process immediately following the state $r_0 = r_s$ is called the 'drainage period'. It is governed by entirely different laws as will be shown later.

14.3.1 Normal treatment

According to equations 14.22 and 14.8, the initial rate of flow through the cake is

$$Q = \frac{dV}{dt} = \frac{\pi \kappa W \rho \omega^2}{\ln (r_B/r_S)} (r_B^2 - r_0^2) \tag{14.37}$$

where V is the volume of the liquid and t is a time coordinate.

During a time increment dt, the volume of liquid 'above' the cake diminishes by

$$dV = 2\pi r_0 W \, dr_0 \tag{14.38}$$

Substituting equation 14.38 into equation 14.37 gives the differential equation

$$dt = \frac{2r_0 \ln (r_B/r_S) \, dr_0}{\kappa \rho \omega^2 (r_B^2 - r_0^2)} \tag{14.39}$$

With the boundary conditions

$$t = 0 \qquad \text{for } r_0 = R$$

and

$$t = \tau \qquad \text{for } r = r_S$$

the solution of equation 14.39 is found to be

$$\tau = \frac{\ln(r_B/r_S)}{\kappa \rho \omega^2} \ln \frac{r_B^2 - R^2}{r_B^2 - r_S^2} \tag{14.40}$$

where τ is the period of time required for the inner surface coordinate r_0 to increase from R to r_S.

14.3.2 Dimensionless treatment

Multiplying equation 14.40 by $\kappa \rho \omega^2$ yields the dimensionless term

$$\psi = \frac{\tau \kappa \rho g C_B}{r_B} \tag{14.41}$$

where C_B is the centrifugal number referred to the radius r_B of the basket as introduced in equation 14.35. ψ is called the 'centrifugal filtration number'. R and r_S can be expressed as fractions of r_B by defining

$$\delta = \frac{R}{r_B} \tag{14.42}$$

and, as in equation 14.26

$$\alpha = \frac{r_{\mathrm{S}}}{r_{\mathrm{B}}} \tag{14.43}$$

In combination with equations 14.41 and 14.40, this gives

$$\psi = \ln \frac{1}{\alpha} \ln \frac{1 - \delta^2}{1 - \alpha^2} \tag{14.44}$$

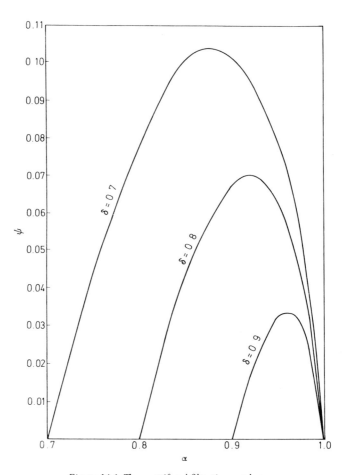

Figure 14.4. The centrifugal filtration number

A graphical representation of this function is shown in *Figure 14.4*. The left-hand starting points of the curves correspond to the condition of having no liquid above the cake, and the end points at $\alpha = 1$ represent the condition of having no cake.

14.3.3 Simplified treatment

If the derivation is started not with equation 14.37 but with the approximate form equation 14.36

$$\frac{dV}{dt} = \frac{2\pi\kappa W}{\ln(r_B/r_S)} \rho g C_B h \tag{14.45}$$

introducing

$$r_B - r_S = \Delta r \tag{14.46}$$

gives

$$\ln\frac{r_B}{r_S} = \ln\left(1 + \frac{\Delta r}{r_S}\right) \tag{14.47}$$

In as much as for commercial filter centrifuges

$$\frac{\Delta r}{r_S} \ll 1 \tag{14.48}$$

so that

$$\ln\left(1 + \frac{\Delta r}{r_S}\right) \approx \frac{\Delta r}{r_S} \tag{14.49}$$

equation 14.45 can be converted to

$$dV = \frac{2\pi\kappa W r_S}{\Delta r} \rho g h C_B \, dt \tag{14.50}$$

On the other hand

$$dV \approx -2\pi r_S W \, dh \tag{14.51}$$

hence combination of equations 14.50 and 14.51 yields the differential equation

$$dt = \frac{-\Delta r}{\kappa \rho g C_B} \frac{dh}{h} \tag{14.52}$$

With the boundary condition

$$t = 0 \qquad \text{for } h = r_B - R$$

the solution of equation 14.52 is found to be

$$t(h) = \frac{\Delta r}{\kappa \rho g C_B} \ln\left(\frac{r_B - R}{h}\right) \tag{14.53}$$

so that the period of time required to lower the surface of the liquid to the surface of the cake becomes

$$\tau = \frac{r_B - r_S}{\kappa \rho g C_B} \ln\left(\frac{r_B - R}{r_B - r_S}\right) \tag{14.54}$$

Multiplying this by

$$\frac{\kappa \rho g C_{B}}{r_{B}}$$

and introducing equations 14.42 and 14.43 yields

$$\psi = (1 - \alpha) \ln \frac{1 - \delta}{1 - \alpha} \qquad (14.55)$$

For the important case of $\delta = 0.8$, corresponding to commercial peeler centrifuges, a comparison of the exact relation equation 14.44 with the approximate formula equation 14.55 is shown in *Figure 14.5*.

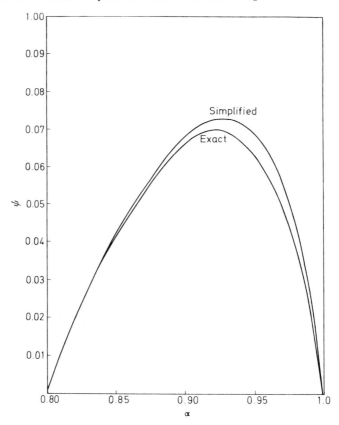

Figure 14.5. Exact and simplified centrifugal filtration numbers for $\delta = 0.8$

The values of the simplified ψ-function are seen to be too high, the maximum deviation from the exact function being of the order of 10%.

14.3.4 Numerical example

A filter centrifuge is characterized by $r_{B} = 1.0$ m and an operating speed of

650 rev. min^{-1}. Its rim radius is $R = 0.75$ m. After being filled to the rim, the radius of the cake surface is $r_s = 0.9$ m. The liquid component of the feed stock has a density of 1000 kg m^{-3}, and the cake has an intrinsic permeability of 1.6×10^{-10} m^4 N^{-1} s^{-1}.

Question: What is the filtration period?

Solution: Application of equation 14.40 yields

$$\tau = 113 \, s$$

14.4 MEASUREMENT OF THE INTRINSIC PERMEABILITY OF A FILTER CAKE IN A CENTRIFUGAL FIELD

It is well known that a filter cake can undergo significant compaction when exposed to centrifugal fields. Consequently, filter cake permeability measurements based on customary vacuum techniques can lead to results which may not be reliably applicable when the filtration process is to be carried out in a centrifuge. In as much as the permeability of a filter cake is proportional to the third power of the cake porosity, it is readily realized that even a small reduction in porosity as a result of compaction can cause a sizeable reduction of the permeability. Hence, for filtration processes in centrifuges, it is desirable to work with a cake permeability determined in the centrifugal field to be employed. Taken to include the viscosity of the liquid to be separated, this property is called the 'intrinsic permeability' of the system, as mentioned already in section 14.2.1.

14.4.1 Measurement with a filter beaker

A simple experimental set-up aimed at determining the permeability of a filter cake in centrifugal fields is shown in *Figure 14.6*. The apparatus consists of a 'filter beaker' suspended in a laboratory centrifuge and observed by means of a stroboscope. The beaker is made from a transparent material such as plexiglass (polymerized methyl methacrylate) and carries scale divisions perpendicular to its axis. At the bottom, the beaker is provided with a filter cloth supported by a wire screen and a perforated cap.

The test procedure involves two operations:

1. The beaker is filled with the suspension to be filtered, and spun at the centrifugal acceleration which will be used for the full-scale operation. This leads to the formation of a cake having undergone a compaction under the same conditions as the separation. The quantity of suspension filtered should be chosen so as to yield a cake thickness exceeding 20 mm.
2. The beaker is refilled with clear filtrate obtained by vacuum filtration or other methods, and spun at the centrifugal acceleration used in 1 while observing the level of the liquid phase by means of a stroboscope and measuring the time required for this level to pass from a scale division close to the top all the way to the surface of the cake.

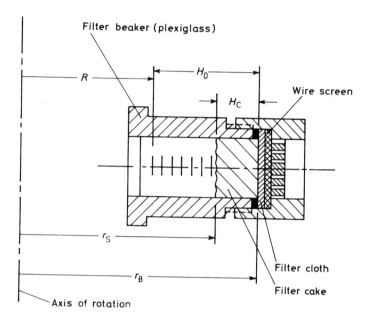

Figure 14.6. The filter beaker

According to the approximate equation 14.54, the period of time required for the level of the liquid to move from an initial radius coordinate R to the radius coordinate r_S of the cake surface is

Using

$$\tau = \frac{r_B - r_S}{\kappa \rho g C_B} \ln \frac{r_B - R}{r_B - r_S} \qquad (14.56)$$

$$r_B - R = H_0 \qquad (14.57)$$

and

$$r_B - r_S = H_C \qquad (14.58)$$

where H_0 is the height of the liquid level 'above' the filter cloth at the start of the time measurement and H_C is the height of the filter cake, equation 14.56 becomes

$$\kappa = \frac{H_C}{\tau \rho g C_B} \ln \frac{H_0}{H_C} \qquad (14.59)$$

14.4.2 Measurement with a filter basket

The intrinsic permeability can also be measured by means of a filter basket. This alternative is adopted when:

1. the plexiglass of which the filter beaker is usually made is attacked by

the liquid phase of the system, e.g. solvents such as esters, ketones, benzene, or halogenated hydrocarbons, and certain concentrated acids such as formic acid, acetic acid, and the like;

2. the permeability is so great that in the filter beaker experiment the level of the liquid phase passes the starting mark before the centrifuge has reached the desired speed;

3. it is desirable to measure the permeability after the cake has been submitted to the action of a discharge peeler which may lead to partial blinding due to compression and particle disintegration.

The experimental set-up required for measuring the permeability with a filter basket is shown schematically in *Figure 14.7*. There is no limitation in terms of size so that the measurement can be carried out with a full-scale production machine.

After the basket has been provided with a cake of appropriate height by introducing a corresponding quantity of suspension through the normal feed stock distributor of the machine (not shown in *Figure 14.7*), the basket is charged with a continuous flow of the pure liquid involved, the latter obtained, of course, in some anterior operation. To introduce this liquid in a smooth fashion, without splashing, it is customary to use a central feed pipe leading behind a slightly submerged ring attached to the rear wall of the basket as depicted in *Figure 14.7*. The feed rate is adjusted to be slightly greater than what can pass the cake, the excess flowing over the rim into a special chamber well separated from the principal stream that has passed the cake. This separation is accomplished by a suitable divider in the casing.

The principal stream is measured by a suitable set-up, usually consisting of an electronic balance, an amplifier, and a recorder so that the weight of the filtrate is determined as a function of time. In steady-state conditions, the slope of the recording divided by ρg yields the volumetric filtrate flow rate Q. In view of equations 14.22 and 14.35 it is found that

$$K = \frac{Q \ln \dfrac{r_B}{r_S}}{\pi \, W \rho g r_B C_B \left(1 - \dfrac{r_o^2}{r_B^2}\right)} \qquad (14.60)$$

14.4.3 General remarks on permeability measurements

Obviously, the measurement of the intrinsic permeability must be carried out at the temperature of the planned process so that, to be of universal applicability, the centrifuge used must permit both heating and cooling, with an appropriate thermostatic control. In addition, if the liquid phase contains any significant amount of dissolved solids (solute), the centrifuge must be operated in an atmosphere saturated with the vapour of the liquid, otherwise evaporation of some liquid could cause some solute to be precipitated due to supersaturation, especially on and in the filter cloth forming the evaporation surface. This precipitation could lead to partial or total blinding of the filter medium. Evaporation also produces a cooling effect increasing the viscosity

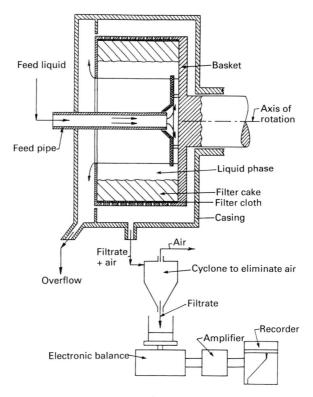

Figure 14.7. Experimental set-up for permeability measurements with a filter basket

of the liquid, usually lowering the solubility of the solid phase, and, in extreme cases, causes ice to form from the water vapour of the surrounding air. It is obvious that all these phenomena would lead to a totally false intrinsic permeability.

In general, the filter cloth should be chosen so as to retain all particles. If, for special reasons, the process envisaged imposes a lower limit on the openings of the filter medium, as in cases where it is intended to use a machine equipped with metal screens having gaps not smaller than 100 μm, then, of course, the cloth to be employed for the permeability measurement must have openings of the same order of magnitude because any much smaller openings would retain particles not retained in the machine, thus leading to a cake and a permeability not commensurate with production reality. In such exceptional cases, actually representing a classification rather than a filtration process, the result of the permeability measurement must be accompanied by a specification of the volumetric solids content of the filtrate.

Frequently, it is useful to measure the permeability as a function of C_B, the resulting graph being called the 'dynamic permeability characteristic', as this dependence may permit optimization of the process. A high centrifugal acceleration will not be advantageous if the cake compaction it causes is found to reduce the permeability to an unacceptable level.

14.4.4 Numerical example

A filter beaker set-up is characterized by $r_B = 0.16$ m and is operated at $C_B = 1000$. With a certain product, the following experimental data are obtained:

$$H_0 = 0.073 \text{ m}$$
$$H_C = 0.027 \text{ m}$$
$$\rho = 1000 \text{ kg m}^{-3}$$
$$\tau = 113 \text{ s}$$

Questions:
1. What is the number of revolutions per minute of the centrifuge?
2. What is the intrinsic permeability of the filter cake?

Solutions:
1. Application of equation 14.35 yields $\omega = 248$ s^{-1} so the number of revolutions must be $n = \dfrac{\omega}{2\pi} = 39.5$ s$^{-1} \triangleq 2370$ rev. min^{-1}.

2. Application of equation 14.59 yields κ 0.25×10^{-10} m^4 N^{-1} s^{-1}.

14.5 CENTRIFUGAL DRAINAGE

In agreement with common terminology, drainage is taken to start at the end of filtration, this moment being defined by the level of the liquid having reached the surface of the cake. Thus, drainage begins with a cake completely saturated with the liquid phase.

14.5.1 The cake model

It would be hopeless to attempt to describe a gas by the highly differing velocities, velocity directions, masses, and spatial as well as temporal distributions of its molecules. Fortunately, this is not necessary as the properties of any gas can be fully described in a simple fashion by employing two easily measured integral characteristics which are known as temperature and pressure, the adjective 'integral' being understood as representing the summary effect of a very large number of molecules.

It is equally hopeless to attempt describing a filter cake by the size, shape, and mutual arrangement of its particles, and by local geometries of its liquid phase[4-8]. The structure of the cake is much too complicated for an approach of this type. On the other hand, the filter cake as a whole can be readily characterized by three easily measured integral properties: the void space of the particle packing without the liquid, the permeability, and the volumetric degree of saturation.

To use these properties for a treatment of the drainage process, it is necessary to represent the cake by a model which is sufficiently simple to prevent getting lost in the great complexity of the many possible cake structures, but which is sufficiently flexible to permit its adaptation to reality. A model satisfying these requirements is a filter cake assumed to consist of a solid mass perforated by cylindrical pores parallel to the acting acceleration, with the diameters of these pores being defined by a continuous spectrum[9,10].

Analytically, the pore diameters are taken to be distributed according to the function

$$g(s) = \frac{dn_s}{ds} \tag{14.61}$$

where dn_s is the number of pores with diameters between s and $(s + ds)$ per unit area perpendicular to the pore axes. Hence

$$g(s)ds \frac{s^2\pi}{4} \tag{14.62}$$

is the porosity contribution due to pores with diameters between s and $(s + ds)$ so that

$$\int_0^\infty \frac{s^2\pi}{4} g(s)ds = \epsilon \tag{14.63}$$

with ϵ as the porosity of the cake. On this basis, a 'Boltzmann-type distribution of pore diameters' can be defined as

$$g(s) = \alpha \frac{s}{\hat{s}^2} e^{-\frac{1}{2}\left(\frac{s}{\hat{s}}\right)^2} \tag{14.64}$$

where α is a factor still to be determined and where $\hat{s}$ represents the 'predominant pore diameter', i.e. the pore diameter for which the distribution function exhibits a maximum. Introduction of equation 14.64 into equation 14.63 leads to the condition

$$\frac{\pi}{4} \int_0^\infty \frac{\alpha s^3}{\hat{s}^2} e^{-\frac{1}{2}\left(\frac{s}{\hat{s}}\right)^2} ds = \epsilon \tag{14.65}$$

Since

$$\int_0^\infty s^3 e^{-ks^2} ds = \frac{1}{2k^2} \tag{14.66}$$

equation 14.65 assumes the form

$$\frac{\pi\alpha}{4\hat{s}^2}\cdot\frac{1}{2}\cdot 4\hat{s}^4 = \epsilon \tag{14.67}$$

hence

$$\alpha = \frac{2\epsilon}{\pi\hat{s}^2} \tag{14.68}$$

so that substituting this expression into equation 14.64 yields the 'Boltz-mann-type distribution of pore diameters' as

$$g(s) = \frac{2\epsilon s}{\pi\hat{s}^4} e^{-\frac{1}{2}\left(\frac{s}{\hat{s}}\right)^2} \tag{14.69}$$

Its dimensionless form

$$g(s)\frac{\pi\hat{s}^3}{2\epsilon} = \frac{s}{\hat{s}} e^{-\frac{1}{2}\left(\frac{s}{\hat{s}}\right)^2} \tag{14.70}$$

is shown in *Figure 14.8*.

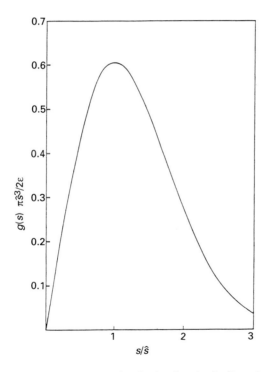

Figure 14.8. The pore diameter distribution function in dimensionless form

Inasmuch as the true distribution of pore diameters is not known, the model must now be adapted to the actual cake. This can be accomplished by expressing $\hat{s}$ in terms of the permeability and the porosity of the cake. If the height of the latter is H and if a pressure difference ΔP is applied to the boundary surfaces, the volumetric flow through a straight pore of diameter s as given by Poiseuille's law is

$$Q_s = \frac{\pi s^4 \Delta P}{128 \eta H} \tag{14.71}$$

where η is the viscosity of the liquid phase. In view of the previous definition, the number of pores with diameters between s and $(s + ds)$ per unit area of cake surface is $g(s)ds$, so that for a cake of area A the fraction of flow due to this small range of pore diameters becomes

$$d\dot{V}_{s/s+ds} = Ag(s) \, Q_s \, ds \tag{14.72}$$

Substituting equation 14.69 into 14.71 yields

$$d\dot{V}_{s/s+ds} = \frac{2\pi \epsilon A \Delta P s^5}{128 \pi \eta H \hat{s}^4} e^{-\frac{1}{2}\left(\frac{s}{\hat{s}}\right)^2} ds \tag{14.73}$$

Hence, the total volumetric flow through all pores is obtained by integrating equation 14.73 from $s = 0$ to $s = \infty$ which leads to

$$\dot{V} = \frac{\epsilon A \Delta P}{64 \eta H \hat{s}^4} \int\limits_0^\infty s^5 e^{-\frac{1}{2}\left(\frac{s}{\hat{s}}\right)^2} ds \tag{14.74}$$

Since

$$\int\limits_0^\infty x^5 e^{-\lambda x^2} \, dx = \frac{-1}{2\lambda^3} e^{-\lambda x^2} (\lambda^2 x^4 + 2\lambda x^2 + 2) \Big|_0^\infty \tag{14.75}$$

and since

$$\lim_{x \to \infty} \frac{x^{2n}}{e^{x^2}} = 0 \tag{14.76}$$

so that

$$\int\limits_0^\infty x^5 e^{-\lambda x^2} \, dx = \frac{1}{\lambda^3} \tag{14.77}$$

equation 14.74 reduces to

$$\dot{V} = \frac{\epsilon A \Delta P \hat{s}^2}{8 \eta H} \tag{14.78}$$

For the same geometry and the same pressure difference, the filtration law gives

$$\dot{V} = \frac{\kappa A \Delta P}{\eta H} \tag{14.79}$$

where κ is the permeability of the cake. Setting equation 14.78 equal to equation 14.79 results in

$$\hat{s} = \sqrt{\frac{8\kappa}{\epsilon}}$$

(14.80)

Thus, the 'predominant pore diameter' of the model can be readily obtained from the permeability and the porosity of the actual cake.

14.5.2 Capillary effects

The liquid in a cylindrical capillary assumes an axially symmetrical curvature determined by the contact angle between the liquid and the solid phase. In the most frequent case of a contact angle smaller than 90°, the liquid surface will be concave in the direction of the acting acceleration.

If the capillary is a radial pore of an annular filter cake in a centrifuge, the pore opening at the larger radius of the cake will be in contact with a filter medium (cloth or screen). Due to the almost flat surface of the latter in a plane perpendicular to the acting acceleration, drainage will lead to the formation of flat liquid lenses on the outermost surface. The curvature of these lenses is negligible as compared with the curvature of the liquid surfaces in the pores so that the liquid in the pores may be considered as having only one curved surface. Hence, according to the well-known theory of Laplace[11], the equilibrium height of a liquid of density ρ in a pore of diameter s under the influence of an acceleration B parallel to the pore axis will be

$$h_o = \frac{4\sigma\cos\alpha}{\rho Bs}$$

(14.81)

where σ is the surface tension of the liquid and α the contact angle between the liquid and the solid phase.

Thus, for the filter cake model described above, it is obvious that for pore diameters small enough to yield

$$h_o = \frac{4\sigma\cos\alpha}{\rho Bs} > H$$

(14.82)

where H is the height of the cake, a drainage process cannot take place at all. In other words, there is a 'critical pore diameter'

$$s^* = \frac{4\sigma\cos\alpha}{\rho BH}$$

(14.83)

below which the pores remain completely filled with the liquid phase.

On this basis, the liquid coalescing around the points of contact between the particles of the real cake and the liquid covering the surface of these particles as a thin film are represented in the model by the liquid in pores which are so small in diameter that their drainage is prevented by capillary forces.

14.5.3 Kinetics of drainage

The analysis of the drainage process begins with the consideration of a single pore of diameter $2b$ and height H, parallel to the acting acceleration B, and initially filled with the liquid phase throughout its length. The diameter of the pore is chosen such that drainage is possible, i.e. the diameter is assumed to be greater than s^* of equation 14.83. Under the influence of the centrifugal acceleration B, the liquid in the pore will be set in motion. After a time interval t, the height of the liquid in the pore is described by $y(t) < H$. There are five forces to be considered:

1. the centrifugal force resulting from the rotation of the basket;
2. the capillary force resulting from the curvature of the upper surface;
3. the rheoligigal force resulting from the viscosity of the liquid;
4. the Coriolis force resulting from the centrifugal movement in a pore;
5. the inertia force resulting from any change in velocity parallel to the pore axis.

Force (4) can be disregarded as it acts in a direction perpendicular to the pore wall, and force (5) is certainly very small as compared to force (1). Consequently, the velocity w is obtained from the differential equation

$$r^2\pi y\rho B = r^2\pi \frac{2\sigma\cos\alpha}{b} - 2\pi r y\eta \frac{dw}{dr} \qquad (14.84)$$

where r is the radius coordinate starting at the pore axis. Integration yields the radial velocity distribution

$$w(r) = \frac{b^2 - r^2}{4\eta y}\left(y\rho B - \frac{2\sigma\cos\alpha}{b}\right) \qquad (14.85)$$

The volume flow between the radii r and $(r + dr)$ during the time interval t is

$$dV_{r/r+dr} = 2\pi r dr w dt \qquad (14.86)$$

Hence, substituting equation 14.85 into equation 14.86 results in

$$dV_{r/r+dr} = \frac{2\pi dt}{4\eta y}\left(y\rho B - \frac{2\sigma\cos\alpha}{b}\right)(b^2 r dr - r^3 dr) \qquad (14.87)$$

Integration between $r = 0$ and $r = b$ leads to

$$dV = \frac{\pi b^4 dt}{8\eta y}\left(y\rho B - \frac{2\sigma\cos\alpha}{b}\right) \qquad (14.88)$$

When this volume element leaves the pore, the height of the liquid column diminishes. This can be expressed by

$$dV = -b^2\pi dy \qquad (14.89)$$

Of course, the actual flow in the pore will be rather complicated as regions close to the pore axis move more rapidly than regions close to the wall. Consequently, the upper surface of the liquid tends towards a radius of curvature which is smaller than the value required by the capillary action. This leads to a radial flow from the wall to the pore axis to counteract any excess concavity. Nevertheless, these phenomena in no way affect the validity of equation 14.89 as an integral concept. Substitution of equation 14.88 into equation 14.89 yields

$$dy = \frac{b^2}{8\eta y} \left(\frac{2\sigma\cos\alpha}{b} - y\rho B \right) dt \qquad (14.90)$$

Introduction of

$$\frac{2\sigma\cos\alpha}{b\rho BH} = \zeta \qquad (14.91)$$

and

$$\frac{b^2 \rho B}{8\eta} = \beta \qquad (14.92)$$

leads to the dimensionless form

$$\frac{\frac{y}{H} d \left(\frac{y}{H} \right)}{\zeta - \frac{y}{H}} = d \left(\frac{\beta}{H} t \right) \qquad (14.93)$$

and integrated to

$$\frac{\beta}{H} t = -\frac{y}{H} - \zeta \ln \left(\frac{y}{H} - \zeta \right) + C_1 \qquad (14.94)$$

As the drainage process begins with $y = H$ at $t = 0$, it is found that

$$C_1 = 1 + \zeta \ln (1 - \zeta) \qquad (14.95)$$

Hence,

$$\frac{\beta}{H} t = 1 - \frac{y}{H} + \zeta \ln \frac{1 - \zeta}{\frac{y}{H} - \zeta} \qquad (14.96)$$

Comparison of equation 14.91 with equation 14.81 shows that

$$\zeta = \frac{h_o}{H} \qquad (14.97)$$

which converts equation 14.96 to

$$\frac{\beta}{H}t = 1 - \frac{y}{H} + \frac{h_o}{H} \ln \frac{1 - \frac{h_o}{H}}{\frac{y}{H} - \frac{h_o}{H}} \tag{14.98}$$

This equation correlates the momentary height, y, of the liquid column in the pore with the time, t, required to reach this state. However, equation 14.98 gives t as a function of y but does not permit y to be expressed as an explicit function of t. Inasmuch as it is this latter form which is needed to derive the moisture content as a function of time, equation 14.98 must be replaced by an appropriate surrogate function. This can be done by an exponential relationship that matches equation 14.98 in both the slope at $t = 0$ and in the asymptote approached by the curve at large values of t. It is readily seen that the curve described by equation 14.98 begins at $y/H = 1$ and that it has an asymptote at $y/H = h_o/H$. Thus, agreement in both the starting point and the asymptote is provided by a surrogate function of the general form

$$\frac{y}{H} = G\, e^{-\frac{\beta}{H}t} + \frac{h_o}{H} \tag{14.99}$$

where G is a still unknown factor which can be determined from the condition of slope equality at $t = 0$. It is found that

$$G = 1 - \frac{h_o}{H} \tag{14.100}$$

so that the surrogate function assumes the form

$$\frac{y}{H} = \left(1 - \frac{h_o}{H}\right) e^{-\frac{\beta}{H}t} + \frac{h_o}{H} \tag{14.101}$$

A graphical representation of equations 14.98 and 14.101 is shown in *Figure 14.9*. As can be seen, the agreement between the exact and the approximate drainage functions is by no means perfect, but for a model concept the deviations are certainly acceptable. Besides, it is to be noted that drainage according to the exact function is somewhat faster than in the case of the surrogate function so that any conclusion based on the latter will be conservative.

Based on these considerations for a single pore, it is now possible to treat the drainage process of the total filter cake. To this end, the pores are subdivided into two groups: one group comprising the pores which do not drain at all, and the other group comprising all the other pores. Hence, the 'cut' between the two groups is defined by the 'critical pore diameter' given by equation 14.85. For a cake of area A and height H, the volume of liquid

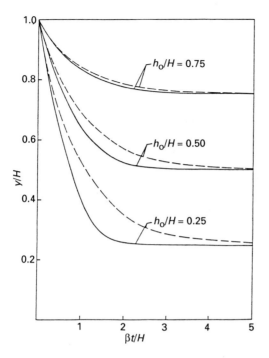

Figure 14.9. Comparison of exact (————) and approximate (----) drainage functions

contained in the pores of the first group is

$$V\bigg|_0^{s*} = \int_0^{s*} g(s) \frac{s^2 \pi}{4} HA \, ds \tag{14.102}$$

In view of equation 14.69, this converts to

$$V\bigg|_0^{s*} = \frac{\epsilon AH}{2\tilde{s}^4} \int_0^{s*} s^3 e^{-\frac{1}{2}\left(\frac{s}{\tilde{s}}\right)^2} \, ds \tag{14.103}$$

Analogously, the liquid in the pores of the second group is found to be

$$V\bigg|_{s*}^{\infty}(t) = \int_{s*}^{\infty} g(s) \frac{s^2 \pi}{4} y(t) A \, ds \tag{14.104}$$

Using equation 14.101, the latter equation assumes the form

$$V\bigg|_{s*}^{\infty}(t) = \frac{\epsilon AH}{2\hat{s}^4} \int_{s*}^{\infty} s^3 e^{-\frac{1}{2}\left(\frac{s}{\hat{s}}\right)^2} \left[\left(1 - \frac{h_o}{H}\right) e^{-\frac{\beta}{H}t} + \frac{h_o}{H}\right] ds \quad (14.105)$$

Since

$$\int_{0}^{s*} s^3 e^{-ks^2} ds = \frac{1}{2k^2} - \frac{ks*^2 + 1}{2k^2} e^{-ks*2} \quad (14.106)$$

equation 14.103 yields

$$V\bigg|_{0}^{s*} = \epsilon HA \left[1 - \left\{\frac{s*^2}{2\hat{s}^2} + 1\right\} e^{-\frac{1}{2}\left(\frac{s}{\hat{s}}\right)^2}\right] \quad (14.107)$$

In view of equations 14.81 and 14.83 and with the substitution

$$\frac{\rho B}{32\eta H} = \theta \quad (14.108)$$

equation 14.105 can be rewritten as

$$V\bigg|_{s*}^{\infty}(t) = \frac{\epsilon AH}{2\hat{s}^4} \left[\int_{s*}^{\infty} s^3 e^{-s^2\left(\frac{1}{2\hat{s}^2} + \theta t\right)} ds - \right.$$

$$\left. s* \int_{s*}^{\infty} s^2 e^{-s^2\left(\frac{1}{2\hat{s}^2} + \theta t\right)} ds + s* \int_{s*}^{\infty} s^2 e^{-s^2\frac{1}{2\hat{s}^2}} ds \right] \quad (14.109)$$

Introducing

$$\frac{1}{2\hat{s}^2} + \theta t = \phi \quad (14.110)$$

and

$$\frac{1}{2\hat{s}^2} = \psi \quad (14.111)$$

equation 14.109 which reduces to

$$V\Big|_{s^*}^{\infty}(t) = \frac{\epsilon AH}{2\tilde{s}^4}\left[\int_{s^*}^{\infty}s^3 e^{-\phi s^2}\,ds - s^*\int_{s^*}^{\infty}s^2 e^{-\phi s^2}\,ds + s^*\int_{s^*}^{\infty}s^2 e^{-\psi s^2}\,ds\right] \quad (14.112)$$

The first integral in the bracket is

$$I_1 = \frac{1}{2\phi}\left(s^{*2} + \frac{1}{\phi}\right)e^{-\phi s^{*2}} \quad (14.113)$$

and the second integral can be written as

$$I_2 = \frac{s^*}{2\phi}e^{-\phi s^{*2}} + \frac{1}{2\phi}I \quad (14.114)$$

where

$$I = \int_{s^*}^{\infty}e^{-(\sqrt{\phi}s^*)^2}\,ds \quad (14.115)$$

The additional substitution

$$\sqrt{\phi}s = u \quad (14.116)$$

leads to

$$I = \frac{1}{\sqrt{\phi}}\int_{\sqrt{\phi}s^*}^{\infty}e^{-u^2}\,du \quad (14.117)$$

Since

$$\int_{\sqrt{\phi}s^*}^{\infty}e^{-u^2}\,du = \frac{1}{2}\int_{-\infty}^{\infty}e^{-u^2}\,du - \int_{0}^{\sqrt{\phi}s^*}e^{-u^2}\,du = \frac{\sqrt{\pi}}{2} - \int_{0}^{\sqrt{\phi}s^*}e^{-u^2}\,du \quad (14.118)$$

equation 14.117 becomes

$$I = \frac{\sqrt{\pi}}{2\sqrt{\phi}} \left[1 - \frac{2}{\sqrt{\pi}} \int_0^{\sqrt{\phi}s^*} e^{-u^2} du \right] = \frac{\sqrt{\pi}}{2\sqrt{\phi}} \left[1 - erf\,(\sqrt{\phi}s^*) \right]$$

(14.119)

where

$$\frac{2}{\sqrt{\pi}} \int_0^x e^{-u^2} du = erf\,(x)$$

(14.120)

represents the 'error function' of x which is available in tabulated form[12,13]. With equation 14.120, equation 14.114 converts to

$$I_2 = \frac{s^*}{2\phi} e^{-\phi s^{*2}} - \frac{\sqrt{\pi}}{\sqrt{4\phi}\,\sqrt{\phi}} \left[1 - erf\,(\sqrt{\phi}s^*) \right]$$

(14.121)

An analogous formulation of the third integral of equation 14.112 leads to

$$V \Big/_{s^*}^{\infty} (t) = \frac{\epsilon A H}{2\hat{s}^4} \left[(s^{*2} + \frac{1}{\phi})e^{-\psi s^{*2}} \right]$$

$$- \frac{s^{*2}}{2\psi} e^{-\psi s^{*2}} - \frac{s^* \sqrt{\pi}}{4\psi\,\sqrt{\phi}} \left\{ 1 - erf\,(\sqrt{\phi}s^*) \right\}$$

$$+ \frac{s^{*2}}{2\psi} e^{-\psi s^{*2}} + \frac{s^* \sqrt{\pi}}{4\psi\,\sqrt{\phi}} \left\{ 1 - erf\,(\sqrt{\phi}s^*) \right\} \Big]$$

(14.122)

In view of equations 14.110, 14.83 and 14.80

$$\psi s^{*2} = \frac{s^{*2}}{2\hat{s}^2} = \frac{\epsilon \sigma^2 \cos^2\alpha}{\kappa B^2 \rho^2 H^2}$$

(14.123)

which is abbreviated by $1/D$ so that

$$D = \frac{\kappa B^2 \rho^2 H^2}{\epsilon \sigma^2 \cos^2\alpha}$$

(14.124)

With equations 14.110, 14.83 and 14.108

$$\phi s*^2 = \frac{\epsilon\sigma^2\cos^2\alpha}{\kappa B^2\rho^2 H^2} - \frac{\sigma^2\cos^2\alpha}{2\eta H^3\rho B}$$

(14.125)

By introducing

$$\mu = \frac{\sigma^2\cos^2\alpha}{2\eta H^3\rho B}$$

(14.126)

and with due consideration of equation 14.124

$$\phi s*^2 = \frac{1}{D} + \mu t$$

(14.127)

The total volume of liquid contained in the cake at the time t is

$$V_R(t) = V\Big/_0^{s*} + V\Big/_{s*}^{\infty}$$

(14.128)

so that introducing equations 14.107 and 14.122 and using equations 14.123, 14.124 and 14.127 leads to

$$\frac{V_R(t)}{\epsilon AH} = 1 - e^{-\frac{1}{D}} + \frac{\sqrt{\pi}\,\sqrt{D}}{2D}\left\{1 - \text{erf}\,\frac{1}{\sqrt{D}}\right\} + \frac{1}{D^2\left(\frac{1}{D} + \mu t\right)}\left[\frac{1}{\frac{1}{D} + \mu t}\right.$$

$$\left. e^{-\left(\frac{1}{D} + \mu t\right)} - \frac{\sqrt{\pi}\left\{1 - \text{erf}\sqrt{\frac{1}{L}} + \mu t\right\}}{2\sqrt{\frac{1}{D} + \mu t}}\right]$$

(14.129)

By combining all constant terms and all terms depending on the time, equation 14.129 can be written in the form

$$\frac{V_R(t)}{\epsilon AH} = \frac{V_\infty}{\epsilon AH} + \frac{A_\Delta(t)}{\epsilon AH}$$

(14.130)

where

$$\frac{V_\infty}{\epsilon AH} = 1 - e^{-\frac{1}{D}} + \frac{\sqrt{\pi}\,\sqrt{D}}{2D}\left[1 - \text{erf}\,\frac{1}{\sqrt{D}}\right]$$

(14.131)

represents the relative content of liquid after an infinitely long drainage time, whereas $V_\Delta(t)/\epsilon AH$ is the fraction of liquid still in the cake at time t but susceptible to drainage by longer processing. Thus, V_∞ is the volume of liquid that cannot be drained, and V_Δ the volume exceeding V_∞. A graphical representation of these terms is given in *Figure 14.10*.

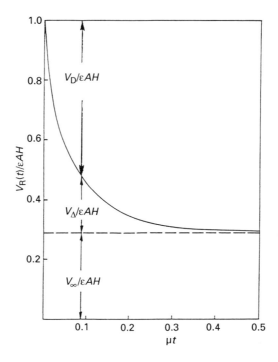

Figure 14.10. The drainage characteristic for D = 10

In the centrifuge industry, this type of diagram is called the 'drainage characteristic' of the cake to be processed. According to *Figure 14.10*, the relative portion of liquid removed from the cake by centrifugal drainage is

$$\frac{V_D(t)}{\epsilon AH} = 1 - \left\{\frac{V_\infty}{\epsilon AH} + \frac{V_\Delta(t)}{\epsilon AH}\right\} = 1 - \frac{V_R(t)}{\epsilon AH} \qquad (14.132)$$

so that

$$V_D(t) = \epsilon AH - V_R(t) \qquad (14.133)$$

A graphical representation of equation 14.133 is known as the 'drainage part of the filtrate flow diagram'. With a set-up such as that shown in *Figure 14.7*, the 'drainage part' is routinely obtained by recording the weight of the filtrate as a function of time, using an electronic balance.

Equation 14.129 shows that the relative volume of liquid in the cake depends only on D, μ and the time t. As the centrifugal acceleration in the cake does not vary significantly, it can be expressed by the constant term

$$B = r_B \omega^2 \qquad (14.134)$$

where r_B is the inner radius of the basket and ω its angular velocity, so that substitution of equation 14.134 into equation 14.124 gives

$$D = \frac{\kappa r_B \omega^4 \rho^2 H^2}{\epsilon \sigma^2 \cos^2 \alpha} \tag{14.135}$$

where D is a dimensionless quantity called the 'centrifugal drainage number'.

According to equation 14.131, the terminal degree of saturation, $\zeta_\infty = V_\infty / \epsilon A H$ is a function of D only. A graphical representation of this important dependence is shown in *Figure 14.11*. As can be seen, the larger the value of D, the smaller the residual moisture content of the cake after an infinitely long drainage. On this basis, D gives a useful indication of what centrifugal treatment can achieve with a given system. In addition, the expression for D according to equation 14.135 permits the effects of the various terms governing the drainage process to be appraised. These terms are:

1. the centrifugal acceleration as the characteristic of the machine;
2. porosity, permeability, and height as characteristics of the cake;
3. density, surface tension, and contact angle as characteristics of the liquid phase.

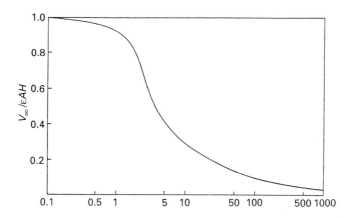

Figure 14.11. *The relative terminal liquid volume in a filter cake as a function of the centrifugal drainage number D*

For a drainage period of finite duration t, the result of the process is given by equation 14.129. As can be seen, in this case the relative residual moisture content depends on three quantities: t, D, and μ. The physical meaning of μ is readily recognized if equation 14.126 is divided by the square of 14.83 leading to

$$\frac{\mu}{s^{*2}} = \frac{\rho B}{32 \eta H} \tag{14.136}$$

so that

$$\mu = \frac{\left(\frac{s^*}{\lambda}\right)^2 \rho B}{8\eta H} \tag{14.137}$$

Inasmuch as the introduction of equation 14.92 into equation 14.101 shows the time dependence of the level of liquid in a pore of radius b to be

$$y(t) = (H - h_o) e^{-\frac{b^2 \rho B}{8\eta H} t} + h_o \tag{14.138}$$

it is seen that μ is the time factor for the level drop in a pore of 'critical diameter' s^*. As there is no drainage in pores with diameters smaller than s^*, μ is the smallest time factor possible.

For obvious reasons, in the given treatment of drainage kinetics exclusive use has been made of the volumetric degree of saturation $\psi = V_R/\epsilon AH$. For practical applications, this term can be readily transformed to the customary gravimetric moisture content, λ, by considering that

$$\lambda = \frac{1 - \nu}{1 + \frac{\rho_S (1 - \epsilon)}{\rho_L \epsilon \psi}} \tag{14.139}$$

where ν is the weight fraction of solids dissolved in the liquid phase, and where ρ_s and ρ_L are the densities of the solid and liquid phases respectively.

14.5.4 Effects of surfactants

It is well known that the surface tension of many liquids can be significantly diminished by suitable additives. Substances capable of producing this effect even when added in very small quantities are called 'surfactants'. With water having a relatively high surface tension, surfactants for water are of particular interest.

According to equation 14.133, lowering the surface tension greatly increases the centrifugal drainage number, D, as the latter is inversely proportional to the square of the surface tension. Thus, in view of *Figure 14.11* it can be expected that the addition of surfactants lowers the terminal degree of saturation, but that the relative effect will depend on the absolute value of D. A maximum improvement in cake drainage by surfactants is to be expected in the range where the curve of *Figure 14.11* exhibits its maximum slope, i.e. for D values between 0.5 and 10, whereas for D values far below 0.1 there will be no effect at all.

Conclusions of this type can be checked by means of experimental data available in the literature[14]. These results refer to centrifugal drainage of quartz sand with particle sizes between 0.2 and 2.0 mm. The cake porosity ϵ was 0.42 and the centrifugal acceleration 6g. After drainage of distilled water from this sytem, the terminal moisture content λ_D as referred to the dry solids was found to be 5.7% by weight. In view of

$$\psi = \frac{\lambda_D \rho_s (1 - \epsilon)}{\rho_L \epsilon}$$

(14.140)

the terminal degree of saturation with distilled water was

$$\psi_\infty \text{ (distilled water)} = 0.154$$

After the surface tension of the water was lowered by the factor of 0.63 by a surfactant 1 and by a factor of 0.51 by a surfactant 2, identical drainage treatment led to terminal moisture contents of 3.2% and 2.7% by weight, respectively, corresponding to

$$\psi_\infty \text{ (surfactant 1)} = 0.088$$

and ψ_∞ (surfactant 2) = 0.074

According to *Figure 14.11*, ψ_∞ = 0.154 (distilled water) leads to $D(H_2O)$ = 38. The reduction in the surface tension by the factors 0.63 and 0.51 corresponds to an increase in the centrifugal drainage number by the factors 2.52 and 3.85 respectively. Starting with $D(H_2O)$ = 38, this results in the theoretical conclusion that for the water modified by the surfactants 1 and 2 the centrifugal drainage numbers were 96 and 146. The terminal degrees of saturation corresponding to these D values according to *Figure 14.11* are exactly the same as the measured values given above, thus confirming the theory with an extraordinary accuracy.

The simple data analysis presented was possible as the surfactants 1 and 2 were anionic so that the contact angle remained unchanged. For cationic surfactants the change in contact angle would also have to be considered.

14.6 FILTER CENTRIFUGES

Filter centrifuges can be subdivided into two fundamentally dissimilar classes, namely continuously and discontinuously fed machines, the crucial distinction being the axial confinement of the feed suspension in the basket as effected by a cake of particles in the continuous machines and by a solid lip in the other cases.

14.6.1 Continuously fed machines

14.6.1.1 Pusher centrifuges

14.6.1.1.1 Design of pusher centrifuges

The most common design employing continuous feeding is the pusher centrifuge. Its principle is illustrated schematically in *Figure 14.12*. The feed stock enters the machine in the axial direction and flows over a distribution cone past a more or less vertical disc on to the filter medium of a cylindrical basket having a horizontal axis.

The cone and the disc rotate at the same angular velocity as the basket and carry out a reciprocating motion in the axial direction. In general, the filter medium consists of a bar-type metal screen, the smallest practical spacing

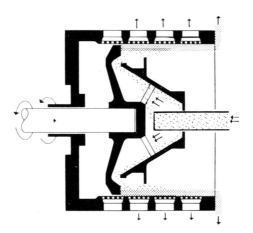

Figure 14.12. Schematic of a one-stage pusher centrifuge

between the bars being 100 μm, thus limiting the application of these machines to the separation of particles larger than this. The bars have a trapezoidal cross-section giving the screens the important feature of being self-cleaning.

The particles retained by the screen during one cycle of the reciprocating motion are pushed by the disc to its right-hand end position, thus moving in turn the solids left there in the preceding cycle. In this fashion, an annular cake is formed and intermittently advanced in axial direction towards the right-hand end of the basket, where the centrifugal acceleration throws the particles into a casing.

The liquid phase passes the screen and leaves the machine by ducts well separated from the solids discharge.

In very small machines, the reciprocating motion is created by an eccentric drive whereas in large machines hydraulic action is used exclusively. In the latter case, the stroke length, generally in the order of a few centimetres, can be adjusted by appropriate positioning of limit switches. The stroke frequency ranges between 20 and 100 min^{-1}.

In as much as the basket has no lip at the discharge end, axial retainment of the feed stock is effected by the annular cake of particles. This imposes an upper limit for the feed rate, the so-called 'overflow limit'. When at any instant, generally during the forward stroke of the disc, the feed stock introduced behind the cake rises to a height exceeding the cake thickness, the suspension will take the lowest path towards the solids discharge end of the basket, rapidly eroding a deep canyon in the cake. The phenomenon depends on four factors:

1. The percentage of liquid in the feed stock.
   If the solids concentration of the feed stock is increased, the overflow limit of the machine will rise, and if the feed stock does not exhibit any free liquid at all, there will be no overflow limit. In the latter case, the maximum processing capacity of the machine increases to what is known as the 'volumetric swallowing capacity', determined by the maximum possible cake height, the stroke length, and the stroke frequency.
2. The permeability of the screen.
   For bar-type screens, the permeability is proportional to the third power of the gap width so that doubling the gap width increases the screen permeability by a factor of 8. Consequently, increasing the gap width raises the overflow limit.
3. The angular velocity.
   Normally, the filtration rate is directly proportional to the centrifugal acceleration so that increasing the angular velocity will raise the overflow limit.
4. The length of the basket.
   The cake thickness is proportional to the basket length. Thus, increasing the basket length (e.g. by removing a pusher disc extension) will raise the overflow limit.

Instead of featuring a simple cylindrical basket, as shown in *Figure 14.12*, pusher centrifuges can be made to have multi-stage screens consisting of two or more steps of successively larger diameter, with alternating steps joining the reciprocating motion of the disc. A two-stage system is depicted schematically in *Figure 14.13*.

Multi-stage machines are particularly suitable for particles forming a soft cake or exhibiting a high frictional resistance to sliding on the screen. They

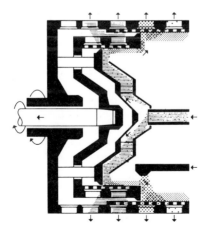

Figure 14.13. Schematic of a two-stage pusher centrifuge

also favour extensive drainage as the transit of the cake over a step leads to a reorientation of the particles. On the other hand, because of the small length of the screens, multi-stage machines have relatively low overflow limits.

All pusher centrifuges lend themselves to washing or extracting the particles by spraying a suitable liquid on the cake. After having passed the basket, this liquid can be collected in separate sections of the casing. An illustration of this process is included in *Figure 14.13*.

The enforced motion of the cake in the centrifugal field and the violent ejection over the edge of the basket against the casing leads to some attrition of the particles but in general this is considered a moderate price for the simplicity and elegance of the process.

14.6.1.1.2 The overflow limit of pusher centrifuges

As described in the preceding section, the feed rate of a pusher centrifuge is limited by the so-called 'overflow limit' which is reached when the feed stock introduced behind the cake rises to a height exceeding the cake thickness. In this case, the suspension will take the lowest path towards the solids discharge end of the basket, rapidly eroding a deep canyon in the cake. To find the parameters governing this phenomenon, the feed process must be studied in some more detail. This process is assumed to begin after the pusher disc has reached its most forward position, having moved the cake by a certain increment. It is assumed that at this stage the inner surface of the cake is a smooth cylinder. The solids configuration existing at this moment is called the 'principal cake'.

The feed process can be logically subdivided into three periods:

I The period covering the backward stroke of the disc. During this interval, the feed stock is introduced into the annular ring formed by the principal cake, the screen, and the retracting disc. The new cake thus formed in this region is called the 'initial layer'. The final phase of this period is shown schematically in *Figure 14.14*.

II The period covering the first portion of the forward stroke as characterized by a continuous deformation of the initial layer and the solids placed thereon, as well as by complete axial immobility of the principal cake. The sum of the deformed initial layer and of the solids placed thereon is called the 'buffer cake'. The final stage of this period is illustrated schematically in *Figure 14.15*.

III The period covering the second portion of the forward stroke as characterized by the axial motion of both the buffer cake and the principal cake. During this period, the buffer cake is no longer being deformed but its height gradually increases up to the height of the principal cake as further feed stock is added. The final phase of this period is depicted schematically in *Figure 14.16*.

Obviously, the transition from period II to period III is defined by the principal cake beginning to move in axial direction. As far as overflow is concerned, the consideration can be limited to period III since this phenomenon will not occur as long as the incoming stream of liquid does not exceed the

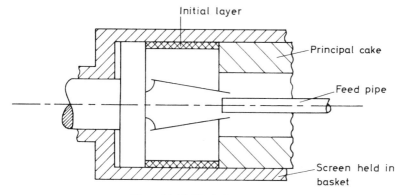

Figure 14.14. *Final phase of period I*

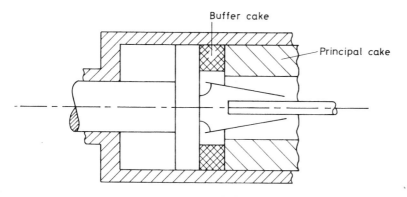

Figure 14.15. *Final phase of period II*

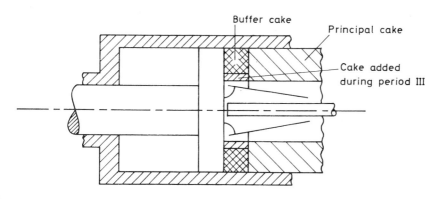

Figure 14.16. *Final phase of period III*

filtration rate permitted by the buffer cake. This condition can be readily formulated.

If $\dot{V}_S$ is the volumetric feed rate of the suspension to be treated, and if c is the volumetric solids content of the feed stock as determined by dividing the apparent volume of solids by the corresponding volume of suspension, the incoming stream of 'free' liquid is

$$\dot{V}_L = (1 - c)\dot{V}_S \tag{14.141}$$

Liquid 'stored' in the interstices between the particles of the cake can be disregarded as it appears on both sides of the balance to be formulated, thus dropping out by cancellation.

Taking into account that in applications of pusher centrifuges it is customary to express the processing capacity in terms of dry weight of solids processed per unit of time, it is appropriate to introduce this rate as

$$\dot{W} = \rho_A g \dot{V}_{Solid} \tag{14.142}$$

where ρ_A is the apparent density of the dry solids (the so-called 'packing density') as expressed in mass per unit of apparent volume taken up by the pack of particles, and $\dot{V}_{Solid}$ is the rate of apparent volume of solids processed.

Since from the definition of c

$$\dot{V}_{Solid} = c\dot{V}_S \tag{14.143}$$

substituting equations 14.143 and 14.142 in equation 14.141 yields

$$\dot{V}_L = \frac{(1 - c)\dot{W}}{c\rho_A g} \tag{14.144}$$

On the other hand, if the filtration resistance of the screen is disregarded as small in comparison with the filtration resistance of the cake, the rate of flow through the cake when the latter has reached its final thickness and when it is not covered by excess liquid can be expressed using equation 14.36 as

$$\dot{V}_F = \frac{2\pi\kappa j Z \rho_L g C_B H_C}{\ln\left[D/(D - 2H_C)\right]} \tag{14.145}$$

where

j is a dimensionless number between 0 and 1 indicating the fraction of the stroke length corresponding to period III
Z is the stroke length
H_C is the final height of the cake
and
D is the diameter of the basket.

The denominator of equation 14.145 can be converted to

$$\ln\frac{1}{1 - (2H_C/D)} \approx \ln\left(1 + \frac{2H_C}{D}\right) \approx \frac{2H_C}{D} \tag{14.146}$$

yielding

$$\dot{V}_F = \pi \kappa j Z \rho_L g D C_B \tag{14.147}$$

The critical overflow condition is obtained by equating equations 14.144 and 14.147 leading to

$$\dot{W}* = \frac{\pi c \rho_A \rho_L g^2 \kappa j Z D C_B}{1 - c} \tag{14.148}$$

where $\dot{W}*$ is the maximum possible rate of dry solids that can be processed before overflow will occur.

As can be seen, $\dot{W}*$ is directly proportional to the permeability of the cake which explains why solids of small permeability cannot be processed on pusher centrifuges at reasonable rates.

14.6.1.1.3 The maximum volumetric swallowing capacity of pusher centrifuges

According to equation 14.148, the function governing the dependence of the overflow limit on the volumetric solids content of the feed stock is

$$F = \frac{c}{1 - c} \tag{14.149}$$

It is called the 'concentration factor'. As for c approaching unity this factor becomes infinite, it is obvious that another limiting condition must prevail in this range.

Consideration of the discharge process shows the volumetric solids output to be

$$\dot{V}_{Solid} = \tfrac{1}{4}[D^2 - (D - 2H_c)^2]\pi Z f \tag{14.150}$$

where f is the stroke frequency.

As the feed stock enters the machine in axial direction from the open end of the basket, thus requiring at least a pipe or a screw conveyor if not a distribution cone as well, the cake height cannot be increased arbitrarily. Introduction of its limiting value $\hat{H}_c$ in equation 14.150 leads to the so-called 'maximum volumetric swallowing capacity' $\hat{V}_{solid}$ of the machine. Thus

$$\hat{V}_{Solid} = D\hat{H}_C\left(1 - \frac{\hat{H}_C}{D}\right)\pi Z f \tag{14.151}$$

and using equation 14.142

$$\hat{W} = \rho_A g D\hat{H}_C\pi Z f \left(1 - \frac{\hat{H}_C}{D}\right) \tag{14.152}$$

14.6.1.1.4 The capacity characteristic of pusher centrifuges

Equations 14.148 and 14.152 can be combined into a single curve, with the understanding that equation 14.152 overrides equation 14.148 for all c-values which cause $\dot{W}^*$ to exceed $\dot{W}$. The resulting graph is called the 'capacity characteristic' of the machine for the product considered. An example is shown in *Figure 14.17*.

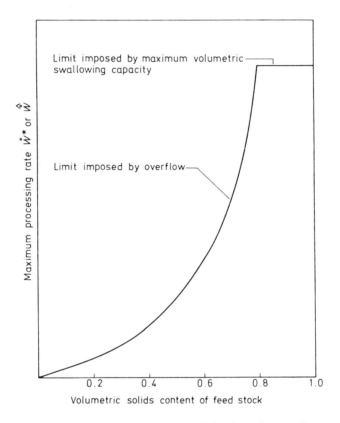

Figure 14.17. The capacity characteristic of a pusher centrifuge

14.6.1.1.5 Conversion procedure for pusher centrifuges

Equation 14.148 can be used to compute the unknown overflow limit of a machine 1 for a volumetric solids content c_1 from a known overflow limit determined experimentally on a machine 2 using the same type of solids but at a volumetric concentration c_2 not necessarily equal to c_1. Formulating equation 14.148 with the indices 1 and 2, respectively, and considering that in view of basic similarities among pusher centrifuges no great error is committed by setting

$$j_1 \approx j_2 \qquad (14.153)$$

it is found that

$$\dot{W}_1^* = \dot{W}_2^* \frac{c_1(1 - c_2)}{c_2(1 - c_1)} \frac{Z_1}{Z_2} \frac{D_1}{D_2} \frac{C_{B_1}}{C_{B_2}} \qquad (14.154)$$

As can be seen, for such conversions it is not necessary to know the many physical properties contained in equation 14.148 as they drop out of the calculation. Thus, a quick overflow test on a laboratory machine and basic handbook data for the geometric and kinematic parameters provide all of the information required to compute the capacity characteristics for other machines.

Needless to say, so far as any capacity characteristic of a pusher centrifuge is governed solely by the feed region, the derivations presented are equally applicable to one-stage and multistage machines.

14.6.1.1.6 Numerical example

A laboratory pusher centrifuge having a basket diameter of 0.160 m and a stroke length of 0.025 m was operated at 3000 rev. min^{-1}. When fed at gradually increasing rates using a suspension having a solids content of 25% by apparent volume, corresponding to 12% by weight, the overflow was found to occur at an input rate of 0.750 m^3 h^{-1}. The density of the suspension was 1200 kg m^{-3}.

Question:
What is the processing capacity of a production machine having a basket diameter of 0.8 m, a stroke length of 0.070 m, and a rotational speed of 1200 rev. min^{-1} if by using a simple thickener the solids concentration of the feed stock is increased to 75%?

Solution:

$c_2 = 0.25$ $\qquad \dot{W}^* = 1060$ N h^{-1} $\qquad c_1 = 0.75$

From equation 14.35

$$C_{B_2} = 800 \qquad C_{B_1} = 640$$

Thus using equation 14.154 yields

$$\dot{W}_1^* = 107\,000 \text{ N h}^{-1}.$$

14.6.1.2 Conveyor discharge centrifuges

Instead of using the pusher action, it is equally possible to effect the axial solids discharge by means of a helical conveyor turning slightly faster than the basket. Such a discharge system can be employed for both cylindrical and conical baskets. The tumbling action of the scroll is claimed to help drainage but crystal breakage is much greater than with the pusher mechanism.

An example of such a machine is shown in *Figure 14.18*. The basket axis is vertical with the large diameter downward, and the feed stock is introduced at the small end. The differential speed of the conveyor controls to some extent the rate at which the solids move through the drainage zone.

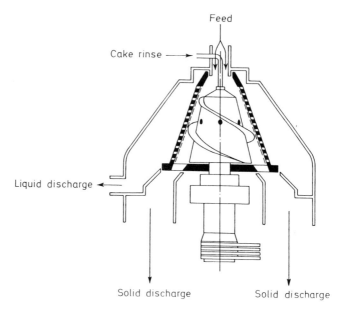

Figure 14.18. Schematic of a conical conveyor discharge centrifuge

14.6.1.3 Centrifuges discharging by mass forces

In machines with conical baskets, the scroll can be replaced by an axial vibratory motion of the screen. The combined action of the axial and centrifugal accelerations leads to a transport of the particles towards the large diameter of the basket.

Finally, if the basket is conical and the angle of the cone is sufficiently large to permit the cake to overcome its friction on the screen, the machine becomes self-discharging. This leads to a centrifuge of impressive simplicity, but it has a great disadvantage in that different products require different cone angles. Besides, the designer is forced to march on a narrow ridge. If the cone angle is too small, the cake will not discharge, and if the cone angle is too large, the residence time of the particles on the screen becomes too short, thus impairing drainage.

Self-discharging centrifuges are primarily used in the sugar industry, where the product is always the same so that the cone angle is well established. A typical example of such a machine is shown in *Figure 14.19*. The basket consisting of a conical metal screen has a vertical axis, with the large diameter being at the top. The feed stock introduced at the bottom and accelerated by a distributor slides over the inner conical surface and is ejected over the upper rim. During its passage over the screen, the sugar is drained of syrup and can be washed with water or steam. Sugar and syrup leave the machine in a downward direction via separate annular compartments.

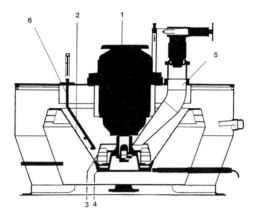

Figure 14.19. Self-discharging sugar centrifuge: 1, motor; 2, basket; 3, feed distributor; 4, screen clamping ring; 5, feed pipe; 6, wash pipe

14.6.2 Discontinuously fed machines

Batch-type filter centrifuges are much older in conception than continuous machines, but they are by no means obsolete. On the contrary, the total number annually put in operation, not even counting household machines, far exceeds the corresponding figure for the continuous types. The reason for this lies in the fact that the predominant number of suspensions to be filtered involves particles considerably smaller than the practical lower limit of 100 μm quoted above for pushers. After all, large crystals are more the exception than the rule.

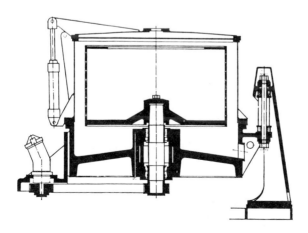

Figure 14.20. Schematic of a three-column centrifuge for manual solids discharge

The $100 \mu m$ limit of pushers is associated with the self-cleaning bar-type metal screens employed in those machines. It is not possible to manufacture such screens with gap widths of the order of 1 μm or less. This size range is the domain of cloths in which particularly fine openings are obtained by employing overlap weaves. As a consequence of this, such filter media are not self-cleaning but rather tend to be plugged up, especially if a cake of small particles is moved over their surface while being pressed against it by strong centrifugal action. Consequently the principal requirements for centrifugal filtration involving very small particles are: use of a cloth as separation agent, and no movement of the cake with respect to the cloth. Obviously, this can only be done in a batch-type machine.

14.6.2.1 Three-column centrifuges

The simplest and most common form of a batch-type machine consists of a cylindrical basket with a vertical axis, a closed bottom, and a partially closed top, the partial closure being called the lip. The basket is suspended on three columns and because of this characteristic it is called a three-column centrifuge. A schematic of this design is shown in *Figure 14.20*.

Generally, the mantle surface of the basket is perforated with a large number of holes covered on the inside by one or more coarse metal screens. The latter serve as a back-up for the filter cloth which can have the shape of a bag to be clamped into the basket. The machine is charged from the top, before or after starting the rotation. The filtrate escapes through the mantle surface and the particles remain on the filter cloth.

In the simple case shown in *Figure 14.20*, the discharge of the solids is effected by stopping the machine and manually removing the cake or replacing the bag.

A more advanced version of the three-column machine, depicted in *Figure 14.21*, allows the solids to be discharged by moving a plough into the cake after having slowed down the basket to a few revolutions per minute. The plough directs the solids towards the axis, where they fall through the bottom openings of the basket. The procedure is fully automated by means of timers, a typical cycle involving the following phases:

1. acceleration to a medium speed;
2. introduction of the feed stock at the medium speed;
3. filtration of the charge at the medium speed;
4. acceleration to a high speed;
5. drainage at the high speed;
6. deceleration to a very low speed;
7. removal of the cake by ploughing at the low speed.

Noteworthily, the plough cannot be moved all the way to the filter cloth as the latter is never perfectly smooth. Consequently, the filter cake is not removed entirely. A thin layer of particles, commonly called the 'residual heel', remains in the basket. Where a residual heel is not acceptable, ploughing can be replaced by a pneumatic cake removal system employing either pressure jets or a vacuum. Obviously, for such a system it is necessary that the cake

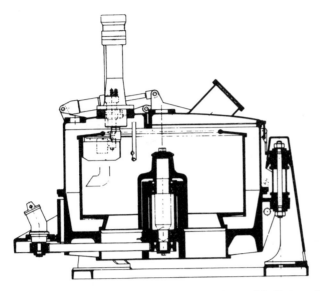

Figure 14.21. Schematic of a three-column centrifuge for a solids discharge by ploughing

disintegrates readily and that the particles are not sticky.

14.6.2.2 Peeler centrifuges

Because three-column machines have a vertical axis, the effect of gravity tends to lead to a non-uniform cake, its lower portion being thicker and comprising larger particles than the upper portion. If washing is required, this non-uniformity of the cake can cause variations in the purity of the solids discharged. In such cases, ti is preferable to employ a machine which has a horizontal axis, as shown in *Figure 14.22*. This type, representing the most highly developed filter centrifuge, allows the cake to be removed at full speed by means of a sturdy knife commonly called a 'peeler'. Because of this feature, the machines are termed 'peeler centrifuges.' In general, peeler centrifuges are operated at constant speed. Thus, they do not require any non-productive periods for acceleration and deceleration. Accordingly, these machines have been found particularly attractive where the filtration and drainage periods are relatively short. If, with a three-column centrifuge, acceleration and deceleration take 3 min while filtration and drainage take 30 min, the non-productive periods are only 10% of the productive time and may be considered acceptable, but if, on the other hand, filtration and drainage take no more than 3 min, then the non-productive periods amount to 100% of the production time, which is utterly uneconomic. In those cases, the peeler centrifuge turns out to be the less expensive alternative, in spite of its higher cost of investment.

Peeler centrifuges are mounted on a heavy platform supported by a system

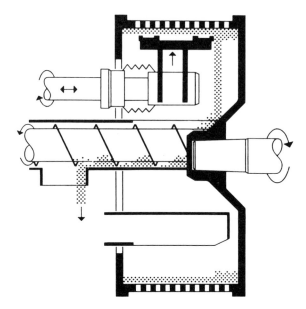

Figure 14.22. Schematic of a peeler centrifuge

of springs and dampers capable of absorbing large imbalances when the machine is operated in the superresonance range.

Imbalances can occur when the rate of filtration is greater than the feed rate, thus leading to a condition where no liquid layer covers the cake during the feed process. Such a liquid layer is required to even out non-uniformities of the cake as caused by variations of the solids input.

Since peeler centrifuges are generally fed at full speed, operation of these machines with a rapidly filtering suspension is readily understood to require powerful motors to permit the feed rate to exceed the filtration rate at any moment of the feed process.

The high speed during the solids discharge may cause considerable breakage of crystals and may lower the permeability of the residual heel by plugging it with fractured fines and by producing compaction as a result of the compressive effects ahead of the peeler. Consequently, it may be necessary to recondition the heel by suitable washes. In addition, compaction of the cake during discharge can be greatly reduced by employing an axially reciprocating knife rather than a full-width blade operated by rotation.

The reciprocating knife may also be required for mechanical reasons. By working like the tool of a lathe, it takes but a small cut and is, therefore, capable of removing very hard cakes where a full-width blade fails. Needless to say, this advantage must be paid for by a longer discharge time.

A particularly sophisticated version of the peeler centrifuge is the so-called 'siphon machine'[15,16] as shown schematically in *Figure 14.23*. In this design, the filter basket is provided with a peripheral enclosure (1) and a narrow lateral chamber (2) equipped with a withdrawal pipe (3). The annular space (4) between the filter medium (5) and the peripheral enclosure (1) is connected

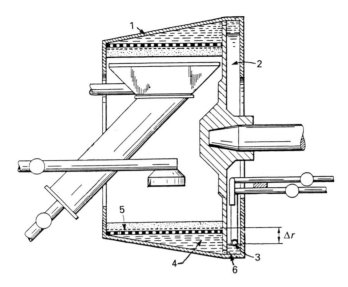

Figure 14.23. Schematic of the 'siphon machine': 1, peripheral enclosure; 2, lateral chamber; 3, withdrawal pipe; 4, space between the filter medium and the peripheral enclosure; 5, filter medium; 6, openings between the lateral chamber and the enclosure space

to the lateral chamber (2) by means of openings (6). Before starting the intended filtration process, the mother liquor of the suspension or some other liquid phase compatible with the system is introduced into the basket or the lateral chamber so as to fill the annular space (4) and the lateral chamber (2) up to a level corresponding to the radius of the filter medium. This level is ensured by an appropriate position of the withdrawal pipe. The suspension to be processed is then fed into the basket while the withdrawal pipe is moved to a greater radius so as to create and maintain a difference, Δr, between the radius of the level in the lateral chamber and the radius of the filter medium.

The consequences of this mode of operation as compared to the conditions in a normal filter centrifuge are illustrated in *Figure 14.24* showing the pressure distribution for the two cases. As the level in the lateral chamber is at atmospheric pressure and as, according to equation 14.5, in a liquid there must be an increase in pressure with increasing radius, the pressure at the filter medium must be below atmospheric pressure. The design is chosen such that this pressure is slightly above the vapour pressure of the liquid so as to prevent cavitation. The result is an increased pressure drop across the cake, thus leading to a faster passage of the liquid phase. Moreover, due to the additional liquid 'below' the filter medium, equation 14.81 now reads

$$h_\mathrm{o} + \Delta r = \frac{4\sigma\cos\alpha}{\rho Bs} \tag{14.155}$$

so that the residual height of liquid in the pores of the cake is decreased by the quantity Δr. Hence, the cake ends up with a lowered residual moisture content.

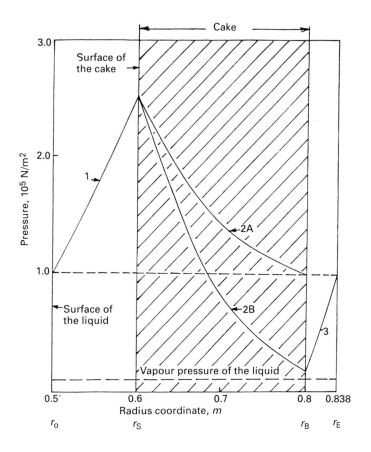

Figure 14.24. Pressure distribution in a 'siphon machine' as compared with the pressure distribution in a normal filter centrifuge. Numerical values are the same as those used in Figure 14.3. Curves 1 and 2A — normal filter centrifuge. Curves 1, 2B and 3 — 'siphon machine'. Curves 1 and 3 — computed using equation 14.5. Curves 2A and 2B — computed using equation 14.23 modified for the different boundary conditions

A further advantage of the 'siphon machine' is the possibility of reconditioning the residual heel. When the filtrate or some other suitable liquid is fed into the lateral chamber, this leads to a backward (centripetal) flow through the cake. In this fashion, the residual heel is transformed to a fluid bed, so that after cessation of the backward flow a new cake is formed from loosely settled particles. Thus, a reduction in permeability due to compaction of the heel by the peeler is eliminated.

With the 'siphon machine', the given improvements in filtration rate and drainage can be augmented by employing a gas overpressure in the interior of the basket. This merely requires surrounding the latter with a sealed housing[15,16].

14.6.2.3 Hanging centrifuges

For the treatment of thick and viscous suspensions liable to cause large imbalances during the feed process, it is advantageous to use a centrifuge with a basket hanging on a long shaft held in a bearing system capable of permitting the axis to deviate from the vertical position. With such a design, imbalances lead to a precessional motion, with the axis describing a cone. In this motion, the basket at the lower end of the axis is deflected from the vertical position so that its huge weight causes a powerful moment that counteracts the deflection. Consequently, the system is self-stabilizing. No other centrifuge has such a characteristic. On the other hand, the lateral freedom of the basket has the obvious disadvantage of leading to vibrations when a discharge plough is moved into the cake. For this reason, the discharge must be effected at very low speed.

Machines of the type described are called 'hanging centrifuges'. They are used primarily in the sugar industry, but they are also used in chemical plants. A typical design is shown in *Figure 14.25*. The feed stock is stored in a container (1) equipped with a mixer driven by the worm gear (2). The feed stock flows via a gate valve (3) and an annular opening surrounding the shaft (4) into the basket (5). The basket is closed at the bottom by a cone (6) which is actuated pneumatically. The basket is driven by a motor (7) connected to the shaft (4) by a flexible coupling (8). The bearing system (9) is held in the support structure (10) by means of a spherical slide surface and rubber buffers. A wash pipe (11) permits the drained cake to be treated with water or steam. The cake is removed by a plough (12) which is actuated hydraulically. During the discharge phase of the process, the cone (6) is lowered in order to allow the solids to slide out. The filtrate is collected in the housing (13) featuring a divider (14) which separates the filtrate outlet from the solids outlet. Of course, the operation of the machine is fully automatic.

14.6.3 Bypass centrifuges

The principal modes of filtration are illustrated in *Figure 14.26*. Conventional filtration corresponds to the case of particle vector 1 which implies that the movement of the dispersed solid phase has the same direction as the filtrate flow, the latter being perpendicular to the filter medium. In general, processes of this type become slow and uneconomic whenever a sizeable fraction of the particles is smaller than 10 μm. In such cases, the principal resistance opposing the flow of the liquid is caused by the conglomerate of particles representing the cake and not by the filter medium. Consequently, the only way of facilitating these filtration processes lies in preventing the particles from blocking the filter medium, the principal possibilities being illustrated by the particle vectors 2 and 3. A well-known example for the case of particle vector 2 is ultra-filtration, where the suspension to be separated flows in a direction parallel to the filter medium, thus preventing the formation of a cake, while permitting the liquid phase to permeate the filter medium. The disadvantage of this process is its inability to yield the solids in a drained state. The suspension undergoes a mere thickening.

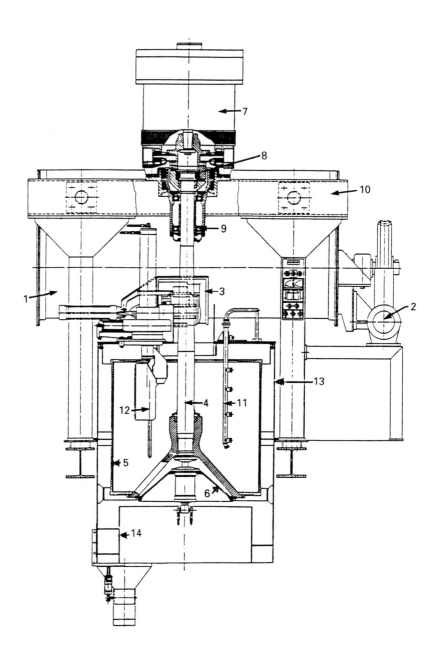

Figure 14.25. Typical design of a 'hanging centrifuge' for the sugar industry: 1, feed container; 2, worm gear; 3, gate valve; 4, shaft; 5, basket; 6, closure cone; 7, motor; 8, coupling; 9, bearing system; 10, support structure; 11, wash pipe; 12, plough; 13, housing; 14, divider

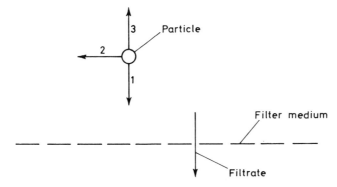

Figure 14.26. Modes of filtration

If drainage of the solids is essential in spite of the particles being very small, the only recourse is the bypass concept. Its principal feature is the formation of a cake by employing mass forces which selectively move the solids away from the filter medium while inducing the fluid to pass through the medium.

As small particles require very high accelerations to be moved rapidly and as the acceleration due to gravity is rather weak and not conducive to directional manipulation, practical applications of the bypass concept are mostly limited to systems based on centrifugal fields.

14.6.3.1 The side filter centrifuge

A simple but highly successful bypass filtration machine is the side filter centrifuge. Its design is shown schematically in *Figure 14.27*. A cylindrical basket rotating on its axis is equipped with filter areas in the cylindrical surface, in the rim, and in the rear wall. The first area is called the mantle filter while the other two areas have come to be known as side filters.

When a suspension characterized by particles heavier than the liquid phase is fed into a basket of this type, the ensuing separation process involves the following more or less simultaneous phenomena:

1. sedimentation in radial direction leading to the formation of an annular cake on the mantle filter;
2. radial flow of liquid through the cake and the mantle filter;
3. axial flow of liquid through the side filters.

When the particles are very small, the build-up of the annular cake will result in a rapid decrease of the radial flow, soon reducing it to a negligible value.

As to the flow through the side filters, it is evidently influenced by any transport of particles to the planes perpendicular to the axis. If the process is to be effective, the centrifugal force acting on the particles must be greater than the force tending to hold them on the side filters. The mathematical formulations defining this condition are available in the literature[17].

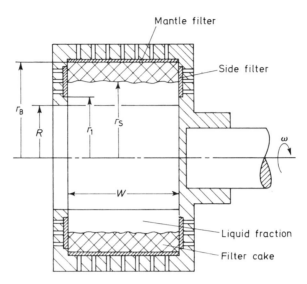

Figure 14.27. Geometry of a side filter basket

14.6.3.2 The planetary centrifuge

Returning to *Figure 14.26* we now consider the most desirable case of particle vector 3 indicating a movement by which the distance between the particle and the filter medium is increased. Disregarding the rare case of particles lighter than the liquid phase, it must be said that a practical realization of this mode of filtration is difficult. It can be accomplished, however, in a periodic fashion, by employing a planetary centrifuge[18] as shown schematically in *Figure 14.28*. The machine consists essentially of two or more baskets A, the axes of which are parallel and arranged on a circle B in equal angular distribution. The basket wall is the filter medium. Each basket rotates around its own axis C which in turn moves along the circle B around a central axis D parallel to C. The angular velocities are chosen to be such that in a certain region E of the basket circumference the centrifugal acceleration due to the rotation around C is more than compensated by the centrifugal acceleration due to the rotation around D so that referred to the basket a centripetal component results. The remaining region F of the basket circumference is under the influence of accelerations having centrifugal components. Thus, when due to the rotation around C a particle located in the centrifugal region F arrives in the centripetal region E, it will be moved away from the filter medium of the basket. As to the operation of a planetary filter centrifuge, the principal features can be readily recognized by considering a gradual increase in speed of the planetary motion ω_1 while the basket is rotating at a constant angular velocity ω_2. The process is illustrated schematically in *Figure 14.29*.

Figure 14.29a represents the initial stage where ω_1 is zero. This is the case of the normal centrifuge where the solids and the liquid form layers concentric with respect to the basket axis. As the latter begins to move, *Figure 14.29b*,

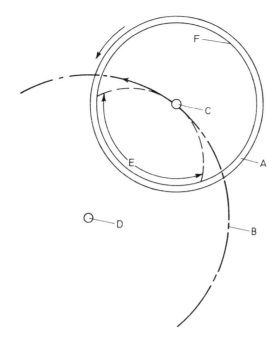

Figure 14.28. Schematic of the planetary centrifuge

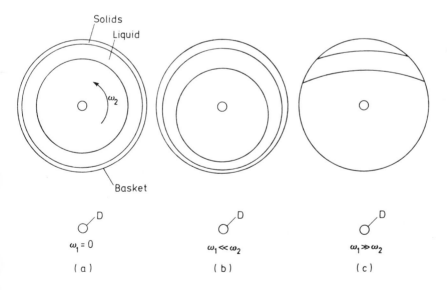

Figure 14.29. Schematic of planetary effects

the layers will no longer be concentric, the region close to D getting thinner as the mass of the basket charge concentrates in the region farther away from D. Finally, if ω_1 is increased to a level where the rotation of the basket becomes negligible as compared to the rotation of the basket axis, *Figure 14.29c*, the charge of the basket will cease to be a ring and will form a cylinder segment. As the liquid covers a greater portion of the filter medium than the solids, it can readily bypass the cake.

In addition, the solids are not permitted to retain liquid by stationary points of contact as the particles are forced to constantly move over the filter medium. Accordingly, drainage is greatly favoured.

Last but not least, the centripetal region is freed of all particles so that plugging of the filter medium can be ruled out.

The theory of the planetary centrifuge is available in the literature[17]. It will not be repeated here as, so far, this machine has been more a subject of laboratory curiosity than of industrial application.

14.7 PRACTICAL ASPECTS OF CENTRIFUGAL FILTRATION

14.7.1 Guidelines for the selection of filter centrifuges

From the description of the various designs, it may have become apparent that a principal criterion for the selection of a particular type of filter centrifuge is the size of the particles to be separated. However, inasmuch as no suspension contains particles of a single size but rather a range of sizes not readily characterized by a single number, and since the filtration behaviour of a suspension is governed not only by the particle size distribution but also by the shape and deformability of the particles as well as by the viscosity of the liquid phase, it is customary to base the selection procedure on the so-called 'intrinsic permeability' of the filter cake as introduced in section 14.2 and discussed in section 14.4, i.e. the original Darcy permeability including the viscosity term but measured in the centrifugal field.

If all lengths are measured in metres, the force in newtons, and the time in seconds, the unit of the intrinsic permeability is $m^4 N^{-1} s^{-1}$.

Once the intrinsic permeability has been determined, a rough selection of the design can be arrived at on the basis of the following rules of thumb:

1. For intrinsic permeabilities above $20 \times 10^{-10} m^4 N^{-1} s^{-1}$: continuously fed machines, usually pushers.
2. For intrinsic permeabilities between 20×10^{-10} and $10^{-10} m^4 N^{-1} s^{-1}$: peeler centrifuges.
3. For intrinsic permeabilities between 10^{-10} and $0.02 \times 10^{-10} m^4 N^{-1} s^{-1}$: three-column centrifuges.
4. For intrinsic permeabilities below $0.02 \times 10^{-10} m^4 N^{-1} s^{-1}$, filtration should be replaced by centrifugal sedimentation.

In the centrifuge industry, it is customary to use centimetres, force kilograms, and minutes so that the respective unit of the intrinsic permeability becomes $cm^4 kg^{-1} min^{-1}$. With this unit, the numbers in the selection rules assume more convenient values[19].

14.7.2 Importance of desliming

Inasmuch as centrifugal filtration is a filtration accompanied, if not largely preceded by, sedimentation, it is essential to pay attention to the characteristics of the latter process. Sedimentation does not lead to a uniform cake unless all particles are of the same size. In practice, suspensions have particles of a more or less wide size distribution. Thus, sedimentation of such particles inevitably leads to a cake with the large particles at the bottom and the fines on the top. This is readily realized by inspecting the cross-section of a cake formed in a filter centrifuge. The phenomenon is commonly known as 'cake stratification'. In view of this fact, the permeability is not constant throughout the cake but has a high value in the bottom layers of large particles and a low value in the top layers of fines. With such a situation, the overall permeability will largely be determined by the top layer of fines. In fact, if the particle size distribution goes all the way down to zero, the top layer of the sediment will be impermeable. For this reason, a suspension to be separated by centrifugal filtration must be free of extremely small particles. If it is not, it must be submitted to a pretreatment leading to the fulfilment of this condition. Such a pretreatment, commonly called 'desliming', can be effected with a hydroclassifier (see chapter 16).

Such apparatus, e.g. a wash cyclone, is capable of eliminating from the suspension all particles smaller than a critical size known as the 'desliming limit', c^*. For a 'normal particle size distribution', as defined in chapter 16, it can be shown analytically[20] that the overall (mean) permeability κ_s of a cake formed by sedimentation is always smaller than the permeability κ_N of a cake made up of randomly distributed particles of the same size spectrum, and it is not surprising that the ratio of these two permeabilities is a function of the desliming limit c^*. This dependence is shown in *Figure 14.30*, where the abscissa is the ratio of c^* and the predominant particle size $\hat{c}$ of the spectrum (the particle size where the distribution function has its maximum).

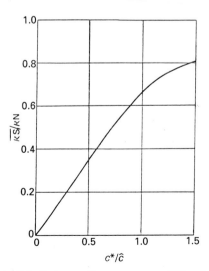

Figure 14.30. The effect of desliming on the mean permeability of a filter cake formed by sedimentation

Desliming is particularly indicated where the particles of the suspension to be treated are formed in a crystallization process. In such cases, the fines fraction of the desliming hydroclassifier is simply recycled to the crystallizer where the small particles are allowed to undergo further growth before they are 'accepted' for the filtration intended. It can be seen from *Figure 14.30* that effective desliming can greatly reduce the cost of investment as it may greatly reduce the number of centrifuges required for the process. In addition, desliming reduces the residual moisture content of the cake as was shown earlier in the discussion of centrifugal drainage.

REFERENCES

1. Darcy, H., '*Les fontaines publiques de la ville de Dijon*', Victor Dalmont, Paris (1856)
2. Zeitsch, K. and Adens, M., 'La filtration centrifuge'. In *La filtration industrielle des liquides*, Part 2, Vol. IV, Chap. XVI, Société Belge de Filtration, Liège (1975)
3. Sommerfeld, A., *Mechanics of deformable bodies*. Academic Press Inc., New York (1950)
4. Batel, W., *Chemie-Ing.-Techn.*, **27**, 497–501 (1955)
5. Batel, W., *Chemie-Ing.-Techn.*, **28**, 343–349 (1956)
6. Batel, W., *Chemie-Ing.-Techn.*, **31**, 388–393 (1959)
7. Batel, W., *Chemie-Ing.-Techn.*, **33**, 541–547 (1961)
8. Mersmann, A., *Verfahrenstechn.*, **5**, 23–28 (1971)
9. Zeitsch, K., 'Proceedings of the International Symposium on Liquid–Solid Filtration', (Antwerp, Belgium, 6–7 June 1978), Institute of Chemical Engineers, Université Catholique de Louvain, Louvain-la-Neuve, Belgium (1978)
10. Zeitsch, K., *Chem. Techn.*, **33**, 456–461 (1981)
11. Laplace, P. S., 'Théorie de l'Action Capillaire'. Supplement au livre dixième de la Mechanique Celeste, Paris (1806)
12. Abramowitz, M. and Stegun, I., '*Handbook of Mathematical Functions*', National Bureau of Standards, Washington (1964)
13. Jahnke, E. and Emde, F., '*Tables of Functions with Formulae and Curves*', Dover Publications, New York (1945)
14. Schubert, H., Meinel, A. and Göll, G., *Chem. Techn.*, **20**, 732–737 (1968)
15. Hultsch, G. and Zeppenfeld, K., Deutsches Patentamt, *Auslegeschrift* 22 60 461 (20 November 1975)
16. Hultsch, G., Zeppenfeld, K., Niedner, P. and Ostermeyer, P., Deutsche Demokratische Republik, Amt fuer Erfindungs- und Patentwesen, *Patentschrift 108 218* (12 September 1974)
17. Zeitsch, K., 'Centrifugal filtration'. *In Solid–Liquid Separation* (L. Svarovsky, Ed.), Chap. 14, Butterworths, London (1977)
18. Zeitsch, K., *Deutsches Patentamt, DOS No. 23 57 775* (1976)
19. Zeitsch, K., *Filtration and Separation,* **13**, 223–226 (1976)
20. Zeitsch, K., *Tech. Mitt. Krupp Forsch.-Ber.*, **42**, 71–76 (1984)

15

Countercurrent Washing of Solids

L. Svarovsky

Department of Chemical Engineering, University of Bradford

15.1 INTRODUCTION

Washing of a particular product (contaminated with a solvent, solute or ultrafine particles) by reslurrying is possible with separator arrangements in simple series connections without recycle, by adding washwater to the underflows in between the stages. Wash liquid requirements can be minimized, however, by countercurrent operation if the additional cost of the necessary equipment can be justified. In this arrangement, as shown in *Figure 15.1*, the feed slurry and the wash liquid move through the separator series in opposite directions, with the wash liquor becoming gradually more contaminated with the mother liquor while the concentration of the mother liquor in the solids stream reduces.

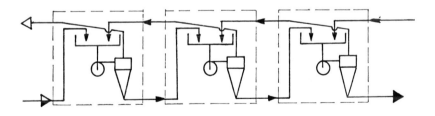

Figure 15.1. Schematic diagram of a countercurrent washing train

The separators in this context may mean any dynamic separators such as hydrocyclones, sedimenting centrifuges or gravity thickeners, but much of the following also applies to countercurrent washing with filters, with or without reslurry. In the case of filters, the washed cake then represents the 'underflow' from the filter whilst the filtrate is the 'overflow'. In the case of dynamic separators, the solids leave the washing train as a slurry in the system underflow whilst most of the contaminated wash liquid leaves through the system overflow.

Systems used in industry may or may not employ intermediate sumps between individual stages, depending on the type of separator used in the train. Gravity thickeners, for example, provide the bulk and the residence time necessary for any mixing or mass transfer to take place. Hydrocyclones usually need sumps between stages, although closed systems without sumps are often found in starch washing, for example.

15.2 MASS BALANCE CALCULATIONS

Mass balance calculations of countercurrent washing systems can be quite involved. Assuming perfect mixing in the feed tanks and for a simple case of the fresh wash liquid containing no solubles, Fitch[1] showed a simple way of working out the algebraic solubles balance. The dotted lines in *Figure 15.1* show rectangles representing the different stages, each containing a separator and its feed tank and pump. The critical information needed for easy calculation of a general case such as that in *Figure 15.1* is the ratio, in each stage, of the quantity of solubles carried in the feedback flow to that carried in the advancing flow. The ratio for perfect mixing in the tank and separator is equal to $1/S_i$ in each stage i, if S_i is the free liquid underflow/overflow split. The algebraic solubles balance can thus be made, starting at the final stage.

An unknown value, x, is assigned to the mass flow rate of the solubles in the advancing flow of washed solids from the final stage (the system underflow). This unknown x will carry through the system and appear as a factor in the solubles balance for all stages in the system. The difference between the advancing flow and the return flow between any two stages in the system must therefore be equal to x. Taking the wash liquid introduced in the final stage to be clean, free of any mother liquor, the return flow from the last stage in *Figure 15.1* would contain x/S_3 of the solubles while the advancing flow, entering stage 3, would contain x more, i.e. $x/S_3 + x$. Application of the method by Fitch[1] to the three-stage system in *Figure 15.1* yields the following equation for the overall loss of the solubles to the system underflow[2]:

$$x/M_s = 1/\{1 + [1/S_1] + [1/(S_1 \cdot S_2)] + [1/(S_1 \cdot S_2 \cdot S_3)]\rho\} \qquad (15.1)$$

or, in terms of the underflow-to-throughput ratios, R_f (again as free liquid ratios):

$$x/M_s = [R_{f1}R_{f2}R_{f3}] / [1 - (1 - R_{f3}) \cdot (R_{f1} + R_{f2})] \qquad (15.2)$$

where M_s is the mass flow rate of solubles in the solids feed.

The amount of wash liquid added to the circuit only affects the dilution of the solubles in the system underflow and overflow and has no effect on the overall washing efficiency.

Another important overall performance parameter is the loss of washwater in the system underflow. This can be calculated from the mass balance for the washwater: Fitch's method (starting this time from the other end of *Figure*

15.1) yields the loss of washwater in the underflow (*w*), for the simple case where no washwater comes with the solids, as

$$w/W = 1/[1 + 1/(S_1 S_2 S_3 + S_2 S_3 + S_3)] \tag{15.3}$$

where W is the wash liquid feed flow rate. It can be seen that this ratio only depends on the splits in each stage and increasing the washwater supply merely dilutes the underflow solids stream. Reducing the washwater supply will therefore thicken the solids underflow stream, but in so doing the separation efficiency of the solids may be diminished causing some solids to leave through the overflow of the particular stage.

In the limit, there is a minimum washwater flow required to carry the solids, and the washwater supply cannot be reduced below this point. For hydrocyclones, for example, the maximum solids concentration in the underflow is of the order of 45% or 50% by volume and enough water must be provided at all stages for the solids to be able to pass through to the system underflow.

The mass balances of simple washing trains with clean washwater and with the solids being dry on entering the system can be calculated manually using the above-described method proposed by Fitch[1]. If some washwater enters with the solids or if the washwater feed contains some solubles, the mass balance becomes more complicated and is best carried out with a computer. An iterative calculation may then be used to introduce an artificial, hidden stage at the front and also one at the back of the train, which would produce the wet solids and dirty washwater feeds to the train as required.

This is the principle behind the programme distributed by Fine Particle Software[3]. The programme has been used to study the countercurrent washing systems and to make the conclusions presented here; it has also been used to compare the predicted performance with experimental tests on washing hydrogel, as reported in the literature[4].

15.2.1 Examples of mass balance calculations

Figure 15.2 sets out three cases of a three-stage countercurrent washing train in which the solids enter dry (no wash liquid with the solids feed) and the wash liquid comes in clean (no solubles in the wash liquid). All three cases are designed to give the same washing efficiency (from 1 g s^{-1} with the feed to 0.008 g s^{-1} in the system underflow) they all have the same minimum amount of liquid in any underflow along the train to carry the solids along (0.261 l s^{-1}), as indicated in *Figure 15.2*.

The three cases are shown at this stage only as a demonstration of how a programme like this can be used to test a system and optimize it quickly. It is also set out to show the effect of the values and the order of the flow ratios on the wash liquid requirement and the position of the thickest underflow in the system. The example will be discussed later, but it can be noted from *Figure 15.2* that, for the case of the R_f ratios increasing along the train, the wash requirement is highest and the system underflow is very dilute (the thickest underflow is in the first stage). As it happens, this case is also very bad for the

Case 1: worst

No. of stages	3			
Solids feed:		Wash:		
Flow rate of wash, $l\,s^{-1}$	0	Flow rate of wash, $l\,s^{-1}$		**3.64 (high)**
Mass rate of solubles, $g\,s^{-1}$	1	Mass rate of solubles, $g\,s^{-1}$		0

Stage No.	R_f		Input		Output	
			Feed	Feedback	Overflow	Underflow
1	0.100	Wash	0.000	2.613	2.352	**0.261**
		Solubles	1.000	0.103	0.992	0.110
2	0.200	Wash	0.261	3.005	2.613	0.653
		Solubles	0.110	0.018	0.103	0.026
3	0.300	Wash	0.653	3.640	3.005	1.288
		Solubles	0.026	0.000	0.018	**0.008**

Case 2: best for wash requirement

No. of stages	3			
Solids feed:		Wash:		
Flow rate of wash, $l\,s^{-1}$	0	Flow rate of wash, $l\,s^{-1}$		**1.52 (low)**
Mass rate of solubles, $g\,s^{-1}$	1	Mass rate of solubles, $g\,s^{-1}$		0

Stage No.	R_f		Input		Output	
			Feed	Feedback	Overflow	Underflow
1	0.180	Wash	0.000	1.450	1.189	**0.261**
		Solubles	1.000	·0.209	0.992	0.218
2	0.180	Wash	0.261	1.507	1.450	0.318
		Solubles	0.218	0.038	0.209	0.046
3	0.180	Wash	0.318	1.520	1.507	0.331
		Solubles	0.046	0.000	0.038	**0.008**

Case 3: best for solids separation

No. of stages	3			
Solids feed:		Wash:		
Flow rate of wash, $l\,s^{-1}$	0	Flow rate of wash, $l\,s^{-1}$		**2.11 (medium)**
Mass rate of solubles, $g\,s^{-1}$	1	Mass rate of solubles, $g\,s^{-1}$		0

Stage No.	R_f		Input		Output	
			Feed	Feedback	Overflow	Underflow
1	0.300	Wash	0.000	2.641	1.849	0.792
		Solubles	1.000	0.417	0.992	0.425
2	0.160	Wash	0.792	2.352	2.641	0.503
		Solubles	0.425	0.072	0.417	0.079
3	0.100	Wash	0.503	2.110	2.352	**0.261**
		Solubles	0.079	0.000	0.072	**0.008**

Figure 15.2. Mass balances for three cases of three-stage countercurrent washing. Note: cases 1 and 3 have the same washing efficiency (0.008 g s^{-1}); all cases have the same minimum wash with solids (0.261 l s^{-1})

separation of the solids, as will be discussed later. Case 1, therefore, must be rejected.

Case 2, of equal flow ratios, is clearly good for wash requirement but the thickest underflow is still at the front of the train rather than at the end of it as is most often required in practice. It could be shown, however, that the point of thickest underflow may be moved to the back end of the train by adding some of the wash liquid with the solids rather than having it all come in at the end of the train.

Case 3 is best from the point of view of solids separation because the separator in the first stage, as the unit responsible for passing solids in the system overflow, operates with dilute underflow, i.e. high separation efficiency. The point of thickest underflow is firmly at the back end of the train and the wash liquid requirement is still fairly low.

The conclusions drawn from the mass balances presented in *Figure 15.2* are discussed further in the following section.

15.3 WASHING TRAIN DESIGN RECOMMENDATIONS

The countercurrent washing trains installed in industry often represent the most complex parts of the overall processes and are usually installed and operated at conditions far from optimum. Experimenting with and optimizing such systems is difficult, particularly in cases where no interstage sumps are used and samples of the intermediate streams cannot easily be taken. It is, therefore, advantageous to use a computer model to find out how such systems work and how they respond to operational or design changes.

The following recommendations may be drawn from the mass balance calculations performed on different possible scenarios encountered in industry. Consulting experience with several working systems in industry confirms the conclusions given below.

15.3.1 Washing efficiency

The washing efficiency of a countercurrent washing train is primarily determined by the values and order of the flow ratios, R_f, used in the train, and by the number of stages employed. It is not affected by the washwater supply rate, irrespective of whether the water comes in with the solids or through the washwater feed at the end of the train; the washwater supply only determines the dilution of the solids in the underflows in the different stages and overall.

15.3.2 Washwater requirement

The washwater requirement of an existing washing train designed for a given washing efficiency and for a given maximum underflow solids concentration anywhere in the train is also determined by the values and order of the flow ratios, R_f, and by the number of stages employed. Fewer stages require lower R_f ratios and higher washwater supplies if they are to achieve the same washing efficiency and to operate at the same maximum underflow solids concentration.

15.3.3 Setting of the flow ratios

It is clear from equation 15.2 that the washing efficiency is maximized by minimizing the flow ratios in all stages. The solubles/wash system cannot be considered in isolation, however, and the transport of the solids along the washing train and the solids separation efficiency both in the individual stages and overall must also be taken into account. The problem is, therefore, one of optimizing a three-phase separation system.

In order to be able to optimize the system, we need to know how the flow ratio affects the separation of the solids. In the case of hydrocyclones, the effect of the ratio is two-fold: increasing R_f leads to improvements in the separation efficiency through the contribution of 'dead flux'; and a further improvement is caused by a reduction in the crowding of the underflow orifice. Both of these effects can be described analytically for certain hydrocyclone geometries[2] and the above-mentioned optimization is, therefore, possible, using the entropy index[5] as a general criterion for the optimization.

It is often the case that the flow ratios are about the same in all stages for some engineering reasons (like in the case study reported by Svarovsky and Potter[4]), but if this is not the case and they can be varied along the train, the following general conclusions may be stated.

The best design of a countercurrent washing train using hydrocyclones is such that the flow ratios decrease along the train, so that:

1. the overall solids recovery is high;
2. the solids concentration is always highest at the end of the train;
3. the wash liquid requirement is low.

The actual design depends on how sensitive the grade efficiency curves of the individual stages are to the changes in the flow ratios. The overall grade efficiency of the train G, which determines the amount of the solids lost in the system overflow, may be calculated from the individual efficiencies of the stages G_i as for a series connection on underflow with overflow recycle[2]. The following equation allows such calculation for a three-stage system as an example:

$$G = G_1 G_2 G_3 / (1 - G_1 - G_2 + G_1 G_2 + G_2 G_3) \qquad (15.4)$$

Total recoveries in the individual units combine into the overall recovery of the whole train in exactly the same way as do the grade efficiencies in the above equation.

15.3.4 Mass transfer from the solids

The mass balance calculations outlined in section 15.2 assume that the mass transfer of the solubles from the solids into the suspending liquid is complete; this is reasonably true if the solids are non-porous and if enough time and shear is provided in the hydrocyclone or in the sump before it for the mass transfer to take place before the solids leave the unit with the underflow.

Mass balance calculations are further complicated if the above assumptions are not true such as in the case of porous materials and where the mixing sumps are not large enough to accommodate the mass transfer. The net result, as shown in the case study given by Svarovsky and Potter[4], is that the liquid in the pores within the individual particles has more solubles than the surrounding liquid and this complicates not only the mass balances but also the analyses of the samples taken from a working washing train.

15.4 APPLICATIONS

Countercurrent separator systems can be found in many diverse industries. One example of the use of such systems is in the production of potato or corn/wheat starch. Washing here usually means removal of one solid from another, i.e. gluten from starch. The gluten must be sheared off the starch particles and hydrocyclones are ideally suited to this duty. The density of the solids is low, however, and this combined with low particle size necessitates the use of small diameter (10 mm) cyclones in parallel arrangements and the number of stages used must be high (eight or nine, up to 24 in extreme cases).

Removal of solubles from the feed solids is achieved using identical washing arrangements, sometimes known as countercurrent decantation. This may be used either when the solids represent the product, like gypsum in the phosphoric acid process, or when the solvent is the product, like the leaching residues (leached uranium, copper, etc.[6–8]). Hydrocyclones here compete strongly with the conventional means of decantation in gravity thickeners but offer lower installation costs and greater ease of control.

A case study of a pilot test series for washing solubles off a solid product is given by Svarovsky and Potter[4] by way of illustration of the problems involved in testing such a system and of the complexity of countercurrent washing arrangements in general.

15.5 CONCLUSIONS

Countercurrent reslurry or cake washing has been used in many areas of chemical, mineral and food processing and some filters, gravity thickeners and hydrocyclones are particularly suitable for this process. Very little is known, however, about the optimum operating conditions and design of such washing trains, and many installations can be much improved if some basic rules and recommendations are followed.

In this chapter we have evaluated the process of countercurrent washing from the point of view of washing efficiency, washwater requirement and separation of the solids, and have given the basic rules for the best operation and design of such systems. A microcomputer model has been used to carry out the mass balances and illustrate the theoretical performance of different systems under different operating conditions.

REFERENCES

1. Fitch, B., 'Countercurrent filtration washing', *Chem. Eng.*, **22**, 119–124 (1962)
2. Svarovsky, L., *'Hydrocyclones'*, Holt, Rinehart and Winston, London (1984)
3. *'Counter-current Washing of Solids'*, Fine Particle Software, Low Shann Farm, Keighley, West Yorkshire
4. Svarovsky, L. and Potter, J. K., 'Counter-current washing with hydrocyclones', *BHRA Conference on Hydrocyclones* (Oxford, 30 September to 2 October 1987)
5. *'Solid–Liquid Separation'*, Continuing Education Course sponsored by the Institute of Chemical Engineers, Bradford (15–18 September 1987)
6. Trawinski, H., 'Counter-current washing of thickened suspensions by repeated dilution and separation by sedimentation', *Verfahrenstechnik,* **8**, 28–31 (1974)
7. Trawinski, H., 'Counter-current washing of leaching products in thickeners and hydrocyclones, including the mathematical calculation', *Aufbereitungs-Technik,* **18**, 395–404 (1977)
8. Chandler, J. L., 'Deep thickeners in counter-current washing', *Inst. Chem. Eng. Symposium on Solid/Liquids Separation Practice and the Influence of New Techniques* (Leeds, 3–5 April 1984), Paper 22, pp. 177–182, Institute of Chemical Engineers, Yorkshire Branch (1984)

16

Problems with Fine Particle Recycling

K. Zeitsch

Consultant, Köln, West Germany

NOMENCLATURE

c	Particle size
$\hat{c}$	Predominant particle size
c^*	Critical particle size of a sedimentation process
c_n^*	Critical particle size of a separation machine featuring the centrifugal acceleration ng
c'	Desliming limit
$\bar{c}$	$= \sqrt{2}\hat{c}$, characteristic particle size employed in the Bennett distribution
D	Depth of the suspension in a settling tank
$f(c)$	Particle size distribution as resulting from the processing of fresh raw material
$F(c)$	Particle size distribution in the steady state of the recirculation process
g	Acceleration due to gravity
$g(c)$	A function defined by equation 16.20
$g_i(c)$	A function defined by equation 16.29
h	Initial height of a particle
$h(c)$	A function defined by equation 16.39
J	Relative fines volume due to recirculation build-up
L	Length of a settling tank
dm_c	Total mass fraction of particles of diameter c
dm_c^{∇}	Deposited mass fraction of particles of diameter c
m_i	Relative mass fraction of group i
m_S	Relative mass fraction of the sediment
n	g-Number = centrifugal acceleration/g
n^*	Exponent in the Bennett distribution
v_c	Settling velocity of a particle of diameter c
v_H	Horizontal velocity
$\dot{W}$	Weight of solids deposited in unit time
z_c	Vertical distance travelled by a particle of diameter c
α	One half of the apex angle of a disc separator

541

ϵ Separation efficiency of a sedimentation process as defined by equation 16.23

η Dynamic viscosity of the liquid phase

κ A quantity defined by equation 16.5

$\Delta\rho$ Density difference between the solid and the liquid phases

τ Residence time of a fluid element in a settling tank

16.1 INTRODUCTION

In numerous industrial processes, suspensions are partially depleted of solids by means of a sedimentation device (settling tank, decanter, separator), thus leading to a thick sludge and a 'residual suspension'. In view of the basic separation principle involved, the latter fraction is commonly called the 'overflow'. When the thick sludge is the valuable fraction (e.g. large crystals), it is often undesirable or even impossible to discard the overflow as it may contain valuable substances, dissolved and undissolved, or represent a serious environmental problem if discharged as waste water. In such cases, it is tempting to recycle the overflow to the stage where the initial suspension is being prepared. In this fashion, no valuable material is lost, the requirement for the liquid phase is greatly reduced, and there is no waste water problem. Unfortunately, overflow recycling is not possible unless certain conditions are satisfied as otherwise fine particles accumulate to the point of rendering the total process unfeasible.

An illustrative example, not to be imitated, is shown schematically in *Figure 16.1* depicting an attempted process for the production of fuel ethanol from corn. Such a process was actually built on a huge scale but failed completely as the requirements for fine particle recycling were overlooked. As indicated by the schematic, the raw material is steeped and, after comminution of the grain followed by liquefaction and saccharification of the starch, the resulting 'mash' is submitted to a fermentation by yeast leading to a suspension of protein, cellulose, and yeast particles in an aqueous solution of ethanol. This suspension enters the first distillation column where all the ethanol and some water are stripped off as the head fraction. The water of this fraction is separated in subsequent columns and returned to the first column so that the sump fraction of the latter comprises all the water contained in the feed. This sump fraction, called 'stillage' or 'slop', represents approximately 95% by weight of the feed and consists essentially of protein, cellulose, and yeast particles suspended in water.

According to the process shown in *Figure 16.1*, this stillage was fed into centrifugal decanters leading to a thick sludge and an overflow. The thick sludge was pumped into screw filter presses for further dewatering, and then entered a thermal drier delivering a marketable fodder. The filtrate of the screw presses was added to the decanter overflow, thus yielding a stream denoted as 'thin stillage'. This suspension of relatively fine particles was to be used for steeping the raw material, together with some fresh water satisfying the balance of the system (water in the corn + fresh water = water lost in the drier).

As it turned out, this process did not work, because fine particles in the 'thin stillage' accumulated without limit so that after a short period of

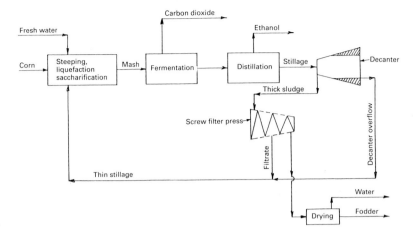

Figure 16.1. Example of a fine particle recycling process which cannot run continuously

operation the mash became so rich in fine particles and thus so thick that the decanters could no longer operate satisfactorily, not to speak of other problems encountered in fermentation and distillation. The plant literally 'suffocated' in fines. When this condition was reached, the process had to be stopped and the entire stillage discarded in order to use fresh water to start a new cycle. Hence, the process originally planned as being continuous degenerated into a stop-and-go operation, totally unacceptable for a large-scale fuel ethanol plant.

The following discussion will show why fine particle recycling processes of the type depicted in *Figure 16.1* cannot be run continuously and how they must be modified to prevent excessive fines build-up.

16.2 THE SEPARATION CHARACTERISTICS OF SEDIMENTATION PROCESSES

To a first approximation, any continuously operating sedimentation device can be represented by a long and flat settling tank with a feed inlet on one side and an overflow on the other side as shown schematically in *Figure 16.2*. For simplicity, the initial distribution of particles at the inlet is assumed to be random, due to an appropriate design of the feed system. In view of the large length-to-depth ratio, the somewhat non-uniform but short flow pattern at the outlet end is negligible so that the flow over most of the length can be considered uniform and parallel to the bottom surface. In this range of uniform flow, the horizontal velocity, v_H, is also uniform, and for a known cross-section perpendicular to the flow and a known feed rate, v_H is a known quantity. Thus, if the length of the tank is L, the residence time, τ, of any fluid element is found to be

$$\tau = \frac{L}{v_H} \qquad (16.1)$$

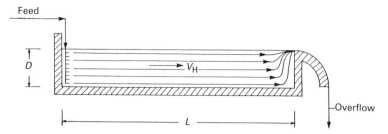

Figure 16.2. Schematic of a continuously operating settling tank

In addition to the horizontal movement, the particles suspended in the feed exhibit a simultaneous vertical movement due to gravity. In a highly diluted system, a spherical particle of diameter c would settle with a velocity described by the law of Stokes as being

$$v_c = \frac{c^2 \Delta \rho g}{18\eta} \tag{16.2}$$

where $\Delta \rho$ is the density difference between the solid and liquid phases, g is the acceleration due to gravity, and η is the dynamic viscosity of the liquid phase.

Although real particles are rarely spherical, this does not reduce the usefulness of equation 16.2 if c is very generally interpreted as a dimension characterizing the settling behaviour of the particle considered. All particle size determinations depend on generalizations; for example, screening is based on the tacit agreement that all particles passing a certain square mesh are considered equal although in reality they may be of greatly differing shapes. On the other hand, in most separation problems, the particles are too small for screening so that their size distribution must be measured by sedimentation analysis. Thus, the general interpretation of c as a characteristic of settling behaviour is very much in agreement with both the measurement and the application of the distribution data for characterizing a sedimentation process.

In view of

$$v_c = \frac{z_c}{\tau} \tag{16.3}$$

where z_c is the vertical distance travelled by a particle of diameter c during the residence time τ, equation 16.2 can be written in the form

$$z_c = \kappa \tau c^2 \tag{16.4}$$

where

$$\kappa = \frac{\Delta \rho g}{18\eta} \tag{16.5}$$

is a constant characteristic of the system. Hence, in any fixed time interval, particles differing in size by a factor of 10 will travel vertical distances differing by a factor of 100.

In practice, the system will not be highly dilute, the settling of each particle being hindered by other particles, but this complication can be disregarded as it does not affect the final result to any significant extent.

With this background, let us consider what happens in a continuously operating settling tank of depth D. After the residence time τ, the bottom of the tank will have been reached by two groups of particles:

1. by *all* particles of the system for which the residence time τ is sufficient to travel a vertical distance D or more, i.e. for which according to equation 16.4

$$\kappa \tau c^2 \geqslant D \qquad (16.6)$$

2. by *some* (not all) particles of the system which are too small to have travelled the distance D but for which

$$\kappa \tau c^2 \geqslant h \qquad (16.7)$$

where h is the initial height, i.e. the height at which the settling process began.

For these two groups of particles, the weights deposited per unit of time are designated by $\dot{W}_1(\tau)$ and $\dot{W}_2(\tau)$, respectively, so that the total weight deposited per unit of time is

$$\dot{W}_S(\tau) = \dot{W}_1(\tau) + \dot{W}_2(\tau) \qquad (16.8)$$

If the total weight of particles introduced into the settling tank per unit of time is $\dot{W}$, division of equation 16.8 by $\dot{W}$ yields

$$m_S(\tau) = m_1(\tau) + m_2(\tau) \qquad (16.9)$$

where m_S, m_1 and m_2 are the relative mass fractions of the sediment and of the two groups.

The mass fraction of incoming particles having diameters between c and $c + dc$ is described by

$$dm_c = f(c)dc \qquad (16.10)$$

where $f(c)$ is the size distribution function of the particle population to be treated. Hence, any fraction with diameters between c_1 and c_2 can be expressed as

$$m \Big/ {}_{c_1}^{c_2} = \int_{c_1}^{c_2} f(c)dc \qquad (16.11)$$

According to equation 16.6 group (1) comprises all particles for which

$$c \geqslant \sqrt{\frac{D}{\kappa \tau}} \qquad (16.12)$$

so that in view of equation 16.11

$$m_1(\tau) = \int_{\sqrt{\frac{D}{\kappa\tau}}}^{\infty} f(c) dc \tag{16.13}$$

As regards group (2), referring only to diameters which are smaller than the value on the right-hand side of equation 16.12, the particles of any narrow size range that are found in the sediment will represent only a part of the total population of this size range because some of its members will have failed to reach the bottom on account of having started too high up. If for a narrow range of diameters between c and $c + dc$ and the given residence time τ the sedimentation path is equal to $D/2$, then at the end of the residence time τ only one half of particle fraction (2) will be found in the sediment, and if the sedimentation path is $D/4$, then only one-quarter of the fraction will have settled. Thus, for each particle size c of group (2) the mass fraction deposited, dm^∇_c, is related to the total mass fraction dm_c, as the sedimentation path of this particle size, z_c, is related to the total depth D of the suspension in the tank so that

$$\frac{dm^\nabla_c}{dm_c} = \frac{z_c}{D} \tag{16.14}$$

In view of equations 16.4 and 16.10, equation 16.14 can be rewritten as

$$dm^\nabla_c = \frac{\kappa\tau c^2}{D} f(c)\, dc \tag{16.15}$$

According to its derivation, dm_c represents the contribution of particles with diameters between c and $c + dc$ to particle group (2) with diameters smaller than

$$c^* = \sqrt{\frac{D}{\kappa\tau}} \tag{16.16}$$

Consequently, the mass fraction m_2 of equation 16.9 is obtained by integrating equation 16.15 over the diameter range $c = 0$ to $c = c^*$ which yields

$$m_2(\tau) = \int_0^{\sqrt{\frac{D}{\kappa\tau}}} \frac{\kappa\tau c^2}{D} f(c)\, dc \tag{16.17}$$

Introducing equations 16.13 and 16.17 into equation 16.9 reveals the sediment fraction after residence time τ to be

$$m_s(\tau) = \int\limits_{\sqrt{\frac{D}{\kappa\tau}}}^{\infty} f(c)dc + \int\limits_{0}^{\sqrt{\frac{D}{\kappa\tau}}} \frac{\kappa\tau c^2}{D} f(c)\, dc \qquad (16.18)$$

In view of equation 16.16, equation 16.18 assumes the form

$$m_s = \int\limits_{c^*}^{\infty} f(c)dc + \int\limits_{0}^{c^*} \left(\frac{c}{c^*}\right)^2 f(c)\, dc \qquad (16.19)$$

Equation 16.19 can be readily interpreted as shown in *Figure 16.3*, the first integral being given by area I and the second by area II, where

$$g(c) = \left(\frac{c}{c^*}\right)^2 f(c) \qquad (16.20)$$

is the function dividing the areas II and III.

Area I represents the coarse particles of the sediment, area II the fine particles embedded in the coarse particles of the sediment, and area III the

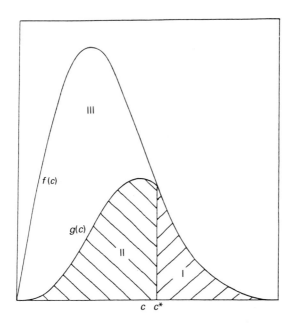

Figure 16.3. Graphical interpretation of equation 16.19

fine particles leaving the system with the overflow. The value c^* is termed the 'critical particle size' of the sedimentation process considered.

The relationships derived for the settling tank are equally valid for a centrifugal decanter, the only difference being that in equation 16.5 the value of g must be multiplied by the 'g number'

$$n = \frac{\text{centrifugal acceleration}}{g} \tag{16.21}$$

so that in view of equation 16.16 the critical particle size assumes the diminished value

$$c_n{}^* = \sqrt{\frac{18D\eta}{\Delta\rho ng\tau}} \tag{16.22}$$

For a separator having conical discs with an apex angle 2α, n must be multiplied by $\cos\alpha$.

The separation efficiency of a sedimentation process is given by

$$\epsilon = \frac{\text{mass fraction of particles in the thick sludge}}{\text{total mass of all particles}} \tag{16.23}$$

On the basis of *Figure 16.3*, this leads to

$$\epsilon = \frac{\text{area I} + \text{area II}}{\text{area I} + \text{area II} + \text{area III}} \tag{16.24}$$

For a normal particle size distribution defined by

$$f(c) = \frac{c}{\hat{c}^2} e^{-\frac{1}{2}\left(\frac{c}{\hat{c}}\right)^2} \tag{16.25}$$

where $\hat{c}$ is the 'predominant particle size' (i.e. the particle size where the distribution function has a maximum), the separation efficiency equation 16.24 is given by

$$\epsilon = \frac{2}{(c^*/\hat{c})^2} \left\{ 1 - e^{-\frac{1}{2}\left(\frac{c}{\hat{c}}\right)^2} \right\} \tag{16.26}$$

This dependence of the separation efficiency on $(c^*/\hat{c})^2$ is illustrated in *Figure 16.4*. Inasmuch as, according to equation 16.16, the value of $(c^*/\hat{c})^2$ is proportional to the reciprocal of the residence time, *Figure 16.4* reflects the dependence of the separation efficiency on the throughput of the separation device. As can be seen, this relationship is by no means linear. Approximate linearity holds only in the range of rather high separation efficiencies, i.e. those in excess of about 80%.

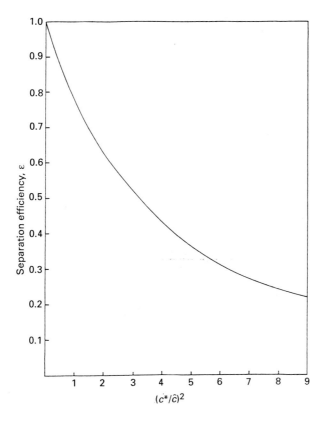

Figure 16.4. Graph to show the dependence of the separation efficiency on $(c^*/\hat{c})^2$

The normal particle size distribution (equation 16.25) is a special case of the well-known Bennett distribution [1]

$$f(c) = \frac{n^*}{\bar{c}^{n^*}} c^{n^* - 1} e^{-\left(\frac{c}{\bar{c}}\right)^{n^*}}$$ (16.27)

if $n^* = 2$ and $\bar{c}^2 = 2\hat{c}^2$.

16.3 UNLIMITED FINES BUILD-UP DUE TO OVERFLOW RECYCLING

Let us consider the faulty process shown in *Figure 16.1*, but, for the sake of simplicity, the screw filter press will be omitted so that the thick sludge goes directly into the drier and the decanter overflow is recycled without being modified by a screw press filtrate. The change in the particle size distribution with time for this case is shown schematically in *Figure 16.5*. After starting the process with fresh water, the size distribution of the particles entering the

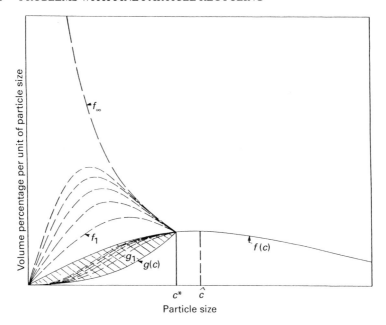

Figure 16.5. Graph to show the change in particle size distribution with time

decanter in the first pass is taken to be $f(c)$, with $\hat{c}$ as the predominant particle size, where $f(c)$ has a maximum, and c^* is the critical particle size of the decanter process. According to equation 16.19 and *Figure 16.3*, the particles removed as thick sludge will be represented by the area under $g(c)$ between $c = 0$ and $c = c^*$, and by the area under $f(c)$ between $c = c^*$ and $c = \infty$. The particles represented by the hatched area between $f(c)$ and $g(c)$ in the range between $c = 0$ and $c = c^*$ remain in the decanter overflow so that recycling them means adding them to a distribution $f(c)$ as formed by the continuous comminution and further treatment of new raw material. Thus, in the second pass the particles in the decanter feed resulting from the newly processed raw material and the first-pass overflow is obtained by adding the hatched area to $f(c)$ which leads to $f_1(c)$. Hence, in the second pass, the decanter feed contains more fines than in the first pass so that more fines become embedded in the coarse sediment. In accordance with equation 16.20, this situation is reflected by the curve

$$g_1(c) = \frac{c^2}{c^{*2}} f_1(c) \tag{16.28}$$

which is somewhat higher than $g(c)$. The particles in the decanter overflow are now represented by the area between $f_1(c)$ and $g_1(c)$. Again, their recycling means adding them to a distribution $f(c)$ which is formed continuously by the comminution and further treatment of new raw material, so that in the third pass the particles in the decanter feed resulting from the newly processed raw material and the second-pass overflow is obtained by adding the area between $f_1(c)$ and $g_1(c)$ to $f(c)$ which leads to $f_2(c)$.

Further recycling results in analogous changes, with distribution functions f_2, f_3, f_4, etc., for the feed, and distribution functions g_2, g_3, g_4, etc., for the embedment, where

$$g_i(c) = \frac{c^2}{c^{*2}} f_i(c) \tag{16.29}$$

Figure 16.5 shows these changes after five passes and the feed distribution function after an infinite number of passes.

For simplicity, the critical particle size c^* was kept constant, which is conservative as in reality an increased hindering of the sedimentation process due to an increased volume of suspended particles would shift c^* towards higher values. *Figure 16.5* illustrates the following.

1. The relative volume of fines increases and imparts a second maximum to the distribution curve.
2. The new maximum continues to increase, and to become more pointed while shifting towards smaller particle diameters. After an infinite number of passes, the maximum becomes infinitely high and will have shifted towards the particle size zero.
3. The relative volume of fines embedded in the coarse sediment also increases; initially this increase is primarily in the area close to the critical particle diameter c^*. As recycling goes on, the embedment curves $g_i(c)$ approach the virgin distribution $f(c)$ but will only reach it after an infinite number of passes when the relative volume of fines has become infinite.

In the steady state, if it could be realized, all fines resulting from the processing of new raw material would have to leave the system together with the coarse sediment. In the range below c^*, this implies the condition

$$g_{\text{steady state}}(c) = f(c) \tag{16.30}$$

According to equation 16.20 this leads to

$$f(c) = (c/c^*)^2 f_{\text{steady state}}(c) \tag{16.31}$$

or, rearranged,

$$f_{\text{steady state}}(c) = f_\infty(c) = (c^*/c)^2 f(c) \tag{16.32}$$

In view of equation 16.25, this shows that for $c \to 0$ the steady-state distribution would go towards

$$\lim_{c \to 0} \left(\frac{c^*}{c}\right)^2 \frac{c}{\hat{c}^2} e^{-\frac{1}{2}\left(\frac{c}{\hat{c}}\right)^2} = \lim_{c \to 0} \frac{1}{c} \left(\frac{c^*}{\hat{c}}\right)^2 e^{-\frac{1}{2}\left(\frac{c}{\hat{c}}\right)^2} = \infty \tag{16.33}$$

Thus, if no special measures are taken, overflow recycling will lead to an unlimited build-up of fines.

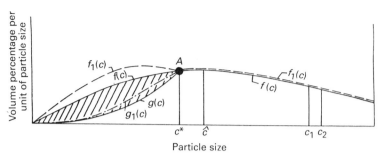

Figure 16.6. Change in the particle size distribution for the case of recirculating half of the overflow

16.4 MEASURES AGAINST FINES BUILD-UP

16.4.1 Partial evaporation

In *Figure 16.6*, as before, $f(c)$ is the size distribution of particles formed by the continuous comminution and further treatment of new raw material. If this distribution is referred to the total stream of particles, the area under the curve between the two particle sizes c_1 and c_2 represents the volume of particles of this size range. Again, c^* is the critical particle size of the decanter so that this machine will separate a sludge consisting not only of all the particles greater than c^* but also of the fines portion represented by the area $0-c^*-A-g(c)-0$. Consequently, the particles in the overflow are represented by the hatched area.

Against this background, let us consider what would happen if, for instance, half the overflow were submitted to an evaporation process leading to a thick syrup and if this syrup were added to the decanter sludge and thermally dried, thus leaving the system. In this case, recirculation to the other half of the overflow is reduced, as represented by half of the hatched area of *Figure 16.6*. In the first recirculation, this group of particles is added to the particles formed from new raw material so that the stream now entering the decanter will have a particle size distribution $f_1(c)$ which is equal to the sum of $f(c)$ and half of the hatched area; in *Figure 16.6*, $f_1(c)$ is shown as a dashed line. Of these particles, the decanter will separate a sludge consisting of all particles greater than c^* and of a portion of the finer particles, the quantity of the latter group now being greater than in the first pass because each infinitesimal particle size range below c^* now has a somewhat higher value. In *Figure 16.6*, this is indicated by the line $g_1(c)$ above $g(c)$. The corresponding solids in the overflow are represented by the area between $f_1(c)$ and $g_1(c)$. Again, one half of this area is deducted for evaporation while the other half is added to the spectrum of particles, $f(c)$, formed from new raw material. In this fashion, the change in the particle size distribution can be obtained by graphical means. This transient process, however, is of little practical importance as the principal point of interest is the question as to whether or not such a partial evaporation of the overflow will lead to a steady-state condition.

Obviously, if a steady-state condition exists, all particles continuously formed from new raw material must somehow leave the system. This equilibrium condition is clearly satisfied for all particles larger than c^* because these particles end up in the decanter sludge. However, the particles smaller than c^* must leave the system either by becoming embedded in the sludge of coarse particles or by leaving the overflow portion by evaporation. Thus, the equilibrium condition for the range of particles smaller than c^* must have the form

$$dV_0 = dV_1 + dV_2 \qquad (16.34)$$

where

$$dV_0 = f(c)\, dc \qquad (16.35)$$

represents the particles formed from new raw material, where

$$dV_1 = (c/c^*)^2\, F(c)\, dc \qquad (16.36)$$

represents the particles embedded in the sludge, with $F(c)$ being the particle size distribution at the steady state, and where

$$dV_2 = p\, [F(c)\, dc - (c/c^*)^2\, F(c)\, dc] \qquad (16.37)$$

represents the particles leaving the system by evaporation, with $p<1$, defining the percentage of overflow not recycled.

Substitution of equations 16.35, 16.36 and 16.37 into equation 16.34 and subsequent division by dc yields the important relationship

$$\frac{F(c)}{f(c)} = \frac{1}{\left(\dfrac{c}{c^*}\right)^2 + p\left[1 - \left(\dfrac{c}{c^*}\right)^2\right]} \qquad (16.38)$$

The above analysis allows the following conclusions to be drawn.

1. If the withdrawal factor p is equal to zero (if there is no withdrawal of overflow for evaporation), then for $c \to 0$ the ratio $F(c)/f(c) \to \infty$, in accordance with the previous statement that a complete recirculation of the overflow results in an unlimited build-up of fines.
2. For finite values of the withdrawal factor p, if $c \to 0$ then $F(c)/f(c) \to 1/p$. Hence, if, for example, one half of the overflow is withdrawn for evaporation ($p = 0.5$), then for very small particles the size distribution in the steady state will be twice as steep as the initial distribution. For all values of $p>0$, the particle size distribution will remain finite, in accordance with the empirical fact that partial evaporation of the overflow will lead to a steady state.
3. If the entire overflow is submitted to evaporation ($p = 1$), then, of course, $F(c) = f(c)$, because in this case the initial distribution remains unchanged.

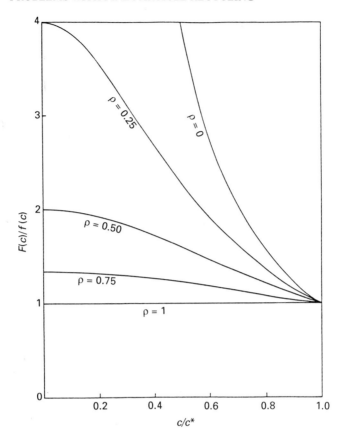

Figure 16.7. Graphical representation of equation 16.38

A graphical representation of the equation 16.38 is shown in *Figure 16.7*. As the stillage overflow of ethanol plants is customarily processed with a withdrawal factor $p = 0.5$, it is seen that this has no serious consequences.

16.4.2 Counterflow classification

Another means of coping with fines build-up is to use counterflow classification which involves freeing the recycle fraction from all particles smaller than a 'cut' c^l. Such a process cannot be accomplished in common sedimentation devices such as settling tanks, decanters, solid bowl peeler centrifuges, or disc separators. While these types of apparatus yield a fine fraction devoid of all particles greater than an upper limit c^*, there is no lower particle size limit for the coarse fraction, as explained in section 16.2, with the fine particles initially in the very vicinity of the sedimentation wall being able to reach this wall and thus becoming embedded in the sediment of coarse particles.

The types of apparatus capable of providing the sharp 'cut' desired for the purpose at hand are called 'hydroclassifiers'. The mode of action of hydro-

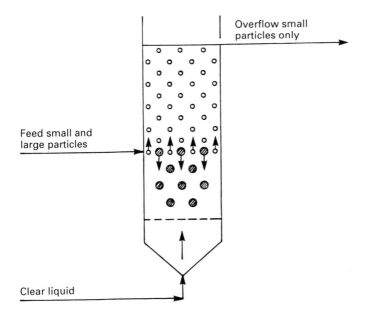

Figure 16.8. Schematic of a hydroclassifier

classifiers is based on the formation of a sediment, but prior to settling the coarse fraction is passed by a counterflow of clear liquid, and is thus 'washed free of fines'. The principle of action of hydroclassifiers is shown schematically in *Figure 16.8* which depicts particles of two sizes in a settling tank featuring an upward flow due to a clear liquid being injected in the bottom. The velocity of the upward flow is chosen to be sufficient to carry the small particles into the overflow but insufficient to prevent sedimentation of the large particles. Consequently, the sediment will be free of small particles.

The removal of particles from a small 'cut' size c' all the way down to zero is called 'desliming'. A typical example of a desliming apparatus is the 'wash cyclone'[2] in which high-pressure water is injected tangentially into the lower part of a hydrocyclone. The resulting displacement flow added to the normal cyclone flow does not permit the fine particles to enter the underflow, thus effecting a sharp classification.

Figure 16.9 gives an illustration of what happens when the overflow of a decanter is deslimed to a particle size c' and then recycled. As before, $f(c)$ represents the particle size distribution resulting from the processing of fresh raw material, $\hat{c}$ being the predominant particle size of this distribution, and c^* being the critical particle size of the decanter, so that the decanter sludge contains all particles greater than c^* and the fines portion is represented by the area under the line 0–g(c)–A. Hence, the overflow particles are represented by the area 0–$f(c)$–A–g(c)–0. A hydroclassifier is assumed to remove from this overflow all particles smaller than c'. This fraction is indicated in the figure by the hatched area. Thus, the particles remaining in the deslimed overflow are represented by the area B–g(c)–A–$f(c)$–C–B. In the first recircu-

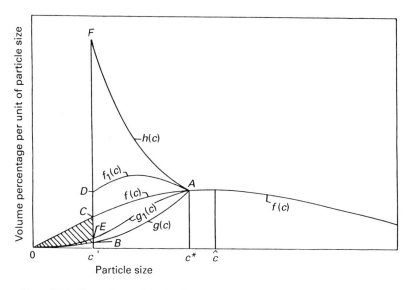

Figure 16.9. Change in particle size distribution for the case of overflow desliming

lation, these particles are added to the particles of distribution $f(c)$ that result from the processing of fresh raw material. Consequently, the suspension now entering the decanter contains all particles greater than c^* and the fines represented by the area under the line $0-C-D-f_1(c)-A$. The sludge separated from this suspension by the decanter contains all particles larger than c^* and the fines represented by the area under the line $0-B-E-g_1(c)-A-f(c)$, so that the corresponding overflow contains the particles represented by the area $0-g_1(c)-A-f_1(c)-D-C-0$. From this overflow, the hydroclassifier will again remove all particles smaller than c', thus leading to a deslimed recycle fraction with particles represented by the area $E-g_1(c)-A-f_1(c)-D-E$.

Continuing this procedure in an analogous fashion yields the change in the particle size distribution by graphical means. After an infinite number of recycles, the suspension entering the decanter will contain all particles larger than c^* and a fines portion represented by the area under the line $0-C-F-h(c)-A$, where, according to equation 16.32

$$h(c) = (c^*/c)^2 f(c) \tag{16.39}$$

Thus, with a hydroclassifier affecting a desliming up to c', the suspension entering the decanter can experience a maximum increase in solids volume as represented by the area $C-A-F-C$. Hence, the 'fines build-up factor' defined by the ratio of the solids volume in the steady state and the initial solids volume is

$$A = (1 + J)/1 \tag{16.40}$$

where J is the area $C-A-F-C$.
In view of *Figure 16.9* and equation 16.39

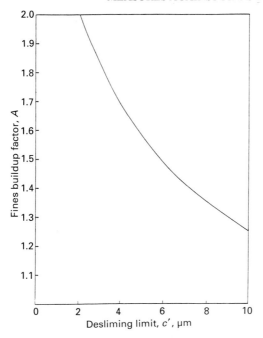

Figure 16.10. The fines build-up factor A as a function of the desliming limit c' for a typical ethanol plant. The initial distribution is represented by the superposition of two normal distributions with $\hat{c}_1 = 30 \mu m$ and $\hat{c}_2 = 1100 \mu m$. The critical particle size is c* = 27 μm

$$J = \int_{c'}^{c*} \left(\frac{c*}{c}\right)^2 f(c) \, dc - \int_{c'}^{c*} f(c) \, dc \qquad (16.41)$$

Introducing equation 16.25, this leads to

$$J = \int_{c'}^{c*} \frac{c*^2}{c^2} \times \frac{c}{\hat{c}^2} e^{-\frac{1}{2}\left(\frac{c}{\hat{c}}\right)^2} dc - \int_{c'}^{c*} \frac{c}{\hat{c}^2} e^{-\frac{1}{2}\left(\frac{c}{\hat{c}}\right)^2} dc \qquad (16.42)$$

resulting in

$$A = 1 + \frac{1}{2} \left(\frac{c*}{\hat{c}}\right)^2 \left[\mathrm{Ei}\left(-\frac{1}{2}\left[\frac{c*}{\hat{c}}\right]^2\right) - \mathrm{Ei}\left(-\frac{1}{2}\left[\frac{c'}{\hat{c}}\right]^2\right) \right]$$

$$- \left[e^{-\frac{1}{2}\left(\frac{c'}{\hat{c}}\right)^2} - e^{-\frac{1}{2}\left(\frac{c*}{\hat{c}}\right)^2} \right] \qquad (16.43)$$

where Ei is the 'exponential integral' which is available in tabulated form[3]. Equation 16.43 gives the fines build-up factor A as a function of the desliming limit c'.

A numerical evaluation of equation 16.43 for a typical ethanol plant is shown in *Figure 16.10*. Whereas for $c' = 0$, the case of recirculating the decanter overflow without desliming, the fines build-up factor A becomes infinite, it is seen that even modest desliming reduces the fines build-up to an acceptable level.

REFERENCES

1. Bennett, J. G., 'Broken coal', *J. Inst. Fuel* **10**, 22–39 1936
2. Dahlstrom, D. A., 'Fundamentals and applications of the liquid cyclone', *Chem. Eng. Progr. Symp. Ser.,* **50**, 41–61 (1954)
3. Jahnke, E. and Emde, F., *Tables of Functions with Formulae and Curves*, pp. 6–8, Dover Publications, New York (1945)

17

Filter Media, Filter Rating

Lawrence G. Loff

Tetko Inc., Briarcliff Manor, New York, USA

17.1 INTRODUCTION

The primary goal of this chapter is to gain a better understanding of filter media and filter ratings. Our definition of filtration will be in reference to liquids and will be the process whereby solids are separated from a liquid by passage through a permeable medium. Particles larger than the passages through the medium are retained, while the liquid (filtrate) passes through. Smaller particles may or may not be retained, depending upon the complexity of the channels through which the filtrate must pass.

The filter medium is installed in the filter over a stationary structure and provides adequate porosity capable of passing liquid through, while retaining solids. The medium is characterized by its filtration properties of permeability and retention. Filtration will not take place without the medium and the overall performance of the filtration system depends directly upon the properties and its interaction with the influent.

Filter media can be natural, synthetic or glass, woven or non-woven cloth: paper, ceramic, cellulosic or synthetic membrane, or, in some cases, woven, etched or sintered metal, depending upon the requirements and/or allowable cost.

Most filtration techniques require the use of a filter medium. The medium can be considered as the heart of the filtration device or machine. It is important to choose a suitable material of construction and to specify correctly the filter medium according to the properties and characteristics of the material to be filtered. The manner of application and attention to maintenance during the use are additional factors of importance. Failure to consider carefully each of these factors relative to each filtration application, will reduce the efficiency of separation and the operational performance achieved.

Prior to discussing any specific types of filter media, I wish to give a small representation of various types of filtration equipment or hardware, which is the vehicle in which the medium operates.

Various types of filtration equipment or hardware employ one of four main classes of driving force: gravity, vacuum, pressure or centrifugal force. The choice of which type of equipment to use in a specific application depends upon many widely differing factors, which will not be detailed in this chapter.

Gravity is simple, and has zero direct running cost. However, the equipment is usually bulky and fairly coarse solids will still contain an appreciable amount of liquid after separation. This may lead to unfavourable capital and running costs overall.

Gravity filters depend upon atmospheric pressure to force the solids through the medium. The slurry is fed into the top of the filter and through the medium, with the clear filtrate emerging from the underdrain. Sand filters, travelling belt filters and rotary drum gravity filters are found in this group.

In vacuum filters, vacuum can be easily produced either by the suction of an ordinary liquid pump or by a gas displacement device, such as a rotary vacuum pump or steam ejector. Depending upon the device, the effect is to find a driving force of up to about 12 p.s.i. (0.81 bar), which in many instances is sufficient to give vastly improved rates of filtration with all except the finest solids.

The vacuum is created behind the filter medium, which causes the atmospheric pressure in front of the filter medium to drive the slurry through the medium, filtering out suspended solids in the process. Types of vacuum filters include: belt, horizontal pan, vertical disc, and drum varieties.

The third type of driving force is pressure. Pressure filters produce a greater output per unit area, so enabling smaller equipment to fit easily into the process circuit, and making the handling of volatile liquids easier. However, it is sometimes difficult for continuous discharge of solids, such as cake. The feed slurry is introduced into the filter under pressure and forced through the filter medium. There are two basic designs used in pressure filtration: plate-type filter presses, whereby a series of plates utilize a filter medium between them; and casing-enclosed pressure filters, which consist of filter medium-covered elements enclosed in a pressure tank.

Another type of pressure filter utilizes two belts which move in the same direction. These belts are normally on top of each other and for a certain distance they share the same rollers with the sludge travelling in between the two belts. These rollers decrease in diameter as the belt travels, and thus the sludge is subjected to a tremendous shear force as it goes over the roller. This assists in the dewatering and filtration process.

Centrifugal driving force takes in two types of separators: cyclones and centrifuges. Cyclones have simple, compact construction with low running costs. However, separating efficiency falls off rapidly for particles below about 10 μm, and solids are discharged as a slurry. Centrifuges give higher separating efficiency even for very fine particles. Discharge of solids is either in cake or slurry form, although with very fine particles continuous discharge is only possible as a slurry. High throughputs are achieved relative to the size of a centrifuge so, hopefully, justifying the high capital cost of this piece of equipment.

The liquid is passed under centrifugal force through a filter medium or perforated plate. Filtration occurs as the liquid passes through the interstices of the solid particles that have built up on the medium surface. Depending upon the degree of separation desired, filter media can be a filter fabric of twilled weave, dutch weave or plain weave in either stainless steel or synthetic fibre.

17.2 FILTER MEDIA — GENERAL

Just as there are vast numbers of different types of filters, the same holds true for filter media, on which the filters are so dependent. I will present as many types of media as feasible, and will concentrate on those that are most commonly used.

An important question on filter media is to ask how fine a degree of filtration it will give. There are three groups of considerations which are raised when attempting to define the degree of fineness of filtration that can be achieved with a given medium. The first is the structure of the medium; the second is the mechanism by which filtration occurs, which is dependent upon the medium and the fluid–particle system to be handled; and the third group is the method of measurement used.

Structural considerations are only relevant for solid media in rigid or semi-rigid form such as porous metals, fabrics, screens or layers of powders (precoats). Filtering characteristics are greatly affected by the size and shape of the holes, as well as by their path through the medium (i.e. straight or tortuous), and whether they vary in size and shape through the depth of the medium. Also, the number of holes per unit area, and their uniformity, affect filtering.

Obvious uniformity is found in simple form with plain-weave metal cloth of light gauge wire. As the gauge of the wire becomes heavier and the weave is changed to a twilled or Dutch-type weave, we have a more elaborate medium that is generally used for filtration. The nature of the holes is more complex and more difficult to recognize with the unaided eye. Woven fabrics become more complicated due to the flexible nature of yarns, and therefore it is more difficult to try to define the size of the hole in a woven fabric. The same is true for media with random structure, such as felts, paper, fibrous and porous material.

The main types of mechanism by which filtration occurs are surface and depth. These are also referred to as classifications of filter media. Other classifications are rigid versus flexible, loose versus integral, and permanent versus disposable media.

An example of surface filtration would be a straining process with wire cloth, wherein all of the solids larger than the media openings are deposited on the cloth surface, while the smaller particles pass through. Depth filtration utilizes the thickness of the medium as well as its surface, primarily to trap particles within the interstices of their internal structure as the fluid travels a tortuous path through the medium. Porous media, felt or the wound type of filter cartridge, are examples of depth media.

Cake filtration occurs after surface filtration where a cake acts as the filter medium. In some instances the cake formed has a lower permeability than the filter medium and thus becomes the limiting factor in the filtration cycle.

The last factor, when attempting to define the degree of fineness of filtration achieved with a given medium, is the method of measurement used. The three criteria by which the performance of a medium is judged revolve around:

1.   the measured data of how small a particle it can stop;
2.   the permeability (the ability of the medium to allow flow);
3.   the relationship between build-up of dirt in the medium and the rate of increase of resistance to flow.

These three criteria are discussed in section 17.9.

17.3 CARTRIDGE FILTERS

Cartridge filters are one of the most widely used types and have the advantage of being simple, with minimal installation cost. Areas of common use for cartridges are in the automobile and aerospace industries for protection of hydraulic and lubricating systems. They are also used in electroplating for filtration of solutions, as well as in the beverage manufacturing field and in chemical processing. Cartridges have had many improvements in manufacturing and pore size control. Some media used in making cartridges, such as porous ceramics, sintered metals, felts and papers, are dealt with separately as we proceed. However, there are certain media that are used only for cartridges because of their physical form. These media are wound cartridge and bonded cartridge, which are discussed with depth filters in the following. There are three basic types of cartridge filters: edge, surface and depth.

The edge filter is of a solid fabrication structure wherein the medium consists of the edges of a stack of specially formed washers or thin discs mounted on a central perforated core (shaft), and held under compression so as to form a continuous cylindrical outer surface. Each washer has grooves or scallops cut into it, such that very fine and carefully controlled gaps occur between the adjacent washers. The gaps are usually sized from 5 μm upwards. Filtration takes place by flow of liquid inward through the narrow gaps between adjacent discs. The filter is essentially a strainer in that particles are stopped on the outside surface. The discs are usually metal but they can be nylon, polypropylene or paper.

Another filter medium in the same category is that of elements made by winding a wire on a cylindrical support. The wire can be round or wedge shaped. Wedge-shaped wire gives wider apertures on one face than the others of adjacent wires wound edgeways on to the perforated support. The micron retention range for the edge filters can range from 1 μm for the paper disc type, from 25 μm for the wire wound, and upwards of 50 μm for the metal disc type. Edge filters can also be cleaned by a back-flushing on paper disc-type or scraping on wire wound or metal disc types. Their cleanability could make them more economical than less expensive media.

The depth cartridge filter holds to the meaning of depth filtration whereby the particles are primarily trapped within the interstices of the internal structure. The two main types of depth cartridges are bonded and wound. Both types offer a simple, compact diposable unit with a high dirt-holding capacity able to remove solid particles from liquids, down to about 1 μm. They have a small surface area but it can be increased by cutting grooves on the outer surface, which also increases life by delaying plugging.

Bonded cartridges are composed of fine, loose fibres of wool, cotton, cellulose, glass and various synthetic materials built into a thick-walled tube by a filtration technique wherein the fibres are formed wet. After dry: the tube is impregnated with resin and cured, forming a light, very porous rigid structure with good dirt-holding capacity. These cartridges are inexpensive because they are self-supporting and do not generally have a central support core. Dirt holding is improved by changing the density of the medium. Bonded cartridges are not cleanable.

Wound cartridges consist of spun stable fibres of wool, cotton, glass and various synthetic products. The fibres can be brushed to raise the nap after each layer is formed, wound on a hollow perforated core until the desired cartridge thickness is achieved. The nap forms the filtering medium by varying the closeness of the winding from the central to the outside layer. The porosity of the medium is determined by control of winding pitch, tension, fibre length and other characteristics. Wound cartridges are similar to bonded filters in that they are inexpensive with large dirt-holding capacity; also, they cannot be cleaned.

In going now to surface filtration cartridges, we find the most widely used media are in thin, sheet-form of cellulose paper or resin-treated paper. They are normally corrugated or pleated to increase useful filter life and filter surface area (e.g. automobile air filters). A single compact cartridge can have up to 2.8 m^2 of surface area. The retention range of this type of cartridge is from 0.5 to 50 μm. Woven materials, primarily woven wire, are also used for this type of cartridge. The bulk of the filtration takes place on the surface but some depth filtration is also involved.

In general, and in summary, cartridge filters have an integral cylindrical configuration made with disposable or cleanable filter media and utilize either plastic or metal structural hardware. Appropriate housings form the filter assembly and for large flows, multiple cartridges are installed in a pressure vessel. Such materials as cotton, wool, rayon, cellulose, fibreglass, polypropylene, acrylics, nylon, asbestos, cellulose esters, fluorinated hydrocarbon polymers and ceramics are commonly employed in the manufacture of disposable cartridge filters. Porous media for cleanable or re-usable cartridges typically employ such materials as stainless steel, monel, ceramics and fluorinated hydrocarbon polymers, as well as exotic metal alloys.

17.4 RIGID POROUS MEDIA

Ceramics and stoneware have good resistance to chemical attack and high temperature. Stoneware is made from certain types of clay rich in silica and is more a sub-class of ceramics, but all ceramics are made from powdered solids by a process that involves kiln temperatures of 1400°C.

The ceramic elements are made in many shapes and sizes in various grades, which correspond to different average pore sizes. One disadvantage is that they are often fragile. However, their fragility is off-set by their low cost and high resistance to corrosion, as well as the wide range of porosities available when compared with available alternatives. Pore sizes can range from 2500 down to 1 μm. Ceramics are widely used in gas filtration, and in separating dust and liquid droplets from gases.

Sintered metals can be produced either from powdered metals or from woven wire. Various mesh-size powders are sintered in a controlled atmosphere at temperatures slightly below their melting point, causing a strong metal-to-metal fusion forming a tortuous path (depth) for filtration, approximately 1.6 mm (1/16 in) thick. Many different metals, such as bronze, various stainless steels and Inconel, can be used. Cutting, bending and welding of sheets into parts is also possible. The cost is usually high but can be off-set by the media's ability to be cleaned for re-use by back-flushing or chemical cleaning.

The woven wire-type, sintered media can be made either with a single layer or several layers of woven wire sintered with the powder. By controlling the particle size of the powder or the wire gauge and weave of the woven metal, the porosity of the sintered product can be made with precision from 400 to under 1 μm.

Keeping our discussion to the different types of porous media, we should not forget to include porous plastics. The range of porous plastics and the shapes (forms) in which they are available, are still growing. A variety of materials such as nylon, polyester polyurethane, polyethylene and fluorocarbon, among others, have gained in popularity. Pore sizes can vary from large holes down to under 1 μm, produced by sintering and foaming techniques. Nucleopore medium, which is a form of porous plastic, is produced by a radiation technique on polycarbonate and polyester which can yield uniform holes in the range 1–10 μm at will. Laser technology is also used to produce membranes of uniform openings at under 1 μm.

The Nuclepore membrane is a thin film of about 6–10 μm thick, containing very fine pores ranging from 0.015 to 14.0 μm of polycarbonate and polyester materials. Membranes of mixed esters of cellulose or PTFE can range between 100 and 200 μm thick with pore sizes of 0.1–5 μm. This medium has found wide use in water filtration, cell culture, and serum filtration among others. Membranes are formed from dissolving the raw material in a mixture of two solvents leading to an evaporation to form the porous structure. Some are formed around woven monofilament fabrics or paper for reinforcement. Their main use has been for liquid and gas filtration where high standards of resolution are needed, such as in water desalination by reverse osmosis or separation by ultra-filtration (separation of colloid substances such as clays or pigments, and relatively high-molecular weight materials as polymers or protein separated from liquids in which they are suspended). For a wide diversity of applications membrane is available in materials other than those mentioned above. Examples are PVDF, polypropelene, nylon, acrylic co-polymer plus others which may be standard or special.

There are two general types of membrane, symmetrical and asymmetrical. Both sides of a symmetrical membrane are the same, therefore, either side can be used in filtration. Symmetrical membranes have very uniform pores, such as a Nuclepore membrane. The sides of the asymmetrical membrane, unlike the symmetrical membrane, are very different. One side has very small openings that branch out through the membrane to form a very large opening on the other side; the performance of the filtration depends upon which side is used. This membrane has a tendency to load up with solids.

17.5 NON-WOVEN MEDIA

Although this section is particularly about non-woven media, much of what we have discussed earlier has, in fact, been non-woven. In this sense of the term, 'non-wovens' are loose assemblages of fibres arranged in short form and physically bonded with a bonding system. They are lighter and thinner than felts with higher permeability, lower retention and greater strength than paper media. In direct contrast to this, I find a paper manufacturer describing his paper-making process for filter paper in the same manner. From studying what both sources are presenting, I believe it to be a matter of how one wishes to interpret 'non-woven'. Purchas (1967) classifies felts, papers, sheets and mats, among other similar media, all in the category of non-woven — so will I. They all basically have random pore structures.

Felts, also known as 'non-woven fabrics', have been used for many years for filtration, but they were formerly made from wool. Once synthetic materials were used in the manufacture of felts, they became very widely accepted and used for many filtration applications. In fact, the word 'filter' is derived from the mediaeval Latin word *filtrum*, which means felt. Felts are now made from a wide range of materials such as olefin, nylon, polyester acrylic and fluorocarbon fibres. These fibres are arranged at random, in mass, with or without a resinous bonding agent (or heat or chemical bond), and are formed into compacted pads which control the thickness of the fibre as well as the porosity and density of the medium. They have fibrous finishes but can be calendered in order to smooth this out.

Felts have many applications in chemical, electrolytic, biological, thermal, cryogenic and foodstuff fields, for gas and liquid filtration. They can be used for filter presses, moulded filter elements, and fabricated cartridges where felt would be formed with the pleated woven wire. Felts can also be used as a drum cover or filter belt for rotary vacuum filters and in the manufacture of filter bags. Synthetic felts also show an excellent particle retention, freedom from blinding (plugging) and resistance to fungus and corrosion. They have dimensional stability and maintain their clean-cut edges in addition to being easy to sew for fabrication. Felts also make efficient gaskets, preventing side leakage between elements on filter press applications. They provide ease of cleaning with hosing or backwashing, which helps to reduce down-time.

An item that we all see throughout our daily lives but do not realize that it is, in one of its forms, a filter medium, is paper. It is a very widely used medium, made by dispersing fibres into a suspension in water. This suspension is then filtered into a mat form, which is compressed and dried. By varying the process, different porosities are achieved. The size of the fibre is important. Cellulose fibres are relatively coarse while glass is finer. Cellulose-based papers have poorer retentive power and are widely used in industrial liquid filtration applications because of lower cost and good mechanical properties. Glass is typically used in laboratory applications.

Both glass and cellulose papers can be impregnated with bonding agents such as melamine, resin and neoprene to increase strength and to modify filtration characteristics. Silicon is also used to give water repellancy for coalescer and separation applications. A very popular end-use for filter paper is for pleated cartridges in automotive and industrial applications where the micron range is from 0.5 to 500 μm. Filter paper is also used on plate and frame presses and pressure leaf presses. Because of its weakness, paper is usually supported by perforated metal, woven wire or synthetic cloth.

Glass fibre paper has workable temperatures up to 500°C, which will exceed that of cellulose paper, membrane and porous plastic materials. This brings glass into competition with asbestos and metallic media. Glass fibre is known for good retention and high flow rates. In the same respect as paper, we have filter sheets and mats which are distinguished as slightly different due to characteristics of thickness that classify them as depth media, whereas paper is primarily a surface medium. Sheets are basically made from asbestos mixed with cellulose for a binder or diatomaceous earth for increased permeability. Other materials such as carbon, lime, synthetic powders, etc. are also used. One face of the filter sheet is usually harder than the other. The harder side is intended for use on the outlet side to prevent fibre loss in the filtrate. Other treatments are also available depending upon the end-use application.

Before describing woven filter media further, it should be noted that there are many filter media available which cannot be detailed in this chapter. The following are some not previously mentioned and they are presented here only in name in order to indicate the vast variety available. Filter sheets are available, formed from diatomaceous earth rather than asbestos. Loose fibre filter media, made from asbestos, cellulose and glass wool, are other types of material available. There are a wide variety of loose particles available as filter media, including precoats, sand or body aids. Body aids are additives to the liquid to be filtered. Lastly, perforated sheets in both synthetic materials and metals are used but have restrictions because hole sizes are generally no smaller than 50 μm. They are commonly used as support or back-up for finer media, or alone as strainers or roughing filters.

17.6 WOVEN WIRE

Woven media can be called woven wire, woven fabric, wire mesh, wire cloth etc. Precision woven wire cloth is a versatile, wear-resistant filter medium which has been widely used for many years and is available in a large variety of weaves made from many metals. The most frequently used metal is stainless steel, in either type 304 or type 316. The basic difference between the two types is the addition of molybdenum to the type 316 for increased corrosion resistance. Both are otherwise 18-8 alloys, i.e. 18% chromium and 8% nickel. Actually type 304 is 18–20% chromium and type 316 is 16–18% chromium.

Woven wire can be made very fine with small wire diameters spaced very close, yielding as many as 635 openings per lineal inch with wires as small as 0.0008 in (0.02 mm) and aperture size of 20 μm. However, this combination, which is woven in a twill square weave of over two wires and under two in each direction, gives a very delicate material for filtering on a large scale. The answer to this situation lies in the weaving of a Dutch weave material, wherein there are a greater number of fine wires in the weft, or width, of the material and a smaller number of heavier wires in the warp, or length. Over one and under one gives a plain Dutch and to achieve a denser cloth the weft wires cross over two warp wires at a time in a staggered arrangement called a 'twilled Dutch'. Both materials are very strong and can be woven with retention down to 1 μm on a twill Dutch and down to 8 μm on a plain Dutch (duplex-twin warp). We will see more details on these weaves a little further on.

Woven wire lends itself easily to fabrication of pressure filter leaves and filter baskets. It is easily applied to filter drums. It is resistant to blinding and is easily cleaned. Stretching or shrinking is non-existent. It usually costs more than non-metallic media, but usually has a long life which helps to off-set the cost along with the other factors above.

Another use for woven wire is for cleanable cartridge filters with nominal ratings from 300 to 1 μm. Although the use of metal media is expensive, the cartridge can be chemically cleaned and re-used. The aircraft and aerospace industries have been big users of these types of elements.

There are several available weaves of woven wire. Some of these weaves lead the product to be called woven fabric. The basic types of weaves of wire cloth, which are woven for filtration purposes, are plain, twilled, plain Dutch, twilled Dutch, plain reverse Dutch, duplex (twin warp) plain Dutch, Betamesh and braided (basket of multibraid)—see *Figure 17.1*.

As mentioned above, a plain weave has each shute or fill wire passing alternately over and under each warp wire. Square or rectangular openings are available. Openings are 'straight through', thus permitting filtrate to pass through the cloth in a perpendicular path. The width of the opening is limited by the wire diameter and the number of wires per lineal inch. Finer meshes with smaller openings are lacking in physical strength.

Twilled weave (plain twill), also has straight through openings. Each shute or fill wire alternately crosses over two and under two warp wires, forming a diagonal pattern. Larger wires can be used for a given mesh size yielding proportionally smaller openings and greater physical strength. For example, if a 0.0055 in (0.14 mm) square opening is required for a particular application, it could be obtained with plain square weave, 100 mesh with 0.0045 in (0.11 mm) wire diameter; or in the twilled weave, 80 mesh with 0.007 in (0.18 mm) wire diameter, which is a stronger fabric; stronger yet is 60 mesh 0.010 in (0.25 mm) wire diameter with 0.0057 in (0.14 mm) openings. Each of the two latter specifications would have longer service life and good wear resistance, but at the expense of a smaller percentage of open area or slower flow rates through the woven wire.

When the twilled weave is woven with multiple wires in both warp and shute, it results in a strong, dense fabric known as braided, basket or multibraid weave. The mesh openings are irregular because the multiple shute wires have a tendency to twist around each other.

In plain Dutch weave, the warp wires are generally larger than the shute or fill wires, and are spread far enough apart so that each shute wire passes alternately over one and under one warp wire and is positioned tightly against the adjacent shute wire. This yields relatively small openings with high strength. There are no straight through openings; they are similar to a triangle in shape and twist through the material on an angle. These materials are mainly rated by particle retention in microns because opening sizes and percentage of open area are difficult to determine. A popular plain Dutch weave is 24 × 110 mesh with 0.010 x 0.015 in (0.25 × 0.38 mm) wire diameters.

Duplex (twin warp) plain Dutch weave is a stronger weave than convential plain Dutch weave with the same micron rating. Although similar to a plain Dutch weave, it has two small-diameter warp wires in place of one large warp wire. This arrangement gives an even stronger material than plain Dutch weave and smaller openings can be achieved. The openings are triangular shaped and twist through the material on an angle, as described for the plain Dutch weave. Duplex weave is rated by particle retention, in microns, as are all the Dutch weaves. Filter ratings generally range from 65 to 8 μm.

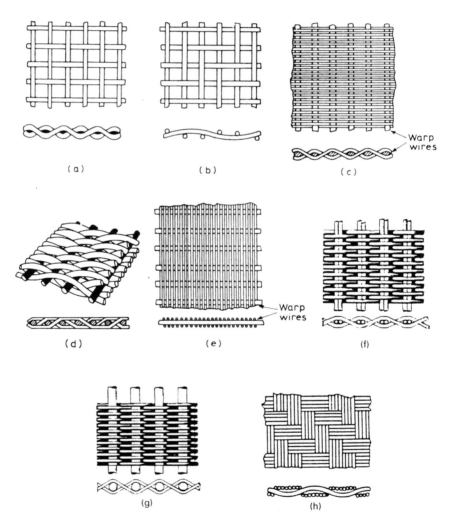

Figure 17.1. Different types of weaves. (a) Plain square weave. (b) Plain twilled weave. (c) Plain Dutch weave. (d) Twilled Dutch weave. (e) Reverse Dutch weave. (f) Duplex (twin warp) plain Dutch weave. (g) Betamesh Dutch weave. (h) Basket (braided or multibraid) weave.

With a twilled Dutch weave the mesh count of the warp wires is much lower than that of the shute. The shute wires pass over two and under two wires in both directions. They are pushed together such that one shute wire is on top of the next one, which is directly under the warp wire. This yields twice as many shute wires for the same wire diameter as in the plain Dutch weave. This is a very strong and dense weave, through which it is almost impossible to see light on a perpendicular projection. Mesh counts range from 20×250 with $100 \ \mu m$ retention to 510×3600 with less than $1 \ \mu m$ retention.

Reverse Dutch weave is a weave where the arrangement of the warp and shute wires is reversed as compared with the plain Dutch or twilled Dutch weave. The greater number of wires closely woven together are found in the warp, and the shute wires have the larger diameter. This yields a very strong material with high rates of flow as compared with twilled Dutch weave with the same micron rating. It has characteristics of good cleanability and resistance to blinding. Mesh counts range from 140×40 with approximately $100 \ \mu m$ retention to 850×155 with $10 \ \mu m$ retention.

The last weave to be mentioned at this point is betamesh Dutch weave. Betamesh is similar to a plain Dutch weave with superior performance characteristics. A larger portion of solids is retained on the surface of the material than with other Dutch weaves. This results in a higher contaminant tolerance and excellent back-flushing properties. Betamesh has more open area and consequently allows higher flow rates than do the other Dutch weaves. The ratio of warp-to-shute wire sizes and the spacing of warp wires have been carefully established to yield precise equilateral triangular openings. Their uniform $60°$ angles are set during weaving, so that internal flow passages of the fabric will always be larger than the surface openings; flow rate is thus improved, and blinding reduced. The surface slots are narrower than the inner triangular openings. Thus, particles are retained on the medium surface before they can plug the inner passages. Betamesh Dutch weave is available in the retention range 81–$10 \ \mu m$.

17.7 WOVEN FABRICS

Woven fabrics are made with natural and synthetic fibres. In some cases there is similarity with woven wire in the fabric construction or job capability. Many synthetics are woven in order to replace as many woven wire applications as possible and/or to replace a similar competing fabric in natural or synthetic fibre itself.

Woven fabrics are made from three different forms of yarn. One is the spun staple yarn, which is made by twisting short lengths of natural or synthetic fibre into a continuous strand; it has good particle retention because of its hairy filaments, and offers excellent gasketing properties. Next are monofilament yarns, which are made of a single continuous filament of synthetic fibre. They allow minimum blinding, good cleaning and excellent cake discharge characteristics. Lastly, multifilament yarns are made by twisting two or more continuous monofilament threads together. They have the greatest tensile strength of any yarn and allow better cake discharge than spun yarns. Cloths woven with multifilament or spun fibre are called 'multifilament fabrics' while materials woven with monofilament fibre are known as 'monofilament fabrics' — see *Figure 17.2* and *Table 17.1*.

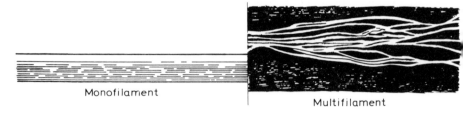

Monofilament

Multifilament

Spun staple

Figure 17.2. Effect of yarn. (Courtesy of Purchas, 1967)

Table 17.1. EFFECT* OF TYPE OF YARN ON CLOTH PERFORMANCE
(From Purchas, 1967)

Maximum filtrate clarity	Minimum resistance to flow	Minimum moisture in cake	Easiest cake discharge	Maximum cloth life	Least tendency to blind
Staple	Monofil	Monofil	Monofil	Staple	Monofil
Multifil	Multifil	Multifil	Multifil	Multifil	Multifil
Monofil	Staple	Staple	Staple	Monofil	Staple

*In decreasing order of preference.

Filter fabrics are mainly woven in four common weaves: plain, twill, plain reverse Dutch and satin. The first three are overlaps from wire cloth weaves but may have slightly different characteristics when put into the framework of woven non-metallic cloth. The weaves will not be detailed again here. The fourth weave is the satin weave where the shute (or warp) fibre passes over several warp (or shute) fibres, then under one in an alternating pattern. *Figure 17.3* shows an over three–under one pattern for the satin weave (three shaft satin); see also *Table 17.2.*

The woven fabric can be supplied either off-loom (greigh goods) or subjected to a finishing process. The two most common to the filtration

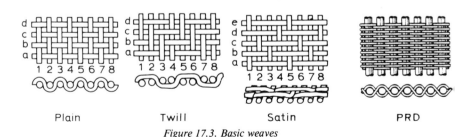

Plain   Twill   Satin   PRD

Figure 17.3. Basic weaves

Table 17.2. EFFECT* OF WEAVE PATTERN ON CLOTH PERFORMANCE
(based on a 40 μm nominal retention fabric of each weave)

Maximum filtrate clarity†	Minimum resistance to flow	Minimum moisture in cake	Easiest cake discharge‡	Maximum cloth life§	Least tendency to blind
Satin	Plain	Plain	Satin	Satin	Plain
Twill	PRD	PRD	Twill	Twill	PRD
PRD	Twill	Twill	Plain	PRD	Twill
Plain	Satin	Satin	PRD	Plain	Satin

*In decreasing order of preference.
†If cake is formed then clarity would be the same.
‡Depends on the texture of the cake (i.e. for sticky cake the order is reversed).
§Can depend on the type of equipment.

industry are heat treatment and calendering. Heat treatment is applied to synthetic cloth in order to stabilize the fabrics and make them more suitable at elevated temperatures. Calendering uses high-pressure, hot rollers whereby the fabric runs between the rollers, which cause the cloth surface to have a smooth polish that improves cake discharge and helps keep the cloth clean while in service.

Filter fabrics are commonly woven from such synthetic fibres as nylon, polyester, polypropylene and occasionally polyethylene. Other thermoplastics can be, or may have been woven, but are not commonly available. Each fibre type has specific characteristics in relation to chemical, thermal and mechanical properties that singles it out as the best candidate for use on the specific application. *Table 17.3* summarizes the use of common fibres.

Even though fibres of the above synthetic materials can be woven multifilament or monofilament, there is a trend towards the use of monofilament filter fabrics. This is so mainly because in years past it was very difficult to weave monofilaments in the fine, low-micron retention areas. Today, monofilament filter fabrics can be woven twilled down to 6 μm and plain reverse Dutch down to 14 μm. Further finishing (shrinking and calendering) can bring the particle removal rating down to the 1 μm area.

17.8 MATERIAL SELECTION

A very important, if not the most important, part of the filtration process is the proper selection of a filter fabric. When selecting a fabric, one must first look at the material from which it is made and take into account that material's chemical, heat and abrasion resistance with regard to the product to be filtered. The next consideration is the fabric weave and how this affects filtrate clarity, blinding characteristics and cake release properties. Multifilaments can yield clearer filtrates but have a greater tendency towards blinding than monofilaments. Where fine solids must be handled and a clear filtrate is required, one of the high-twist multifilament yarn, twilled-type weaves frequently provides the best compromise between clarity and blinding tendencies. This is not to say that some of the very fine monofilament calendered fabrics would not be a good choice. Assumptions are not the

Table 17.3. PHYSICAL PROPERTIES OF FIBRES

	Softening point, °C	Estimated safe temperature, °C	Specific gravity	Moisture regain %*, 70°F (21.1°C), 65% r.h.	Wet breaking tenacity**, g density^{-1}	Wet breaking elongation†, %	Resistance to wear
Acetate	204–229	99	1.32	6.3–6.5	0.8–1.0	35–50	Fair
Acrylic	NA	143	1.17	1.5	1.8–3.0	34–60	Good
Fluorocarbon	275	204	2.10	0	0.5	52	Fair
Glass (multifilament)	732–877	302	2.50	0	6.7	5.3–5.7	Poor
Modacrylic	NA	77	1.35	2.5	1.5–2.4	45.65	Fair
Nomex	—	218	1.38	—	4.1	14	Excellent
Nylon (6 or 6.6)	170–235	116	1.14	2.8–5.0	3.7–6.2	20–47	Excellent
Nylon 12	155–170	90	1.01	0.8	—	—	Excellent
Polyester	220–240	149	1.38	0.4	3.2–4.2	21–35	Excellent
Polyethylene							
Low density	107–113	71	0.92	0	1–3	20–80	Fair
High density	116–102	102	0.95	0	3.5–7	10–45	Fair
Polypropylene	141–149	121	0.91	0.01–0.1	3.5–5.0	14–30	Fair
PVDF	143–149	141	1.78	< 0.1	—	—	Good
Rayon	NA	99	1.54	11	0.95–1.1	24–28	Poor
Saran		77	1.70	0.1–1.0	1.5	15–30	Fair

NA—Not applicable to these fibres.

*That percentage by weight of moisture gained by an oven-dried textile product in a standard atmosphere of 70°F (21.1°C) with 65% relative humidity.

**The stress at which a fibre breaks.

† The increase in length or deformation of a fibre as a result of stretching as a percentage of the original length.

‡Difficult to ignite. Does not propagate flame. Does not melt. Decomposes at 499°C.

Table 17.4. QUESTIONNAIRE FOR FABRIC SELECTION

Make of filter. . . . . Type and size. . . . .

Filter operating data:
1. Minimum and maximum pressure employed. . . . .
2. Temperature at which solution to be filtered is maintained . . . .
3. Kind and concentration if liquid is acid, and pH value if known . . . .
4. Nature and percentage if liquid is alkaline or caustic, and pH value if known . . . . - . . . .
5. Chemical composition or nature and density of liquid . . . .
6. Chemical composition or nature and specific gravity of solids. . . . .
7. Physical characteristics of solids—i.e., crystalline, granular, slimy, colloidal, etc. . . . .
8. Retained on mesh

	40	80	100	150	200	300	400	+500
Percentage.	%	%	%	%	%	%	%	%

9. Maximum particle size that can be permitted to pass through filter. . . . .
10. Relative proportion by weight of solids to liquid. . . . .
11. Is cake difficult to discharge? . . . .
12. Is a filter aid used? If so, state kind . . . .
13. Would there be any advantage to increasing the acidity or causticity of the solution as now fed to your filters? If so, what would be the desired acid or caustic content? . . . .

Present type of filter medium . . . .

Average operating period of present filter medium:
1. Average total time it remains in place . . . .
2. Average total operating time obtained (if filter is not operated on a 24 hour daily basis) . . . .

Reason for removing present filter medium:
1. Mechanical failure ☐ ☐
2. Chemical corrosion ☐ ☐
3. Medium becomes blinded ☐

Remarks. State any special conditions not specified that are present in your process and that you feel affect the filter medium you are now using:

Source: 'Filter Media', *Chemical Engineering*, 1963 (Oct. 14).

proper way to make a fabric selection. Bench scale and/or actual testing on the filter is what should be done. We will discuss this aspect of selection further on.

In many cases, the filter fabric selection is based on a compromise rather than on locating the Utopian material that will do everything required at an economical price. The user of the medium must decide whether he is primarily interested in the filter rate or clarity, and whether the initial price of the material or life is important.

It is important to note that suppliers of filter media can make qualified recommendations based on experience, but no one can guarantee the performance of a filter fabric for an end-use until it is actually applied to the equipment for the application (proven). When selecting a filter fabric, a supplier looks for certain characteristics of the filtration problem in detail. Basic questions should be asked to narrow the choice down fairly accurately. Important points would be the filter type and size, temperature of the product being filtered, pH of the product being filtered, physical characteristics of the product (i.e. crystalline or granular), distribution of particle size and the current problem with existing media (blinding or mechanical failure). These, along with other questions about the wash cycle, should be answered to help achieve the best possible material to replace an existing medium and to obtain greater productivity for the filtration equipment. Just because a piece of equipment arrives at the plant with a certain material on it does not necessarily mean that it is the best material for the job. The operator should always be on the look-out for a 'better' medium based on the results he obtains from his existing media. *Table 17.4* shows a typical questionnaire for fabric selection.

The three methods for testing filter media are laboratory tests, pilot plant and full-scale tests. The laboratory tests usually consist of a Buchner funnel, in which the filter medium is placed in the mouth of the funnel and a sample of the slurry is poured on top of the medium. The slurry is then drawn through the medium and such factors as filtrate clarity and time can be checked. Laboratory tests are quick and economical, but often deceptive. The pilot plant, or small operating filter, is an improvement over the laboratory process as it gives results that are more likely to be reproduced in production.

The most reliable method of test as well as the most expensive, is to conduct a full production test. However, when testing, it is far better to test the entire filter—for example, a full press load of cloths on a plate and frame press, rather than one or two cloths, as these cloths might be doing the work of the others in the press, yielding no specific indication as to their actual capacity. The same situation applies to the disc filter where each sector of an entire disc should be covered, or even on the drum of a caulked-in fabric where the entire drum should be covered.

From the above, it is clear that the operator has many cloth selection factors to consider in order to obtain optimum performance. Often the only way is by trial and error, guided by broad general principles. Once the optimum cloth is found, the following benefits are obtained: a clean filtrate with no loss of solids by bleeding, an economic filtration time (production rate), an easily discharged filter cake, no deterioration of the medium by sudden or gradual blinding and an adequate cloth life.

17.9 FILTER RATING

There are many questions which centre around testing and rating methods of filter fabrics. Manufacturers generally do not use identical rating methods by following any one established test procedure. No universal standard seems to be followed. Any successful effort to standardize test methods lies in the distant future. However, current rating methods make it difficult for uniform procurement as well as causing frustration to the user in selecting the optimum medium. Suppliers make qualified recommendations based on their own experience but cannot really guarantee the performance of a filter medium for a given end-use until it has been thoroughly tested and found satisfactory.

Some common filter rating characteristics that are considered prior to testing the media are as follows: nominal filter rating in microns, absolute filter rating in microns, bubble point (inches water gauge) and air permeability. The nominal and absolute filter ratings can be determined by computations through the bubble point test. Other considerations that are not often used are: mesh count per inch, fibre diameter, cloth thickness, weight of material, tensile strength and water permeability.

Nominal ratings are intended to describe an average pore size for filtration performance. This would give a reading as to what size particles, in general, the cloth will retain. The absolute rating is intended to measure the largest single pore rather than an average pore. There are several tests to determine nominal and average pore size but a frequently used test is the bubble-point test described in the next paragraph. Other tests (see Cole, 1975), which are not detailed here, include filtration efficiency and bead transmission.

Bubble-point tests are, as already mentioned, frequently used to estimate pore sizes and distributions. The tests consist of immersing the test specimen in a wetting liquid and then displacing the liquid from the pore structure with gas or air pressure. The first bubble through is a measure of the largest pore or absolute rating of the specimen. These tests are reproducible but do not give much information about the general pore structure.

Mean flow ratings, sometimes called 'boil-all-over bubble points', attempt to indicate a defined average pore size which can be more meaningful than the 'average' pore size alone. Once the first bubble appears as the largest pore, or absolute rating, the air pressure is slowly increased. The filter will then, fairly suddenly, 'boil all over'. At this point we have the mean flow rating. We can directly associate mean flow rating and nominal rating as the same in this regard, as the key objective is the average pore size. However, they are distinguished as different, depending on which test method is used. For example, flow permeability test for nominal rating and bubble-point test for mean flow rating.

The last factor presented here to rate filter media, is that of air permeability. This is the measurement of a filter medium's ability to pass air. Air permeability is stated in CFM, the number of cubic feet of air that could pass through a square foot of medium per minute at a given pressure rating to determine air flow. The standard equipment used for this rating in the USA is the Frazier differential pressure air permeability machine. The air permeability test is based on Frazier method number 5450, equivalent to Federal Test Method Standard number 191 and ASTM D737-75(80). The test pressure is

0.01806 p.s.i. and the inches of water column is 0.5 in.

17.10 SUMMARY

It should be remembered that the filter medium is the keystone of any filtration system, and for the filter to give an optimum performance the medium used must be the best one available for the purpose. We should be aware that by changing the filter medium, a user can perhaps significantly alter the economics of his process.

The users of filter media continue to benefit from new developments in materials technology. Manufacturers respond to the challenges of various applications by making continued improvements in the quality of more traditional materials, in addition to the new products they may offer. Many monofilament filter fabrics were virtually unavailable ten years ago. New fabrics are continually being added, and those already available are refined by calendering, shrinking, etc., to meet new applications or to improve performance for existing applications.

One main advantage that filter media users have is that there are several qualified suppliers, who are willing to work with users on potential applications. It is a two-way street. Suppliers need users to test, analyse and report back, while users need suppliers to make these materials available by working with the user to pin-point the best possible material. Confide in your supplier—it will help improve your operation.

BIBLIOGRAPHY

Bosley, R., 'Vacuum filtration equipment innovations', *Filtration and Separation,* March/Apr., 138 (1974)

Cole, F. W., 'Filter ratings—an alternative to "black art"', *Filtration and Separation,* Jan./Feb., 17 (1975)

Corte, H., 'Why can paper be used as a filter medium?', *Filtration and Separation,* Jan./Feb., 42 (1980)

Dean, J. H., 'Nonwoven west-laid filter media', *Filtration and Separation,* Nov./Dec., 669 (1972)

Dickey, G. and Bryden, C., *Theory and Practice of Filtration,* Reinhold Publishing, New York (1946)

Deitrich, H. and Gurtler, H. C., 'A new textile for high temperature filtration of dust and gases', *Filtration and Separation,* July/Aug., 403 (1974)

Dyson, N. W., 'Cartridge filtration for processing edible oils', *Filtration and Separation,* March/Apr., 167 (1979)

Franks, E. H., *Tetko Training Manual* (1972)

French, R. C., 'Filter media', *Chem. Eng.,* Oct. 14 (1963)

Gale, R. S., 'Control of sludge filter operation', *Filtration and Separation,* Jan./Feb., 76 (1975)

Goeminne, H., de Bruyne, R., Roos, J. and Aernoudt, E., 'The geometrical and filtration characteristics of metal-fibre filters—a comparative study', *Filtration and Separation,* July/Aug., 351 (1974)

Hutto, F. B., 'What the filter man should know about filter aid filtration', *Filtration and Separation,* March/Apr., 164 (1975)

Jensen, K. E., 'Concepts of fabric filtration for air pollution control', *Filtration and Separation,* May/June, 254 (1969)

Johnson, P. R., 'Submicron filtration with cartridges', *Filtration and Separation,* July/Aug., 352 (1975)

Lloyd, P. J. and Ward, W. S., 'Filtration applications of particle characterization', *Filtration and Separation,* May/June, 246 (1975)

Nickolaus, N., 'What, when and why of cartridge filters', *Filtration and Separation,* March/Apr., 155

Penderson, G. C., 'Fluid flow through monofilament fabrics', *Filtration and Separation,* Nov./Dec., 586 (1974)

Pointon, C. W. and Giles, J. W., 'Industrial screening filters with special reference to cartridge filters', *Filtration and Separation,* May/June, 259 (1974)

Purchas, D. B., *Industrial Filtration of Liquids,* CRC Press, Cleveland (1967)

Purchas, D. B., 'Art, science and filter media—I', *Filtration and Separation,* May/June, 253 (1980)

Purdy, A. T., 'The structural mechanics of needlefelt filter media', *Filtration and Separation,* March/Apr., 134 (1980)

Redmon, O. C., 'Improvements in filters and separators for jet fuel', *Filtration and Separation,* May/June, 241 (1969)

Rushton, Albert and Griffiths, P. V. R., 'Role of the cloth in filtration', *Filtration and Separation,* Jan./Feb., 81 (1972)

Rushton, Albert and Rushton, Alan, 'Size and concentration effects in filter cloth pore bridging', *Filtration and Separation,* May/June, 274 (1972)

Schweitzer, P. A. (Ed.), *Handbook of Separation Technology for Chemical Engineers, Part 4,* McGraw-Hill, New York (1979)

Shoemaker, W., 'Filtration fundamentals, filter media, filter rating', *Filtration Symposium* (1974)

Shoemaker, W., 'The spectrum of filter media', *Filtration and Separation,* Jan./Feb., 62 (1975)

Shoemaker, W., 'The industrial filtration market for non-wovens', *Filtration and Separation,* May/June, 252 (1979)

Snow, C., 'Toward improving standards in nonwoven filter media', *Filtration Engineering,* Sept./Oct., 23 (1974)

Squires, B. J., 'Fabric filter plants for cleaning gases from non-ferrous metal furnaces', *Filtration and Separation,* May/June, 277 (1974)

Suttle, H. K., 'The proportions and properties of particles', *Filtration and Separation,* May/June, 272 (1972)

Tarala, F. E., 'The particle size war', *Filtration Engineering,* May/June, 7 (1973)

'Tenth Magdeburg Conference', *Filtration and Separation,* Jan./Feb., 28 (1973)

Tetko Catalog, No. 1000, Tetko Inc. (1974)

Tetko Betamesh Brochure, Tetko Inc. (1981)

Tetko Catalog, No. 3000 WC, Tetko Inc. (1985)

Tetko Catalog, No. 3000 S, Tetko Inc. (1986)

Thomas, C. M., 'Filter media for filter pressing applications', *Filtration and Separation,* Nov./Dec., 629 (1975)

Ward, A. S., 'Filter media development and innovation', *Filtration and Separation,* Jan./Feb., 61 (1973)

Wrotnowski, A. C., 'Felt filter media', *Filtration and Separation,* Sept./Oct., 426 (1968)

18

Methods for Limiting Cake Growth

L. Svarovsky

Department of Chemical Engineering, University of Bradford

18.1 INTRODUCTION

This is an area of solid–liquid separation which has enjoyed considerable interest recently, although the basic concept is an old one. In conventional cake filtration, a cake is allowed to form on the filter medium and this gradually increases resistance to flow, leading to increasing time or pressure requirement for filtration of a given volume of slurry.

The flow rate of the liquid through the medium can be maintained high if no, or little, cake is allowed to form on the medium. Most of the methods used for such limitation of cake growth lead to mere thickening of the slurry. The cake can be prevented from forming by hydraulic or mechanical means and such operation can be carried out over either a part or the whole of the filtration cycle. The methods for limiting cake growth vary in the extent of their commercial exploitation; their general classification is as follows:

1. removal of cake by mass forces (gravity or centrifugal) or electrophoretic forces tangential to or away from the filter medium;
2. mechanical removal of the cake by brushes, liquid jets or scrapers;
3. dislodging of the cake by intermittent reverse flow;
4. prevention of cake deposition by vibration;
5. cross-flow filtration by moving the slurry tangentially to the filter medium so that the cake is continuously sheared off.

The above-listed methods are briefly described in the following, with particular reference to known commercial exploitation whenever applicable. As most of the concepts are relatively new, a greater use of references than usual is made to enable the reader to follow up any topic of interest.

18.2 REMOVAL OF CAKE BY MASS FORCES

This is a method of limiting cake growth through the action of gravity, centrifugal or electrophoretic forces, which act on the particles either tangentially to, or away from, the filter medium. *Figure 18.1* shows the method schematically on a gravity 'side filter' which has a vertical filter medium. The filtrate flow is essentially horizontal and this results in a horizontal drag force on the particles. Additionally, the particles are subjected to a gravity force pulling them down. This effect takes place both in the main body of the suspension where it leads to gravity settling and in the cake that forms on the face of the medium. Whether or not the particles deposited on the medium will fall off depends on the relative magnitude of the forces and the angle of friction between the particles and the filter medium. In the case of a gravity system, if the filter were to be operated continuously, this would only work with relatively coarse particles with a high mass-to-surface ratio. It is quite practicable, though, in an intermittent

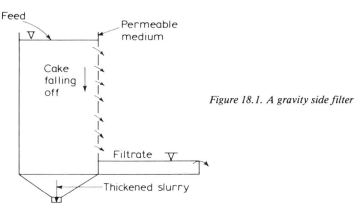

Figure 18.1. A gravity side filter

system where the cake is first formed in a conventional way and the feed is then stopped to allow gravity removal of the cake. Henry *et al.*[1] described and studied a similar system of pressure filtration of 2.5–5 μm particles (in the neutralized acid mine drainage water) in vertical hoses where a pressure shock associated with relaxing the pressure in the hose was used to aid cake removal.

While a system similar to that shown in *Figure 18.1* may be impractical under gravity, it is used in a centrifugal field as the 'bypass filtration'. Zeitsch[2] describes the side filter centrifuge, which combines the conventional centrifugal filtration on the cylindrical basket with filtration through the sides, as sometimes incorporated in peeler centrifuges. He laid down the theoretical basis of the side filtration process and showed that the side filters reduce the filtration time to less than 35% of the normal value. To take the concept a step further, Zeitsch also described and patented[3] a planetary filter centrifuge, which has two or more baskets rotating about their own axes and both spinning on arms around a parallel central axis. This arrangement, under certain operating conditions, gives rise to centrifugal forces on the particles which keep a portion of the medium free of particles and thus allow the filtrate to bypass the cake. While this is certainly an attractive concept it remains to be seen whether it will find wider application in practice. Another method of keeping particles away from the filtration medium is by using electrophoretic forces in 'electrokinetic filtration'[4]. This is achieved by application of a d.c. voltage between two electrodes, one on either side of the filter medium. The electrical charge on the upstream electrode is opposite to that carried on the surface of the particles (zeta potential). The rate of cake formation is reduced by electrophoresis and at the same time the rate of permeation is increased by electro-osmosis occurring inside the filter cake. The theoretical basis of electrokinetic filtration has been studied and examined experimentally in Japan[4]; for a given electric field strength E the filtrate volume V is related to electric power consumption W_p by an empirical equation[4]:

$$V = a \cdot W_p^n \tag{18.1}$$

where a and n are experimental constants. There is an optimum field strength because cake resistance decreases and electric power consumption increases with increasing E.

18.3 MECHANICAL CAKE REMOVAL

This is a method by which the cake is prevented from forming by the mechanical action of brushes, scrapers or liquid jets. Tiller[5] showed theoretically that the capacity of a drum or belt filter can be increased by using brushes to keep the cake off a portion of the available filtration surface, with the remaining portion used for normal filtration of the thickened surry. The reduction in filter area available for cake formation need not lead to reduction in solids handling capacity because the thicker slurry would give faster cake build-up. This method has not caught on in practice, with the exception of the scraping action of the turbine rotors used in the American version of the dynamic filter (see section 18.6).

18.4 DISLODGING OF CAKE BY REVERSE FLOW

This is a method of limiting cake thickness by intermittent back-flushing of the filter medium, as reported by Muira[6] and Brociner[7], for example. Conventional vacuum or pressure filters can be used with some modifications for the effects of the forces of the back-flush. Thus filtering through thin cakes in short cycles can be used.

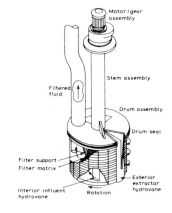

Figure 18.2. Ebclear filter. (Courtesy of Applied Products Corp.)

One commercial application of this principle is the Ebclear filter (Applied Products Corp.), shown schematically in *Figure 18.2*. This consists of a rotating filter basket under vacuum, with particle deposition on the outside surface. The filtration flow is also assisted by interior fixed vanes, and periodic cake removal is accomplished through the hydrodynamic action of an external fixed vane which momentarily reverses the flow. The dislodged cake flies off due to the centrifugal acceleration of more than 20 *g*. Units up to 5 m³ h⁻¹ are available and can be simply dipped into an existing tank or a pond.

18.5 PREVENTION OF CAKE DEPOSITION BY VIBRATION

Vibration is another means of preventing formation of a dense filter cake. An example of a commercial application of this principle is the tubular pressure filter reported elsewhere[8]. It consists of elements of three 38 mm (1.5 in) diameter tubes enclosed in a cylindrical housing, with the elements being pneumatically vibrated. The high-frequency, low-amplitude vibration leads to slurry agitation, cuts premature blinding and helps the dislodging of the solids during backwash which is used to follow the filtration cycle. The filter is designed for pressures up to 10 bar and flow rates up to 1.3 m³ h⁻¹, its applications include filtration of paper-coatings, colloidal gels and ceramic slips.

Sawyer[9] has patented a similar pressure filter, also tubular, in which the filter blanket in the form of a membrane is constantly cleaned by a sonic sinusoidal wave. This is induced by a transducer which is affixed to the tubular wall at an antinodal point and which causes cavitation and continuous cake removal within the annular filtration chamber. This invention is specifically related to the membrane process of filtration, such as ultra-filtration reported in the following section.

18.6 CROSS-FLOW FILTRATION

This is another, and certainly the most popular, method of limiting cake growth in which high relative velocities of the suspension with respect to the filter medium, and parallel with it, are induced (*Figure 18.3*). The slurry forces in the flow close to the medium continuously remove a part or the whole of the cake and mix it with the remaining suspension. As more and more of the filtrate is removed from the slurry, the latter gradually thickens and may become thixotropic. Surprisingly, a higher solids content in the final slurry can sometimes be obtained than with the conventional pressure filters

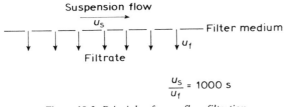

Figure 18.3. Principle of cross-flow filtration

(by as much as 10 or 20%). Kaspar *et al.*[10] suggested a range of velocity gradients from 70 to 500 s[-1] which, depending on the suspension properties, are necessary to prevent cake formation and to keep the thickening slurry in a fluid state.

Depending on how the relative velocity of the fluid and the filter medium is achieved there are two basic methods of cross-flow filtration: by using rotating elements (which may or may not carry the filter medium) in a stationary vessel, or by pumping the slurry through pipes made of permeable medium.

18.6.1 Cross-flow filtration with rotating elements

The first literature reference to this principle is in a patent by Morton[11]. His was a cylindrical rotating filter element mounted in a cylindrical pressure vessel, and the suspension was pumped into the annulus between the rotor and the stator. This is just one possible arrangement; the British Patent Specification by Kaspar *et al.*[10], first filed in 1964, covers virtually all possible arrangements in this category. *Figure 18.4* gives an example of one principle

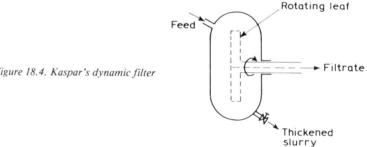

Figure 18.4. Kaspar's dynamic filter

of a 'dynamic' filter, which comprises a pressure vessel filter with a hollow shaft and a hollow filter leaf attached to the shaft. The slurry is continuously fed at a constant rate into the vessel via the inlet; the filtrate passes through the rotating filter medium and out through the hollow shaft while the thickened slurry is discharged via a control valve that is used to maintain optimum slurry concentration.

As can be seen from the principle in *Figure 18.4*, the cake removal by fluid shear is aided by spinning off the cake under the centrifugal action. Kaspar *et al.*[10] also described alternative arrangements with stationary filtration media and rotating discs to create the shear effects, rotating cylindrical elements, etc. and also showed how such filters can be used for cake washing.

Since its conception by Kaspar *et al.* there have been several published reports of developments of the dynamic filter. Most European designs (*Figure 18.5*) pass the slurry through a series of stages which have both stationary and rotating filter surfaces. Such filters have been found, for the same moisture content of the final slurry, to be 5–25 times more productive[12] (in terms of mass of dry cake per unit area and time) than filter presses; in several cases the moisture content with the dynamic filter was lower and the productivity was 5–10 times greater than with a filter press. The maximum

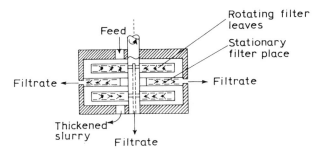

Figure 18.5. European dynamic filter (only two stages shown)

productivity was achieved with peripheral disc speeds of 2.8–4.5 m s⁻¹.

The so-called 'axial filter', developed in the Oak Ridge National Laboratory[13], is remarkably similar to Morton's and Kaspar's dynamic filter in that the filter leaf is in the tubular form and the outer shell is also cylindrical. An ultra-filtration module based on this principle has also been described more recently[14]. Unlike the European dynamic filters referred to in the previous paragraph, however, this filter is not suitable for scale-up because it poorly utilizes the available space. The Escher–Wyss pressure filter

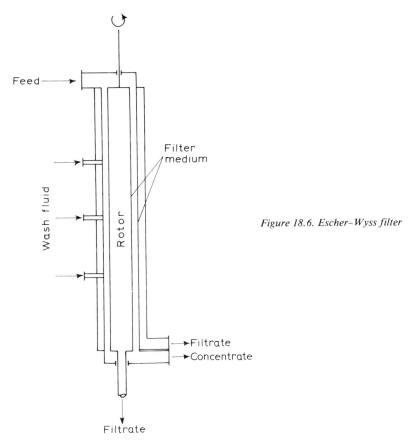

Figure 18.6. Escher–Wyss filter

described recently[15] (identical copy of this paper also appeared subsequently in *Filtration and Separation*[15]) takes the idea of axial filtration a step further: in addition to the cylindrical rotating filter surface, it also provides a stationary outer cylindrical filter surface (*Figure 18.6*) so that the suspension passes through a narrow annulus between the two, and filters through both cylinders. Strategically placed washwater entry points allow continuous and efficient washing (equivalent to reslurrying) if needed. This filter is particularly suitable for intrinsically viscous, plastic or thixotopric slurries.

There are several general disadvantages that apply to all dynamic filters with rotating filter elements. The filter medium is subjected to a centrifugal force and this imposes special requirements on its fastening and stretch resistance. The filtrate is collected into a rotating element radially inwards, i.e. against the centrifugal force, and discharged through the central hollow shaft—this reduces filtering velocities and makes the filter relatively complicated. Finally, any solids that may penetrate through the medium can accumulate inside the rotating element due to centrifugal settling.

These disadvantages can be avoided by replacing the rotating filter elements by solid elements, so as to prevent or reduce cake formation.

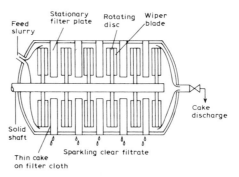

Figure 18.7. Artisan continuous filter. (Courtesy of Artisan Industries Inc.)

The American version of the dynamic filter, known as the 'Artisan continuous filter' (*Figure 18.7*), uses turbine-type elements as rotors and filtration medium only on the stationary surfaces. This arrangement gives less filtration area per volume of the pressure vessel (productivities up to 5 times those with conventional machines are claimed) but provides a better control of cake thickness. At low speeds the cake thickness is governed by the clearance between the scraper and the filter medium on the stationary plate while at higher speeds part of the cake is swept away and a thin layer remains to act as the actual medium. Tiller *et al.*[16–18] published results of their work with this type of filter, producing cakes with 1 mm thickness using 3 mm clearance. They found that the cake formed on the medium was generally stable, thus giving high filtration rates over long periods of time. The precoat-type cake did not blind with time and there was generally no evidence of size selectivity of the process (with the exception of conventional filter aids, which are preferrentially picked up by the rotating fluid). This was attributed to inertial bypassing of the cake by larger particles and re-entrainment of the smaller ones. Good control of the agitator speed and operating pressure was recommended in order to maintain a balance between the overall filtration and the slurry flow rates, so that the critical mud concentration and excessive torque are avoided.

The energy going into the agitation converts into heat and thus the extruded cake may be quite hot, 50–80°C. This makes the final product adaptable to drying. Little information exists about power requirements but in a discussion following one paper (reference 18, p. 598) a favourable comparison with a solid bowl centrifuge, used with the same calcium carbonate slurry, is quoted.

Murkes[19] has reported tests with the same 'American' dynamic filter, except that he eliminated cake formation altogether by spinning the discs faster. No visible abrasion of the medium was observed while the filtration velocities could be maintained high even at relatively low pressures, below 5 bar. Murkes showed clearly the effect of speed and number of rotor vanes on filtration velocities and specific energy input. His results show, for any number of rotor vanes, an increase in specific energy input (kWh per m³ of filtrate) with rotor speed; this increase can be minimized by optimization of the number of vanes, but not eliminated. This does not justify the use of higher speeds in terms of running costs but a saving can be made in capital expenditure because a smaller filter can be used for the same capacity.

18.6.2 Cross-flow filtration in porous tubes

This is a method of limiting cake growth by pumping the slurry through porous tubes at high velocities, so that the ratio of the axial flow velocity to filtration velocity through the tube walls is of the order of thousands. This is in direct analogy with the now well-established process of ultra-filtration applicable to much finer solids, which itself borders with reverse osmosis on the molecular level. It is therefore appropriate to review briefly the latter and then to follow on with ultra-filtration and, finally, with the relatively recently explored cross-flow filtration in porous tubes. The reader is also referred to chapter 21 for a more detailed treatment of this subject.

18.6.2.1 Analogy with reverse osmosis

The reverse osmosis process, also known as 'hyperfiltration', is based on the passage of solvent molecules through a dense membrane from a concentrated solution to a dilute one. As this process is opposed by osmotic pressure, the pressure drop across the membrane must be higher to overcome it. *Figure 18.8* shows how simple the process is in principle. The solution is

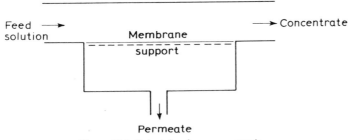

Figure 18.8. Principle of reverse osmosis

pumped over a membrane held on a permeable support and it is split into the solvent (permeate) and concentrate streams.

The osmotic pressure which has to be overcome depends on the concentration of the solution and on the nature of the solutes. As the osmotic pressure increases with concentration, the rate of permeation of the solvent diminishes as the concentrate thickens and it is not economically viable to aim at above a certain concentration[20] (5% for NaCl and 30% for a sugar solution, for example). The usual operating pressures are 40–50 bar giving solvent flux anywhere between 2 and 40 l m^{-2} h^{-1}, depending on the solution and the membrane used. Most membranes are made of cellulose acetate or its close relatives and they usually allow a limited solute passage in order to obtain reasonable permeate flow rates.

Although known for over two decades, reverse osmosis has only been accepted commercially in the last ten years, mainly due to the development in membrane technology. It is now used on a large scale in portable water treatment, for desalination of sea water, for concentration of sulphite-spent liquor and generally as a pre-concentration step before evaporation or thermal processing. There is a wide energy margin in favour of reverse osmosis as compared with evaporation[20] but the final concentrations economically achievable are limited. The types of equipment used are the same as in ultra-filtration that follows.

18.6.2.2 Ultra-filtration

There is no distinct limit between reverse osmosis and ultra-filtration but the latter employs lower pressures of no more than 10 bar (seldom above 6 bar) and more open membranes for separation of large molecules and ultra-fine, sub-micron solids. Henry[21] tried to make a clear distinction between ultra-filtration and cross-flow filtration by defining the former as retention of only dissolved species from solutions (as opposed to retention of particulate material from suspensions) but this has not caught on in practice, mainly because of the fact that dissolved and undissolved (ultra-fine) solids are often separated together. Thus, ultra-filtration is used for example for the concentration of proteins from low-cost dairy byproducts, or in the separation of emulsified oil and suspended solids from waste waters. In such

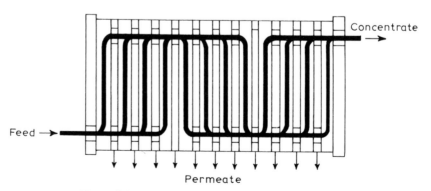

Figure 18.9. The DDS ultra-filtration module—internal flow

processes a cake or a layer of gel would form on the membrane and reduce the filtration rates if it were not for the cross-flow characteristics of most designs, in which the suspension flows at high speed across the membrane surface and prevents cake build-up.

The original reverse osmosis and ultra-filtration system was based on traditional plate and frame filter designs, in which modules were made up of spaces and membrane-covered plates mounted in a sandwich fashion, pressed together. *Figure 18.9* gives a schematic diagram of such a module by DDS, which is probably the only major company still using this design. This has been largely superseded by the tubular, hollow fibre, spiral and leaf-type systems.

The simplest configuration is the tubular system, shown schematically in *Figure 18.10*. It usually consists of stainless steel membrane support tubes with porous glass fibre backing tubes inside. The very frail membranes are coated on to the inside of the fibre tubes, and several of the complete tubes are connected in parallel in modules which resemble, both internally and externally, a long and thin single-pass shell and tube heat-exchanger. The tubes are commonly 13–26 mm in diameter, easy to clean physically and chemically, and are simple to remove. However, the modules occupy a large area and the liquid hold-up in them is high.

The hollow fibre systems are based on very small-bore tubes cemented at both ends in a wide plastic tube, which also serves as a receiver for the filtrate.

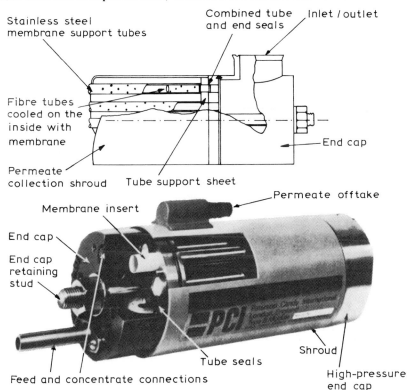

Figure 18.10. Tubular ultra-filtration module. (Courtesy of Miller)

A large membrane area can be packed into a small volume, and the liquid hold-up is also reduced. The main disadvantages of the hollow fibres are in their susceptibility to blocking due to the small internal diameters, and difficulties with cleaning.

The spiral and leaf systems both use two rectangular sheets of membrane placed one on top of the other, separated by a porous sheet and sealed together on three sides. In the spiral systems, the fourth sides are separately sealed along a length of a tube which forms a header to receive the permeate from the membranes. With spacers laid on top of the two membranes, the whole sandwich is rolled around the header tube to form a spiral which is sealed into a tubular housing, and the liquid is pumped from one end to the other across the membrane surfaces. In the leaf systems, the double-sheet leaves are attached to a square section metal header common to several leaves so that a whole section resembles a comb. The feed passes along several such sections in series through the teeth of the comb. Both the spiral and the leaf systems have a high surface-to-volume ratio but they can create problems with abrasive materials, are difficult to clean and susceptible to blocking, particularly where the feed contains fibrous matter.

Whatever system is used, high fluid velocities over the membranes must be used. This, coupled with the small rate of permeation through the membrane, makes it impracticable to use a one-pass arrangement and, consequently, the fluid must be recirculated. This is done either on a batch or on a continuous 'feed and bleed' basis. The latter, continuous operation often incorporates two or more 'feed and bleed' stages with the concentration increasing from the first stage to the last.

In application, ultra-filtration is mostly applied to effluent treatment but the major advantage is using ultra-filtration for recovery of a valuable component which is difficult to recover by other means. Thus details of applications have been published[22] on recovery of electrophoretic paint, protein extraction from cheese whey or from skimmed milk, pre-concentration of starch and enzymes, textile de-sizing, latex concentration, recovery of oil from oil–water emulsions or waste machining and metal-rolling emulsions, recovery of lanolin from wool-scouring effluent, recovery of lignosulphonates from spent sulphite liquor in pulping mills, and concentration of impurities in alkaline cleaning-bath effluents. Ultra-filtration is also used for concentration of viruses. The process of ultra-filtration is still undergoing rapid development and it is certain to have a healthy future.

18.6.2.3 Filtration in porous pipes

Having shown the principles of ultra-filtration it is now simple to extend the idea to filtration of more conventional suspensions with coarser particle size, and this is the cross-flow filtration in porous pipes. Essentially, the same equipment and types of modules as in ultra-filtration can be used but thicker flow channels or greater-diameter hollow fibres are necessary. It seems, however, that porous pipe systems represent the most practicable approach here. Zhevnovatyi[23] was first to report the use of porous tubes for cross-flow filtration, in his case for separation of hydrated alumina and red bauxite.

Dahlheimer et al.[24] published an investigation on concentration of kaolin clay with woven fibre hoses, and Krauss[25] reported on sewage filtration in permeate hose pipes with the addition of activated carbon. The latest investigation is that of Rushton et al.[26], who studied the effect of filtrate velocity on the slurry velocity required for particle transport and examined the separation of particles with porous stainless steel tubes.

If cross-flow filtration in porous tubes is carried out at constant pressure, the permeate flux drops, initially approaching steady state, which ideally would result in a constant rate at which cake deposition and stripping are equal and the resistance of the precoat layer is unchanged. In practice, however, the rate decreases slowly with time and back-flushing[26] or even chemical treatment[27] must be used for clearing the pipes.

The most important parameter in designing the porous tube filtration systems is the ratio of filtrate velocity to pumping velocity because the latter determines the shear rates at the wall and these, in turn, control the rate of cake stripping. In contrast to reverse osmosis and ultra-filtration of large molecules, the permeate flux here is much less dependent on the concentration in the pipe flow and thus higher final slurry concentrations can be achieved.

Theoretical investigations at low permeate fluxes[21] predict the permeate flux to increase exponentially with shear rate

$$J \sim \gamma^m \qquad (18.2)$$

The shear rate γ for circular pipes in Newtonian fluids is:

$$\gamma = \frac{8u}{d} \qquad (18.3)$$

where u is the average piping velocity and d is the diameter of the pipe. This shows that if the pipe diameter is reduced, a given shear rate or permeate flux can be achieved at lower flow velocity.

Rushton et al.[26] derived empirical equations relating the slurry velocity u_s to the permeation velocity u_f in the form:

$$u_s = a + bu_f \qquad (18.4)$$

for $MgCO_3$ where constants a and b depend on particle size.

The possibility of using conventional precoating techniques in cross-flow filtration in tubes has also been suggested[26]. With respect to the filtration velocity, Krauss[25] suggested that tubular filtration systems should obtain filtration velocities in excess of 4.72×10^{-5} m s^{-1} to be economically viable.

18.6.3 Conclusion

In conclusion to cross-flow filtration as a unit operation, it is probably the most exciting development in solid–liquid separation yet to be fully explored. Its advantages are in high filtration rates due to minimized particle deposition on the medium without a strong effect of particle size on performance, and in the absence of chemical additives or filter aids in the products.

REFERENCES

1. Henry, J. D. Jr, Lui, A. P. and Kuo, C. H., 'A dual functional solid–liquid separation process based on filtration and settling, *A.I.Ch.E.*, **22**, 433–441 (1976)
2. Zeitsch, K., 'Centrifugal filtration', ch. 14 in *Solid–Liquid Separation*, ed. by L. Svarovsky, Butterworths, London (1977)
3. Zeitsch, K., Dos. No. 2357 775, Deutsches Patentamt, München
4. Yukawa, H., *et al.*, 'Studies of electrically enhanced sedimentation, filtration and dewatering process', in *Progress in Filtration and Separation*, ed. by R. J. Wakeman, pp. 83–112, Elsevier, Amsterdam (1979)
5. Tiller, F. M., 'Improving filtration rates through delayed cake formation', *29th Res. Conf. Filtration and Separation*, Nagoya, Japan, Soc. Chem. Engrs (Japan), pp. 150–153 (1972)
6. Muira, M., 'Water filtration with the reversible flow filter', *Filtration and Separation*, **4**, 551–554, 556 (1967)
7. Brociner, R. E., 'An improved vacuum leaf thickener for china clay slurries', *Filtration and Separation*, **9**, 562–565 (1972)
8. Ronningen-Petter Div., Dover Corp., *Chem. Eng.*, **80**(6), Mar. 19 (1973)
9. Sawyer, H. T., 'Dynamic self-cleaning filter for liquids', US Patent No. 4158 629, June 19 (1979)
10. Kaspar, J., Soudek, J. and Gutwirth, K., 'Improvements in or relating to a method and apparatus for dynamic filtration of slurries', British Patent No. 10577 015
11. Morton, C. D., 'Method of filtering', US Patent No. 1762 560, June 10 (1930)
12. Malinovskaya, T. A. and Kobrinsky, I. A., 'The separation of finely divided suspensions in a dynamic filter', *Mechanical Liquid Separation (9th Symp.)*, Magdeburg, Oct., pp. 47–51 (1971)
13. Irizarry, M. M. and Anthony, D. B., Ornt-Mit- 129, Oak Ridge National Laboratory, 28 April (1971)
14. Hallström, B. and Lopez-Leiva, M., 'Description of a rotating ultra-filtration module', *Desalination*, **24**, 273–279 (1978); also in: 'Membranes: desalination and waste water treatment', *Proc. 13th Annual Desalination Conf.*, Jerusalem, 8–12 Jan. (NRD 8/78) (1978)
15. Tobler, W., 'Dynamic filtration—the engineering concept of the Escher–Wyss pressure filter, *Escher-Wyss News*, 2/1978-1/1979, pp. 21–23; also in: *Filtration and Separation*, **16**, 630–632 (1979)
16. Tiller, F. M. and Cheng, K. S., 'Delayed cake filtration', *Filtration and Separation*, **14**, 13–18 (1977)
17. Bagdasarian, A., Tiller, F. M. and Donovan, J., 'High-pressure, thin-cake, staged filtration', *Filtration and Separation*, **14**, 455–460 (1977)
18. Bagdasarian, A. and Tiller, F. M., 'Operational features of staged pressure, thin cake filters', *Filtration and Separation*, **15**, 594–598 (1978)
19. Murkes, J., 'Cake-free filtration in a rotary, high shear filter', *The Second World Filtration Congress*, Filtration Society, London (1979)
20. Pepper, D., 'Reverse osmosis for preconcentration', *The Chemical Engineer*, **339**, 916–918 (1978)
21. Henry, J. D., Jr, 'Cross-flow filtration in recent development', in *Separation Science*, **11**, ed. by N. N. Li, Chemical Rubber Co., Cleveland, Ohio (1972)
22. Bailey, P. A., 'Ultrafiltration—the current state of the art', *Filtration Separation*, **14**, 213–219 (1977)
23. Zhevnovatyi, A. I., 'The thickening of suspensions without cake formation', *Intl J. Chem. Eng.*, **4**, 124 (1964)
24. Dahlheimer, J. A., Thomas, D. G. and Krauss, K. A., 'Application of woven fiber hoses to hyperfiltration of salts and cross-flow filtration of suspended solids', *I.E.C. Proc. Des. Dev.*, **9**, (1970)
25. Krauss, K. A., *Eng. Bull, Purdue University*, **145**, 1059 (1974)
26. Rushton, A., Hosseini, M. and Rushton, A., 'Shear effects in cake formation mechanisms', pp. 149 –158. In *Proc. Symp. on Solid–Liquid Separation Practice*, Yorkshire Branch of the I. Chem. E., Leeds, 27–29 March (1979)
27. Zhevnovatyi, A. I., *Zhurnal Prikladnoi Khimii*, **48**, 334 (1973)

19

Flotation

R. J. Gochin

Royal School of Mines, Imperial College, London

NOMENCLATURE

a_s	Air-to-solids ratio	mg mg^{-1}
A_1, A_2	Concentrations of dissolved air	mg m^{-3}
c, c_E, c_0	Solids concentration; weight solids per weight liquid	—
d, D	Diameters	m
$E(t)$	Residence time distribution function	—
ΔG	Free energy change per unit area of surface	J m^{-2}
k	First-order reaction rate constant	s^{-1}
k_s	Solubility of air in water at STP	mg m^{-3}
n	Number of flotation cells in series	—
N	Impeller speed	Hz
P	Pressure (above atmospheric)	kPa
$P*$	Pressure	atmospheres
P_a	Power	kW
Q, Q_a, Q_r	Flow rates	m^3 s^{-1}
R_B, R_c	Proportion of suspended solids recovered	—
t	Time	s
$\bar{t}, \bar{t}_c$	Mean retention times	s
v_L	Limiting downward velocity	m^3 m^{-2} day^{-1}
V	Flotation cell volume	m^3
W_a	Weight of air dissolved in water	mg l^{-1}
γ_{GL}	Gas–liquid interfacial tension	—
θ	Gas bubble–solid surface contact angle measured through the liquid	—

19.1 INTRODUCTION

Flotation is a process in solid–liquid separation technology whereby solids in suspension are recovered by means of their attachment to air bubbles. As such, it is the antithesis of solid–liquid separation by sedimentation. In flotation, aggregates of particles and bubbles with densities less than the suspension itself, rise to the surface and are removed.

The term 'flotation' is a generalization for a number of processes known collectively as 'adsorptive bubble techniques'. These can be broadly divided into two groups: foaming and non-foaming separation methods. The first category requires the presence of surface-active agents to generate a relatively stable foam or froth, which acts as a carrier fluid during the removal of the floated particles; the second does not. Both groups are further sub-divided and Lemlich[1] has provided an excellent description of the various types. Examples of both foaming and non-foaming techniques are found in solid–liquid separation technology.

It must be emphasized that if flotation is used to separate solids from liquids, then the objective is almost invariably to remove all the solids. This contrasts with the selective froth flotation techniques used in the minerals industry to concentrate certain species of particles in a finely ground ore. Several authors have summarized the complex chemical balances required to float mineral particles selectively[2–4].

In a suspension of a single species it may be relatively easy to find a set of conditions to float most of the solid. However, if the suspension is made up of many species, differing in their surface chemical and electrical properties, then the simultaneous flotation of all species becomes much more difficult. One solution is to add a coagulating solid, such as the ferric or aluminium salts used in the water industry. Here, the precipitation of hydrated salts of very high surface area compared to the naturally occurring particles means that the flotation is largely governed by the properties of the single (known) precipitated solid. In all systems the controlling step for flotation to proceed is bubble–particle attachment, which in turn is determined by the surface chemical and hydrodynamic parameters that exist in the flotation cell. The properties which have been shown to be most important in determining the success of flotation are solid hydrophobicity, bubble-to-particle size ratio and the degree of turbulence in the suspension. These areas will be described in more detail below.

19.2 HYDROPHOBICITY AND FLOTATION

Almost all naturally occurring solid particles and most inorganic chemical precipitates have surfaces with a strong affinity for water (hydrophilicity) and they are invariably unfloatable. Therefore the process of attachment of these wetted solids to air bubbles involves the displacement of a water film from the solid surface. However, displacement can only be spontaneous if the forces acting between the particle surface and water are sufficiently weakened. This can be accomplished by surface-active agents (collectors) which act at the solid–water interface and diminish the particle surface–water interactions. The particle surfaces are rendered, at least in part, hydrophobic, i.e. more amenable to the displacement of the wetting film by a bubble. Collectors are usually long-chain hydrocarbon molecules containing polar groups. The molecule adsorbs on to the solid surface via the charged group with the hydrocarbon chain presented to the aqueous phase. They may be added to the suspension or be present from natural sources, or as the result of a previous industrial process.

For bubble–particle attachment to occur in practice both thermodynamic and kinetic conditions must be fulfilled. The thermodynamic condition for the equilibrium attachment of a gas bubble to a surface is controlled by the hydrophobicity as shown by the Young–Dupré equation[5], which relates free energy to interfacial tension through the contact angle thus:

$$\Delta G = \gamma_{GL}(\cos\theta - 1) \qquad (19.1)$$

where

ΔG is the free energy change per unit area of surface
γ_{GL} is the gas-liquid interfacial tension
θ is the gas bubble–solid surface contact angle measured through the liquid.

Equation 19.1 shows that only for $\theta > 0$ (i.e. a degree of hydrophobicity) is the attachment of an air bubble to a surface thermodynamically feasible. For good flotation to occur, contact angles over 30 degrees are usually required. *Figure 19.1* illustrates the relationship between θ and hydrophobicity.

(a)   (b)

Figure 19.1. Hydrophobicity and contact angle. (a) Low hydrophobicity. (b) Good hydrophobicity

It should be noted that equation 19.1 applies to a flat surface and represents the maximum free energy change; surface geometry and bubble deformation may alter ΔG in reality, although hydrophobicity is still a necessary condition.

The kinetic factors which intervene in a flotation system arise from the physical processes involved in particle–bubble attachment. These can be summarized as:

1. collision between bubble and particle;
2. thinning of the liquid film between the bubble and particle;
3. rupture of the liquid film;
4. rapid expansion of the air meniscus over the particle so that a stable attachment has been achieved.

The bubble–particle collision stage is mainly a hydrodynamic process and is highly dependent on particle and bubble sizes[2]. Electrical effects between the charged surfaces of particles and bubbles may also have an influence[6,7].

The time required for the thinning and rupture of the liquid film between the colliding particle and bubble to take place is known as the induction time.[8] Measurements on moderately hydrophobic solids have shown this to be of the order of a few milliseconds[9]. After collision and bubble rupture, the probability that a stable bubble–particle aggregate will be formed is a function of contact angle, relative particle and bubble sizes, and particle density. Hence the overall probability that flotation will occur may depend on a large number of factors. However, the controlling steps are the rate of collision of particles and bubbles, few collisions, little flotation and the rapid establishment of a contact angle, i.e. sufficient hydrophobicity.

19.3 BUBBLE GENERATION IN FLOTATION SYSTEMS

All flotation systems require the generation of a swarm of bubbles to levitate the solid phase. Gas bubbles can be formed by three main methods:

1. *Mechanical means*—As in dispersed air flotation cells where bubbles of about 1 mm are formed by an agitator, or by passing air through a porous plate or nozzle.
2. *By nucleation of gas from solution*—This can be achieved either by applying a vacuum to the suspension (vacuum flotation), or by saturating water with air under pressure and injecting it into the suspension (dissolved air flotation), or by air supersaturation induced by dissolving air into a stream as it flows down a shaft (~ 10 m) and its subsequent release as the stream is returned to ground level (micro-flotation). These methods all produce fine bubbles of around 60 μm diameter.
3. *By electrolysis of the aqueous phase (electroflotation)*—This procedure requires sufficient conductivity but will generate very small bubbles, less than 50 μm, with a minimum degree of turbulence.

The actual mean bubble size produced by each method will also depend on the presence of surfactants active at the air–water interface.

In solid–liquid separation dispersed air, dissolved air and electroflotation systems have been economically viable on a large scale, and further discussion will be limited to these. Of these, the dissolved air flotation method has proved the more attractive because of its small bubble size relative to cost[10]. Indeed, since the removal of coarse particles from suspensions by filtration or sedimentation is relatively cheap and technically simple, flotation has found use mainly for the separation of fine and colloidal particles. In these regions the relationship between the particle distribution by size and bubble numbers and sizes is critical to efficient operation.

19.4 PARTICLE SIZE AND FLOATABILITY

The efficiency of recovery of particles of differing size has been the subject of numerous studies. Traher and Warren[11] have summarized the findings for the froth flotation of mineral particles. Recovery of particles over 200 μm and below about 10 μm was found to be poor. Particles of an intermediate size range show the best recoveries, although the optimum size range is not the same for all minerals. For a bubble of given size the reduced flotation of large particles is obviously due to mass increases (particle diameter cubed) overcoming adhesion forces, which depend on particle diameter. The inefficiency with small particle sizes is not well understood but hydrodynamic conditions between a relatively large bubble and a small particle are usually unfavourable to direct collision[7,12].

The overall collection efficiency of a bubble (E_c) can be defined as the product of collision and attachment efficiencies:

$$E_c = E_1 \cdot E_2 \tag{19.2}$$

Considering a spherical bubble of diameter d_b rising vertically at constant velocity and a spherical particle of diameter d_p, Reay and Ratcliff[13] predicted for $d_b < 100 \ \mu$m that

$$E_1 \propto \left(\frac{d_p}{d_b}\right)^2 \tag{19.3}$$

Anfruns and Kitchener[7] confirmed this relationship for a model system using quartz and bubbles between 0.5 and 1.1 mm diameter. Jameson *et al.*[14] correlated theoretical and experimental capture data and showed that flotation rate was proportional to $(d_p)^2/(d_b)^{2.7}$ for bubbles between 0.6 and 1.0 mm. All studies indicate that for a given air consumption the flotation of fine particles is enhanced by the presence of small bubbles. This is due to their larger number density and longer residence time in the pulp increasing the probability of collision. It is also evident from equation 19.3 that if an increase in particle size can be achieved at a given bubble size then flotation should be further enhanced. This leads in practice to the use of coagulating and flocculating aids in the flotation of fine particles. Indeed, there is some evidence in mineral flotation, where collectors have been used to induce hydrophobicity, that the collector itself has coagulated fine particles and increased their flotation rate[15,16]

19.5 BUBBLE–PARTICLE AGGREGATION

The collision of a bubble with a particle, induced by fluid turbulence, is only one possible mechanism by which bubble–particle contact might be achieved. The four possible ways of contacting the two phases are:

1. *Collision*—Brought about either by
   (a) turbulence induced in the suspension, or by
   (b) a 'filtering-out' effect as rising bubbles meet a sedimenting slurry.
2. *Entrapment*—Bubbles are trapped in a floc structure during its formation.
3. *Precipitation*—Air from a supersaturated solution precipitates directly on to a particle or floc.

Method (2) is only feasible with aggregated solids whereas the others can operate on dispersed particles. The 'filtering-out' of air bubbles by a sedimenting suspension is really only a type of turbulent collision and hydrophobicity will still be required for stable adhesion of bubbles to occur[15,17]. However, this may well be the best mechanism where high turbulence is to be avoided, such as with weak alum or ferric coagulation.

The flotation cell must be arranged such that there is a population of rising bubbles over the whole tank area. This may be difficult to achieve in practice except with electrolytic bubble generation.

The entrapment procedure and direct precipitation are attractive in that they avoid the need for a collision between a bubble and a particle. However, the entrapment method still requires both that the solids be aggregated and that there is a high probability that a suitable bubble (or bubbles) is in the vicinity of the coagulating or flocculating suspension. This is not likely in practice unless large numbers of very small bubbles are present. It also demands that bubbles are generated before or during a flocculation stage and this may interfere with the low-shear hydrodynamic conditions necessary for good floc formation. Vuuren[18] has described the use of air induced into a pump before the coagulation of sewage and its subsequent flotation.

The precipitation process is where bubbles are formed *on* the solid surfaces, although a degree of hydrophobicity is still required to ensure adequate adhesion of the bubble during its growth. This phenomenon takes place during micro- and vacuum-flotation but only when pre-existing gas pockets (Harvey nuclei) occur on the particles[19]. This is probably true in almost all solid–liquid separation processes except perhaps where the main solid phase has been formed under water and out of contact with gases, as for instance in the coagulation of clays in river waters by iron or aluminium hydrated oxide precipitation for potable water production[20,21].

In dissolved air flotation, the normal practice is to release most of the air as micro-bubbles by pressure-release devices and not to add the supersaturated water itself to the suspension[21]. Although some small degree of supersaturation exists in dissolved air, dispersed air and electroflotation cells, its relative importance in bubble–particle attachment is not known but is thought to be small. Hence both precipitation on solids and entrapment in flocs as they form offer attractions in avoiding the collision stage; however neither of these systems has been developed commercially to any degree. Industrially, solid–liquid separation processes which utilize flotation still rely in the main on achieving collision between a small bubble and hydrophobic areas on particles or flocs.

19.6 MACRO-KINETIC MODEL OF FLOTATION

The complexity of the microprocesses that occur in a flotation system has made the establishment of a suitable design model from first principles almost impossible. Flint[22] has provided a good summary of the problems involved. However an approach that has given some success is to regard a flotation system as analogous to a chemical reactor. Much experimental evidence has been accumulated to show that the *overall* rate of removal of solids by flotation is a first-order process[4] of the form:

$$x(\text{particles}) + y(\text{bubbles}) \rightarrow z(\text{floated aggregates})$$

If the concentration of bubbles is constant, and assuming that the solid removed represents a small volume, then the rate of change in solids concentration is proportional to the concentration, c, thus

$$\frac{dc}{dt} = -kc \tag{19.4}$$

The rate constant k is a lumped parameter involving physical, chemical and surface properties of the system, and its value must be found experimentally.

Assuming that this reaction takes place in a continuous flotation machine operating at the steady state, the hydrodynamics involved can be approximated by using the concepts of residence time distribution[23]. The suspension is regarded as consisting of a large number of identical 'fluid elements' in which the 'reaction' (the removal of solids) is taking place. Obviously, the extent to which solid is removed in any particular element is proportional to the length of time it stays in the machine—its residence time. As a result of flow patterns induced by machine design, all elements need not stay in the system for the same time period, hence a residence time distribution function, $E(t)$, can be derived to describe the overall movement of fluid through the reactor. $E(t)$ is a normalized function such that

$$\int_0^\infty E(t)\,dt = 1$$

The shapes of some residence time distributions are shown in *Figure 19.2*.

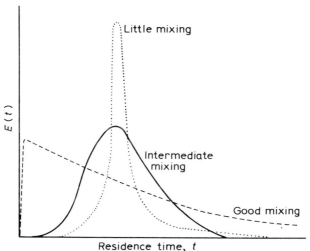

Figure 19.2. *Typical residence time distribution functions*

Then, at the steady state in a flotation cell of volume V with fluid flow rate Q, the concentration of solids leaving the cell can be written as

$$\left\{\begin{array}{l}\text{mean concentration}\\\text{of solids in the}\\\text{exit stream}\end{array}\right\} = \sum \left\{\begin{array}{l}\text{conc. of solids}\\\text{remaining in an element}\\\text{of age } t \to t + dt\\\text{leaving the reactor}\end{array}\right\}$$

$$\times \left\{\begin{array}{l}\text{fraction of exit stream}\\\text{which consists of elements}\\\text{of age } t \to t + dt\end{array}\right\}$$

or

$$c = \int_0^\infty c_E \cdot E(t)\,dt \qquad (19.5)$$

where c_E is obtained from the solution of equation 19.4 which is $c_E = c_o \exp(-kt)$ for a concentration of solids in the feed of c_o.

Most flotation systems with a high degree of turbulence have been shown to approximate to perfect mixing[4] where it can be proved that $E(t)$ is given by

$$E(t) = \frac{Q}{V}\exp\left(-\frac{Qt}{V}\right) = \frac{\exp\left(-\frac{t}{\bar{t}}\right)}{\bar{t}} \qquad (19.6)$$

where $\bar{t}$ is the mean residence time, V/Q. Hence

$$c = \int_0^\infty c_o \exp(-kt) \cdot \frac{\exp\left(-\frac{t}{\bar{t}}\right)}{\bar{t}}\,dt \qquad (19.7)$$

The solution of equation 19.7 is

$$\frac{c}{c_o} = \frac{1}{1 + k\bar{t}} \qquad (19.8)$$

The flotation recovery of solids is then given by

$$\text{recovery} = \frac{c_o - c}{c_o} = \frac{k\bar{t}}{1 + k\bar{t}} \qquad (19.9)$$

In the case of some dissolved air systems and most electroflotation cells the degree of fluid turbulence is much lower and the hydrodynamics are probably close to plug flow. In plug flow there is no distribution of fluid element residence times, all elements remain in the cell for the same time period, i.e. $\bar{t}$. Then equation 19.5 becomes

$$c = c_E \int_0^\infty E(t)\,dt = c_E \times 1$$

or

$$c = c_E = c_o\,e^{-k\bar{t}} \qquad (19.10)$$

Hence the flotation recovery for plug flow is given by

$$\frac{c_o - c}{c_o} = 1 - e^{-k\bar{t}} \qquad (19.11)$$

It is evident from equations 19.9 and 19.11 that theoretically a plug flow flotation cell is more efficient than a perfectly mixed system, provided that k and $\bar{t}$ are the same. Conversely, a smaller plug flow tank will give the same recovery as a larger perfectly mixed system[24]. In practice it is very difficult to achieve plug flow conditions, although the trend towards this state in dissolved air cells may account for some of their superiority in fine-particle flotation compared to dispersed air systems.

19.7 FACTORS IN PLANT DESIGN

19.7.1 General considerations

The chemical reactor models of flotation using first-order kinetics allow the volume of the cell to be calculated if k and Q are known. Laboratory batch experiments can be conducted to find k by measuring the amount of solid floated over various times from samples of the suspension[4]. If such an experiment has given a solids flotation recovery of R_B in t minutes, then k can be found from the solution to equation 19.4 for the batch test

$$R_B = \frac{c_0 - c}{c_0} = 1 - e^{-kt} \tag{19.12}$$

An estimate of the size of a large-scale continuous plant can now be made using the reactor models. For instance, consider the following two alternative designs.

19.7.1.1 A single perfectly mixed flow cell (PMF)

Continuous machine, fluid mean residence time $= \bar{t}_c$

Continuous machine solids recovery $= \dfrac{k\bar{t}_c}{1 + k\bar{t}_c} = R_c$

If it is required that $R_c = R_B$ then

$$1 - e^{-kt} = \frac{k\bar{t}_c}{1 + k\bar{t}_c}$$

or

$$\frac{\bar{t}_c}{t} = \frac{e^{kt} - 1}{kt} \tag{19.13}$$

The ratios of $\bar{t}_c/t$ given by equation 19.13 for some possible laboratory batch experimental results are shown in *Table 19.1*. It can be seen that as a

Table 19.1. LABORATORY FLOTATION RECOVERIES AND MEAN RESIDENCE TIME OF CONTINUOUS EQUIPMENT

Laboratory recovery* (R_B)	kt**	$\bar{t}_c/t$†
0.85	1.9	3.0
0.90	2.3	3.9
0.95	3.0	6.3
0.99	4.6	21.5

* As a proportion of total solids present.
** From equation 19.12. † From equation 19.13.

higher solids removal efficiency is required, the single continuous machine must have an increasingly large mean residence time compared to the time taken in the laboratory test. This is a reflection of the inefficiencies caused by having a distribution of fluid residence times in the plant machine.

19.7.1.2 A single plug flow cell (PF)

Continuous machine solids recovery $= R_c^1 = 1 - e^{-k\bar{t}_c}$

If it is required that $R_c^1 = R_B$ then

$$1 - e^{-kt} = 1 - e^{-k\bar{t}_c} \text{ and } t = \bar{t}_c$$

Hence, in this case the mean residence time in the continuous plant is the same as the time required to achieve a given recovery in a laboratory batch cell, which confirms the superior efficiency of plug flow reactors. Examination of equation 19.13 and *Table 19.1* reveals the extra volume required in a perfectly mixed cell compared to one that exhibits plug flow. The ratio $(\bar{t}_{PMF})/(\bar{t}_{PF})$ is identical to the third column of *Table 19.1*.

Fluid flows that are intermediate between perfectly mixed flow and plug flow can be modelled in a similar fashion[23, 24]. Of course, in reality k is only a parameter that averages the influence on flotation of a number of factors. King[25] has described distributed parameter flotation models that allow the use of varying values of k according to particle size, chemical environment and changes in bubble dimension.

Experience has shown that it can be dangerous to scale-up directly from laboratory batch experiments. Whenever possible, the suspension should be further tested in a continuous pilot plant to confirm the value of the flotation rate constant, k, and to determine other parameters such as air flow rate, fluid velocity, etc. These will be discussed below under the individual flotation methods.

19.7.2 Dissolved air flotation (DAF)

In dissolved air flotation water is saturated with air under pressure and is passed into a circular or rectangular flotation tank through a nozzle or valve. The resultant pressure reduction releases air from solution as small bubbles, which collide with the dispersed phase and carry it to the surface. Hydrophobicity may be obtained by surfactant addition or may be a pre-existing condition due to impurities in the water. While some turbulence is induced during the injection of the bubble swarm, it is usually very much less than the shear involved in dispersed air flotation. This makes DAF very useful when loosely bound flocs are to be floated. Three principal modes of operation of DAF systems are recognized:

1. full-stream pressurization;
2. split-stream pressurization;
3. recycle-stream pressurization.

Figure 19.3 illustrates these variations schematically. In full pressurization systems the pressure required to provide a given air-to-solids ratio is a minimum. However, a much larger pressurization plant is needed compared to the split- or recycle-stream methods. Both full and split procedures can be difficult to operate where flocculation or coagulation before flotation is important, since the passage of the flocs either through the saturator or through the injection nozzle may give rise to excessive break-up. Recycling

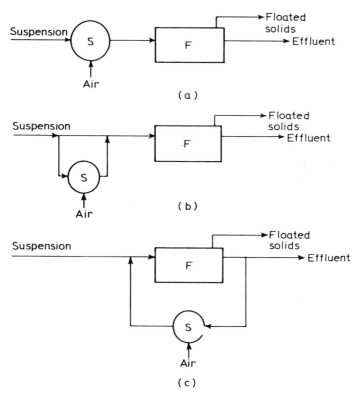

Figure 19.3. Dissolved air flotation systems. (a) Full stream pressurization. (b) Split stream pressurization. (c) Recycle pressurization. S—saturator; F—flotation cell

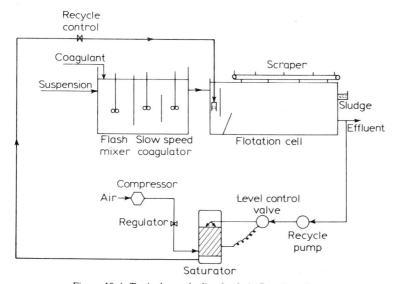

Figure 19.4. Typical recycle dissolved air flotation plant

part of the effluent through a saturator (method 3) avoids this problem, although it does increase the total fluid flow to the flotation cell.

The choice between the three systems will be governed by the flocculation stage (if any) and the volume and flotation characteristics of suspension to be treated. A typical recycle system used to treat solids-laden water and employing 5–12% clarified product recycle, is shown in *Figure 19.4*. In practice, industrial wastewaters have required recycles of effluent varying from 5 to 50%[17,26].

Floated product removal is usually by a mechanical scraper which slowly traverses the surface of the flotation unit. There must be a balance between the time allowed for drainage of water from the accumulated sludge and its mechanical properties. Generally when floating fine solids, sludges of 2–10% solids can be produced with sufficient solids loading and drainage time. However, if the sludge is left too long on the surface it may become de-aerated and brittle, and so break up during scraping thus leading to solids re-entering the water. A number of authors have discussed the problems of solids removal and several systems using paddles or chain scrapers have been described[10,26,27].

There are four main design parameters involved in practical dissolved air systems: air-to-solids ratio, hydraulic loading, saturator characteristics and injection nozzle performance.

19.7.2.1 Air-to-solids ratio (a_s)

If Henry's law is assumed to be valid, then the theoretical weight of air available for flotation at atmospheric pressure, W_a, is proportional to the saturation pressure, P, at which air has been dissolved in water, thus

$$W_a = k_s P \tag{19.14}$$

At the saturation pressures employed in water treatment systems (280–520 kPa) the values of W_a range from about 30 to 75 mg air per litre of saturator feed. An estimate of the theoretical flotation power of this dissolved air can be seen from the following calculation.

Assume the saturator gives 50 mg air dissolved per litre. This is equivalent to approximately 42 cm^3 of air per litre. In a recycle system with 10% recycle of raw water, the buoyancy available from this air per litre of wastewater is about 4 g. If the wastewater contains 4 g l^{-1} of solids then there should be enough air available to just lift these particles to the surface. The theoretical value of a_s needed is then

$$\frac{50 \times \dfrac{1}{10}}{4000 \times \dfrac{9}{10}} = 0.0014$$

In practice, the air-to-solids ratios actually required vary from 4 to 100 times the theoretically calculated value and usually fall within the range 0.007 to 0.7. This discrepancy is due to one or more of the following factors:

1. Inefficiencies in the saturator leading to lower quantities of air being dissolved than indicated by Henry's law. This will be discussed later.
2. The amount of air actually released as bubbles by the injection nozzle is less than 100%, the value depending on parameters of the system. A well-designed nozzle should give at least 90% of the dissolved air as bubbles.
3. The bubble size distribution produced by the nozzle is not suitable for the solids to be floated. For instance, a significant proportion of the air may be taken up by a few large bubbles.
4. The probability aspects of bubble–particle collision ensure that some bubbles never collide effectively with solid particles. This is particularly detrimental at low solids loadings. The degree of hydrophobicity will also influence the adherence of those bubbles that do impact.
5. In order to achieve reasonable total fluid retention times in flotation systems, the bubble–particle agglomerates must rise fairly rapidly. This means that excess air, over that just required to achieve uplift, must be attached to particles.

The value of the air-to-solids ratio required in a particular situation can only be determined by continuous pilot plant test work at various hydraulic loadings.

Once the optimum air-to-solids ratio has been determined by test work, a simple calculation allows the relationship between the recycle rate and the saturator operating pressure to be calculated. For a solids suspension flow rate of Q m^3 s^{-1} containing c_i mg m^{-3} of solids, the following can be defined:

saturator discharge, dissolved air concentration, A_1 mg m^{-3}
saturator feed, dissolved air concentration, A_2 mg m^{-3}
saturator pressure, P^* atmospheres
recycle flow rate, R m^3 s^{-1}

If it is assumed that A_2 is the solubility of air at one atmosphere then, from Henry's law,

$$A_1 = k_s(P^* - 1) \tag{19.15a}$$

The amount of air released in unit time in the cell is given by

$$\text{air released} = Rfk_s(P^* - 1) \qquad \text{mg} \tag{19.15b}$$

where f is the efficiency of the injection nozzle at releasing dissolved air as bubbles.

In unit time this amount of dissolved air is able to contact with a weight of solids given by

$$\text{solids inflow} = Qc_i \qquad \text{mg}$$

Hence, the air-to-solids ratio is

$$a_s = \frac{Rfk_s(P^* - 1)}{Qc_i} \tag{19.16}$$

or

$$R = \frac{a_s Qc_i}{fk_s(P^* - 1)} \tag{19.17}$$

Hence the most advantageous values of R and P^* can be derived from an economic analysis.

19.7.2.2 Hydraulic loading

The hydraulic loading is a measure of the flow rate through a DAF machine and hence of the fluid mean residence time. To facilitate comparisons with sedimentation methods the hydraulic loading is usually expressed in units of m^3 water per m^2 horizontal cross-sectional area of the flotation cell per day. Values found in practice vary from 30 to 400 $m^3 m^{-2} day^{-1}$. Also defined for a particular system is the limiting downward velocity, v_L. This is the value of the hydraulic loading at which bubble–particle agglomerates are just carried out of a flotation cell with the effluent. Obviously a plant must work at values of hydraulic loading less than v_L. However the bouyancy of a particle–bubble agglomerate, and hence v_L, depends primarily on the quantity of air attached to the particle and hence on the hydrophobicity and the air-to-solids ratio. Bratby and Marais[27] have shown for a coagulated wastewater that at a_s values between 0.001 and 0.4 the value of v_L increased from 40 to 600 $m^3 m^{-2} day^{-1}$, the relationship having the form

$$v_L = k_1 a_s^{k_2} \tag{19.18}$$

k_1 and k_2 being constants of the system.

At a given hydraulic loading there is normally an optimum a_s value above which no increase in flotation recovery is obtained. Presumably the surface chemical and hydrodynamic conditions only allow a certain total volume of gas to adhere to a particle. Rees *et al.*[21] give a good example of this phenomenon in the treatment of river water for the removal of turbidity.

In the case of high suspended solids content the surface loading rate, expressed as kg suspended solids per m^2 cross-sectional area of the flotation cell per day, also becomes important. Within the constraint of a limiting downward velocity, surface loading rates normally range from 5 to 50 kg m^{-2} day^{-1}. Higher values of surface loading can lead to problems of bubble–particle attachment stability and turbulence in the floated product zone decreasing flotation efficiency.

19.7.2.3 Saturator and nozzle performance

The two main methods of air dissolution in water are by sparged or packed saturator. In a sparged air system, compressed air is bubbled through water in a pressure vessel by means of a gas diffuser. With the packed system, water is distributed under pressure over some form of proprietary packing in the saturator. The sparged system has efficiencies of 60–70% of the packed saturator and is rarely used except in laboratory situations[21,27]. Surface areas of packing vary such that the water loading rate on the packing is in the range 400–2000 $m^3 m^{-2} day^{-1}$. The actual value will depend upon the volume of water to be saturated, its temperature and the pressure used.

The solubility of air in water is also a function of temperature, being about 35 mg l^{-1} at 0°C and 19 mg l^{-1} at 30°C and 1 atmosphere. Hence, saturators operating on process streams or wastewaters at temperatures above ambient will have to be commensurately larger or the recycle rate increased.

After saturation the dissolved air in solution is made available for flotation by precipitation over a pressure-reduction device such as a nozzle or valve. The bubble size distribution obtained and the quantity of air released from solution are critical to flotation performance and, hence, so is the design of the nozzle. Simple nozzles using orifice plates of varying thickness and open area provide efficient and cheap devices. Micro-bubble clouds representing over 90% of the available air have been achieved. Multi-nozzle combinations have also been used to obtain a better distribution of air in a flotation cell. However only experiments with the actual system to be used can give realistic design data. Urban[28] has investigated the theoretical aspects of orifice plate nozzles while Takahashi et al.[29] and Hyde[17] describe the performance of practical systems.

19.7.2.4 Outline design procedure

The following is a suggested design procedure:

1. Determine the solids concentration, solids composition and size distribution in the water to be treated.
2. Use a simple laboratory saturator and glass vessel to investigate the rate of flotation. Hyde[17] and Bratby and Marais[27] describe suitable systems which use a manually operated plug valve to release known quantities of supersaturated air into the suspension contained within a graduated measuring cylinder (*Figure 19.5*). Samples of unfloated solids are

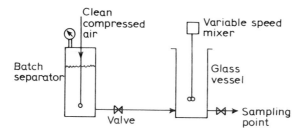

Figure 19.5. Laboratory DAF test apparatus

removed from near the base of the cylinder, and solids recovery is calculated by comparing the residual concentration with the initial value.
3. If coagulation or flocculation is required to increase the rate, determine optimum conditions using standard jar tests[17,20].
4. Use laboratory batch test work to determine the flotation kinetics at various air-to-solids ratios and the rise velocity of particle–bubble agglomerates. Study the influence of collector additions if necessary.
5. Calculate the saturator characteristics to give an acceptable value of a_s and at a given recycle rate if a recycle system is to be used.
6. Design a continuous pilot plant using the flotation kinetics and rise velocities to determine hydraulic loading, v_L and $\bar{t}$.

7. Determine the true characteristics of flotation by test work on the continuous pilot plant and scale-up to full size.

19.7.3 Electroflotation

Electroflotation closely resembles dissolved air flotation except that the bubbles produced are generally smaller and the turbulence in the cell is less. Instead of using a saturator and injection nozzle to develop a bubble swarm, an electroflotation cell uses the passage of a direct current between electrodes. In water, the migration of hydrogen and hydroxyl ions to the electrodes leads to the evolution of hydrogen and oxygen gas. The rate at which gas is evolved depends upon the current density, usually expressed as ampere per unit area of electrode surface. At a given voltage between the electrodes, the magnitude of the current is dependent upon the concentration of ions in solution and the spacing between the electrodes. Many industrial waters have low inherent conductivity, and inorganic salt or sulphuric acid additions may be used to improve current flow. The bubbles generated by this method are very small, often 30 μm or less in diameter.

Particle flotation has been found to be influenced by pH gradients set up in the vicinity of the electrodes: the anode region (oxygen evolution) tends to become more acidic than the bulk suspension, and the cathode (hydrogen evolution) more alkaline[30]. However, this effect must depend upon the mixing prevalent in the cell and its overall importance is unknown. Certainly most applications of electroflotation have required prior coagulation of the solids phase to give acceptable efficiency despite the small diameter of the bubbles.

19.7.3.1 Design parameters

Electroflotation cells consist of a single tank with horizontal electrodes placed at or near the bottom. Some designs incorporate devices to clear any settled solids from the base of the cell while others have diaphragms to

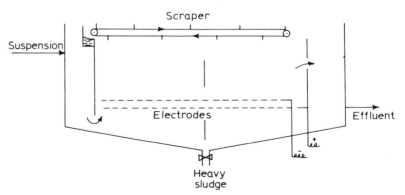

Figure 19.6. Electroflotation cell

separate the oxygen and hydrogen gas flows. A schematic diagram of a typical cell is shown in *Figure 19.6*.

The design criteria involved in the kinetics of flotation by electrolysis include those described earlier for dissolved air flotation, i.e. air-to-solids ratio, hydraulic loading and rise velocity. Acceptable values of these parameters are similar to those used in DAF[30] and the design procedure is as given in section 7.2.4.

A suitable laboratory electroflotation system has been described by Sandbank *et al.*[31]. A simple glass tank houses a pair of stainless steel mesh electrodes. A variable d.c. voltage is supplied from the mains via a rectifier and rheostat.

Electrodes have been made from many substances including mild steel, aluminium and graphite. However, severe corrosion problems have limited practical systems to stainless steel or equivalents. The upper electrode is usually in the form of a grid and is placed 6–100 mm above the lower electrode, commonly a solid sheet or grid. Voltages used are 5–15 V, creating current densities of 40–600 A m^{-2} and giving rise to gas evolutions of about 0.3 m^3 m^{-2} day^{-1}.

In general larger bubbles, but less total gas volume, are produced by lower voltages. One of the advantages of an electrolytic system is that bubble production is easily initiated and controlled. However, supersaturation ensures that bubbles continue to appear for some time after stopping the current.

19.7.3.2 Coagulation

If solids aggregation is required to promote flotation rate, a coagulation or flocculation stage may be added prior to the flotation cell, as with recycle dissolved air flotation. In a few cases the aggregating agent may be added to the cell during flotation.

Two possible methods of inducing coagulation of a suspended phase by the electrolysis itself have been described: electrode dissolution and collision induced by electrophoretic velocities. If an anode is constructed of, say, iron or aluminium then the ions released into solution due to electrolytic dissolution form effective coagulants. However, the overall costs of such sacrificial electrodes is usually higher than using inorganic aluminium or ferric salt additions. More successful is orthokinetic coagulation brought about by the movement of charged particles in an electric field. The velocity of movement of a charged particle (the electrophoretic velocity) towards an oppositely charged electrode in a stable electrical field depends upon the charge-to-mass ratio of the particles. Hence due to the distributions of particle types, sizes and surface chemical properties that usually exist, velocities will vary and collisions will occur, leading to coagulation. However, the time required for significant agglomeration to be achieved by this method normally means that either the electrodes must be far apart or the fluid residence time in the flotation tank must be increased. Both lead to extra costs and coagulation or flocculation of the suspended solid phase, if required, is invariably carried out by separate chemical addition prior to, or during, flotation.

19.7.4 Dispersed air flotation

Dispersed (or induced) air flotation systems usually generate bubbles that are significantly larger than those produced in dissolved air systems. These bubbles are also larger than most of the particles or aggregates to be floated. Consequently the beneficial influence of a high particle-to-bubble-diameter ratio is lost and the kinetics of flotation must be enhanced by improving either (or both) the collision frequency or surface hydrophobicity. The turbulence induced by the bubble-generating mechanism can provide a higher collision frequency. Additions of surfactants active at the solid–air–water interface can improve hydrophobicity considerably and dosages up to 50 p.p.m. are common where environmental considerations permit. Frothers may also be used to promote a smaller bubble size and a more stable froth. In practice, the use of additional turbulence and hydrophobicity can give high flotation rates and recoveries of solids provided that the particles are not too small. If very fine particles are involved flocculation as well as good hydrophobicity will be required to give acceptable recoveries.

Since hydrolysing metal salts cannot normally withstand the shear involved in dispersed air flotation, polymer flocculants are often used to aggregate particles. These can be used in conjunction with coagulants; for instance, where the solids have a high negative surface potential, an inorganic calcium or magnesium salt may be added to lower this potential. This will reduce the amount of cationic polymer flocculant required and may even make the use of less costly non-ionic or anionic varieties possible.

The total air flow in dispersed air flotation exceeds that for dissolved air flotation because of the larger bubble size. Although the capture efficiency of these large bubbles is low unless enhanced by collectors and flocculants, once a particle adheres to a bubble it rises rapidly to the surface. Hence the overall fluid residence time within a flotation cell using dispersed air is often less than that involved in a dissolved air system. This may give capital cost savings although if polymer flocculants and collectors are required, operating costs can be higher.

A type of effluent recycle cell has been developed in which air is entrained in recycled water. This water is pumped into the cell through Venturi-type nozzles and the air is released as a bubble cloud somewhat finer than that generated by conventional dispersed air cells[32]. This system may represent a compromise between dispersed and dissolved air cells.

19.7.4.1 Flotation kinetics

The kinetics of dispersed air cells are well described by the perfectly mixed flow, first-order kinetic model given in equation 19.8. In practice, the loss of efficiency due to short-circuiting associated with perfectly mixed flow is compensated by using several cells to give the overall residence time. It can be seen from equation 19.8 that if n individual flotation machines were arranged in series, each machine being identical, then as long as k remains constant, the overall solids recovery R_c^n is given by

$$R_c^n = 1 - \cfrac{1}{\left(1 + \cfrac{k\bar{t}_c}{n}\right)^n} \qquad (19.19)$$

Hence if the recovery in a batch flotation test were R_B and it is required that $R_B = R_c^n$ then

$$1 - e^{-kt} = 1 - \left[\frac{n^n}{(n + k\bar{t}_c)^n}\right] \qquad (19.19a)$$

This expression can be re-arranged to give

$$k\bar{t}_c = n\left[\exp\left(\frac{kt}{n}\right)\right] - n$$

or

$$\frac{\bar{t}_c}{t} = \frac{n}{kt}\left[\exp\left(\frac{kt}{n}\right) - 1\right] \qquad (19.20)$$

Equation 19.20 can be evaluated for values of n at a given laboratory batch solids recovery. For example, if $R_B = 0.9$ then $kt = 2.3$ and the ratio $\bar{t}_c/t$ for values of n can be calculated. *Table 19.2* shows that as the number of

Table 19.2. CONTINUOUS RECOVERY IN A SERIES OF MACHINES ($R_B = R_c^n = 0.90$)

No. of machines, n	$\bar{t}_c/t$
1	3.9
2	1.9
3	1.5
5	1.3
7	1.2

flotation machines in the series increases, there is a substantial drop in the overall retention time required to achieve a given recovery. It can also be seen that more than five cells in series produces few gains in reduced residence time. There will be a balance between the capital and operating costs of a single unit compared to a chain of smaller cells. Jowett and Sutherland[24] have discussed the factors that determine the most favourable value of n in a particular industrial situation; in wastewater treatment this is unlikely to be more than four[32].

19.7.4.2 Cell design

Dispersed air flotation cells normally consist of a rectangular or circular tank with a stirrer mechanism. The agitator provides turbulence to suspend solids and aid particle–bubble collision frequency. It also mechanically breaks up the air into small bubbles. This dual role has given rise to some complex agitator designs. However, the most common systems usually involve a vein-type rotor revolving within a fixed stator. Air is pumped down the rotor shaft or may be ingested by the action of the impeller–stator assembly. A fluid recirculation system ensures intimate contact between the suspension and air

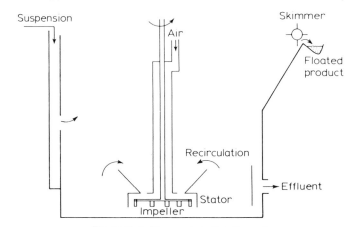

Figure 19.7. Dispersed air flotation cell

bubbles. *Figure 19.7* shows a typical design. The cell has a quiescent zone above the agitator mechanism into which the solid-laden bubbles rise. The froth or scum layer formed on the surface is removed by rotating skimmer paddles. Individual cell volumes up to 50 m³ are available.

Since a dispersed air cell is essentially a two-phase fluid pump, the main scale-up parameters are related to impeller speed and diameter. Degner and Treweek[33] have reported on comparisons between the dimensional analysis of such systems and hydrodynamic studies on actual cells. They have investigated the following relationships between impeller speed N and diameter D and liquid recirculation flow Q_r, air ingestion rate Q_a and power draw P_a:

$$Q_r \propto (DN)^{0.2} \tag{19.21}$$

$$Q_a \propto D^{2.4} \ln(N) \tag{19.22}$$

$$P_a \propto D^{3.7} N^{1.8} \tag{19.23}$$

These relationships apply to a particular cell geometry and to a given rotor submergence. Hence for a given rotor, an increase in speed will increase both the power drawn and the air ingestion rate but will have little influence on fluid recirculation. However, equations 19.21–19.23 have been used to scale-up from laboratory and pilot plant test results once suitable flotation rates have been achieved by adjustment of the surface chemical environment.

In practice, the air flow rate and the power required are related to cell volume. For larger cells these values are approximately 0.6 m³ min⁻¹ air per m³ and 1.3 kW per m³ cell volume, respectively.

19.7.4.3 Design procedure

The design procedure is similar to that for dissolved air and electroflotation, in that laboratory batch experiments lead to pilot plant work and hence to full-scale design. The procedure normally follows the route outlined below:

1. Laboratory batch test work is carried out using the 0.5 to 3 l cells

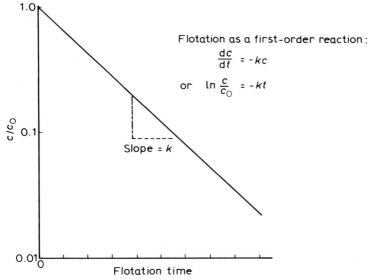

Figure 19.8. Determination of flotation rate constant

available commercially (Denver and Wemco manufacture suitable devices[4]). The value of k is found by plotting recovery versus flotation time graphs as shown in *Figure 19.8*.

2. Flotation rates are optimized using flocculants and/or collectors if necessary.
3. Continuous pilot plant tests are carried out to confirm chemical conditions and the value of k.
4. The retention time is calculated using equation 19.20 for a given solids recovery, and total cell volume is found from $\bar{t} = V/Q$.
5. Air flow rates and impeller speeds are derived for the full-scale system from pilot plant data using equations 19.22 and 19.23

19.8 RECENT DEVELOPMENTS

Dissolved and induced air flotation continue to find applications in solid–liquid separation, particularly in the provision of potable water and for the treatment of industrial wastewaters. Electrolytic flotation, however, largely remains a laboratory technique as the costs of electrodes and power are prohibitive.

A comprehensive review by Zabel[34] of dissolved air flotation for water treatment in the UK provides data on testing, plant design and costs. Flotation tanks of around 100 m^3 active volume with retention times of 5–15 min are in operation. Attention has been drawn to the efficiency of air saturation systems which contribute some 50% to the overall power costs. Experience has shown that packed saturators with polypropylene rings give almost 100% saturation at loadings of up to 2000 $m^3 m^{-2} day^{-1}$ and operating pressures of 400 kPa. The need to undertake comprehensive testwork including pilot plants has also been stressed[35].

Air requirements for the removal of suspended solids have generally been given a_s values of 0.03–0.4 with the degree of hydrophobicity of the solids playing a key role in reducing air demand[36,37]. However, it is now clear that sub-micrometre hydrophilic particles may be collected by microbubbles due to heterocoagulation between oppositely charged bubbles and particles[38]. Note that in the absence of surfactants all gas bubbles are negatively charged in waters of pH > 3. This severely limits the application of this mechanism.

Two reviews have given good descriptions of the potential of electroflotation and of its engineering and economic deficiencies[39,40]. The development of dimensionally stable anodes of platinum or iridium coated titanium has increased useful electrode life to 2–3 years but at substantial increase in costs. Power requirements are highly dependent on suspension conductivity and can contribute greatly to operating costs. At best, values of 2 MJ m^{-3} have been reported but generally the energy requirement is 30–100 times this value. Such power consumptions (15–50 kWh m^{-3}) are an order of magnitude above dispersed and induced air systems[41]

REFERENCES

1. Lemlich, R., *Adsorptive Bubble Separation Techniques*. Academic Press, London (1972)
2. Gaudin, A. M., *Flotation*, McGraw-Hill, New York (1957)
3. Somasundaran, P. and Grieves, R., 'Advances in interfacial phenomena of particulate-solution-gas systems; applications to flotation research', *A.I.Ch.E., Symp. Series,* **150** (1975)
4. Fuerstenau, M. C. (Ed.), *Flotation*, A. M. Gaudin Memorial Volume, American Institute of Mining Engineers, New York (1976)
5. Fuerstenau, D. W. and Raghaven, S., 'Some aspects of the thermodynamics of flotation', in ref. 4, pp. 21–65 (1976)
6. Derjaguin, B. V. and Dukhin, S. S., 'Theory of flotation of small and medium sized particles', *Trans. I.M.M.,* **70**, 221 (1961)
7. Anfruns, J. and Kitchener, J., 'The rate of capture of small particles in flotation', *Trans. IMM,* **86**, C9–15 (1977)
8. Laskowski, J., 'Particle–bubble attachment in flotation', *Min. Sci. Eng.,* **6**(4), 223 (1974)
9. Eigeles, M. A. and Volova, M. L., 'Kinetic investigations of the effect of contact time, temperature and surface condition on the adhesion of bubbles to mineral surfaces', *Proc. 5th Int. Min. Proc. Cong.,* p.271, IMM, London (1960)
10. Lundgren, H., 'Theory and practice of dissolved air flotation', *Filtration and Separation,* **24** (1976)
11. Traher, W. and Warren, L., 'The flotability of very fine particles—a review', *Int. J. Min. Proc.,* **3**, 103 (1976)
12. Anfruns, J., 'The flotation of small particles', PhD Thesis, London University (1976)
13. Reay, D. and Ratcliff, G., 'Removal of fine particles from water by dispersed air flotation: effects of bubble size and particle size on collection efficiency', *Can. J. Chem. Eng.* **51**,179 (1973)
14. Jameson, G., Nam, S. and Moo Joung, M., 'Physical factors affecting recovery rates in flotation', *Min. Sci. Eng.,* **9**, 103 (1977)
15. Solari Saavedra, J. A., 'Selective dissolved air flotation of fine mineral particles'. PhD Thesis, London University (1980)
16. Rubio, J. and Kitchener, J., 'New basis for selective flocculation of slimes', *Trans. IMM,* **86**, C97–100 (1977)
17. Hyde, R., 'Water clarification by flotation', *J. Amer. Water Works Assn,* **69**, 369 (1977)
18. Vuuren, L., 'Advanced purification of sewage using a combined system of lime softening and flotation', *Water Res.,* **1**, 463–474 (1967)
19. Hemmingsen, E., 'Cavitation in gas-supersaturated solutions', *J. Appl. Phys.,* **46**, 213–218 (1975)
20. Ives, K. J., 'The scientific basis of flocculation'. *Nato Advanced Study Series,* Sijthoff and

Noordhoff, Alphen aan den Rijn, The Netherlands (1978)
21. Rees, A., Rodman, D. and Zabel, T., *Water Clarification by Flotation—5*, Tech. Report TR114, Water Research Centre, Medmenham, UK (1979)
22. Flint, L. R., 'Factors influencing the design of flotation equipment', *Min. Sci. Eng.,* **5,** 232–241 (1973)
23. Himmelblau, D. and Bischoff, K., *Process Analysis and Simulation. Deterministic Systems,* Wiley, New York (1968)
24. Jowett, A. and Sutherland, D., 'A simulation study of the effect of cell size on flotation costs', *The Chemical Engineer,* Aug., 603–607 (1979)
25. King, R., 'Simulation of flotation plants', *Trans. AIME,* **258,** 286–293 (1975)
26. Rovel, J., 'Experiences with dissolved air flotation for industrial effluent treatment', *Proc. Conf. on Flotation for Water and Waste Treatment,* Paper 3, Water Research Centre, Medmenham, UK (1976)
27. Bratby, J. and Marais, G., 'Dissolved air (pressure) flotation', *Water S.A.,* **1**(2) (1975)
28. Urban, M., 'Aspects of bubble formation in dissolved air flotation', PhD Thesis, University of London (1978)
29. Takahashi, T., Miyahara, T. and Mochizuki, H., 'Fundamental study of bubble formation in dissolved air pressure flotation', *J. Chem. Eng. Japan,* **12,** 275–280 (1979)
30. Kuhn, A., 'Electroflotation—the technology and waste treatment applications', *Chem. Processing,* June, 9–12 (1974); July, 5–7 (1974)
31. Sandbank, E., Shelef, G. and Wachs, S., 'Improved electroflotation for the removal of suspended solids from algal pond effluents', *Water Research,* **8,** 587–592 (1974)
32. Degner, V. and Winter, M., 'Recent advances in wastewater treatment using induced-air flotation', *Water—1978. A.I.Ch.E., Symp. Series,* **75,** 119–133 (1979)
33. Degner, V. and Treweek, H., 'Large flotation cell design and development', in *Flotation,* A. M. Gaudin Memorial Vol., ed. by M. C. Fuerstenau, American Inst. Min. Eng., New York (1976)
34. Zabel, T., 'The advantages of dissolved air flotation for water treatment', *J. Am. Water Work Ass.,* **77,** 42–47 (1985)
35. Bratby, J., 'Batch flotation tests: how useful are they', *J. Water Poll. Con. Fed.,* **55,** 110–113 (1983)
36. Gochin, R. J. and Solari, J., 'The role of hydrophobicity in dissolved air flotation', *Water Res.,* **17,** 651–657 (1983)
37. Travers, S. and Lovett, D., 'Dissolved air flotation for abbattoir wastewater', *Water Res.,* **20,** 421–426 (1986)
38. Fukui, Y. and Yuu, S., 'Collection of sub-micron particles by electroflotation', *Chem. Eng. Sci.,* **35,** 1097–1105 (1980)
39. Mallikarjunan, R. and Venkatachalam, S., 'Electroflotation—a review'. In *Electrochemistry in Mineral and Metal Processing* (P. Richardson, Ed.), The Electrochemical Society Inc., Pennington, New Jersey (1984)
40. Matis, K. and Gallius, G., 'Dissolved air and electroytic flotation'. In *Mineral Processing at a Crossroads* (B. Wills and R. Barley, Eds), Martinus Nijhoff, Boston (1986)
41. Eberts, D., 'Flotation—choose the right equipment for your needs', *Canad. Min. J.,* 25–33 (1986)

20

The Selection of Solid–Liquid Separation Equipment

H. G. W. Pierson

Pierson & Company (Manchester) Ltd., Bozeat, Wellingborough, Northants

20.1 INTRODUCTION

The selection of solid–liquid separation equipment is dependent on such a vast number of different factors, and the choice is so very wide, that it is clearly impossible to give a formula whereby one can select the right piece of equipment without any factor of error.

This chapter should, therefore, be seen as an attempt to assist the chemical engineer in his short-listing exercise, for which it is probably more important to know which pieces of equipment are not suitable. In most cases it appears that more time and money is spent in finding that a certain piece of equipment is unsuitable rather than in finding which piece is suitable. In the end this may lead a company to turn to old, well-known, well-tried principles, thereby often settling for second best.

We would feel that we have achieved something if this chapter enables a chemical engineer to spend most of his time and budget on those pieces of equipment which are likely contenders rather than the reverse.

20.2 SEDIMENTATION OR FILTRATION?

If one ignores impingement, it would be true to say that all mechanical solid–liquid separation systems are based on one of two principles: sedimentation or filtration (see *Figure 20.1*).

With sedimentation one must remember that substantially the only driving force is the difference in specific gravity between the phases. The apparent difference can be increased by, for instance, applying a centrifugal force or by increasing the mass of individual particles through flocculation. but the basic difference is a given factor which cannot be altered. Sedimentation is, therefore, a process which, from the onset, can have limiting

Settling systems
Normally continuous and cheaper than filters

Sometimes clear } little design control
Sometimes cloudy }

Almost always wetter than filter

Filter systems
Either continuous or batch

Almost always drier than settling

Almost always clear

Design control possible

Figure 20.1. Schematic illustration of settling and filtration

factors which are beyond the design control. Sedimentation systems on the other hand tend to be relatively cheap, and are ideally suitable for continuous, and certainly automatic, operation.

Filtration, on the other hand, gives an almost unlimited amount of design control since it relies on a man-made septum, so chosen that it will retain those particles which must be separated. Filtration systems, however, are far less suitable for continuous production, sometimes not even for automatic production, and tend to be more expensive per volume treated than sedimentation systems.

From this it follows that one must first of all look at sedimentation. In this respect it is important to note that even if sedimentation does not achieve exactly the separation required it may still be a valuable first step in a process and may be followed by filtration. A typical example of this is the well-known clarification of water in water works. Here sedimentation with or without flocculation is first carried out, and the still somewhat hazy overflow is then passed through polishing filters. To achieve the end result purely by sedimentation or purely by filtration would be prohibitively expensive. The combined procedure, however, is both practical and reliable.

Since filtration and separation are still to a very large extent an art rather than a science it is important that right from the beginning one obtains a 'feel' for the material. We suggest therefore that before taking any actual measurements one observes the sample and memorizes how it behaves. Does the material form a precipitate, does it spontaneously flocculate, do the eddy currents travelling upwards disturb the flocs, or do they simply move them out of the way, does a small stirring action upset the whole pattern, etc.? These somewhat nebulous facts are very important in the final choice since it will enable one to decide whether or not the specific mechanical details of otherwise identical pieces of equipment are acceptable or not.

20.3 SEDIMENTATION EQUIPMENT

Figures 20.2, 20.3, 20.4 and *20.5* show the settling times for different sludge volumes in the various types of sedimentation systems. It is important to note that the equipment listed is merely representative of whole groups. These figures are, therefore, meant purely as a short listing. Having selected a few likely groups one will then have to go further into specific design and mechanical details to see which particular type of any one group would be most suitable. That this is a task which can only be carried out successfully by an expert with an intimate knowledge of the pieces of equipment hardly requires further labouring.

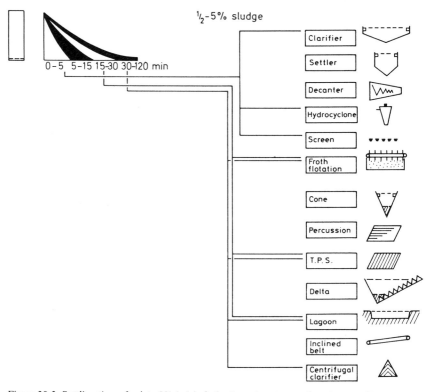

Figure 20.2. Settling times for $\frac{1}{2}$ to 5% (v/v) sludge in various types of equipment. Decanter—In this instance only a centrifugal decanter is meant. Cone—Typically a deep cone settlement system without vertical side walls. Percussion—Percussion screens or shaking tables, while originally mainly used for the classification of solids, can act as very efficient simple liquid–solid separators. Delta decanters—A truely triangular sedimentation system with a screw conveyor taking away the sludge. Particularly suitable for very fast settling coarse material

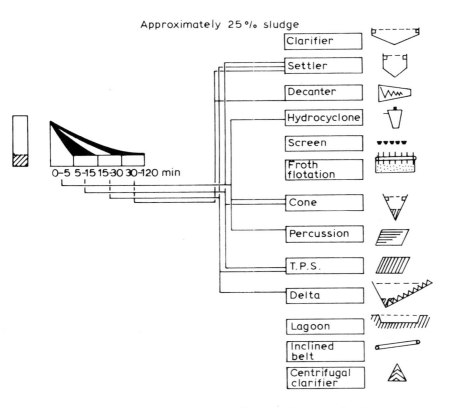

Figure 20.3. Settling times for a 25% (v/v) sludge in various types of equipment

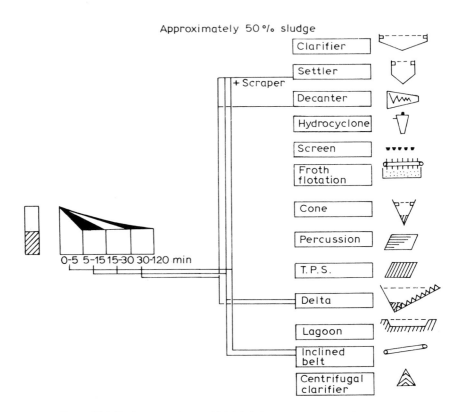

Figure 20.4. Settling times for a 50% (v/v) sludge in various types of equipment

Of the various groups shown two should be mentioned specifically, as they are somewhat anomalous. First of all the screen is not a sedimentation system *per se* but is listed here since there is a good chance that a slurry which behaves as indicated would also be amenable to screening, in which case this may be a preferred method. Similarly, the inclined belt, being somewhat of a hybrid, could be considered by some to be a filter using its own solid mass as a medium. The same, of course would apply for inclined vibrators.

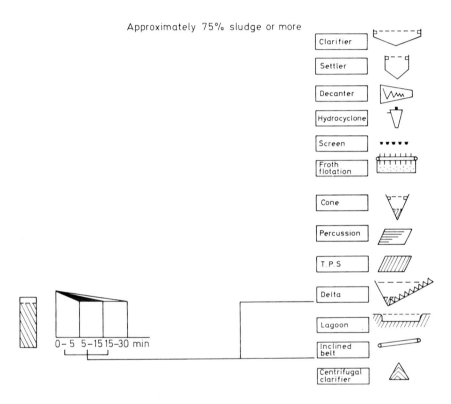

Figure 20.5. Settling times for a 75 % (v/v) sludge in various types of equipment

Whereas an expert would be able to design and calculate the performance of most pieces of equipment from laboratory tests a word of warning should be sounded about centrifugally operated systems, which include hydrocyclones. In our experience these pieces of equipment are best tested on a pilot scale rather than on a laboratory scale.

20.4 FILTRATION EQUIPMENT

Filtration is a very much more complex process than sedimentation since the large degree of design control that one has also, of course, allows a far greater number of parameters to be varied. For a true evaluation it will, therefore, be necessary to have access to a reasonably well equipped filtration laboratory. For the purpose of this chapter, however, we will review purely the short-listing and basic evaluation and assume that only the most simple laboratory equipment is available.

A prerequisite for any filtration tests is the Buchner funnel with filter flask and the filter leaf. In addition a good vacuum source is required. This should preferably be better than the simple water tap system.

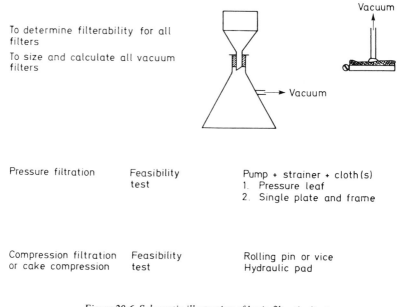

To determine filterability for all filters

To size and calculate all vacuum filters

Pressure filtration   Feasibility test   Pump + strainer + cloth(s)
1. Pressure leaf
2. Single plate and frame

Compression filtration or cake compression   Feasibility test   Rolling pin or vice
Hydraulic pad

Figure 20.6. Schematic illustration of basic filtration tests

Since all filtration techniques depend on passing a slurry through a septum at a pressure differential, it is clear that a basic vacuum test will in most cases at least establish filterability. It is clearly a complete method for determining the possibility of using vacuum filters and even for pressure filters the basic parameters can be plotted (*Figure 20.6*).

To establish what effect greater pressure differentials have one will have to rely on proper test equipment and, of course, the same would apply to compression filtration.

We stress that, before starting any filtration tests, the following basic principles must be adhered to:

1. *Filter medium*
   Use only the material which will be used under plant conditions.

2. *Filter cloth*

   If cloth is to be used remember that the wetter the cake the smoother and possibly thinner, the cloth should be.

3. *Actual tests*

   (a) Sample must be representative.

   (b) The filter area should never be less than 15 cm diameter.

   (c) Test against actual cake thicknesses.

On the last point it is appreciated that formulae exist which should enable one to extrapolate from a thin cake tested under laboratory conditions to a thicker cake expected on a full scale plant. In our opinion these formulae are virtually useless.

Filters can be subdivided into numerous different groups.

A first distinction is the difference between those filters which filter through the actual cloth, paper, or screen etc. and those which filter through the cake. The cake filters have the advantage that one can normally use a support which is less prone to blinding or bleeding than for the non-cake filters, but, of course, a prerequisite is an adequate amount of solids in the feed.

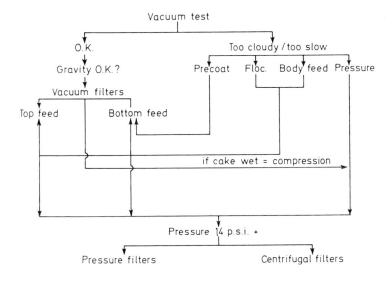

Figure 20.7. Schematic illustration of filter selection from vacuum tests

A further obvious distinction is between vacuum filters (negative pressure) and pressure filters (positive pressure) (see *Figure 20.7*). As the vacuum filters are limited to a maximum practical pressure differential of about $6.8-8.2 \times 10^4 \text{ Nm}^{-2}$ (10 to 12 p.s.i.), theoretically they have a lower capacity. Since most pressure filters, however, rely on a thicker cake and, therefore, a greater cake resistance, the net result is not always so much in favour of pressure filtration. It is important, therefore, actually to check these figures and not assume higher throughput with a higher pressure.

Regarding the actual cake, it must be borne in mind that the way in which the solids are applied to the screen determines to a large extent the ultimate pressure differential. It is, therefore, necessary to select a screen and a feeding mechanism to suit the slurry conditions. Blinding occurs in practically all filters. If the blinding occurs in the cake itself then only chemical treatment of the slurry prior to filtration or varying the pressure differentials can cure this problem. If the blinding occurs in the support screen, cloth or paper then there may be some merit in using a support with a pore size larger than that of the smallest fraction, and allowing the largest fraction first to form a bed, recycling the runnings which will contain the fine particle fraction, through the cake (*Figure 20.8*).

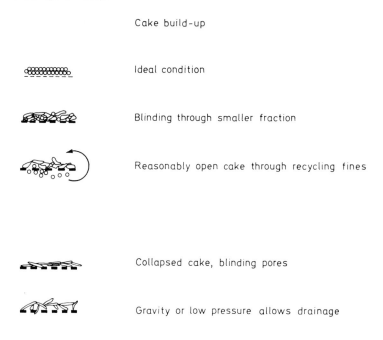

Figure 20.8. Types of cake build-up

When the vacuum tests have been carried out and the results are known, one can use *Figure 20.7* to select likely pieces of equipment. If the vacuum tests are very successful one must first look at gravity filtration and see whether or not this would be adequate. If not, one must calculate the possibilities of using either top feed or bottom feed vacuum filters and work out the necessary costs, capital and running costs, for the particular problem.

If the vacuum tests are not successful because the rate is too slow or, if the tests are successful but the equipment for vacuum filtration would be too large or too costly, or simply because filtration is not achieved, then one has to look at four possible alternatives. These are: a bottom feed vacuum filter with precoating; flocculating the feed or adding body feed, either for bottom feed or top feed vacuum filters; or pressure filtration in all its ramifications, either with or without flocculation or body feed. In certain cases one can even look at precoating pressure filters. In this respect, of course, pressure filters can equally well be centrifugal filters, or compression filters.

Since the solids are often of importance, not only as an end product but also as a costly item to dispose of if they are not wanted, many filters are being classified by the form in which they dispose of the cake. A very broad breakdown is given in *Table 20.1*. In the table the arrows indicate whether or not the cake discharge tends to be a dry or a wet cake.

Table 20.1. FILTER SELECTION BY CAKE QUALITY

Dry cake	Wet cake/sludge	Thin sludge/slurry
Vacuum filters (Precoat rotary)*	←———— Pressure leaf	Sand bed
Filter presses	Candle ———→	Mixed media
Spinning leaf ———→	Gravity ———→	Screens
Disposable belt*		Laminar flow
Cartridge		Edge
Centrifugal filter		Reverse osmosis
Compression filter		

*Solids cannot normally be recovered

No mention of filters would be complete without reference to cake washing. Particularly with the burden of effluent costs becoming greater every day and the cost of water rising, more and more companies are being forced into better and more efficient methods of cake washing. In this respect the filter press is particularly important at present since historically it has been used extensively for cake washing applications. Whereas a filter press will give a very well washed cake in most cases it will only do so at the cost of a tremendous amount of water. Possibly for this reason many companies are looking at continuous vacuum filters and, specifically, rotary vacuum filters, for cake washing.

One must remember the great difference between a rotary drum filter and a rotary belt filter. As shown in *Figure 20.9* the loss factor, or, in other words, the area of the drum which is not used for filtration is very much greater for a belt filter than for a drum filter. This has led to many mistakes being made since rotary vacuum filters are quoted in square feet of filter area, by which the manufacturer means the total area of the drum. As can be seen from *Figure 20.9* for an identical drum circumference and face the drum filter offers a far greater effective filter area than the rotary vacuum belt filter. Or, in other words, for a given effective filtration area the rotary belt filter will have to be a larger machine than the drum filter.

A further point which is often ignored with rotary filters is the necessity to have a certain ratio between pick-up time and filtration time. *Figure 20.10* shows, for example, that in a case where the pick-up time is 60% of the total time an ordinary rotary vacuum drum or belt filter could run at, for instance, a 48% loss factor. This could only be reduced if one uses a machine which has submerged bearings, which is very much more expensive. Since the ratio tends to vary somewhat between various makes it must be stressed

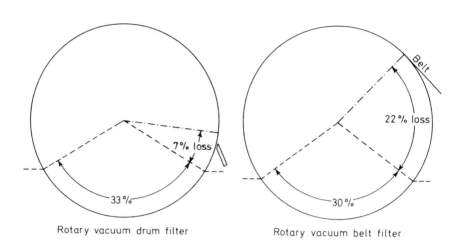

Rotary vacuum drum filter   Rotary vacuum belt filter

Figure 20.9. The loss factor in two types of rotary vacuum filter

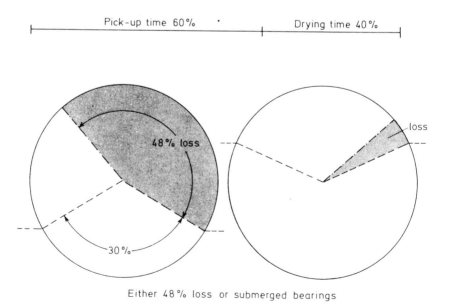

Either 48 % loss or submerged bearings
Rotary vacuum drum filter

Figure 20.10. Reduction of the loss factor in a rotary drum filter with submerged bearings

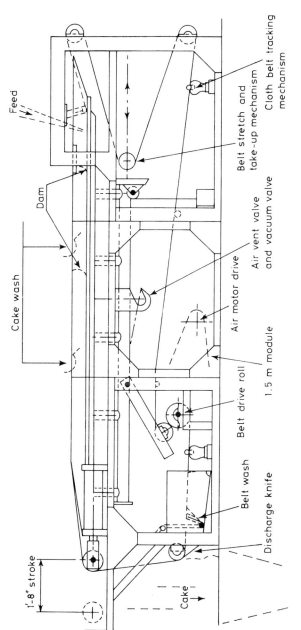

Figure 20.11. Horizontal belt filter

that these loss percentages may not be absolute but should be taken only as an indication.

A rotary vacuum filter, therefore. requires very careful selection and a very careful study before a decision is made. If one wishes to wash with a rotary vacuum filter then the problem is even more aggravated since the natural configuration of the filter does not present a very good surface for washing. This, in itself, can again alter the ratio and it is not uncommon to find that one is reduced either to using two or three filters, or to using machines which are vastly larger than the theoretically required filter area.

These problems do not exist in the horizontal belt filter, of which in the last few years quite a number of very successful machines have been introduced. This filter is basically constructed as an endless belt so it does not have the same dead time factors, is not affected by differences between cake formation and drying time, and in all cases presents a far better washing surface than a rotary filter (see *Figure 20.11*). If one wishes to take cake washing to the ultimate efficiency then one must consider countercurrent washing, which is virtually only possible on a belt filter or any other type of filter which has a continuous horizontal filter surface such as, for instance, the rotary tipping pan (see *Figure 20.12*).

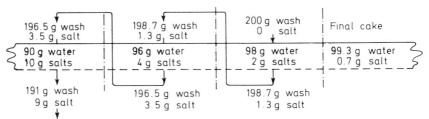

Figure 20.12. *Countercurrent washing on a horizontal filter*

With a growing interest in energy conservation, it is obvious that more and more filters will be required which produce cakes of the greatest possible dryness. For this purpose membrane pressure filters tend to be the best solution. Semi-automatic and automatic units are available. Since they all rely on a flexible membrane the limiting factor appears to be more in the choice of rubber than anything else. It is important to remember, however, that if cake washing has to be carried out prior to ultimate drying the membrane press can be a disadvantage since the very compacting of the cake which frequently occurs quite often prevents effective cake washing.

This chapter attempts to do nothing more than to give a very broad outline of the possibilities, and above all the limitations, of different groups of equipment. We hope that it will enable the chemical engineer to make, relatively quickly, an approximate choice of equipment in order to assess whether or not a given project is financially effective. The ultimate selection and final design of the complete plant can only be made after a very detailed study of the problem as a whole and of the individual pieces of equipment which are short-listed.

21

Membrane Separation

E. H. Stitt

Department of Chemical Engineering, University of Bradford

NOMENCLATURE

C	Solute concentration (mol m^{-3})
C_p	Permeate concentration
C_R	Retentate or bulk solution concentration
C_T	Total molar concentration of solution (molar density)
C_w	Concentration at membrane surface
D	Diffusion coefficient
$\mathbf{J}$	Solvent permeate flux
$\mathbf{J}_s$	Solute flux in the permeate
K	Distribution coefficient of solute between solvent and membrane phases
k	Mass transfer coefficient
N_A, N_B	Solute, solvent flux in reverse osmosis
P	Applied hydrostatic pressure
R	Apparent rejection coefficient
Re	Reynolds number
Sc	Schmidt number
Sh	Sherwood number
T	Temperature
t	Membrane thickness
$\mathbf{U}$	Cross-flow velocity
W	Pure water permeability
x	Mole fraction (see *Figure 21.4* for subscripts)
y	Distance from membrane surface
α	Real rejection coefficient
Γ	Virial coefficient
ΔP	Hydrostatic pressure difference
δ	Whitman diffusive film thickness
ϵ	Membrane porosity
μ	Dynamic viscosity
ν	Kinematic viscosity
Π	Osmotic pressure
ϕ	Concentration polarization modulus ($\phi = C_w/C_R$)

21.1 MEMBRANE SEPARATION PROCESSES

A wide variety of separations may be achieved using membranes. A membrane is best considered as a thin film interposed between two fluid phases, the selective permeation through which is governed by particle or molecular size, chemical or physical affinity to the membrane material and/or the mobility of the permeating species within the membrane. Separation is an entropy reducing process and, as such, work has to be done on the system – either chemically, electrically or mechanically. The means of energy input is to provide a driving force for mass transfer which may be, respectively, a concentration, potential or pressure gradient. The nature of the driving force is a useful crude method of membrane process classification and *Table 21.1* lists a number of operations according to this rationale. In the present context, it is the pressure-driven processes that are of interest.

Table 21.1. MEMBRANE SEPARATION PROCESSES CLASSIFIED BY DRIVING FORCE

Description	Driving force	Operation
Electrically driven	Potential (voltage) gradient	Electrodialysis
Chemically driven	Concentration (chemical potential) gradient	Dialysis Liquid membranes Pervapouration
Pressure driven	Pressure gradient	Hyperfiltration Ultrafiltration Microfiltration

21.2 PRESSURE-DRIVEN MEMBRANE SEPARATIONS

Membrane filtration is now widely considered as an area of rapid expansion and of great potential within the chemical and process industries. The term itself encompasses a number of separation processes varying from solvent purification, through separation of solutes according to molecular size, to filtration of fine or colloidal suspensions. Despite this wide domain, many of the underlying principles governing its •operation remain essentially unchanged. It is, however, useful in the first instance to define terms and thus gain a perspective of the regions of operation. Membrane filtration is generally considered to consist of three types of process:

1. *reverse osmosis* (or hyperfiltration)—the separation of a solvent (typically water) from its dissolved solutes;
2. *ultra-filtration*—the fractionation of dissolved molecules according to size;
3. *microfiltration*—the filtering of fine or colloidal suspensions.

The range of these operations is shown in *Figure 21.1* in terms of the smallest particle retained. As shown, the boundaries between reverse osmosis and ultra-filtration and between ultra-filtration and microfiltration are not distinct. This is especially so in the latter case where macromolecules and particles may be of similar size.

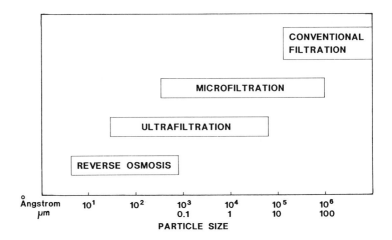

Figure 21.1. Ranges of membrane filtration techniques

Membrane filtration operations employ 'cross-flow' rather than 'dead-end' flow as used in conventional filtration. The difference between these modes of operation is shown in *Figure 21.2*. In cross-flow filtration, solution flow is tangential to, rather than perpendicular to the filtration medium. The shear forces generated by fluid motion across the membrane surface inhibit the build-up of a cake. The cross-flow velocity and the applied pressure used are dependent upon the nature of the separation required. In general, the cross-flow velocity increases from reverse osmosis to microfiltration. Higher pressures, however, are required for reverse osmosis in order to overcome osmotic pressure.

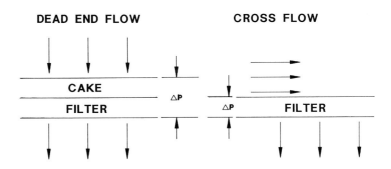

Figure 21.2. Dead-end versus cross-flow filtration

21.3 REVERSE OSMOSIS (HYPERFILTRATION)

Although this chapter is ostensibly concerned with ultra-filtration and microfiltration using membranes, it is useful to introduce some of the concepts using well-developed reverse-osmosis theory. Although many facets of operation undergo a change in the progression from reverse osmosis to microfiltration, the underlying theory remains the same and is least complex for reverse osmosis. The principal difference lies in the fact that reverse osmosis relies substantially upon chemical and physical interactions with the membrane to achieve separation, microfiltration is essentially a sieving operation.

If water and an aqueous solution are separated by a semi-permeable membrane, the solvent will pass through the membrane in a direction such that the concentration difference across the membrane is reduced: this process is osmosis and is shown in *Figure 21.3a*. The driving force for this transport process is termed the 'osmotic pressure', which may be evaluated from, for example, van't Hoff's equation:

$$\Pi = C \cdot R \cdot T \left[1 + C \cdot \frac{\Gamma}{2} \right]^{2} \qquad (21.1)$$

If, however, a pressure is exerted upon the more concentrated solution, the osmotic driving force is effectively reduced. If the applied pressure, P, is equal to the osmotic pressure, no nett driving force exists, and only equimolar counter diffusion will occur—see *Figure 21.3b*. When P exceeds Π, then the solvent passes in the reverse direction due to the nett pressure driving force—see *Figure 21.3c*. Thus the solution becomes more concentrated and 'reverse osmosis' is obtained. The term 'reverse osmosis' is in fact a misnomer and it is for this reason that the operation is sometimes referred to as 'hyperfiltration', because while the solvent flow is reversed with respect to 'normal osmosis', the direction of solute leakage is unchanged.

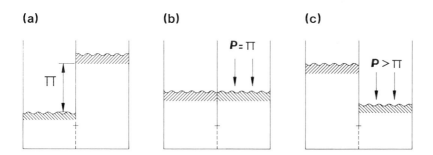

Figure 21.3. Principles of reverse osmosis: (a) osmosis; (b) equimolar counterdiffusion; (c) reverse osmosis

In a reverse osmosis unit the applied pressure must be sufficient to overcome the osmotic pressure in addition to the hydraulic resistance of the membrane. The osmotic pressures corresponding to the salt concentrations prevailing at the membrane surface are substantial and operating pressures in the range 80–100 bar are not uncommon. Despite this, a considerable number of large, industrial-scale units are in operation for desalting brackish waters and sewage treatment. Some typical applications of reverse osmosis are listed in *Table 21.2*.

Table 21.2. APPLICATIONS OF REVERSE OSMOSIS

Desalination of brackish water and seawater
Production of pure water: process water for electronics industries;
                           boiler feed make-up;
                           for air conditioning plants and dialysis
Treatment for dilute pulp and paper effluents
Treatment for textile dyehouse effluents
Treatment of metal finishing wastewaters
Concentration of fruit juices and sugar solutions
Treatment of wine and beer: tartar removal;
                            modification of alcohol content
Dewatering cheese whey

21.3.1 Reverse osmosis transport theory

A number of models of varying complexity and attention to fundamentals exist for describing hyperfiltration. The basic model, derived by Kimura and Sourirajan[1], relies upon characterizing the membrane by two parameters:

1. pure water permeability, W;
2. solute transport parameter $D_{AM}/K \cdot t$

where D_{AM} is the diffusivity of the solute in the membrane, K is the distribution coefficient of the solute between the solvent (aqueous) solution and the membrane and t is the effective membrane thickness.

The model is based upon a mechanism wherein one of the constituents is preferentially absorbed onto the membrane surface and transferred through the membrane pores under the influence of the applied pressure. This is often referred to as the 'preferential sorbtion–capillary flow' mechanism. Other models have been proposed based upon dissolution of the solute and subsequent diffusion (see Mazid [2]).

The mechanism gives rise to a number of basic transport equations. Referring to *Figure 21.4*, it may be shown that[1]:

$$N_B = W\,[P - \Pi\,(x_{A2}) + \Pi\,(x_{A3})] \tag{21.2}$$

$$= \frac{D_{AM}}{K\,t}\,C_T\,\frac{(1 - x_{A3})}{x_{A3}}\,(x_{A2} - x_{A3}) \tag{21.3}$$

$$= k C_T\,x_{B3}\,\frac{(x_{B1} - x_{B2})}{(x_A)_{LM}} \tag{21.4}$$

$$N_A = \frac{D_{AM}}{K t}\,C_T\,(x_{A2} - x_{A3}) \tag{21.5}$$

where

$$(x_A)_{LM} = \frac{(x_{A2} - x_{A1})}{\ln\left[\dfrac{x_{A2} - x_{A3}}{x_{A1} - x_{A3}}\right]}$$

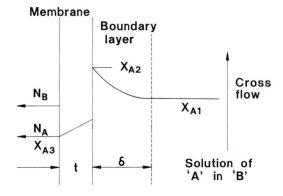

Figure 21.4. Nomenclature for reverse osmosis

Each of the equations 21.2 to 21.4 considers a particular aspect of solvent (water) permeation. Equation 21.2 represents the resistance due to osmotic pressure, equation 21.3 the resistance within the membrane itself, and equation 21.4 accounts for the mass transfer within the diffusional liquid boundary layer. Note that the latter equation includes what is, in effect, the 'drift factor', commonly used in gas absorption calculations, to account for diffusion through what is ostensibly (in mass transfer terms) a stagnant film. Equation 21.5 considers the rate of solute leakage through the membrane.

The important conclusion from these equations in the present context is that for a given membrane and feed condition, performance in terms of permeate flux, N_B, and apparent solute rejection $(x_{A1} - x_{A3})/x_{A1}$, is governed by the mass transfer coefficient, $k (= D_{AB}/\delta)$ and the solute concentration at the solution–membrane interface, x_{A2}. Both these factors are essentially determined by the behaviour of the boundary layer at the membrane. The above observation is not limited to reverse osmosis but is equally applicable to ultra-filtration and microfiltration.

The preferential sorption–capillary flow mechanism infers that the rate of transfer of a solute through the membrane is dependent upon the chemical interaction between the solute and the membrane as much as it is a function of the physical restriction of pore size. The same conclusion may be drawn from the solution–diffusion model. It is well acknowledged that this is indeed the case for reverse osmosis. In ultra-filtration, it has been observed that solute permeation is not dependent upon pore size geometry alone. In certain cases permeation of molecules larger than pore size has been noted. Much current research into reverse osmosis and ultra-filtration is concerned with the effects of surface chemistry and surface force parameters upon permeation rates and the rejection characteristics of the membrane.

21.3.2 Concentration polarization

It was noted that the performance of a given membrane system is dependent upon the solute concentration at the membrane–solution interface, C_W. This is itself subject to a phenomenon called 'concentration polarization' which results in wall concentrations substantially higher than those in the bulk phase.

Consider the system shown in *Figure 21.5*. The solvent flow within the boundary layer may be resolved into two component vectors: the cross-flow flux, U, parallel to the membrane surface and the permeate flux, J, normal to the membrane. Both these vectors carry with them a solute concentration, C. Considering the normal component, on arrival at the membrane the solvent permeates through but the solute is substantially rejected. Thus, at the start-up, the solute flux, JC, leads to a build up of solute at the membrane surface. A diffusional flow of solute from the membrane to the bulk phase ensues. The diffusion within the boundary layer is a slow process compared with the bulk flow and thus, a relatively high wall concentration is attained before back-diffusion equals bulk flow. This is a slightly simplified visualization of the process because the solute flow-back into the bulk solution is usually enhanced by forced and/or natural convection effects resulting from the retentate cross-flow. The mass transfer within this boundary layer is conveniently represented and characterized by conventional mass transfer coefficients.

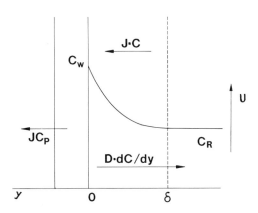

Figure 21.5. Concentration polarization

The effects of the enhanced solute wall concentration can readily be seen by reference to equations 21.2 to 21.5. It results in an increased osmotic pressure resistance, a decrease in solvent flux and a decrease in separation performance. The control of concentration polarization is therefore essential if good separation and product rates are to be attained. The limiting step is the diffusion of the solute away from the membrane which is itself determined by the fluid flow boundary layer. The control of concentration polarization is possible by minimizing boundary layer thickness or by inducing turbulence which acts to increase the mass transfer coefficient, k, in equation 21.4. Thus, enhancement of performance in terms of product purity and flux is possible by means of correct fluid management techniques.

21.3.3 Effect of bulk concentration

The bulk concentration of the retentate or feed stream has considerable influence upon the permeate rate and its purity. This may readily be seen with reference to equations 21.2 to 21.4. Assuming the same hydrodynamic conditions and feed-stream pressure to prevail, and therefore the mass transfer coefficient, k, is constant, then clearly a higher wall concentration, x_{A2}, will be obtained at increased bulk concentration, x_{A1}. This results in an increased osmotic resistance to permeation and thus decreased permeation rates (equation 21.2). An increased wall concentration will also lead to increased transfer of the solute through the membrane (equations 21.3 and 21.5) and thus reduced permeate purity.

 These general observations apply equally to ultra-filtration and microfiltration. Thus, for all membrane filtration operations, increases in feed concentration lead to a reduction in both permeate product rate and purity. In ultra-filtration and microfiltration the situation may be made worse by changes in feed stream rheology. The viscosity of macromolecular solutions and of particle suspensions can be very sensitive to concentration. Consequent increases in viscosity will lead to decreased Reynolds numbers and mass transfer coefficients and thus to an exacerbation of concentration polarization.

21.4 FLUID MANAGEMENT

In all the theoretical approaches to membrane filtration, expressions include a mass transfer coefficient for transport to the bulk feed stream from the membrane surface. This is itself a function of boundary layer behaviour which in turn also controls the degree of concentration or particle polarization and flux stability and decay (membrane fouling): fluid management is the general term used for their collective treatment. Data is presented in the literature by a number of workers for reverse osmosis[3], ultra-filtration and microfiltration[4-7].

 The mass transfer–heat transfer analogies well-known in chemical engineering, have enabled application of heat transfer coefficient correlations to the evaluation of the mass transfer coefficient. For laminar flow Leveque's solution for convective heat transfer analogy gives:

$$Sh = 1.62 \, [Re \cdot Sc \cdot d_h / L]^{0.33} \qquad (21.6)$$

for $100 < Re \cdot Sc \cdot d_h / L > 5000$, which is equivalent to:

$$k = 1.62 \, [UD^2 / d_h L]^{0.33} \qquad (21.7)$$

where k is the mass transfer coefficient, U the mean fluid linear velocity, D the solute diffusivity, d_h the channel equivalent hydraulic diameter and L its length. A more general expression is:

$$k = 0.816 \, [\dot{\gamma} / L \, D^2]^{0.33} \qquad (21.8)$$

here $\dot{\gamma}$ is the fluid shear rate at the membrane surface. For rectangular channels $\dot{\gamma} = 6U/b$ (where b is the channel height) and for circular tubes $\dot{\gamma} = 8U/d$.

Examination of these equations indicates that any fluid management technique which increases the fluid shear rate will increase the flux. This may be achieved by the use of increased cross-flow velocity or by decreasing the channel height.

For tubulent flow, the Dittus–Boelter heat transfer correlation has been successfully applied. For mass transfer this becomes:

$$\text{Sh} = 0.023 \, \text{Re}^{0.8} \, \text{Sc}^{0.3} \qquad (21.9)$$

$$k = 0.023 \, (U^{0.8} \, D^{0.67}/d_h^{0.2} v^{0.47}) \qquad (21.10)$$

where v is the kinematic viscosity. As for laminar flow, increases in cross-flow velocity or a decrease in channel height yield an increased mass transfer coefficient and thus improved permeate flux.

The above observations have given rise to the use of 'thin channel' ultrafiltration techniques wherein low channel heights and high cross-flow velocities are employed to maintain good mass transfer coefficients. Spiral wound and hollow fibre modules are exemplary of this approach. Typical channel heights and linear velocities vary according to the nature of the separation. For example, reverse osmosis uses cross-flow velocities up to 1 m s^{-1}. In microfiltration channel heights will rarely be greater than 5 mm, while linear velocities in excess of 2 m s^{-1} are normally used. Typical permeate fluxes vary in a similar fashion. In reverse osmosis fluxes of 3 to 15 $\mu\text{m s}^{-1}$ (1–5 gallon $\text{ft}^{-2} \text{ day}^{-1}$) are common, while in microfiltration fluxes of above 3 mm s^{-1} (1000 gallon $\text{ft}^{-2} \text{ day}^{-1}$) can be obtained.

Some laboratory research workers have proposed the use of turbulence promotors, pulsatile flow and the like to reduce boundary layer thickness and thus reduce concentration polarization. These ideas have not, however, gained acceptance on a commercial scale. The thin channel technique remains prevalent as a means of mass transfer enhancement.

In microfiltration, a number of reports exist where electric fields have been employed to promote migration of the particles away from the membrane[8-10]. Above a critical voltage gradient, electrophoretic effects cause a bulk flow of the particles away from the membrane, thus reducing particle polarization and enhancing flux. While this approach has yielded encouraging results on a laboratory scale, it poses considerable scale-up problems.

21.5 MEMBRANE MORPHOLOGY AND PRODUCTION

A wide variety of membranes is commercially available with large variations in structure, form and materials. The type of membrane used will depend to a large extent upon the applications. Fundamental differences exist between membranes for reverse osmosis and microfiltration usage.

21.5.1 Microfiltration membranes

Microfiltration membranes are made from a wide variety of materials. Originally the use of cellulose esters and cellulose nitrate was prevalent but nowadays the range includes polymers such as nylon, polyvinylchloride (PVC), co-polymers of PVC and acrylonitrile, polyproylene, polysulphones and polytetrafluoroethylene. Different methods of production have evolved for different polymers. In general, microfiltration membranes are isotropic; i.e. their pore size and porosity are independent of the depth within the membrane.

Polypropylene, and PTFE membranes can be made simply by stretching a thin sheet of the polymer, under controlled conditions, to yield a thin microporous film. In many cases this is then bonded onto a porous support in order to improve mechanical properties. Resistance to flow caused by the support will be small compared with that of the membrane itself. The thickness of the membrane (film) is usually in the order of 100–250 μm.

An alternative method of microfiltration membrane preparation is 'track-etching'. A thin dielectric film is bombarded with massive, charged atomic particles. This produces a series of narrow radiation-damaged trails called 'tracks'. The polymer film is then etched by a solvent which selectively dissolves the damaged material leaving cylindrical, straight pores. This is in contrast to the tortuous path of creating pores using casting or stretching. These latter methods do, however, produce membranes of greater porosity. Pore size in track-etched membranes is controlled by the temperature and residence time in the etching bath. An upper limit to membrane thickness of 15 μm is imposed by the method of production. This method has been used, principally for the production of polycarbonate membranes which are then bonded onto a mechanically strong support.

Most pairs of polymers will separate into two liquid phases when placed in the same solvent. A number of compatible pairs are, however, known. When a film is cast from a pair of compatible, non-complexing polymers and the solvent removed by evaporation, the polymers in the resulting dry film are frequently partially separated. If one of the polymers is leached from that film then this will result in a microporous film of the second polymer which can be used as a microfiltration membrane.

Most membranes are prepared using casting techniques. The oldest and still most important is known as 'phase inversion'. The polymer is dissolved in a solvent or solvent mixture and cast as a thin film (0.2–0.4 mm) onto a suitable surface. This film is exposed to a controlled environment which promotes loss of the solvent from the film. This results initially in precipitation of the polymer. As further solvent loss occurs, the precipitate to solution phase ratio increases, phase inversion ensues and the polymer precipitate becomes the continuous phase. Solvent loss continues, and the polymer phase gelates thus setting the membrane structure. The remaining solvent represents the eventual voids or pores in the finished membrane. The film shrinks during the process and final thicknesses are of the order of 100 μm.

Essentially, three variants of the phase inversion method exist and these differ in the driving force used to obtain solvent loss and/or precipitation of the polymer. In the 'dry process', the film is cast onto the surface and exposed to a controlled atmosphere in which the solvent evaporates. In the 'wet process', the film cast onto its surface is plunged into a bath of antisolvent, i.e. a liquid which is miscible with the solvent but in which the polymer is insoluble. Several applications use acetone as the solvent and water as the antisolvent. Solvent loss is obtained by the counter diffusion of the two solvents. Finally, in the 'thermal process', a saturated solution is cast at an elevated temperature and allowed to cool. Decreasing solubility as the temperature falls leads to successive precipitation, phase inversion and gellation.

In the phase inversion method, the properties of the membrane obtained are controlled by correct choice of solvent and any secondary solvents, use of additives, the molecular weight of the polymer and its concentration in the casting solution, in addition to control of the environment in which casting and the rest of the process are carried out. Some modification is possible by subsequent heat treatment.

Inorganic membranes, also referred to as ceramic and mineral membranes, were developed in France for use in the nuclear industry but have recently been declassified and are now commercially available. They are an exciting development and are well suited to cross-flow microfiltration. They consist of a sintered silica or alumina support, or substrate, with, typically, a 15 μm or greater pore size. On to this is deposited a layer of, typically, γ-alumina from a sol or suspension which forms the active layer. After deposition this is compacted by pressure, dried and baked to give a typical thickness for the active layer of 1–10 μm. The pore size of the membrane is controlled by regulating the particle size distribution in the sol. If required, several layers of decreasing particle size may be used to prevent penetration into the substrate. The thickness of the active layer is simply regulated by varying the time allowed for particle deposition. Inorganic membranes are almost exclusively of tubular form, 'skinned' internally and are currently available with pore sizes down to less than 0.2 μm.

Organic and polymer membranes have a limited temperature stability and, in some cases, especially with cellulosic materials, are particularly sensitive to chemical attack. Advantages may lie with inorganic membranes in that they can be steam cleaned, and in many cases, are essentially chemically inert. They are, however, expensive and the relatively large flow channels in the available equipment may lead to high pumping requirements.

21.5.2 Reverse osmosis membranes

The above methods of production are not applicable for reverse osmosis membranes. The higher working pressures require high levels of rigidity and mechanical strength. Furthermore, the small pore size would lead to intolerably slow permeation in isotropic membranes such as those described above. Thus, for reverse osmosis an antisotropic, or asymmetric, membrane is used which comprises a skin of less than 1 μm thickness, of very fine pores, overlaid on a porous support which provides the mechanical strength. The technology of producing such membranes was developed by Loeb and Sourirajan at the University of California, Los Angeles (USA), around 1960 (see Loeb[11]). The technique is essentially a modification of the phase inversion method described above. The solution is cast on to a surface at a sub-zero temperature. The volatile solvent evaporates more rapidly from the solution–air interface than from the interior of the film. Owing to this differential rate of solvent loss, the polymer density increases at the film surface and a skin layer forms while the underlying solution is still fluid. The pore size and porosity may be modified by annealing subsequent to complete gelation. Loeb and Sourirajan worked with cellulose acetate but other polymer membranes have been produced successfully by this method. A typical Loeb–Sourirajan type membrane will have a dense skin of 0.2–0.5 μm thickness and a 50–100 μm thick porous substructure. The support layer has pores 0.1–1.0 μm in diameter whereas the skin has pores estimated to be in the order of 10 Å. The sublayer offers negligible resistance to flow but provides integral support for the skin which could not otherwise be handled.

More recently, workers have sought greater flexibility in polymer selection by developing composite membranes. These are formed by casting a dense thin film (0.02–0.8 μm) on another microporous substrate. The membrane thus obtained is structurally similar to Loeb–Sourirajan type membranes but can be prepared from materials not amenable to procedures described above.

21.5.3 Ultra-filtration membranes

The Loeb-Sourirajan type membrane developed for reverse osmosis applications also fits the requirements of ultra-filtration. Isotropic membranes had previously been available but the advent of the skinned membrane provided for higher permeate fluxes and superior resistance to plugging by retained solutes. This latter feature is a result of the increasing pore diameter with depth within the membrane.

Asymmetric ultra-filtration membranes are cast using techniques similar to the phase inversion technique described above. The retentivity (pore size or nominal molecular weight cut-off) and porosity are adjusted by varying the composition of the casting solution and by using different polymer solvents. Such membranes have been made from a number of polymers including cellulose acetate, polycarbonate, polyvinylchloride, polyamides, modacrylic co-polymers, polysulphones and halogenated polymers (e.g. PVDF).

Recent developments in ceramic membranes are leading to their availability in the ultra-filtration range of pore sizes. The smaller particle sizes required to obtain the decreased pore diameters mean that less inert materials have to be used than for the microfiltration membranes. They are inherently of an anisotropic structure and pore sizes as low as 40 Å have been cited (ca. 70 000 molecular weight cut-off (MWCO)).

21.6 EQUIPMENT

A typical membrane filtration plant is built up from a number of modules. Scale-up would consist essentially of the addition of further modules to obtain the required membrane area. There are therefore two aspects to consider. Firstly the nature of the modules and the manner in which the membrane surface is presented. Secondly, the configuration of the associated process plant and the manner in which the liquors are passed through the modules.

The theoretical and practical considerations hitherto lead to a number of general rules of thumb for the design of process plant. In order to minimize concentration polarization, it is desirable to maintain high surface shear rates. This requires high cross-flow velocities and small flow channel heights. For economic reasons, however, it is also desirable to maintain a reasonable through-flow pressure drop of the feed stream and to maximize the membrane area per unit volume of module. It has further been shown (section 21.3.3) that it is beneficial to minimize the mean bulk retained solute concentration within the equipment. Fortuitously, many of these objectives are mutually compatible.

21.6.1 Module construction

There are a number of ways of presenting a membrane filtration surface. It is important in unit design to maximize the specific membrane area and to facilitate flow regimes that will inhibit concentration polarization. The principal configurations are given in *Table 21.3*. For each of these configurations, however, different structures are possible.

Table 21.3. MEMBRANE CONFIGURATIONS

Flat sheet	Hollow fibre
Plate and frame	Spiral wound
Tubular	Cartridge

A plate-and-frame unit may be of a similar form to a plate heat exchanger with a series of parallel membrane sheets and feed flow in alternate channels (*Figure 21.6a*). The membrane sheets may be rectangular or eliptical. An alternative embodiment is shown in *Figure 21.6b* where the membranes are in the form of discs and the feed fluid flows through alternate channels, perpendicular to the feed conduit.

The construction of tubular units usually resembles that of a shell and tube heat exchanger with the feed liquor passing through the tubes and the permeate being withdrawn from the 'shell side'.

Hollow fibres are essentially tubular membranes with a very fine bore, usually less than 1 mm, frequently less than 50 μm for reverse osmosis applications. A typical unit is of 'shell and tube' type construction. Depending upon the flow configuration, the fibres can be 'feed through' or a 'U-tube' with the fibres bent double (*Figure 21.7*). The flow regime employed depends upon the application.

For reverse osmosis the feed is fed to the shell side. The use of fibre bores of the order of 50–100 μm precludes the use of tube side flow. For ultra-filtration and microfiltration applications, lower specific membrane areas are required and wider bore fibres may be used. The use of tube side flow now permits high Reynolds numbers to be attained at comparatively low linear velocities; this does, however, lead to high pressure losses and may result in a reduced transport driving force at the effluent end of the unit. Two modes of

(a)

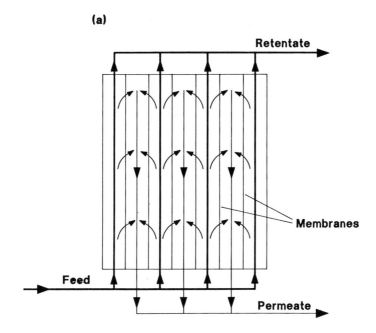

(b)

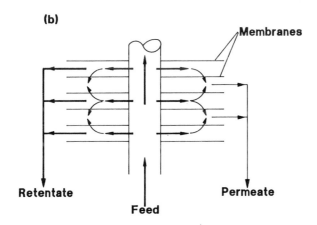

Figure 21.6. Plate-and-frame type modules

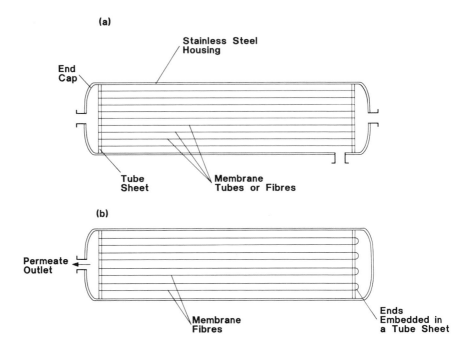

Figure 21.7. Tubular and hollow fibre modules: (a) feed through (tube side flow); (b) 'U tube' (shell side flow)

shell side fluid flow have been proposed: axial flow, where feed inlet and retentate outlet are at opposite ends of the unit; radial flow, where an axial distributor is used to give a uniform outward flow. With asymmetric membranes, the choice between tube side and shell side flow is predetermined by whether the membrane is 'skinned' internally or externally.

Spiral wound modules consist of membrane sheets sandwiched between a porous support and an open spacer (*Figure 21.8*); this is wound 'swiss roll' fashion to give an integral unit. The feed liquor flows through the unit axially. The permeate passing through the membrane flows through the porous substrate and is recovered from the axial tube. It is possible that correct design of the membrane spacer mesh will promote turbulence and thus improve performance. It does, however, lead to high pressure drops and can give fouling problems when particulates are present in the feed stream.

A number of membrane manufacturers supply their products in the form of cartridge filters. The configuration of the membrane within these units may be hollow fibre or spiral wound. An alternative presentation is that of pleated membranes where additional membrane area is achieved by folding the membrane.

Ceramic membranes are supplied as hexagonal multi-channel elements with 'tube side' feed flow. Each element contains 19 tubes of 4 or 6 mm diameter. These elements are mounted in steel modules, each of which may contain 1, 7 or 19 elements.

The configuration of the membrane within the module is substantially a

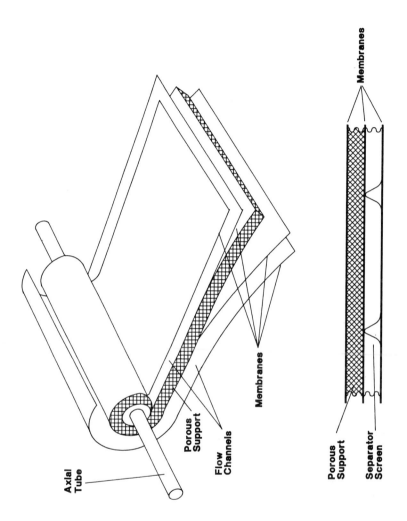

Figure 21.8. Spiral wound module

function of the supplier. Each manufacturer tends to have its own preferred, and in some cases proprietary, design. Thus, for the user, the problem is to select a suitable membrane and evaluate the required membrane area, usually on the basis of pilot-scale work. The selection of a particular membrane, and thus its manufacturer, may in effect dictate the module configuration. For reference, the relative merits of the principal module types are given in *Table 21.4.*

Table 21.4. COMPARISON OF PRINCIPAL MODULE DESIGNS

Module type	Membrane surface Model volume (m² m⁻³)	Capital cost	Operating cost	Flow control	Ease of in-place cleaning
Tubular	25–100	High	High	Good	Good
Plate and frame	400–600	High	Low	Fair	Poor
Spiral wound	800–1000	Very low	Low	Poor	Poor
Hollow fibre (or capillary)	600–1200	Low	Low	Good	Fair
Inorganic	300–600	High	Medium	Good	Very good

21.6.2 Process plant

Having considered the configuration of the membrane modules available it is now useful to look at the approaches to overall plant design. A number of basic flowsheet types are common and these are discussed below.

Figure 21.9 shows a basic flow diagram for a single membrane filtration unit. In many cases, for continuous plants, more than one module may be required. In such cases, where once-through flow is to be used, a tapered plant may be used in order to maintain the feed flow rates at the required linear values (*Figure 21.10*). This configuration is commonly used in large-scale reverse osmosis and desalination plants where the permeate is the desired product.

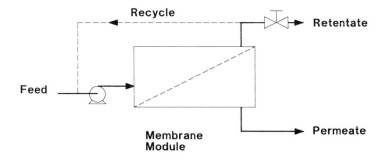

Figure 21.9. Single-stage membrane filtration

For batch operation, the feed liquor is placed in a feed tank from where it is withdrawn, passed through the membrane modules and either recycled to the inlet of the module (*Figure 21.11*) or to the holding tank (*Figure 21.12*): these are referred to as 'closed-loop' and 'open-loop' operation. Advantages of the closed-loop approach include reduced pumping requirements and, if an in-line prefilter is used, only the make-up requires filtration. Conversely, disadvantages of closed-loop operation include lower time-averaged fluxes and, for shear sensitive retained products, repeated passage of the same aliquot may result in degradation. Operation is continued until the required volume reduction is attained. In cases where the charge volume is greater than the capacity of the feed tank, continued topping-up may be practiced until the entire charge has been added. The use of top-up, however, leads to a reduced time-averaged flux.

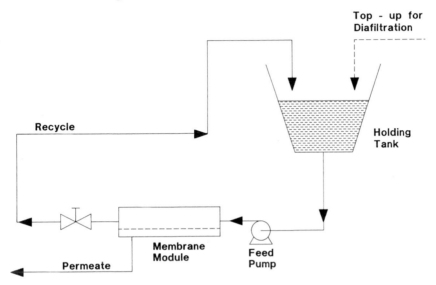

Figure 21.12. Batch ultra-filtration and diafiltration, open-loop operation

In ultra-filtration and microfiltration, a common requirement is that the concentrated, retained product should have a high level of purity. In such cases a given volume of additional water (or solvent) is added gradually to the feed tank and continued processing helps flush through the permeable impurities (*Figure 21.12*). This practice, variously referred to as 'dilution mode' or 'diafiltration', is equally applicable to continuous processing and is analagous to cake washing in conventional filtration.

For continuous operation in ultra-filtration and microfiltration, repeated passage across the membrane is usually required. The use of recycle to obtain an increased retentate product concentration is therefore common practice. A single-stage continuous plant, as shown in *Figure 21.9*, is inefficient due to the high average retentate concentration in the modules: the concentration being essentially nearly equal to the required retained product concentration. For this reason, in order to obtain higher average fluxes, multi-stage systems can be used (*Figure 21.13*) with each stage having its own closed-loop recycle. In this configuration, each stage operates at successively higher retentate

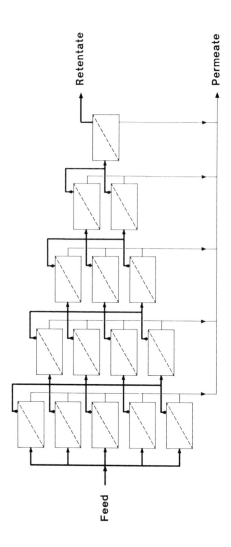

Figure 21.10. Tapered desalination plant

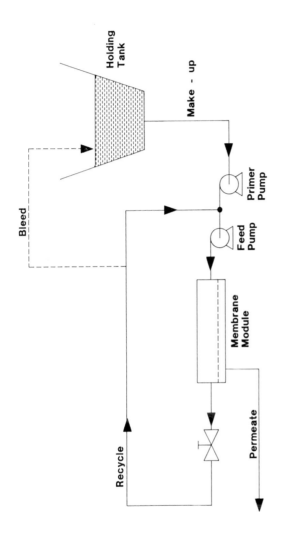

Figure 21.11. Batch ultra-filtration, closed-loop operation

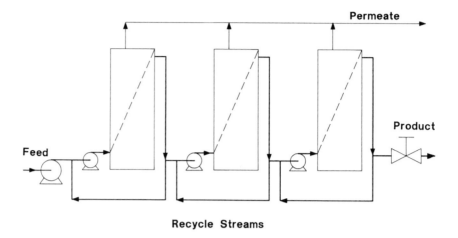

Figure 21.13. Multi-stage process with recycle

concentrations, with only the last stage operating at a retentate concentration close to that of the product. Continuous operation is inherently less efficient than batch processing due to higher average concentrations and the associated reduction in time-averaged fluxes. The use of multi-staging, as shown in *Figure 21.13*, allows this efficiency gap to be closed. A three-stage plant will typically attain over 80% of the efficiency of a batch plant in terms of membrane area utilization.

21.7 ULTRA-FILTRATION

Ideally, ultra-filtration is a screening process and a membrane's permeability to a given solute is dependent substantially upon the molecular dimensions of the solute in relation to the pore size. In many practical situations, however, interaction between the solute and the membrane bears considerable influence upon the membrane's permeability and its rejection characteristics. This is particularly prevalent in the treatment of polymeric solutions and of macro-biomolecules (proteins, enzymes etc.). Under such circumstances, the behaviour of a given membrane can vary widely from one solute to another. There is currently no means of predicting this and pilot tests are invariably required. The situation may be complicated further by the macromolecular solution becoming non-Newtonian at the concentrations prevailing near the membrane surface. Thus the fluid properties in the critical boundary layer may vary widely from those in the bulk phase.

It is usual to describe an ultra-filtration membrane by its 'nominal molecular weight cut-off' (MWCO), which indicates the smallest molecular size that will be retained by the membrane. Since, for structurally similar macromolecules, molecular weight and size have an approximately linear relationship, the MWCO represents a crude method of describing the pore diameter. In most ultra-filtration membranes there is a relatively wide distribution of pore sizes and thus components of a similar size cannot be separated. Unfortunately, there is no specified standard for the determination of MWCO. Variations occur in both the methods and materials used. The solutes most commonly used are dextrous or polyethylene glycols and the given level of rejection is frequently 90%. Thus two membranes of the same quoted MWCO from different manufacturers may have quite disparate characteristics.

Molecular shape and rigidity have an important influence on a membrane's steric rejection. The retention of linear, flexible molecules is normally an order of magnitude less than that for a globular or cyclic molecule of the same molecular weight. As a rule of thumb, for molecules of a similar shape, a molecular weight ratio of about 10 is required to obtain an effective separation. For components of a different molecular shape, however, this rule of thumb should be used with caution and the effects of shape on retention borne in mind.

Despite the above, ultra-filtration is gaining widespread usage as a means of molecular separation and for the concentration of macromolecular solutions. Examples of applications are given in *Table 21.5*. For the present, at the considerable risk of oversimplification, it will be assumed that ultra-filtration is ostensibly a screening operation; i.e. solvent flow through the membrane is limited by hydraulic resistances rather than by solvent–solute-–membrane interactions and that osmotic pressure effects may be neglected.

Table 21.5. APPLICATIONS OF ULTRA-FILTRATION

Recovery of electrophoretic paints from rinsing water
Treatment of emulsified oily wastes
Recovery of sizing agents in the textile industry
Concentration of latex emulsions prior to drying
Treatment of various aqueous effluents: printing industry;
                                                                tanning industry
Dairy and food industry: recovery of proteins from cheese whey;
                                     concentration of milk proteins for cheese production;
                                     fractionation of animal blood;
                                     concentration of soy protein isolates;
                                       gelatine manufacture
Downstream separations in biotechnology
Pharmaceutical industry

21.7.1 Solvent transport

The membrane is treated as a collection of capillary pores and conventional filtration equations may be applied. The solvent flux, **J**, is given by:

$$\mathbf{J} = K_1/\mu \cdot \Delta P/x \qquad (21.11)$$

where μ is the solvent viscosity, ΔP the hydrostatic pressure difference across the membrane, x the effective membrane (skin) thickness and K_1 is the hydraulic permeability given by:

$$K_1 = \epsilon \, d^2/20 \qquad (21.12)$$

where ϵ is the membrane porosity and d the mean pore hydraulic diameter. It may be noted that solvent flux, $\mathbf{J}$, is dependent upon the square of pore diameter. As such, most of the flow occurs through the pores at the upper end of the pore size distribution.

21.7.2 Solute transport

Transfer of a dissolved species through the membrane is normally considered in terms of a 'leaky' membrane type model. The solute transport flux is defined as:

$$\mathbf{J}_s = C_R \cdot (1 - R) \tag{21.13}$$

where C_R is the concentration of the solute in the bulk solution and $(1 - R)$ is the fraction of solvent passing through pores large enough to allow passage of solute molecules. Clearly, the solute flux may be expressed in terms of solvent permeation:

$$\mathbf{J}_s = \mathbf{J} \cdot C_p \tag{21.14}$$

where C_p is the solute concentration in the permeate. Therefore, R, the 'rejection coefficient' of the membrane for the given solute is defined as:

$$R = 1 - C_p / C_R \tag{21.15}$$

In the absence of a build up of retained species at the membrane surface, the flux is a linear function of applied pressure and the rejection coefficient, or retentivity, is independent of pressure.

21.7.3 Concentration polarization

The concept of concentration polarization was introduced for reverse osmosis in section 21.3.2. It was indicated that the effects are deleterious to unit performance in terms of permeate flux and purity. These general observations also hold for ultra-filtration, although the overall situation is somewhat more complex. *Figure 21.14* shows a typical family of curves for permeate flux as a function of applied pressure. It is observed that beyond a given ΔP, the flux attains a limiting value. This phenomenon is called 'pressure independent ultra-filtration' and the corresponding ultra-filtration rate is termed the limiting flux. This behaviour is also prevalent in microfiltration. In order to explain the behaviour, the exact nature of concentration polarization needs to be established.

At steady state, the solute flux towards the membrane must be balanced by permeation through the membrane and the back-diffusional solute flow. Referring again to *Figure 21.5* for nomenclature:

$$JC = J C_p - D \frac{dC}{dy} \tag{21.16}$$

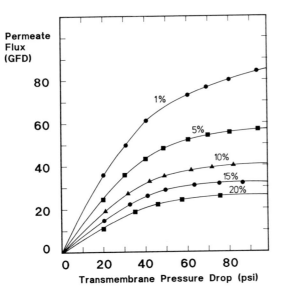

Figure 21.14. Ultra-filtration of styrene butadiene polymer latex with constant cross-flow velocity. (Data from Porter[5])

Assuming complete rejection of the solute,

$$\mathbf{J} \, C = -D \, \frac{\mathrm{d}C}{\mathrm{d}y} \tag{21.17}$$

and integrating with appropriate boundary conditions

$$\mathbf{J} = \frac{D}{\delta} \cdot \ln \, (C_w/C_R) \tag{21.18}$$

Defining the mass transfer coefficient as $k = D/x$, the solvent flux is given by:

$$\mathbf{J} = k \cdot \ln \, (C_w/C_R) \tag{21.19}$$

where C_w and C_R are the solute concentrations at the membrane surface and in the bulk retentate stream respectively. Thus, the 'polarization modulus', $\Phi = C_w/C_R$, increases exponentially with solvent flux and in a similar fashion to that of solute diffusivity. This means that concentration polarization is particularly severe with high permeability membranes and high molecular weight retained solutes, as in ultra-filtration. The hypothetical boundary layer thickness, x, is determined by the fluid properties and by flow adjacent to the membrane surface.

The analagous expression to equation 21.19 for a 'leaky' membrane, i.e.

one which permits passage of some of the solute, is derived in a similar fashion from equation 21.16:

$$\mathbf{J} = k \cdot \ln \left[\frac{C_w - C_p}{C_R - C_p} \right] \tag{21.20}$$

21.7.4 Rejection coefficients

The magnitude of concentration polarization, characterized by the ratio $\Phi = C_w/C_R$, will influence the retentivity of the membrane as defined by equation 21.15. As a result of solute polarization, the membrane sees a solute concentration of C_w rather than C_R. Correspondingly, rejection is based upon the high concentration and apparent retentivity is reduced. Thus, as was the case for reverse osmosis, concentration polarization adversely affects both permeate flux and membrane separatory performance. In order to clarify the situation it is convenient to define:

R, the apparent rejection coefficient

$$R = \frac{C_R - C_p}{C_R} = 1 - \frac{C_p}{C_R} \tag{21.15}*$$

and α, the real (or intrinsic) rejection coefficient

$$\alpha = \frac{C_w - C_p}{C_w} = 1 - \frac{C_p}{C_w} = 1 - \frac{C_p}{\phi C_R} \tag{21.21}$$

Ideally, α is independent of process conditions other than wall concentration, and is a fundamental property of the membrane. In reality, however, because α tends to be influenced by interactive effects, surface forces and membrane fouling, it is frequently sensitive to pressure and permeate flux. At the extreme, the rejection characteristics can become entirely controlled by the polarization layer. Such behaviour is not uncommon, particularly with biological and food products, and is referred to as 'gel controlled ultra-filtration', as distinct from membrane control. This concept of 'gel control' has also been applied to aspects of permeation rate data interpretation.

Some workers have used the fact that a gel layer upon a membrane may be used to control permeation successfully. This has led to the development of dynamic membranes (see Thomas[12], for example). These are, however, only of limited interest in the current context.

21.7.5 Limiting flux

The phenomenon of limiting flux and pressure independent ultra-filtration described earlier, may be interpreted in a number of ways. The oldest, simplest and still probably most favoured conception is described by the 'gel polarization model'[4]. This assumes that the solution at the membrane surface attains a saturation concentration and results in the precipitation of a slime or 'gel' layer on the membrane. The pressure and flux corresponding to the onset of pressure independent permeation are coincident with incipient gelling. Gel concentration represents the maximum attainable solute concentration and a further increase in flux (or pressure) results in a thickening of the gel layer, a corresponding increase in resistance to permeation, and a return at steady state to the initial flux.

According to this model, therefore, the concentration at the membrane surface remains constant throughout the pressure independent filtration domain. This concentration is termed the 'gel concentration', C_G, and equation 21.19 becomes:

$$\mathbf{J}_\infty = k \cdot \ln (C_G/C_R) \qquad (21.22)$$

This relatively simple model has proved surprisingly effective in predicting performance in the pressure independent regime[4-6]. The model predicts that the limiting resistance to flow is the gel layer. Enhancing the mass transfer coefficient, k, will aid diffusion away from the membrane. Thus gel layer thickness is reduced, resistance to permeation is decreased and higher fluxes are attained. Increasing the applied pressure alone, however, has a zero net effect.

It is now widely accepted that this model is not entirely satisfactory and that, even if the wall concentration does remain constant, its value appears to have no physical significance such as that implied by the gel model. There is evidence to suggest that, at the high concentrations prevailing at the membrane surface, osmotic pressure may play a significant role even for very high molecular weight solutes[13]. Trettin and Doshi[14] use this to explain limiting flux by equating osmotic pressure to applied pressure. This approach is not, however, generally favoured because it neglects the effects of solute–membrane interactions. It is now more widely accepted that performance is governed by a combination of hydraulic resistance within the boundary layer and that due to interfacial effects (adsorptive fouling and pore blinding). Numerous attempts to model ultra-filtration using these ideas are to be found in the literature, see for example[15-20], and have recently been reviewed[21].

In conclusion, the nature of concentration polarization and the effects on membrane performance of solute–membrane interaction are still not entirely understood. This, however, has not impeded widespread use of ultra-filtration as a separation process in industry. In general, well run pilot trials on the given system give quite sufficient information for reliable design of the full scale unit.

21.8 CROSS-FLOW MICROFILTRATION

The theory of microfiltration is less well developed than that for reverse osmosis or ultra-filtration primarily because the comportment of the microparticles near the membrane surface is not well understood. Particle polarization occurs in similar fashion to retained solute build up at the membrane in ultra-filtration. The cross-flow is not totally efficient in removal and flux decline occurs due to the formation of a cake or due to pore blockage.

Experimental data indicate that, as with reverse osmosis and ultra-filtration, it is the polarized layer which dominates performance. By its very nature, cross-flow microfiltration cannot be used for complete solid–liquid separation. The cross-flow configuration determines that it can achieve only the concentration of slurries or suspensions. Examples of existing applications are given in *Table 21.6.*

Table 21.6. APPLICATIONS OF MICROFILTRATION

Removal of suspended matter from effluent waters:
  ZnS in television screen production
  recovery of metallic colloids
Clarification of process streams:
  Vinegar
  Fruit juices
Concentration of slurries and suspensions:
  Harvesting of bacterial cells
  Sterile filtration

Figure 21.15 shows the effects of applied pressure and particle concentration upon permeate flux for a typical system. A tendency towards a limiting flux is again apparent. This led early workers to attempt to model the system using the gel polarization model, with the gel concentration corresponding to the non-void fraction of close packed spheres. This approach, however, underpredicts fluxes by, typically, one or two orders of magnitude[4,5]. In addition, in laminar flow, flux dependencies upon cross-flow velocity by an exponent ranging from 0.5 to 1.3, have been observed[5,7]. This is not in accordance with the predicted value of 0.33 (equation 21.7). Porter[5] attributed the apparently enhanced mass transfer to the radial migration of the

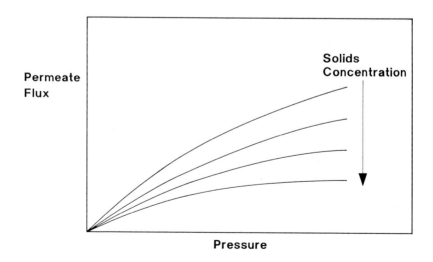

Figure 21.15. Typical dependencies of flux in cross-flow microfiltration

particles in the fluid flow shear field, i.e. a tubular pinch effect, and experimental evidence exists that this phenomenon indeed occurs[22]. Other ultra-filtration models have also been applied to microfiltration modelling, but these attempts have generally met with only limited success. Efforts to evaluate hydraulic resistances using conventional cake filtration theory for the 'gel' layer have also, in general, been unsuccessful. Other recent efforts have employed models, for example, based upon the systematic blinding of pores[15], upon hydrodynamic consideration of sediment transportation in pipes[23] or flow through channels with porous walls.

The complexities of suspension/membrane systems currently preclude the general simulation of cross-flow microfiltration and the universal correlation of cross-flow velocity, particle size distribution and membrane clogging/particle polarization. For given cases and systems, mathematical models may, however, be developed, although predictive simulation is not as yet possible. In any case, empirical design based upon pilot trials may be accomplished satisfactorily.

21.9 FLUX STABILITY AND DECAY

Prevailaing operating conditions and the nature of the feed stream determine the long-term stability of flux and the subsequent decay due to membrane fouling. In many commercial reverse osmosis and ultra-filtration installations the flux may remain constant for many months and sometimes a number of years. The stability of the flux has an important bearing upon plant running costs. The rate of flux decline and the consequent frequency of shutdown for cleaning or membrane replacement are direct functions of the severity and nature of fouling. The causes of fouling are several, as are the means of prevention. These are discussed briefly below.

In reverse osmosis and, to a lesser extent in ultra-filtration, membrane compaction due to the applied pressure leads to flux decline. This is a creep process wherein the skin of an asymmetric membrane grows in thickness by amalgamation with its porous substrate. This process is irreversible and leads to a logarithmic decay of flux. This is especially significant in reverse osmosis plants, where membrane resistance is controlling and can lead to an annual rate of decline of 15–50% depending on the membrane material and the pressure applied.

With anisotropic membranes, internal pore plugging does not appear to be a problem due to the open pore structure of the substrate; it is surface fouling which is the major cause of flux decline. In reverse osmosis the precipitation of various oxides and other salts on the membrane surface leads to a decay in permeation rates. This may, however, be obviated by suitable pretreatment or by good fluid management techniques. The real problem arises when species attach themselves to the membrane irreversibly and are not subject to back-diffusion from the membrane surface.

Organic macromolecules, colloids and various organisms are among the major propagators of fouling. Bacteria may grow on the membrane in some cases producing a 'slime', resulting in a catastrophic reduction in flux. This is a particular problem with plants·that are run on a discontinuous basis. Interaction between the membrane and the foulant may affect the rate of flux decay. In such cases, changing the chemical nature of the membrane may ameliorate the problem. In general, however, the techniques used to arrest flux decline rely upon minimization of solute deposition or removing deposits already formed.

The principal methods for preventing fouling are pretreatment of the process fluid and fluid management. Pretreatment can take any of several forms. The feed stream may be treated to remove the fouling species; precipitation, filtering and activated carbon are commonly used here. Alternatively, the influent may be made to dissolve the foulant; an example of this is the injection of acid to prevent calcium carbonate formation in reverse osmosis. The third method is to use a dispersant, such as hexametaphosphate, to create a suspension of the foulant. Biological growth may be inhibited by continuous addition of chlorine, formaldehyde or the like. Fluid management considerations are essentially the same as for polarization control. A high fluid shear rate will help to prevent deposition and assist removal of foulants upon the membrane surface.

Pretreatment can be omitted if the foulants can readily be removed by periodic flushing or cleaning of the unit. This process itself is dependent upon the nature of the foulant, the type and configuration of the membrane and its resistance to chemical cleaning agents. Two distinct modes of cleaning operation exist: back-flushing and recycling. In the former, permeate is forced back through the membrane and thus lifts off or loosens the cake or gel accumulated on the membrane surface. It is important that the back-flush liquid contains no suspended matter which might damage or foul the membrane substrate. In the recycling mode, the permeate outlet ports are closed off and the feed stream is allowed to recirculate. Permeate is continuously produced resulting in a build-up of pressure on the permeate side of the membrane. Ultimately, the permeate side will attain a pressure approximately equal to the mean feed side pressure. The cross-flow may now effectively clean the membrane since there is no bulk flow through the membrane.

Recycle-type cleaning operations are often enhanced by the addition of suitable chemical cleaning agents. The aggressiveness of the cleaning agent is limited by the chemical resistance of the membrane. It is rarely possible to restore fluxes to 'as new' values. However, good cleaning can result in fluxes of greater than 90% of the 'as new' value being attained. One of the ways in which inorganic membranes offer potential advantages is that far more severe cleaning methods, such as stream or moderately concentrated acid or alkaline solutions, can be used.

21.9.1 Fouling of microfiltration membranes

While the above comments apply equally to microfiltration membranes (except for compaction) there are certain characteristics of microfiltration which deserve special attention. Although flow hydrodynamics exercise considerable control over deposit formation on the membrane surface, the cross-flow is not entirely efficient in particle removal. As a result, fluxes undergo a substantial reduction over time due to increasing hydraulic resistance. In practical terms, two extreme possibilities exist as regards the propagation and level of fouling (*Figure 21.16*):

1. following the initial rapid decrease, a steady state is achieved where the filtrate flux remains sufficiently high;
2. filtrate fluxes cannot be stabilized and decline to unacceptable levels.

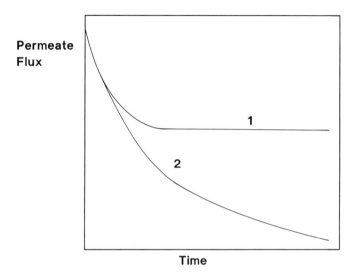

Figure 21.16. Stabilization and decline of cross-flow microfiltration fluxes

In the latter case, periodic back-flushing through the membrane may be practised in order to maintain the desired time averaged flux (*Figure 21.17*). The frequency and duration of the back-flush are dependent on the severity

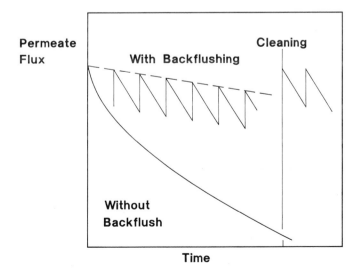

Figure 21.17. Use of back-flush to maintain flux in cross-flow microfiltration

of the problem. Typical practice is a back-flush duration of 1–2 s at intervals of 2 min to 1 h. The use of such periodic back-flushing requires that the membrane is able to withstand pressure gradients in both directions. For this reason, the use of capillary or tubular membrane modules has gained preference.

The use of back-flushing to restore flux levels is widespread despite the fact that returning hard-won filtrate to the feed steam should be something of an anathema. The method represents, however, a simple solution to a common processing problem and allows fundamental questions to be bypassed. The reasons for the gradual build up of the particulate deposits on the membrane which lead to flux decline are several and include the following[24].

1. Inertia deposition: the drag forces exerted on the particles within the laminar sublayer may be small compared to the inertia forces arising from filtrate flux. The particles are thus carried to the membrane surface and plug or blind the pores.
2. Pore plugging and blinding: the probability of membrane clogging is a function of the size distribution of the pores and of the particles. In both cases, a broad distribution increases the propensity to internal and surface plugging.
3. Depth filtration: the smallest particles may readily enter the pores. As they pass through the membrane they may be captured by the same surface forces deliberately exploited in deep bed filtration. This causes constrictions in the pores and leads to other and larger particles being trapped and ultimately to the pore becoming blocked.
4. Particle surface force effects: surface forces at the membrane surface, and between particles and solutes almost certainly play an important

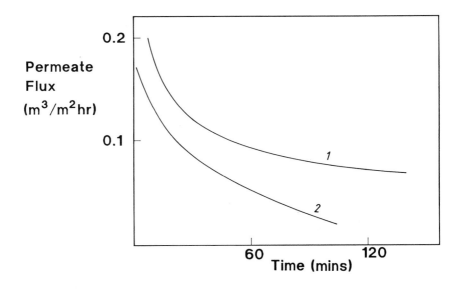

Figure 21.18. The adverse effect of macromolecules in cross-flow microfiltration: 1, bentonite (1 kg m⁻³); 2, as (1) plus Dextran (0.14 kg m⁻³). (Data from Milisic and Ben Aim[24])

role in surface, rather than internal, fouling. The adverse influence of macromolecules often noticed in microfiltration (*Figure 2.18*) is attributed to this type of effect.

To what extent these effects can be encountered by correct system design is uncertain. It may be that back-flushing will remain a necessary evil.

21.10 CONCLUSIONS

From the number of uncertainties expressed above and the lack of comprehensive design equations, the use of membrane filtration might not appear an attractive proposition. This is deceptive. Experience has shown that membrane filtration is effective and the growing number of installations and applications is testament to its efficacy. The nature of the equipment, and its modular construction lead to reliable scale-up procedures providing a correctly managed pilot-scale study has been conducted. In the course of such a pilot study the following factors may beneficially be considered.

1. Membrane: its constitution, pore sizes, and porosity and its configuration in the test module.
2. Cross-flow velocity: higher cross-flow rates lead to decreased polarization and fouling and thus improved fluxes.
3. Applied pressure: increased pressure will, in most cases, increase fluxes but may also increase fouling and separatory performance may deteriorate.
4. Solute concentration: the process plant should be designed such that average concentrations are minimized, thus allowing increased fluxes and separation and, possibly, reduced fouling as a bonus.
5. Temperature: increased temperature should lead to improved solution rheology (lower viscosity, higher diffusivities). Problems may arise with the temperature tolerance of the product and of the membrane itself.
6. Ionic environment: in certain systems ionic strength and pH may play a critical role, particularly in biological process streams.
7. Pretreatment: the performance of the membrane unit may be considerably improved by simple pretreatment of the process stream to eliminate the major propagators of fouling.

REFERENCES

1. Kimura, S. and Sourirajan, S., *Am. Inst. Chem. Eng. J.,* **13**, 497 (1967)
2. Mazid, M. A., *Sep. Sci. Technol.,* **19**, 357 (1984)
3. Gutman, R. G., *Inst. Chem. Eng. Symp. Ser.,* **51**, 71–81 (1977)
4. Blatt, W. F., Dravid A., Michaels, A. S. and Nelson, L. In *Membrane Science and Technology*, (Flinn, J. E., Ed.), pp. 47–97, Plenum Press, New York (1970)
5. Porter, M. C., *Ind. Eng. Chem., Product Res. Dev.,* **11**, 234 (1972)
6. Porter, M. C. In (Turbak, A. F., Ed.), *ACS Symp. Ser.,* **153**, 407–448 (1981)
7. Henry, J. D. In *Recent Developments in Separation Science*, Vol. II (Li, N. N., Ed.), pp. 205–225, Chemical Rubber Co. Press, Cleveland, Ohio (1972)
8. Henry, J. D., Lawler, L. F. and Kuo, C. H. A., *Am. Inst. Chem. Eng. J.,* **23**, 851 (1977)
9. Lo, Y. S., Agidaspow, D. and Wasan, D. T., *Sep. Sci. Technol.,* **18**, 1323 (1983)

10. Wakeman, R., *Chem. Eng.,* **426,** 65 (June 1986)
11. Loeb, S., In (Turbak, A. F., Ed.), *ACS Symp. Ser.,* **153,** Chap. 1 (1981)
12. Thomas, D. G. In *Reverse Osmosis and Synthetic Membranes* (Sourirajan, S., Ed.), pp. 295–312, National Research Council of Canada, Ottawa (1977)
13. Goldsmith, R. L., *Ind. Eng. Chem. Fundam.,* **10,** 113 (1971)
14. Trettin, D. R. and Doshi, M. R. In (Turbak, A. F., Ed.), *ACS Symp. Ser.* **154,** 374–409 (1981)
15. Le, M. S. and Howell, J. A., *Chem. Eng. Res. Des.,* **62,** 373 (1984)
16. Wijmans J. G. *et al., J. Membrane Sci.,* **22,** 117 (1985)
17. Reihanian, H. *et al., J. Membrane Sci.,* **16,** 237 (1983)
18. Bruin, S. *et al., Desalination,* **35,** 223 (1980)
19. Belfort, G. and Altena, F. W., *Desalination,* **47,** 105 (1983)
20. Mattiasson, E. and Sivik, B., *Desalination,* **35,** 59 (1980)
21. Van den Berg, G. B. and Smolders, C. A., *Filtration Separation,* **25,** 115 (1988)
22. Bauser, H., Chimiel, H., Stroh, N. and Walitza, E., *J. Membrane Sci.,* **11,** 321 (1982)
23. Belfort, G., Weigand, R. J. and Maher, J. T. In *Reverse Osmosis and Ultrafiltration,* (Sourirajan, S. and Matsuura, T., Eds), *ACS Symp. Ser.,* **281,** 383–401 (1985)
24. Milisic, V. and Ben Aim, R., *Filtration Separation,* **23,** 28 (1986)

22

High Gradient Magnetic Separation

J. H. P. Watson

Institute of Cryogenics, The University, Southampton

22.1 INTRODUCTION

In this chapter the theory and some of the applications of high gradient magnetic separation are discussed. High gradient magnetic separation is a process in which magnetizable particles are extracted onto the surface of a fine ferromagnetic wire matrix which is magnetized by an externally applied magnetic field. The process, which is used to improve kaolin clay, was developed for and in conjunction with the kaolin industry in the USA[1-3]. This process allows weakly magnetic particles of colloidal size to be manipulated on a large scale at high processing rates. There are, in addition to the clay industry, a large number of potential applications in fields as diverse as the cleaning of human bone marrow, nuclear-fuel reprocessing, sewage and waste-water treatment, industrial-effluent treatment, industrial and mineral processing and extractive metallurgy.

There are three ways in which magnetic separation can be achieved:

1. where the difference in magnetic properties between the particles to be separated is sufficiently large;
2. where the material, although not sufficiently magnetic, can be attached to something which is sufficiently magnetic for separation to be achieved;
3. when the magnetic ions to be separated are in solution, a chemical or, as in a case described below, a biochemical treatment is required to produce a magnetic precipitate which can either be extracted itself or attached to a magnetic particle[4,5].

In this chapter we will first review important aspects of the theory of high gradient magnetic separation and then will concentrate on three applications of the technology which have been of particular interest to the author: the magnetic treatment of kaolin clay; the description of a biomagnetic separation process in which micro-organisms can be used to decontaminate water systems from toxic heavy metal pollutants; and the use of superconducting magnets to lower the cost of magnetic separation.

Electromagnets in conjunction with an iron circuit have been used to

Figure 22.1. A typical high gradient magnetic separator in use in the brightening of clay for use in the ceramics industry. The pole cap, which weighs 30 t, has been removed to reveal the 2 mm diameter separation region which contains the matrix and in which the 2 T field is applied

generate a magnetic field in an air gap. Field gradients are produced by shaping the poles or by using secondary poles. Secondary poles consist of pieces of shaped ferromagnetic material introduced into the air gap. The magnetic induction produced in the air gap in an iron circuit is limited to about 2 T if the separation zone is reasonably large as compared with the volume of the iron in the magnetic circuit[2]. The magnetizable particles processed by these machines are separated by being deflected by the magnetic field configuration or they are captured and held by the secondary poles. The particles are released from the secondary poles either by switching off the magnetic field or by removing the secondary poles from the field mechanically. With particles which are large or strongly magnetic, separation can be accomplished with electromagnets which consume modest amounts of electric power.

Figure 22.1 shows a typical high gradient magnetic separator as presently in use in the kaolin industry. The system is an iron-bound solenoid, weighing about 200 t. The magnetic field of 2 T (20 kG) is produced in a cylindrical hole by copper coils weighing 60 t, approximately 2 m in diameter and 0.6 m long. Into this space are packed pads of ferromagnetic stainless-steel wool with voidage of 95%. The power consumption is approximately 500 kW, although the newer machines operate nearer 300 kW. The kaolin slurry is pumped through the matrix and, when the field is applied, magnetic particles are attracted to and held on the matrix. This process will be discussed in more detail in section 22.3.

22.2 THEORY OF HIGH GRADIENT MAGNETIC SEPARATION

Magnetic separation is achieved by a combination of a magnetic field and a field gradient which generates a force on magnetizable particles such that paramagnetic and ferromagnetic particles move towards the higher magnetic field regions and the diamagnetic field particles move towards the lower field regions. The force, $\mathbf{F}_m$, on a particle is given by:

$$\mathbf{F}_m = \chi \, V_p \, (\mathbf{B}_o \nabla \mathbf{B}_o)/\mu_o \tag{22.1}$$

where χ is the magnetic susceptibility of the particle with volume V_p, $\mathbf{B}_o$ is the applied magnetic field, $\nabla \mathbf{B}_o$ is the field gradient and μ_o is the constant $4\pi \times 10^{-7}$ h m^{-1}.

The fundamental element in the capture process is the interaction between a small magnetizable particle, usually paramagnetic, in a uniformly applied magnetic field[4]. Consider a ferromagnetic wire of radius a and a saturation magnetization M_s placed axially along the z-axis as shown in *Figure 22.2*. A uniform field H_o large enough to saturate the wire is applied in the x-direction. Paramagnetic particles of susceptibility χ and volume V_p (= (4/3)πb^3) and density ρ_p are carried past the wire by a fluid of viscosity η moving with a uniform velocity V_o in the negative x-direction.

The flow around the wire is treated in the hydrodynamic approximation in which the fluid can be considered frictionless. Equations can be derived describing the particle trajectories in terms of cylindrical co-ordinates r_a, θ and z_a. These are related to x and y by $x/a = r_a\cos\theta$ and $y/a = r_a\sin\theta$. The

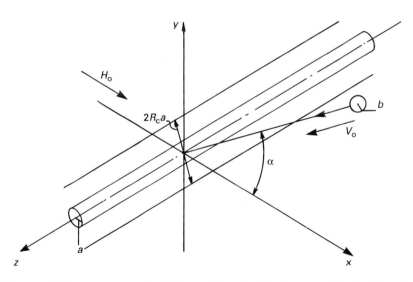

Figure 22.2. Scheme to show the basic filter configuration. The ferromagnetic wire is of radius a and a magnetization, M_s, is placed axially along the z-axis; a uniform field H_o large enough to saturate the wire is applied in the x-direction; paramagnetic particles of susceptibility χ and volume $V_p = (4/3) \pi b^3$ and density ρ_p are carried past the wire by a fluid of viscosity η moving with a uniform velocity V_o at an angle α to the negative x-direction (in this example $\alpha = 0$)

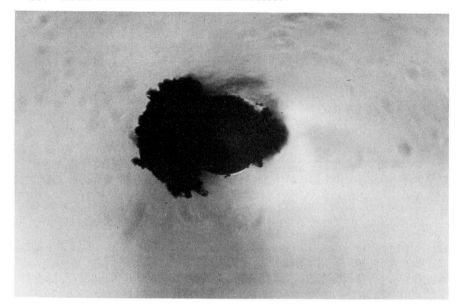

Figure 22.3. Cross-section through a nickel wire (25 μm diameter). Flow is horizontal left to right parallel to the applied magnetic field. Paramagnetic particles are captured on the front of the wire. With a Reynolds number of 5, as used here, small numbers are captured at the back of the wire. The two regions of the wire which are attractive to paramagnetic particles are separated by two repulsive regions

magnetic permeability of free space μ_o is given by $\mu_o = 4 \times 10^{-7} \, \text{H m}^{-1}$. Far away from the wire the particles move at the same velocity as the fluid. It is desired to calculate the effective capturing area that the wire presents to the moving stream. The capturing area/unit length of wire is $2R_c a$ where R_c is the capture radius. This is illustrated in *Figures 22.2 and 22.3*. The equations of motion can be derived by setting the Stokes' viscous drag, $\mathbf{F}_D$, on the particle to the magnetic force, $\mathbf{F}_m$

$$\mathbf{F}_D = 6\pi\eta b \, (\mathbf{V}_p - \mathbf{V}_o) \tag{22.2}$$

where $\mathbf{V}_p$ is the particle velocity.

$$(\mathbf{F}_m)_r = -6\pi\eta b V_m \, [K/r_a^5 + \cos{(2\theta)}/r_a^3] \tag{22.3a}$$

$$(\mathbf{F}_m)_\theta = -6\pi\eta b V_m \sin{(2\theta)}/r_a^3 \tag{22.3b}$$

where $(\mathbf{F}_m)_r$ and $(\mathbf{F}_m)_\theta$ are the two non-zero components of the magnetic force $\mathbf{F}_m$ written in Stokes' form. Here $K = M_s/2H_o$ is the coefficient which multiplies the short-range term and V_m is the *magnetic velocity* and is given by

$$V_m = (2/9) \, (\mu_o \chi \, b^2 M_s H_o / \eta a) \tag{22.4}$$

where M_s and H_o are in Å m^{-1}, all lengths are in metres and η is in Pa-s. (Water at 20°C has $\eta = 10^{-3}$ Pa-s). Examination of equation 22.3a reveals

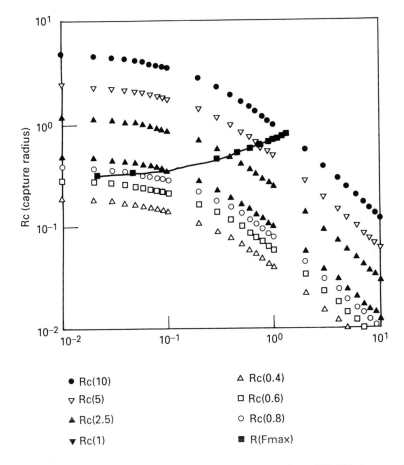

- ● Rc(10)
- ▽ Rc(5)
- ▲ Rc(2.5)
- ▼ Rc(1)
- △ Rc(0.4)
- □ Rc(0.6)
- ○ Rc(0.8)
- ■ R(Fmax)

Figure 22.4. The capture radius, $R_c(V_m/V_o)$, versus f for various values of (V_m/V_o). f_{max} according to the force-balance model[6,7] is shown as a solid line and is labelled as R (fmax)

there are two regions on the wire that are attractive to and two regions which are repulsive to paramagnetic species. This is illustrated in *Figure 22.3*.

The equation of motion is obtained from

$$\mathbf{F}_m + \mathbf{F}_D = O \qquad (22.5)$$

Far away from the wire the particles move at the same velocity as the fluid and it is desired to know the thickness of the band which will be captured by the wire or the effective capturing area the wire presents to the stream $2R_c a$ per unit length of wire, where R_c is called the *capture radius* and is illustrated in *Figure 22.4* for our purposes where V_m/V_o is approximately 1, then

$$R_c = V_m/2V_o \qquad (22.6)$$

As capture proceeds, the capture efficiency becomes reduced as the capture radius is reduced. If all the particles that were magnetically induced to collide with the wire were retained, the capture radius has the form[6,7]

$$R_c = (V_m/2V_o) [1/(1 + 4f)] \tag{22.7}$$

where f is the relative volume of the captured material, i.e. the volume of captured material divided by the volume of the wire. This expression, although approximate, is suitable for this discussion. The equation for R_c is plotted against f in *Figure 22.4* for a number of values of V_m/V_o.

The question of the stability of the captured material must now be considered. In order for a particle that has been captured to be retained, the value of $(\mathbf{F}_m)_\theta$, the tangential magnetic force, must be greater than the tangential fluid drag force $(\mathbf{F}_d)_\theta$. Using this model, the maximum value of f can be calculated as

$$f_{max} = (1/\pi) (V_m/V_o)^{2/3} \int_{3\pi/4}^{\pi} [-\cos \theta]^{2/3} d\theta - 1/4 \tag{2.8}$$

f_{max} is plotted in *Figure 22.4* and may be interpreted as the point on the R_c versus f line beyond which further material cannot be collected, because beyond that value of f, $(F_m)_\theta < (F_d)_\theta$. This means that the capture radius R_c is fairly constant as the captured material builds up to f_{max} after which $R_c = 0$. This theory predicts that particles with $V_m/V_o < 0.62$ cannot be retained although they can strike the wire, but the probability of them striking the wire decreases with increasing f as $1/(4f + 1)$.

The filtering performance of the magnetic stainless-steel wool has been evaluated in the form of deep beds where the wire wool occupies about 5% of the volume, but for special applications knitted meshes are often used, as shown in *Figure 22.5*. It was found that the number of particles coming out of the filter per unit volume of slurry, N_{out}, compared with the number of particles per unit volume of slurry going in, N_{in}, for a filter of length L, is given by[4,5]

$$\ln (N_{out}/N_{in}) = -4FR_cL/3\pi a = -L/L_o \tag{22.9}$$

where F is the fraction of space occupied by the matrix and where $L_o (= 3\pi a/4FR_c)$ is a characteristic length. This expression is only approximately correct when F or R_c is large due to random overlap of capturing areas in the filter.

It is important to realise that this equation only applies to the initial application of the slurry to the filter because as the volume of magnetics builds up, the rate of particle capture decreases. The presence of the captured volume, V, alters the flow around the wire until at some particular volume of magnetics, $V_{max} = F L A_c f_{max}$, where A_c is the cross-sectional area of the canister of length L.

The solution of the magnetic separator equations is very complex and has been studied in the literature. When the attraction between the matrix and the particles is strong, the separator equations can be simplified[7]

$$\ln [N_{out} (L,t)/ N_{in}] = -L/L_o + N_{in}V_oV_pt/Ff_{max}L_o \tag{22.10}$$

where V_o is the velocity of the slurry, V_p is the particle volume and t is time. This equation can be used to estimate the separator performance.

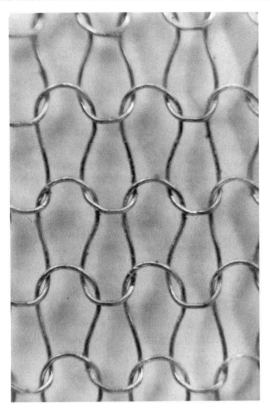

Figure 22.5. For applications where smooth wires are needed, knitted wires are a useful matrix. The magnetic field should be applied to this section of the mesh, oriented perpendicular to the page.

22.3 THE MAGNETIC PROCESSING OF A TYPICAL CERAMIC CLAY

22.3.1 Introduction to the magnetic processing of clay

The clay-processing capacity of a separator depends upon two factors, first, the type of clay being processed and second, the degree of beneficiation required. The measure of beneficiation is determined by the increase in *fired* brightness produced by the processing and this increase is produced by a fall in the iron concentration in the clay. Two clay feed systems have been considered: the axial feed system in which the clay moves parallel to the axis of the solenoid; and the radial feed system in which the clay is passed down a central tube which feeds the clay radially outwards through the matrix where it is collected in an annular space in which it again flows axially out of the solenoid. Magnetic separation is a cyclical process consisting of a feed part and a washing part. Some magnetic separators operate with two reciprocating canisters in which the dead time is that required to exchange the canister, a time of 10 s is considered adequate for this process, as will be discussed in section 22.5. The results show that the production rate of the reciprocating canister system is always appreciably higher than for the stationary canister and particularly so for the radial feed system[9].

The main purpose of magnetically processing clay is to increase the brightness of the clay which is accomplished by a reduction in the iron concentration of the clay. The iron in kaolinite is in substitutional sites usually occupied by aluminium and is usually in the ferric state. The micaceous components present (usually in the form of mixed minerals rather than free mica) contain iron in the ferrous and ferric state which substitutes for aluminium. In the mica, potassium replaces sodium and the potassium concentration can be of interest particularly in ceramic bodies where it can alter the firing characteristics. Potassium is removed because of its association with iron. In the absence of iron, kaolinite is diamagnetic but iron is present in sufficient quantity to give the clay a paramagnetic volume susceptibility (average) of about 10^{-4} (SI).

It has been found that for a given matrix, stainless steel with a strand radius of about 75 μm, normally Type 430, and occupying about 5% of the canister, the beneficiation of the clay depends on two things: (i) an extraction factor, $A = B_0/\eta V_0$, where B_0 is the applied field (Tesla), V_0 is the velocity of the clay through the matrix (m s^{-1}) and η is the viscosity of the clay (Pa-s) normally about 10^{-3} Pa-s; and (ii) on the amount of clay fed through the separator. For a given clay density, the volume of the clay is measured in terms of the canister volume containing the matrix. The objective of experimental work is to establish curves of integrated or accumulated average brightness, $B(J)$, versus the number of canister volumes, J, of clay processed for a given clay density with the extraction factor A. The integrated or accumulated average brightness, $B(N)$, is the total brightness of all the clay in the first N canister volumes treated, rather than the B_N achieved in the Nth canister.

If an objective of five units of brightness gain are desired, then with A_1, N_1, canister volumes can be passed, with A_2, N_2 canister volumes, and so on. Each of these options has a different production rate, with the applied field at its maximum V_0 can be chosen to be maximum consistent with the value of A. Values of $A = 10^5$ normally give the best results. This approach has been adopted here. The objective then is to choose the values of A_x and N_x which achieve a required accumulated average $B(N_x) = B_r$ where B_r is the required value. When this is done the clay velocity V_0 can be established. The viscosity can be determined if $\bar{\omega}$ the density of the clay is known, then using the specific gravity of kaolin of 2.65, the volume fraction c_p of the clay in the slurry can be determined as $c_p = \bar{\omega}/2.65$. The viscosity η of the slurry is $\eta = \eta_0 (1 + c_p)^{5/2}$ where η_0 is the viscosity of water (10^{-3} at 20°C). With the field B_0 at the maximum value $V_0 = B_0/\eta A_x$. If T is the time taken to deliver one canister volume of clay to the system, the feed part of the cycle lasts $N_x T$.

In order to determine the production rate it is also necessary to know the fraction of clay which passes through the separator: this is called the recovery R. The larger the value of A the lower is the recovery. The fraction 1-R is the fraction retained in the separator as magnetics.

22.3.2 Magnetic separation configurations

In the theory of magnetic separation developed for particle systems of single

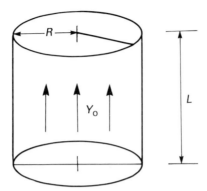

Figure 22.6. A simple axial canister which usually contains 430 stainless wool of medium grade, strand radius ca. 75 μm, saturation magnetization 1.7 T, matrix filling ca. 5%

size, the number of particles escaping depends upon the length of the canister L. It is found that the curves of accumulated average brightness, $B(N)$, versus the number of canister volumes, N, apply to all shapes of canister of various size provided that the volume of clay is measured in terms of the actual canister volume used. This is an extremely surprising result but this property can be used to produce feed systems with very much greater throughput within the space provided by the canister. In this work the axial and the radial feed systems are considered. Both feed systems can be used as stationary or reciprocating canister separators. There is no doubt about this thoroughly surprising result in the processing of English clays even when applied to canisters 2 cm long and to canisters 1 m long.

22.3.2.1 Axial feed system

The axial feed system is shown in *Figure 22.6*.

In the experimental set-up the canister is cylindrical in shape (20 cm long, 2.62 cm i.d.). The applied field is 5 T and is uniform throughout the canister. The production rate, P, is given by

$$P = \bar{\omega}V_oA \cdot R\delta \qquad (22.11)$$

$$\delta = N/(N + 1 + D/T) \qquad (22.12)$$

where $\bar{\omega}V_o$ is the weight of dry clay fed to the system per m^2 and A is the area over which this feed is applied. In the case of the reciprocating canister the dead time, D, is simply the time taken to shunt the filled canister out and replace it with the clean one. The important quantity that determines the efficiency of the system is the duty factor, δ, given by $\delta = N/(N + 1 + D/T)$ and the production rate, P, is directly proportional to δ. The time for one canister is given by $T = L/V_o$, the length of the canister being L (m). In the calculations the dead time, D, is determined to be either 2 or 3 min for the stationary canister. For the reciprocating canister it is assumed that $D = 10$ s, i.e. 10 s is the time taken to switch the canisters.

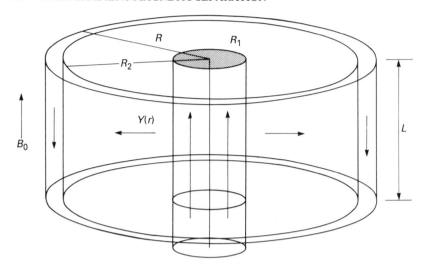

Figure 22.7. Radial flow canister system. Flow is radial through the matrix between R_1 and R_2. The cross-sectional area between R_1 and R_2 is the same as the cross-sectional area of the inner tube

22.3.2.2 Radial feed system[9]

The radial feed system is shown in *Figure 22.7*. Extensive work on the radial canister system indicates that the best results are obtained when $R_1/R = 0\cdot35$. Then R_2 is chosen so that the cross-sectional area $\pi(R^2 - R_2^2) = \pi R_1^2$. Thus the velocities in the central feed tube and the outer return passage are equal. This ensures that the dynamic heads are balanced inside the feed tube and in the outer space. The canister volume is $\pi(R_2^2 - R_1^2)L$. The value of the extraction factor, A, is now a function of radius so that the average extraction factor A^* is given by

$$A^* = A_1(R_2 + R_1)/2R_1 \qquad (22.13)$$

where $A_1 = H_0/\eta V_1$ where V_1 is the velocity of the fluid in the radial direction at R_1. The curves of brightness versus canister volume passed established for an axial feed system at extraction factor A applies to the radial canister if A^* is assigned the value A. If A is known, then V_1 can be determined. If in the axial system the velocity is V_0, then in the radial case $V_1 = V_0(R_2 + R_1)/2R_1$. The feed rate is $2\pi R_1 L \cdot \bar{\omega} V_1$ and the time T for one canister is $T = (R_2^2 - R_1^2)/2R_1 V_1$.

22.3.3 Experimental results

22.3.3.1 Ceramic clay properties

The properties of a ceramic clay are given in *Table 22.1*. It is found that the brightness is a linear function of the iron concentration, so that if the percentage of Fe_2O_3 is known then the brightness can be obtained.

Table 22.1. CERAMIC CLAY PROPERTIES

Particle size composition, %	
$< 2\ \mu m$	55 ± 2
$> 10\ \mu m$	5 ± 2
Brightness	81 ± 0.5
Composition, %	
Fe_2O_3	0.76 ± 0.03
K_2O	2.6 ± 0.02

22.3.3.2 Clay preparation and experimental conditions

The clay is often dispersed in water with N 40 Dispex. The resulting clay has a pH of 8.5–9. The clay is fed through the matrix at a constant velocity V_0 in a background magnetic field of 5 T, which produces the saturation magnetization in the matrix. As the clay comes through the matrix it is collected in 2-l volumes for analysis. As the clay flow comes to an end, water is pumped into the system without interrupting the flow or changing V_0. When this is completed, the applied magnetic field is switched off and the magnetic components of the clay are collected for analysis. From the results the iron concentrations and recoveries are used to obtain the average iron concentration of the accumulated clay. If the iron concentration in the n sample is $Fe(n)$, then the average iron concentration of the accumulated clay is given by

$$\sum w(n)Fe(n)/\sum w(n)$$

where $w(n)$ is the dry weight of clay recovered in the nth sample. The process should be run at a slurry velocity of $V_0 = 3.0\ \text{cm s}^{-1}$.

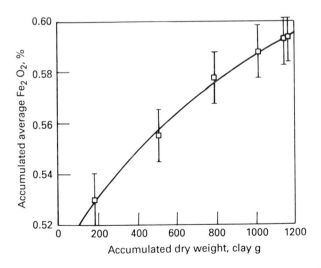

Figure 22.8. The accumulated average percentage of Fe_2O_3 versus the accumulated dry weight of clay (g). A second-order polynomial has been fitted with a correlation coefficient $R = 1$. The error bars are the standard error of the data

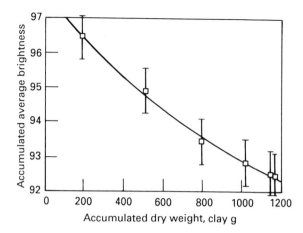

Figure 22.9. The accumulated average brightness versus the accumulated dry weight of clay (g). The second-order polynomial has been fitted with a correlation coefficient of R = 1. The error bars are the standard error of the data.

Figures 22.8 and *22.9* show plots of the accumulated dry weight of clay versus the average percentage of Fe_2O_3 and the accumulated average brightness respectively. In the example used in these figures the total feed was 1227 g and the magnetic species comprised 58 g, which indicates a product recovery of 95.3%. The percentage of Fe_2O_3 in the magnetics was 3.46 and the percentage of K_2O was 6.6, which represents a 4.6- and a 2.5-fold increase over the feed concentrations respectively. As expected, the magnetics are considerably coarser than the feed with 20% of the particles being $>10~\mu m$, 43% being $<5~\mu m$ and 17% being $<2~\mu m$.

22.3.4 Comparison of magnetic fractions

As shown in *Figure 22.10*, the amount of iron and potassium again fall as the value of V_o increases, except for potassium at 3 cm s^{-1} which is probably experimental error. The relative values of potassium and iron are very interesting as the large decrease in potassium is not accompanied by a correspondingly large decrease in iron. However, *Figure 22.10* is extremely interesting as it shows that the iron in the magnetics is most concentrated at 3 cm s^{-1}. This means that at 3 cm s^{-1} the concentration in the product is at the lowest value and, therefore, the brightness gain should be highest at 3 cm s^{-1}. This is confirmed by the experimental results, for example in *Figure 22.11* in which the initial brightness shows a peak at 3 cm s^{-1}.

It is also very interesting to examine the loss of iron by adding together the iron in the product and the magnetic fraction for comparison with the total iron in the feed. The overall iron loss is a few per cent in each run. It is often suspected that there may be iron leached from the product if, say, a high pH is used in flocculation by accident. An analysis, as shown in *Table 22.2*, indicates these fears are not justified in each case. For example, in this case, if all the iron lost were added to the product it would produce a fall in brightness of 1.95 units, on the other hand this 4% iron loss is within the experimental error in the brightness versus per cent Fe_2O_3 correlation curve.

Table 22.2. THE LOSS OF IRON RECOVERY

This is done by comparing the iron content of the sum of the iron in the product and the magnetic fractions with the iron content of the feed

	Weight, g	Fe_2O_3, %	Weight Fe_2O_3, g
Product	1169	0.594	6.94
Magnetics	58	3.46	2.01
Product + magnetics			8.95
Feed	1227	0.76	9.32
% Fe_2O_3 lost			4

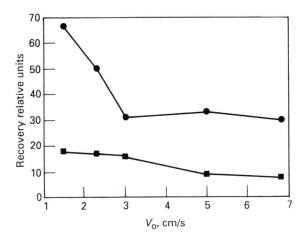

Figure 22.10. The recoveries of Fe_2O_3 (■) and K_2O (•) in the magnetics as a function of V_o

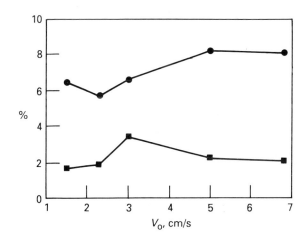

Figure 22.11. The percentage of Fe_2O_3 (■) and K_2O (•) in the magnetics versus V_o

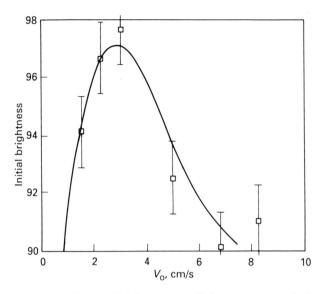

Figure 22.12. The initial brightness versus V_o for a typical ceramic clay

22.3.5 Initial brightness

As shown in *Figure 22.12*, the initial brightness peaks at around 3 cm s^{-1} and, although the best velocity varies from clay to clay, this figure is typical of the brightness achieved as a function of velocity. The reduction in brightness gain at low velocity is due substantially to mechanical entrainment. The initial brightness is limited to the clean matrix condition; however, the other conclusion refers to the whole so it is very likely that for this clay the best overall velocity is that of 3 cm s^{-1}.

22.4 BIOMAGNETIC SEPARATION PROCESS FOR HEAVY METAL IONS FROM SOLUTION

22.4.1 Bioaccumulation

There is a large literature on the role of micro-organisms in metal recovery from solutions[10]. The metal may be taken up intracellularly or by metal deposition on to the surface of the micro-organism. More recently[11] an elegant technique has been worked out in which glycerol-3-phosphate is the carbon source. This substrate does not enter the cell and is cleaved by an extracellular phosphatase to give a high concentration of phosphate anions at the cell surface. These phosphate ions react with metal ions present in solution to form a precipitate on the cell surface of the micro-organism. In the work reported here the scavenging of uranyl ions from solution was studied with two micro-organisms grown on glycerol-3-phosphate to induce a phosphatase activity in accordance with the procedure described by Macaskie and Dean[11]. The cells were then incubated with glycerol-3-phosphate in the presence of uranyl ions. Uranyl ions are strongly paramagnetic so the coated organisms behave like a paramagnetic material in a magnetic field. The cells can then be successfully concentrated by high gradient magnetic separation (HGMS). When $V_m \geq V_o$, the capture of the particles is very strong and the separator behaves like a filter[8]. Of the micro-organisms studied, *Candida utilis* and *Bacillus subtilis* strains behaved in the manner observed by Macaskie and Dean[11] in the *Citrobacter*, that is they picked up significant amounts of material from solution.

22.4.2 Experiments and discussion

Two types of micro-organisms were used in the HGMS tests: *Candida utilis* and *Bacillus subtilis*. The micro-organisms were grown in a flask as a batch culture using a glycerol-3-phosphate rich medium. At mid-logarithmic growth they were collected by centrifugation, resuspended in saline at 4°C and allowed to stand for one week at 4°C. The cells were then transferred into a buffer containing differing concentrations of uranyl acetate for 18 h at 20°C. The micro-organisms were then separated from the buffer solution, again by centrifugation. At this stage a weight increase of the cells was observed, presumably corresponding to the uptake of uranium by the cells.

A number of HGMS tests were performed using *C. utilis* and *B. subtilis* resuspended in saline solution at pH 7. In the tests a magnetic field of 5 T was used to magnetize a stainless-steel wool matrix (wire diameter 150 μm). A fluid flow velocity of 2 cm s^{-1} through the matrix was used. Batches containing 1 g of the *C. utilis* and the *B. subtilis* micro-organisms were passed through the separator and more than 98% of the material was retained and later recovered as the magnetic fraction. It is possible, as the collection was so high, to assume that $V_m \geq V_o$ and thus obtain a minimum estimate for χb^2 from equation 22.4. If b^2 can be estimated, then a minimum value of c can be obtained. If it is further assumed that the uranium is in the form of uranyl ions and the effective Bohr magneton is the same as for other uranyl compounds, then the number of ions associated with each cell can be calculated. If the above assumptions are valid, the numbers obtained for the quantity of adsorbed material is very much a minimum estimate. Using $M_s = 1.7$ T, $H_o = 5$ T, $m_o = 3.98 \times 10^6$, $h = 10^{-3}$ Pa-s and $a = 75 \times 10^{-6}$ m. Then $V_m = 2.004 \times 10^{13} (cb^2)$ m s^{-1}. The dimensions of the micro-organisms were determined by examining electron microscope pictures of the micro-organisms. The results are shown in *Table 22.3* for *C. utilis* and *B. subtilis*. The results in *Table 22.3* were calculated assuming $V_m = V_o$ which means that the value of χ and, consequently, the weight of material accumulated are very

Table 22.3. COMPARISON BETWEEN *C. UTILIS* AND *B. SUBTILIS*

	b_{eff}, μm	X, min	Weight of UO$_2$/ weight of cell
C. utilis	1.4 ± 0.1	(5.1 ± 1.2) × 10^{-5}	0.44 ± 0.2
B. subtilis	1.7 ± 0.15	(2.5 ± 0.8) × 10^{-5}	0.31 ± 0.2

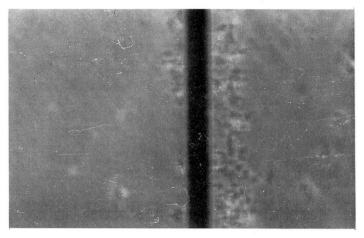

Figure 22.13. A single wire (25 μm diameter) magnetized in a direction parallel to the paper and perpendicular to the axis of the wire. A considerable quantity of magnetic micro-organisms have been captured

much minimum estimates. This assumption is reasonable because the extraction of the micro-organisms by the matrix is almost complete.

Nevertheless, a large amount of material has been captured compared to the wet weight of the cell. In view of the relatively better uptake of material by *C. utilis*, a more detailed series of experiments were carried out using the single-wire technique[12] in order to determine the magnetic properties of the individual *C. utilis* micro-organisms. A method similar to the one used here was first used by Paul *et al.*[13] to determine the magnetic properties of red blood cells. The method used here employs a flow cell in which a 25 μm diameter ferromagnetic wire, in this case a nickel wire, is incorporated. An electromagnet provides the background field. A phase-contrast microscope, a video camera, a video recorder and a television monitor are used to observe the capture and retention of the cells by the ferromagnetic wire. The flow cell consists of an aluminium housing in which a channel of 0.5 mm depth incorporated a 25 μm diameter nickel wire positioned at the mid-point of the channel (*Figure 22.13*). The complete assembly is sealed by two microscope cover glass slides. The single-wire cell is placed between the poles of the electromagnet (maximum field 1.3 T) with the wire perpendicular to the magnetic field. The micro-organisms flow through the cell parallel to the wire and are observed through a phase-contrast microscope (objective magnification × 20).

By measuring the successive positions of the micro-organisms as they approach the wire it is possible to determine the value of V_m or the micro-organism. The values of V_m obtained were consistent with the values given in *Table 22.3*. It was found for the case of uranium that the mean susceptibility measured by this method was $(8.35 \pm 3.51) \times 10^{-5}$ and this corresponds to a particle loading of 0.72 compared with the value of 0.44 in *Table 22.1*. These values are in agreement with the minimum values estimated by magnetic separation of the cells and shown in *Table 22.3*. Following the work with uranyl ions, a batch of *C. utilis* cells were prepared from a solution containing ferrous ammonium sulphate in the same molar concentration as the uranyl acetate used in the previous case. *Figure 22.14* shows the results for the sample of measured cells as a graph of log probability versus magnetic susceptibility. The cells produced in this way are considerably more magnetic than for the uranyl ion case. The mean susceptibility of the cells was $(4.5 \pm 1.3) \times 10^{-4}$ which corresponds to a mass loading of 0.46, although there are cells present with a loading of more than twice this value.

If the magnetic velocity is greater than the fluid velocity, then strong capture occurs. If a horizontal line is drawn in *Figure 22.14*, corresponding to a particular value of V_m, then all the micro-organisms above this line would be strongly captured if a flow velocity $V_0 = V_m$ were chosen. The point of intersection of this horizontal line with the log probability curve indicates the fraction of particles which can be removed at fluid velocity $V_0 = V_m$. If a separator is constructed in which a field of 5 T is used to magnetize stainless-steel wires instead of nickel wires then values of V_m can be increased by a factor of 15. From a practical point of view, this means that the fluid velocity though this matrix can be 15 times greater while the same extraction is maintained. The release of phosphate ions at the cell surface as shown by Dean and co-workers[11] and ourselves gives rise to precipitates of metal phosphates on the cell surfaces of micro-organisms. It is now of interest to estimate the minimum obtainable concentration of metal ions remaining in solution after the biomagnetic extraction process has been completed. Calculations have shown that very low ultimate levels are possible for materials of interest to the nuclear industry. For example, for Am^{3+} the ultimate level is 2.43×10^{-19} p.p.m., for UO_2^{2+} the level is 5.38×10^{-9} p.p.m. and for PuO_2^{2+} the level is 4.37×10^{-9} p.p.m.

As mentioned in the introduction, a number of micro-organisms produce sulphide and the growth of *Desulfovibrio* strains in media containing sulphate ions together with Fe^{2+} ion give rise to iron sulphide precipitates[14]. In conjunction with Professor Brown of Heriot-Watt University, we have been able to show that organisms do precipitate sulphides on their cell surface and we have been able to scavenge cells using HGMS after preparing the cells by this method and the sulphide preparation route has produced similar susceptibilities to the phosphate preparation route. This has been observed in a simple magnetic separation test and measurements have also been made on the micro-organisms using the single wire cell. The ultimate levels to which heavy metal ion concentrations can be reduced appears to be even lower than for the phosphates. For example, for Co^{2+} it is 1.77×10^{-16} p.p.m. and for Hg^{2+} 8.02×10^{-43} p.p.m. In experimental work on industrial effluents from the Netherlands, the Hg concentration has been reduced from 2000 p.p.b. to 2 p.p.b. in a single stage.

22.4.3 Conclusions on the biomagnetic separation process

Effluents from a number of industrial processes contain low but toxic concentrations of metal ions in solution. The process described in this paper allows the removal of metal ions down to very low concentration; furthermore, it allows the collection of these materials in a highly concentrated

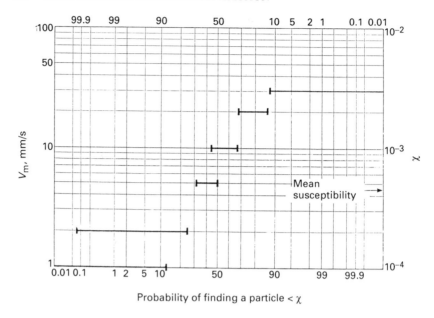

Figure 22.14. Log probability versus V_m and χ for ferrous ions with C. utilis

material form. Therefore, this process will have applications in the mineral-processing industry in addition to the treatment of effluents from nuclear and other industrial plants.

22.5 SUPERCONDUCTING MAGNETIC SEPARATORS

22.5.1 Introduction

For many years magnetic separation has been used by the mineral processing industries on a large scale and it is by this industry where magnetic separation using superconducting magnets has stimulated most interest. The high values of applied field available from superconducting magnets can be used to increase V_m and this can be used to

1. increase extraction with high values of V_m/V_o; or
2. increase the processing rate keeping V_m/V_o constant.

In HGMS the ferromagnetic wire occupies ca. 5% of the space. In order to quantify the amount of material passing through the separator, it is convenient to use the volume of the canister, less the volume of the matrix, as the unit of volume, V_c. The HGMS process is a cyclical one; it has a feed part and a cleaning part, i.e. when the quality of the effluent is inadequate the matrix must be cleaned. In order to clean the canister it is usually necessary to reduce the field to zero and flush the captured material from the matrix. If N_0 canister volumes of slurry can be fed before flushing, the processing rate, P, can be written as

$$P = \rho A_c V_0 R N_0 / (N_0 + 1 + D/\tau) \qquad (22.14)$$

where ρ is the dry weight of material per unit volume of slurry, A_c is the cross-sectional area of the canister, D is the dead time (i.e. the time between stopping and re-starting the feed), τ is the time for one canister volume ($\approx L/V_0$, where L is the length of the canister), and R is the fraction of material passing through the separator. In order that the process is efficient, it is necessary that $N_0 \gg D/\tau$. When using HGMS two possibilities exist:

1. systems which switch off the field to flush;
2. systems which leave the field on.

22.5.2 Systems switching the field off

22.5.2.1 Introduction to switched systems

In this approach the magnetic field is switched off in order to clean the system. This approach has been adopted with conventional iron-bound solenoid systems in clay processing leading to a dead time of approximately 200 s giving a duty factor of ca.75%. There are now a number of outstanding projects using this type of system, particularly notable is the system produced by Eriez Magnetics for Huber Corporation, a large USA clay producer. The electrical power cost inflation in the USA now makes this system extremely attractive, as *Table 22.4* shows.

Table 22.4. ESTIMATED OPERATING COSTS ($US) OF HGMS FOR CLAY PROCESSING

	Conventional			Superconducting	
	1975	1986	1988	1986	1988
Depreciation per					
10 000 h	20	20	20	30	43
Magnet power	12	30	15	3	3
Pumping and					
and cleaning power	4	10	10	10	10
Labour	5.5	16.5	18	16.5	18
Maintenance	4	10	11	10	11
Total ($ h^{-1})	45.50	86.50	74	69.50	85
Relative cost					
per tonne	1.00	1.90	1.52	1.42	0.88

22.5.2.2 Eriez Magnetics first system

The Eriez Magnetics system has the following characteristics:

Canister	213.36 cm diameter, 51 cm deep
Magnet size	391.16 cm^2, 210.82 cm deep
Magnet weight	227 t
Ramp time	1 min, maximum
Helium usage	2 l per cycle at 1 min ramp time
Liquifier capacity	20 l hl^{-1} from gas at 300–4.2 K
Compressor power	47.8 kW (nominal)
Field strength	2 T
Status	On line from 8 May 1986

Eriez Magnetics and Huber Corporation must be congratulated on this venture and I look forward to hearing more about the operation of the system in the future. This machine involves the least technical risk, but until recently the savings in power could not have justified the capital cost of the superconducting coils which simply replaced the conventional coils. *Table 22.1* shows that this is no longer the case.

There are a number of other systems of this type, but they differ in that they do not use an iron yoke and operate at higher magnetic fields.

22.5.2.3 Eriez Magnetics second system

Based on the success of the first superconducting system J. M. Huber have ordered a second 2 T superconducting separator, 3.05 m in diameter, which will have twice the capacity of the earlier separator. As shown in *Table 22.1*, the processing cost per tonne should be appreciably less than for both the first superconducting system and the improved 1988 conventional separator.

22.5.2.4 Performance of switch-on switch-off systems

These systems perform well provided the duty factor can be kept high, i.e. $N_o/(N_o + 1 + D/\tau) \approx 1$. This is the case for the processing of USA clays at 2 T or for water-pollution clean-up. This system becomes difficult when the following situations occur.

1. When V_o is increased with field, τ can be reduced so that if $D/\tau \gg 1$, all the increase in the processing rate due to the increase in V_o is lost. This occurs in situations when large amounts of material are removed in, for example, mineral processing.
2. Another problem associated with a decrease in τ is due to the increasing frequency with which the field must be switched on and off as there is helium boil-off when this occurs. With the Eriez machine this amounts to 8 l h^{-1} of liquid helium, which almost doubles the helium requirement. If this machine were operated on the same job but at 4 T, the boil-off would be 16 l. The power costs and liquifier costs, therefore, increase rapidly as τ decreases.
3. If the machine has simply had coils replaced with superconducting coils

the cost of the machine is considerably higher so that cost savings due to electrical power must at least be greater for economic viability.

Clearly, for mineral processing, switch-on switch-off systems at high field are not economically viable. Another solution must be sought in these cases. It should be also pointed out that the machines that result from the alternative solution should be much more attractive economically even in situations where switch-on switch-off systems are viable.

22.5.3 Reciprocating canister systems[15]

22.5.3.1 Introduction to reciprocating systems

These systems were pioneered by English Clays Lovering Pochin (ECLP), a large clay producer in the UK, and its inception was largely due to the vision

Figure 22.15. The first reciprocating superconducting separator. The hydraulic ram which is used to drive the canister backwards and forwards along the axis of the superconducting magnet can be seen at the top of the figure

of Professor Clark, who was Director of Research and Development at that time. The first reciprocating superconducting magnetic separator is shown in *Figure 22.15*. In the reciprocating canister system (RCS)[15], one superconducting magnetic uses two separate canisters. When one canister is separating the other is out of the field being cleaned. Then the clean canister is returned to the field while the other is moved out of the field for cleaning. The cycle then continues. The canister train must be magnetically balanced in order that the canisters can be moved in and out of the field without using large forces. This is achieved by arranging things so that when the transfer takes place the amount of iron in the coil remains constant. Under these conditions the superconducting magnet can be left at high field continuously. Consequently, this means that a helium liquefier requiring less power can be used. The dead time can be reduced to a few seconds in this situation; therefore, feed times of less than 1 min can easily be achieved while maintaining a high duty factor and fields as high as 5–8 T can be used. This means that the processing rate of the RCS can be a factor of 8–10 times higher per unit area of canister than the switch-on switch-off type of system, as shown in *Figure 22.16*. For a given separation problem, the RCS is appreciably smaller in size and the low boil-off, resulting from the field not being switched, means a smaller helium liquifier requiring less power can be used.

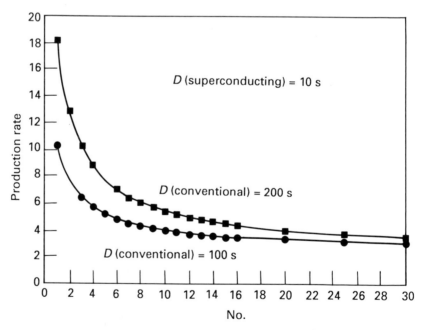

Figure 22.16. Comparison of the production rates of a superconducting system (D = 10 s, field 5 T) with a conventional system (D = 100 s and D = 200 s, field 2 T)

22.5.3.2 ECLP pilot plant

The construction of the pilot-scale RCS took place after extensive trials with various types of clays originating from Cornwall, UK. The system was pilot scale in the sense that the superconducting magnet system was 12.7 cm clear bore with a maximum field of 5 T. The helium liquifier was full scale, using turbo-expanders which were specially cut to provide 2 l h^{-1} rather than the normal 18 l h^{-1} of the machine with properly cut turbines. All aspects of full-scale reciprocator were tested, and valves were operated under pressure in proper sequence for two years. Cyclical matrix compression tests were done. A radial feed system[9] was developed and tested which gave a three-fold increase in the production rate. Some of this work has been described by Riley and Hocking[16].

22.5.3.3 Czechoslovakian RCS for clay processing

A very important and ambitious programme to build and test a supercon-ducting RCS is underway in Czechoslovakia[17]. This machine which was designed and built in Czechoslovakia is intended for the purification of kaolin clay. The magnet has a horizontal working chamber of 56 cm diameter and 1.32 m in length. The field strength of the system is 5 T. The liquid helium consumption is 3.5–5.6 l h^{-1} depending on the current. The helium is supplied by two liquefiers each discharging into a 500-l Dewar which supplies the magnet. This system has been under test since mid-1985.

22.5.3.4 Cryogenic Consultants Limited (London)

Cryogenic Consultants Limited (CCL) have produced a mobile RCS system[18] employing an internal closed-cycle refrigerator, with a power consumption of 14.7 kW, designed to run for 6 months between services. The magnet operates at 5 T in an active separation volume of 26.6 cm diameter and 50 cm long. The magnet is surrounded by an iron yoke to produce a low field around the system which allows the saturated matrices to be cleaned much closer to the magnet and, consequently, the canister train is reduced in length. It is estimated that the off-balance forces on the canister train are 0.5 t. A clay company took delivery of one of these systems in January 1989 and has operated succesfully since.

22.5.3.5 Superconducting separator requirements

A number of interesting projects have been instituted recently which greatly improve the prospects for superconducting separation, namely, the develop-ments by Eriez magnetics, the Czech and the CCL programmes. Here we have a commercially operating system which will provide an answer to the question always posed by properly sceptical process engineers, namely: 'Where can I see one in operation?' This question is very sound as no production engineer wants to operate the first system of a new technology.

Again clay seems to have taken the lead in progressing the technique of magnetic separation. The applications suggested for HGMS have been enormous, but so far little has happened outside of clay. Progress to other applications will depend on collaboration between the HGMS expert and a partner who is an expert in the proposed field of application. This cannot be stressed enough.

Finally, what is needed is an HGMS which is continuous and cannot block easily when wood chips, ferromagnetic material or other unexpected things happen to arrive at the matrix.

REFERENCES

1. Kolm, H. H., *U.S. Patent No. 3567026* (1971)
2. Marston, P. G., Nolan, J. J. and Lontai, L. M., *U.S. Patent No. 3627678* (1971)
3. Oder, R. R. and Price, C. R., Brightness beneficiation of kaolin clays by magnetic treatment, *Tech. Assoc. Pulp Paper Ind.,* **56**, 75 (1973)
4. Watson, J. H. P., 'Magnetic filtration', *J. Appl. Phys.,* **44**, 4209 (1973)
5. Watson, J. H. P., 'Applications of and improvements in high gradient magnetic separation', *Filtration Separation,* **16**, 70–74 (1979)
6. Luborsky, F. E. and Drummond, B. J., *IEEE Trans. Magn.,* **11**, 1969 (1975)
7. Luborsky, F. E. and Drummond, B. J., *IEEE Trans. Magn.,* **12**, 463 (1976)
8. Watson, J. H. P., *IEEE Trans. Magn.,* **14**, 240–245 (1978)
9. Watson, J. H. P., *British Patent No. 1530296* (Czech No. 205022)
10. Kelly, D. P., Norris, P. R. and Brierley, C. L. In *Microbial Technology Current State, Future Prospects* (Bull, A. T., Ellwood, D. C. and Ratledge, C., Eds), pp. 2263–2308, Society for General Microbiology Symposium **29** (1979)
11. Macaskie, L. E. and Dean, A. C. R., *J. Gen. Microbiol.,* **130**, 56–62 (1984)
12. Watson, J. H. P. and Rassi, D., In *Proc. of MINTEK 50* (Haughton, L. F., Ed.), pp. 335–340, The Council for Mineral Technology, Randburg, 2125 South Africa (1985)
13. Paul, F., Melville, D. and Roath, S., *IEEE Trans. Magn.,* **15**, 989–991 (1979)
14. Freke, A. M. and Tate, D., *J. Biochem. Microbiol. Technol. Eng.,* **111**, 29–39 (1961)
15. Windle, W., *British Patent No. 1469765* (Czech. No. 180638 DIV P); *British Patent No. 1599824; British Patent No. 1599825*
16. Riley, P. W. and Hocking, D., *IEEE Trans. Magn.,* **17**, 3299, (1971)
17. Kaiser, Z., Vycydal, P., Fojtek, J., Smrz, S., Kouba, M. and Suma, J., *Proc. Magnetic Tech. Conf. MT-9, Zurich* (1985)
18. Lam, K., Stadtmuller, A. A. and Good, J., *Proc. Mag. Tech Conf. MT-9, Zurich,* p. 317 (1985)

23

Particle-Fluid Interaction, Thermodynamics of Solid-Liquid Separation
Part I — Particle-Fluid Interaction

L. Svarovsky

Department of Chemical Engineering, University of Bradford

NOMENCLATURE

A	Projected area of a particle facing the flow; or the face area of a packed bed
C, c	Solids concentration
C_1	Minimum solids concentration at which zone settling occurs
C_D	Particle drag coefficient
$f(\epsilon)$	Function of voidage (voidage function)
F_D	Drag force on a particle
g	Gravity acceleration
G	Separation factor (number of gs)
K	Permeability of a packed bed
L	Depth of a packed bed
m	Particle mass
Q	Volumetric flow rate
R	Radius of particle position; or medium resistance
R_o	Initial radius of particle position
Re	Reynolds number for the flow in a packed bed
Re_p	Particle Reynolds number
S_o	Volume-specific surface of the bed
t	Time
u	Particle–fluid relative velocity
u_c	Terminal settling velocity in a centrifuge
u_g	Terminal settling velocity under gravity
v	Superficial velocity in a packed bed
V	Filtrate volume collected
w	Angular speed
x	Particle size

x_c	Top size limit in a centrifuge
x_g	Top size limit under gravity
x_{sv}	Mean particle size in a packed bed
α	Specific cake resistance
Δp	Static pressure drop across a packed bed
ϵ	Voidage of a packed bed
μ	Liquid viscosity
ρ	Liquid density
ρ_s	Solid (particle) density
τ	Particle relaxation time

23.1 INTRODUCTION

This chapter reviews some fundamentals on which much of the rest of this book is based. Particle–fluid interaction (Part I), whether in suspensions or in packed beds, plays an important role both in sedimentation and in filtration. Thermodynamics underlie it all, of course, and Part II of this chapter gives an overview of the practical help that thermodynamics can offer in this subject.

23.2 MOTION OF PARTICLES IN FLUIDS

In this first section, the fundamentals of particle–fluid interaction are reviewed, with particular emphasis on the concepts used in scale-up of equipment in particle–fluid separation.

If a particle moves relative to the fluid in which it is suspended, there exists a force opposing the motion, known as the drag force. Knowledge of the magnitude of this force is essential if the particle motion is to be studied. The conventional way to express the drag force, F_D, is according to Newton:

$$F_D = C_D \cdot A \cdot \frac{\rho \, u^2}{2} \tag{23.1}$$

where u is the particle–fluid relative velocity, ρ is the fluid density, A is the area of the particle projected in the direction of the motion and C_D is a coefficient of proportionality known as the drag coefficient. Newton assumed that the drag force is due to the inertia of the fluid and that C_D would then be constant.

Dimensional analysis shows that C_D is generally a function of the particle Reynolds number:

$$\frac{u \cdot x \cdot \rho}{\mu} = Re_p \tag{23.2}$$

(where x is particle size) and the form of the function depends on the regime of the flow. *Figure 23.1* shows this relationship for rigid spherical particles in a log–log plot. At low Reynolds numbers, under laminar flow conditions when viscous forces prevail, C_D can be determined theoretically from

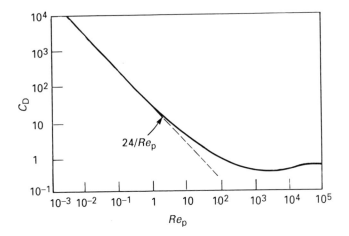

Figure 23.1. Drag coefficient versus particle Reynolds number for spherical particles

Navier–Stokes' equations and the solution is known as Stokes' law:

$$F_D = 3\pi\mu u x \tag{23.3}$$

This is an approximation which gives best results for $\text{Re}_p \to 0$; the upper limit of its validity depends on the error which can be accepted. The usually quoted limit for Stokes' region of $\text{Re}_p = 0.2$ is based on the error of approximately 2%.

Elimination of F_D between equations 23.1, 23.2, and 23.3 gives another form of Stokes' law:

$$C_D = \frac{24}{\text{Re}_p} \quad (\text{Re}_p < 0.2) \tag{23.4}$$

and this is shown in *Figure 23.1* as a straight line. For Reynolds numbers greater than about 1000, the flow is fully turbulent with inertial forces prevailing and C_D becomes constant and equal to 0.44 (Newton's region). The region in between $\text{Re}_p = 0.2$ and 1000 is known as the transition region and C_D is either described in a graph or by one or more empirical equations.

In solid–fluid separation, the greatest concern is with the fine particles which are most difficult to separate. This means that Re_p is low, almost inevitably less than 0.2 due to low values of x and u and, therefore, only the Stokes region need be considered here.

23.2.1 Gravity settling at low concentrations

A single particle settling in a gravity field is subjected primarily to drag force, gravity force and buoyancy, which have to be in equilibrium with the inertial force, i.e.

$$m \, \frac{du}{dt} = m \cdot g - m \, \frac{\rho}{\rho_s} g - F_D$$

(23.5)

inertial   gravity   buoyancy   drag
force   force   force

(taking the downward forces as positive), where m is the particle mass, g is the acceleration due to gravity, ρ_s is particle density and t is time.

If equation 23.5 is divided by m and Stokes' law in equation 23.3 is introduced for F_D, the following equation of motion is obtained:

$$\frac{du}{dt} = g \, \frac{\rho_s - \rho}{\rho_s} - \frac{u}{\tau}$$

(23.6)

where the constant $\tau = \dfrac{m}{3\pi\mu x}$ is known as the particle relaxation time and is for spherical particles

$$\tau = \frac{x^2 \rho_s}{18\mu}$$

(23.7)

Equation 23.6 can be solved with the following result:

$$u(t) = \frac{\rho_s - \rho}{\rho_s} \, \tau \cdot g \, \left(1 - \exp\left(\frac{-t}{\tau}\right)\right)$$

(23.8)

This is an exponential relationship which with increasing time, t, approaches

$$u_g = \left(\frac{\rho_s - \rho}{\rho_s}\right) \tau \, g$$

(23.9)

and this is known as the terminal settling velocity under gravity. Equation 23.9 can also be written as:

$$u_g = \frac{x^2 \, (\rho_s - \rho) \, g}{18 \, \mu}$$

(23.10)

by elimination of τ using equation 23.7.

It can be shown that the terminal velocity is in the case of fine particles approached so fast that in practical engineering calculations the settling is taken as a constant velocity motion and the acceleration period is neglected. This is shown in the following example.

Example 23.1

Determine the time in which a solid particle of 5-μm diameter and a density of

2500 kg m^{-3} will reach 99% of its terminal settling velocity when falling under gravity from a stationary position, in water of viscosity $\mu = 0.001 \text{ Ns m}^{-2}$.

Solution

Assuming Stokes' law, equations 23.8 and 23.9 give:

$$\frac{u(t)}{u_g} = \left(1 - \exp\left(\frac{-t}{\tau}\right)\right) \tag{23.11}$$

and this has to be equal to 0.99 to find the time t_{99}. Particle relaxation time is, from equation 23.7,

$$\tau = \frac{(5 \times 10^{-6})^2 \times 2500}{18 \times 0.001} = 3.472 \times 10^{-6} \text{ s}$$

and the time from equation 23.11 for $u(t)/u_g = 0.99$

$$t_{99} = -\tau \ln 0.01 = 0.016 \times 10^{-3} \text{ s}$$

Checking Re_p for Stokes' region requires knowledge of the instantaneous particle velocity at the time $u(t_{99})$ and, from equation 23.8, this is

$$u(t_{99}) = 0.99 \times u_g$$

$$= 0.99 \frac{\rho_s - \rho}{\rho_s} \tau \cdot g$$

$$= 0.99 \frac{1500}{2500} \times 3.472 \times 10^{-6} \times 9.81$$

$$u(t_{99}) = 20.23 \ \mu\text{m s}^{-1}$$

and the particle Reynolds number at t_{99} is then from equation 23.2:

$$\text{Re}_p (t_{99}) = \frac{20.23 \times 10^{-6} \times 5 \times 10^{-6} \times 1000}{0.001} = 1.012 \times 10^{-4} < 0.2$$

This is, of course, well within the Stokes' region. Note that for a particle ten times larger, i.e. $50 \ \mu\text{m}$, both t_{99} and $u(t_{99})$ increase by a factor of 10^2 for 1.6×10^{-3} s and $20.23 \times 10^{-4} \text{ m s}^{-1}$ respectively, and $\text{Re}_p = 0.1012$ which is still in the Stokes' region.

It is obvious from this example that the times needed for acceleration to a velocity which is very close to the terminal settling velocity are very short indeed and negligible for practical purposes.

Equation 23.10 is the basis of simple scale-up calculations for non-flocculent systems within the restrictions of Stokes' law. It can also be applied to non-spherical particles provided that the size is measured by sedimentation which leads to the equivalent Stokes' diameter.

It should be pointed out that there is also a bottom limit of particle size (usually about 1 μm) below which Stokes' law cannot be assumed to apply in gravity sedimentation. This is due to Brownian diffusion which becomes significant below 1 μm and leads to settling rates lower than those predicted by Stokes' law.

23.2.2 Hindered settling at high concentrations

As the concentration of the suspension increases, particles get closer together and begin to interfere with each other. If the particles are not distributed uniformly the overall effect is a net increase in settling velocity because the return flow due to volume displacement will predominate in particle sparse regions. This is the now well-known effect of cluster formation which is only significant in nearly mono-sized suspensions. With most practical, widely dispersed suspensions clusters do not survive for long enough to affect the settling behaviour and, as the return flow is more uniformly distributed, the settling rate steadily declines with increasing concentration. This is referred to as hindered settling and can be theoretically approached from three starting premises:

1. as a Stokes' law correction by introducing a multiplying factor;
2. by adopting 'effective' fluid properties for the suspension, different from those of the pure fluid;
3. by determining the bed expansion from a modified Carman–Kozeny equation.

The three approaches above yield essentially identical results in the form

$$\frac{u_p}{u_g} = \epsilon^2 f(\epsilon) \tag{23.12}$$

where u_p is the hindered settling velocity of a particle, u_g is the terminal settling velocity of a single particle as calculated from Stokes' law (equation 23.1), τ is the volume fraction of the fluid (voidage), and $f(\epsilon)$ is a 'voidage function' which for Newtonian fluids have different forms depending on the theoretical approach adopted. The differences between the available expressions for $f(\epsilon)$ are not great and are frequently within experimental accuracy. The most important forms are as follows:

(i) from the Carman–Kozeny equation (see section 23.3.1)

$$f(\epsilon) = \frac{\epsilon}{10 (1 - \epsilon)} \tag{23.13}$$

(ii) from Brinkman's theory[1] applied to Einstein's viscosity equation[2]:

$$f(\epsilon) = \epsilon^{2.5} \tag{23.14}$$

and (iii) from the well-known Richardson and Zaki equation[3]

$$f(\epsilon) = \epsilon^{2.65} \tag{23.15}$$

For irregular or non-rigid particles (e.g. flocs) the Einstein constant (2.5) and the Richardson and Zaki exponent (2.65) can be considerably larger than for spheres.

Strictly speaking, the above correlations apply only to the cases when flocculation is absent, such as with coarse mineral suspensions. Suspensions of fine particles, due to the very high specific surface of the particles, often flocculate and therefore show different behaviour. With increasing concentration of such suspensions at a particular concentration C_1, an interface can be observed and this becomes sharper at $C > C_1$. The slurry is then said to be in the zone settling region. The particles below the interface, if their size range is not more than 6:1, settle 'en masse', i.e. all settle at the same velocity irrespective of their size. There are two possible reasons for this: either the flocs become of similar size and settle at the same velocity or the flocs are locked into a loose plastic structure and thus sweep down as a 'web'. Interestingly enough the settling rates of the interface (and of the solids below it) of many practical suspensions can still be described by the Richardson and Zaki equation but the value of u_g must be determined by extrapolating the experimental log–linear plot for $\epsilon = 1$. The value of this intercept has in fact been used for the indirect size measurement of the flocs. The slope of the plot determines the exponent.

The concentration C_1 at which zone settling can first be observed depends very much on the material and its state of flocculation and no guidance can be given on this. Addition of flocculation (or dispersing) agents drastically changes this concentration and its value can only be determined by experimental evaluation. If the concentration is increased further still, a point is reached when the flocs become significantly supported mechanically from underneath as well as hydraulically and the suspension is then known to be in compression or compression settling. The solids in compression continue to consolidate and the rate of consolidation depends not only on the concentration but also on the structure of the solids which, in turn, depends on the pressure and flow conditions. This is a very complex problem closely related to cake filtration and the mechanical squeezing of cakes. For intermediate concentrations between those of zone settling and fully established uniform compression there is sometimes also a phenomenon called channelling which particularly occurs in slowly raked large-scale thickeners. Under these conditions a coarser structure of pores becomes interconnected in the forms of channels.

Most authors who have studied the consolidation process of solids in compression have used the basic model of a porous medium with point contacts which yields a general form of the mass and momentum balances. This must be supplemented by a model describing filtration and deformation properties. Probably the best model to date is that of Kos[4], which uses two parameters to define the characteristic behaviour of suspensions. His model can potentially be applied to four processes: sedimentation, thickening, cake filtration and expression.

23.2.3 Centrifugal sedimentation at low concentrations

Particle terminal settling velocity under centrifugal acceleration can be calculated from Stokes' law by substituting g in equation 23.1 by Rw^2 (where w is the angular speed and R is the radius of particle position) and the same assumptions apply as under section 23.2.1. There is one important difference here however: particle motion is no longer at constant velocity but as the particles move radially outwards they are continuously accelerated (Rw^2 increases). Similarly as under gravity, particle inertia is neglected and the particles are always assumed to move at their respective terminal velocities; this is an acceptable assumption in most practical cases. The top size limit of acceptability of Stokes' law is reduced here due to higher accelerations. It can be shown that if a separation factor G is defined as a number of gs, i.e.

$$G = \frac{Rw^2}{g}$$
(23.16)

then the top size limit of Stokes' law, x_c, for a centrifuge can be calculated from the limiting size under gravity x_g

$$x_c = x_g / \sqrt[3]{G}$$
(23.17)

Under gravity acceleration the distance covered by a particle in a given time is simply $U_g t$; under centrifugal acceleration (as the motion is accelerated) the following expression is obtained from integration of the terminal velocity expression

$$u_c = \frac{dR}{dt} = \frac{x^2 \, \Delta \rho \, Rw^2}{18\mu}$$
(23.18)

from R_0 (the initial radius) to R (radius after time t):

$$\frac{R(t)}{R_0} = \exp \frac{x^2 \, \Delta \rho \, w^2 \, t}{18\mu}$$
(23.19)

The settling time of a given particle size x from R_0 to R can be calculated from the above equation.

23.2.4 Centrifugal sedimentation at high concentrations

Centrifugal sedimentation at high concentrations is not hindered as much as in the case of gravity sedimentation because higher concentrations exist within the settling suspension only for a fraction of the total settling time rather than for all the time as under gravity. This is due to the continuous dilution effect: as particles settle radially they spread out and thus the concentration at a given radius falls steadily with time. However, the settling is still hindered, if only for some time, from the start and this has to be taken into account as it prolongs the settling times. Surprisingly, this effect was neglected until relatively recently, when Baron and Wajc[5] published a full study of the effect. They concluded that some settling in the centrifugal field leads to much the same behaviour as under gravity, i.e. to the formation of a visible interface between the supernatant liquid and the settling suspension, and that the concentration below the interface is the same right down to the sediment level. However, this concentration is not constant but decreases with time because of the above-mentioned dilution effect, until it comes out of the zone settling regime and the interface ceases to be visible. The total settling times are naturally much longer than those obtained at lower concentrations. The analysis derived by Baron and Wajc is much too complicated to be reported here, but the important feature of it is that it uses the experimental settling flux curve obtained under gravity which, in principle, allows scale-up of sedimenting centrifuges from gravity experiments. Further development in this area is anticipated.

23.3 FLOW THROUGH PACKED BEDS

23.3.1 Carman–Kozeny equation and its limitations

Flow through packed beds under laminar conditions can be described by a model in which the flow is assumed to be through capillaries whose surface equals that of the solids comprising the bed. The capillary volume is set equal to the void volume of the bed. The model leads to the well-known Carman–Kozeny equation[6] as follows:

$$\frac{Q}{A} = \frac{\Delta p}{\mu L}\frac{\epsilon^3}{5\ (1-\epsilon)^2 S_o^{\ 2}} \tag{23.20}$$

where Q is the volumetric flow rate, A is the face area of the bed, L is the depth of the bed, Δp is the applied pressure drop, ϵ is the voidage of the bed (porosity), S_o is the volume-specific surface of the bed, and μ is the liquid viscosity.

The constant 5 is valid for lower porosity ranges and particles not too far from spherical in shape; it generally depends on particle size, shape and porosity. Unfortunately, although the above equation has been found to work reasonably well for incompressible cakes over narrow porosity ranges, its importance is limited in cake filtration because it cannot be used for most practical compressible cakes. Its value is mostly in illustrating the high sensitivity of the pressure drop to the cake porosity and to the specific surface of the solids. A major criticism of the Carman–Kozeny equation is that it incorrectly tends to infinity when the voidage approaches 1.

23.3.2 Darcy's law and the basic cake filtration equation

Darcy's law[7] combines the constants in the last term of equation 23.20 into one factor, K, known as the permeability of the bed, which is a constant

$$K = \frac{\epsilon^3}{5\ (1-\epsilon)^2 S_o^{\ 2}} \tag{23.21}$$

for incompressible solids, but for compressible cakes it depends on the applied pressure, the approach velocity, and the concentration and, therefore, it presents serious problems in cake filtration testing and scale-up. There are some materials such as highly flocculated beds for which the above linear relationship between the face velocity and pressure drop does not hold and the flow is then called non-Darcian.

The modern filtration theory tends to prefer the Ruth form of Darcy's law as:

$$\frac{Q}{A} = \frac{\Delta p}{\mu R}$$ (23.22a)

where R is known as the bed resistance. In cake filtration the bed resistance consists of the medium resistance in series with the resistance of the deposited cake (assuming no penetration of solids into the filtration medium) and the general filtration equation is then written as:

$$\frac{Q}{A} = \frac{\Delta P}{\alpha \, \mu c \dfrac{V}{A} + \mu R}$$ (23.22b)

where α is the specific cake resistance, μ is the liquid viscosity, c is the solids concentration in the feed, V is the filtrate volume collected since the commencement of filtration, and R is the medium resistance. This equation is the basis of cake filtration analysis. The feed liquid flow rate and filtrate volume, V, are usually assumed to be related as,

$$\frac{dV}{dt} = Q$$ (23.23)

There is a hidden assumption in the above relationships that the volume of the solids and the liquid retained in the cake is negligible. This is reasonable at low concentrations but can lead to errors at higher solids concentrations and moisture contents of cakes. The usual way to correct for this is by using a 'corrected' value of the concentration, c, in equation 23.22b.

23.3.3 Alternatives to the Carman–Kozeny equation

As was briefly pointed out in section 23.3.1, the Carman–Kozeny equation does not work well towards the limit of $\epsilon = 1$; Carman[6] himself stated that the equation should not be used for $\epsilon > 0.8$. Several researchers have attempted to derive a model with a more realistic and general outcome. Perhaps the most significant attempt is that by Rudnick[8] who used a free-surface cell model by Happel[9] in which each particle is assumed to be a sphere at the centre of a cell, the volume of which is such that the porosity of each cell is the same as that of the bed. If the tangential stresses at the boundaries of adjoining cells are set to zero, an exact solution of the general Navier–Stokes' equations exists, assuming that the inertial terms are negligible.

The solution leads to an equation similar to the Carman–Kozeny equation (23.20) except that the term

$$\frac{\epsilon^3}{(1 - \epsilon)^2}$$

is now replaced by a more complicated one of ϵ, as follows:

$$10 \; \frac{3.45(1-\epsilon)^{1/3} \; + \; 4.5(1-\epsilon)^{5/3} \; - \; 3(1-\epsilon)^2}{3(1-\epsilon) \; + \; 2(1-\epsilon)^{8/3}} \tag{23.24}$$

The two equations compare favourably in the lower porosity range, say below 0.6, but the Rudnick equation predicts correctly finite flow rates through very highly porous beds.

23.3.4 Ergun equation

With increasing Reynolds number, the flow in some of the passages in the bed becomes turbulent, the Carman–Kozeny equation is no longer valid and must be corrected by adding a turbulent term. In turbulent flow, the pressure drop is no longer proportional to the flow velocity but increases with the square of the velocity. As a part of the flow in the bed, in the smaller passages, is still laminar, the overall pressure drop is a sum of two components, laminar and turbulent.

The laminar term is derived from the Carman–Kozeny equation (equation 23.20)

$$\frac{\Delta p}{L} \; = \; 5 \; \frac{Q}{A} \; \mu \; S_o^2 \; \frac{(1-\epsilon)^2}{\epsilon^3} \tag{23.25}$$

whilst the turbulent term, derived using a similar model, is

$$\frac{\Delta p}{L} \; = \; 0.292 \; \left(\frac{Q}{A}\right)^2 \; \rho \; S_o \; \frac{1-\epsilon}{\epsilon^3} \tag{23.26}$$

The two terms combined lead to the well-known Ergun[10] equation in the following form:

$$\frac{\Delta p}{L} \; = \; 4.17 \; \frac{Q}{A} \; \mu \; S_o^2 \; \frac{(1-\epsilon)^2}{\epsilon^3} \; + \; 0.292 \; \left(\frac{Q}{A}\right)^2 \; \rho \; S_o \; \frac{1-\epsilon}{\epsilon^3} \tag{23.27}$$

where the constant of 5 in the Carman–Kozeny term becomes 4.17 in the Ergun equation. Equation 23.27 is often written in a form where specific surface S_o is replaced by $6/x_{sv}$ where x_{sv} is the mean surface volume diameter of the particles making up the bed. This step is not always necessary, however, and it involves an assumption that the particles are spherical; it is best, whenever possible, to use the actual (measured) specific surface of any real particle system.

The relative importance of the two terms in equation 23.27 depends on the Reynolds number which describes the flow in the passages in the bed. This may be defined using x_{sv} as the mean particle size in the bed:

$$\mathrm{Re} \; = \; \frac{v x_{sv} \rho}{\mu} \tag{23.28}$$

where v is the superficial velocity equal to Q/A with A being the face area of the bed. For Re less than about 1, the second (turbulent) term in equation 23.28 becomes negligible compared with the first, whilst for Re greater than about 1000 the turbulent flow term dominates.

Highly turbulent flow in fixed beds is only possible, however, if the bed is constrained so that the individual grains within it cannot move. If the bed is not constrained and the flow through it is upwards, the bed becomes fluidized.

REFERENCES

1. Brinkman, H. C., *Appl. Sci. Res.,* **A1**, 27 (1947); **A1**, 81 (1948); **A2**, 190 (1949)
2. Einstein, A., *Ann. Phys. Leipzig,* **19**, 289 (1906); **34**, 591 (1911)
3. Richardson, J. F. and Zaki, W. N., *Chem. Eng. Sci.,* **3**, 65 (1954)
4. Kos, P., *Second World Filtration Congress* (London, 18–20 September 1979), pp. 595–603
5. Baron, G. and Wajc, S., 'Fluidized settling in centrifuges, Synopsis 686', *Chem. Eng. Technol.,* **4**, 333 (1979)
6. Carman, P. C., *Flow of Gases through Porous Media*, Butterworths, London (1956)
7. Darcy, H. P. G., *Les Fontaines Publiques de al Ville de Dijon*, Victor Dalamont (1856)
8. Rudnick, S. N., 'Fundamental factors governing specific resistance of filter dust cakes', S. D. Thesis, Harvard School of Public Health, Boston, MA (1978)
9. Happel, J., 'Viscous flow in multiparticle systems', ~~Assoc. Ind. Chem. Eng.~~, **4**, 197 (1958)
10. Ergun, S., *Chem. Eng. Prog.,* **48**, 93 (1952)   *A I che*

Particle-Fluid Interaction, Thermodynamics of Solid-Liquid Separation
Part II — Thermodynamics of
Solid-Liquid Separation

L. Svarovsky

Department of Chemical Engineering, University of Bradford

NOMENCLATURE

A	Helmholtz free energy
c	Volumetric concentration of the solids (as a fraction)
E_s	Entropy index
g	Gravity acceleration
H	Total settling height
h	Height of the settled sediment
k	Boltzmann constant
N	Total number of particles in the system
N_1	Number of particles
N_2	Number of sites in the fluid
R_f	Underflow-to-throughput ratio, by volume
S	Overall entropy
S_{susp}	Entropy of the suspension
s	Specific entropy, per volume of suspension
T	Absolute temperature
U	Internal energy
V_p	Volume of a particle
V_o	Volume of the supernatant liquid in a settled system
V_1	Volume of the packed sediment
W	Number of distinguishable arrangements (multiplicity)
x	Particle size (diameter of a sphere)
Δ	Denotes a difference in the variable
$\Delta\rho$	Density difference between that of the solids and of the liquid

23.4 INTRODUCTION

The science of thermodynamics is widely used in systems involving heat or chemical change. This part of chapter 23 sets out to show that even in a cold, non-reacting system such as solid–liquid separation, thermodynamics can be very useful. The concepts of entropy and free energy can be used in:

1. developing criteria for separation efficiency;
2. obtaining a criterion to indicate whether the separation is to occur at all;
3. making estimates of sediment porosity.

The state of the art and applications in the above-mentioned aspects of solid–liquid separation are reviewed and an outline of further work and development is given.

The use of thermodynamics is well accepted and indeed essential in those areas of particle technology which involve heat and/or flow of gases. Properties of a dust-laden gas, for example, deviate from those of a perfect gas as a consequence of the finite particle volume. Internal energy, enthalpy and specific heats of a particulate system, needed in applications involving heat, are the suitably weighted means of the respective properties of the constituents (except that enthalpy of the particulate phase suspended in a gas also depends on the pressure). Another example[1] is the effect of the presence of particles on both the equilibrium and 'frozen' velocities of sound in a gas.

In contrast to the above, very little use is made of thermodynamics of 'cold' systems which do not involve flow of gases such as solid–liquid separation, yet the principles of thermodynamics underlie this subject too and can be made use of in practice.

23.5 SOME NOTES ON ENTROPY

The concept of entropy as a general measure of the disorder of a system is, of course, highly relevant in solid–liquid separation because the degree of restoration of the order (the reduction in entropy) describes the success of the separation process. Fundamentally, entropy is a measure of the way in which the total energy of the system is distributed amongst its constituent atoms.

In order to be able to evaluate the relative reduction in entropy during or after the separation process, we have to be able to calculate the entropy of the suspension as a function of the solids concentration. The necessary relationship may be derived in analogy with a molecular model for an ideal solution as used in chemical thermodynamics[2], summarized in the following. A very small solid particle is treated simply as if it were a large molecule.

Assuming, in the first instance, mono-sized particles, completely interchangeable with pockets of fluid of the same size without affecting the internal energy of the system, the number of distinguishable arrangements (the so-called 'multiplicity') is

$$W = \frac{(N_1 + N_2)!}{N_1! \, N_2!} \tag{23.29}$$

where N_1 and N_2 are the number of sites of particles and fluid respectively.

From molecular thermodynamics, the entropy, S, is proportional to the natural logarithm of the multiplicity W, i.e.

$$S = k \ln (W) \qquad (23.30)$$

where k is really the Boltzmann constant but we shall take it as 'a constant' for the time being as we are merely after a relative measure of entropy at this stage.

Stirling's approximation for equation 23.29, used with equation 23.30, gives the entropy of a suspension

$$S_{susp} = - k \left[N_1 \ln \left(\frac{N_1}{N_1 + N_2} \right) + N_2 \ln \left(\frac{N_2}{N_1 + N_2} \right) \right] \qquad (23.31)$$

In the above equation, S is used for S_{susp} because it is an extensive property for a given number of sites $N_1 + N_2$. If we now divide the right-hand side of equation 23.31 by $k (N_1 + N_2)$ we obtain entropy in arbitrary units, say specific per unit volume of suspension, s:

$$S_{susp} = - \frac{N_1}{N_1 + N_2} \ln \left(\frac{N_1}{N_1 + N_2} \right) - \frac{N_2}{N_1 + N_2} \ln \left(\frac{N_2}{N_1 + N_2} \right) \qquad (23.32)$$

For the mono-disperse system, the concentrations by number in the above equation are the same as the volume, so we can write

$$S_{susp} = - c \ln c - (1 - c) \ln (1 - c) \qquad (23.33)$$

where c is the concentration of particles by volume, as a fraction. Better still, if the maximum entropy (at $c = 0.5$) is to be equal to 1:

$$S_{susp} = - c \ln_2 c - (1 - c) \ln_2 (1 - c) \qquad (23.34)$$

where $\ln_2$ denotes logarithm to the base 2, i.e. $\ln_2(x) = 1.44 \ln (x)$.

Equation 23.34 is the basis of the use of the 'entropy index' defined in the following section; it should be borne in mind, however, that, strictly speaking, equation 23.34 is derived for a mono-sized suspension of fine particles only.

23.6 ENTROPY INDEX

In an ideal world, solid–liquid separation separates the solids and the liquid in a suspension into a stream of dry solids going one way and a stream of pure liquid going the other way. The entropy of the whole is reduced from a given starting value according to the initial solids concentration c in equation 23.34, down to zero ($S_{susp} = 0$ for $c = 0$).

In a real world, we always have to accept some liquid with the separated solids (in the filter 'cake' or in the system 'underflow') and some, usually fine, solids with the liquid (in the 'filtrate' of 'overflow'). In other words, neither the separation of the solids nor that of the liquid is perfect. In evaluating the efficiency of a separation process, therefore, both the separation of the solids and the separation of the liquid must be considered. In most practical applications, the emphasis on either of these two is different and they are kept separate: in thickening, for example, the goal is the separation of the liquid and the (complete) separation of the solids is secondary. In solids recovery or liquid clarification, the completeness of the separation of the solids has priority over the separation of the liquid.

If we wish to consider both the separation of the solids and that of the liquid with equal emphasis, entropy gives us an ideal tool. The efficiency of the overall separation can thus be evaluated as a fraction or percentage decrease in the entropy of the system, taking the initial entropy of the suspension (from equation 23.34) as being 100%. This definition of separation efficiency has been called the 'entropy index' and has been used widely in the Russian scientific literature [3,4].

Mathematically, the entropy index is defined as

$$E_s = \frac{S_{feed} - (S_{underflow} + S_{overflow})}{S_{feed}} \tag{23.35}$$

where S is the total entropy in the respective streams, with the origin $S = 0$ for complete separation. In terms of our specific entropy s, i.e. the entropy per unit volume of slurry, the values of entropy must be weighted by their respective volumetric contributions, and the above equation then becomes

$$E_s = \frac{S_{feed} - R_f \, s_{underflow} - (1 - R_f) \, s_{overflow}}{s_{feed}} \tag{23.36}$$

where R_f is the underflow-to-throughput ratio by volume and the values of s are evaluated from equation 23.34.

The entropy index as defined in equation 23.36 is potentially very useful in the fundamental evaluation of any separation processes, not just in solid–liquid separation. Besides the Russian references [3,4], Ogawa et al.[5] derived the same entropy index (but using mass fractions rather than volumetric ones) from information theory and proposed its use for the evaluation of any separation process.

The use of entropy in particle technology is not restricted to evaluations of separation efficiency. A unit-less definition of entropy virtually identical with that in equation 23.6 but in terms of probability of events, as used in information theory and proposed by Shannon[6], has been applied in geology to histograms and cumulative curves by Sharp and Fan[7] to define a sorting index for particle size analysis. Sharp[8] applied the information entropy to define a measure of parity between the mean and the standard deviation of a distribution whilst, more recently, Full et al.[9] used it to define optimum class intervals of histograms or frequency plots. Entropy also has a great potential in the evaluation of the quality of mixing in powder mixing applications which are really the reverse of separation or classification.

There are two problems, however, with using equations 23.30 and 23.34 blindly, as many authors seem to do. Firstly, the multiplicity W in equation 23.30 is the total number of combinations if the phase space were to be discrete, i.e. the particle centres constrained to lattice points. For a real, continuous-phase space, W should be replaced[10] by a full $3N$-dimensional configurational integral (N being the number of particles) which has no simple solution mainly due to the lack of an adequate equation of state for liquids. Secondly, and notwithstanding the first point, equation 23.34 applies to mono-sized systems only but seems to be applied indiscriminately to all two-phase systems, including polydisperse ones. Strictly speaking, one should go back to equation 23.30 and consider the effect of the presence of particles of different sizes on their respective 'free volumes' or configuration combinations. This is bound to have some bearing on the entropy index as the size distributions of the solids in the feed, overflow and underflow are so very much different in most solid–liquid separators. More work is needed in this area.

23.7 CRITERION OF SEPARATION

Besides its use in assessing separation efficiency, entropy, in conjunction with internal energy, can also be used in deriving a criterion of separation[11,12]. This is in analogy with chemical thermodynamics: the value of the criterion will decide whether separation will take place spontaneously or not at all. Only a summary of the derivation is given in the following.

The Helmholtz free energy, ΔA, for a non-flow process, is calculated from the change in internal energy, ΔU, and the change in entropy, ΔS

$$\Delta A = \Delta U - T \Delta S \tag{23.37}$$

where T is the absolute temperature.

For a spontaneous process A decreases, so the condition for separation to take place is

$$- T \Delta S < - \Delta U \tag{23.38}$$

(as both changes are normally negative), i.e. if we define

$$\text{separation criterion} = \frac{\Delta U}{T \Delta S} \tag{23.39}$$

the separation will take place if the value is greater than 1 and it is impossible if the value is less than 1.

The problem now is how to calculate the changes in entropy and in internal energy for a specific case and arrive at values which are not merely relative (in arbitrary units such as in the previous section) but absolute, measured from a standard state. Attempts have been made to calculate the values of the criterion for some special cases[11,12] and only a relatively simple case of gravity settling is described here for illustration.

In the calculation of the change of entropy, ΔS, the following assumptions are made:

1. mono-disperse system;
2. complete separation;

3. negligible liquid in sediment;
4. analogue of 'regular solution', without excess free energy of mixing of particles and liquid molecules;
5. dilute suspension.

The result is[11]

$$\Delta S = - k N \ln (V_o / V_1) \qquad (23.40)$$

where N is the number of particles, k is the Boltzmann constant, V_o is the volume of the supernatant liquid and V_1 is the volume of the packed sediment.

For mono-disperse, non-interacting particles, the change in internal energy, ΔU, is equal to the change in the potential energy of the settling system. Taking the heights from *Figure 23.2* (H is total settling height, h is the height of the sediment after separation), the change in internal energy can be calculated as

$$\Delta U = - N V_p \Delta\rho \, g \, (H/2 - h/2) \qquad (23.41)$$

where V_p is the single particle volume and $\Delta\rho$ is the density difference between the solids and the liquid.

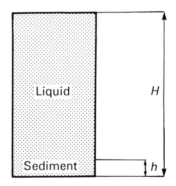

Figure 23.2. Definition of settling heights for equation 23.41

Assuming that V_1/V_o is approximately equal to the volumetric concentration of the solids (for dilute suspensions) and taking a spherical particle of diameter x, the sedimentation criterion may be calculated from equations 23.39, 23.40 and 23.41:

$$\text{sedimentation criterion} = \frac{\pi \, g \, \Delta\rho \, x^3 \, H \, (1 - c)}{12 \, k \, T \ln (1/c)} \qquad (23.42)$$

It is rather surprising to find that the value of the criterion which decides whether the settling will or will not take place depends on the absolute value of the settling distance, H. Taking an example of a 1-μm particle settling through 20 cm at $\Delta\rho = 1600$ kg m^{-3} and 20°C, the value of the criterion is 436663. To bring this value below 1, one would need a particle less than 0.013 μm in size, which would not settle under gravity in the above system; this is quite a credible value considering that it has been derived solely from thermodynamics without any input from the theories of settling or diffusion.

The separation criterion may also be derived for polydisperse systems[11] in terms of summations, and this makes it possible to establish the likelihood of the separation of any particular fraction of the particle population.

In general, the evaluation of the separation criterion is subject to much the same criticisms as is the entropy index. In this case, however, the problem is even more difficult due to the fact that absolute changes in entropies and internal energies are needed. More work is needed to establish these quantities for particulate suspensions, with the inevitable particle–particle interactions and excess free energies of mixing. Textbooks on thermodynamics are remarkably silent on the subject of thermodynamics of particulate systems and suspensions.

23.8 ESTIMATES OF SEDIMENT POROSITY

Reasoning very similar to that in deriving the separation criterion can also be used for making estimates of sediment porosity. The liquid content in the sediment is not neglected this time, however, and the equilibrium condition of the separation criterion equal to 1 is used[11], i.e. when no more settling can take place in the sediment. This is the least researched area, however, requiring much more realistic assumptions about the thermodynamics of real, polydisperse systems than in the cases of the entropy index and separation criterion. The only study on this subject known to the author[11] does not include the effect of the interaction energies of the particles with one another, an omission hardly acceptable for particles in close proximity in the sediment. The potential in this area is great, however, and a serious study is needed.

23.9 CONCLUSIONS

The thermodynamics of suspensions is a subject much neglected in textbooks, yet there are many potential uses, particularly in solid–liquid separation. A review of the state of the art has revealed at least three important applications where thermodynamics can make a real and practical contribution. The available literature shows a sizeable Russian lead but their initial proposals and ideas need fine tuning. It is hoped that this review will stimulate further work in this area.

REFERENCES

1. Rudinger, G., *Fundamentals of Gas-Particle Flow, Handbook of Powder Technology*, Vol. 2, Elsevier, Amsterdam (1980)
2. Pitzer, K. S. and Brewer, L., *Thermodynamics*, McGraw-Hill, New York (1961)
3. Agranonik, R. Ya., Shishmakov, S. Yu. and Shamanaev, Sh. Sh., 'Application of the

entropy index to evaluate the efficiency of separator operation', *Chem. Petrol Eng.,* **19,** 227–229 (1984)

4. Sulla, M. B. and Fikhtman, S. A., 'Application of the entropy index to evaluate the efficiency of thickening equipment', *Vodosnabzhenie i Sanitarnaya Tekhnika,* **11,** 11–13 (1972)
5. Ogawa, K., Ito. S. and Kishino, H., 'A definition of separation efficiency', *J. Chem. Eng. Jpn.,* **11,** 44–47 (1978)
6. Shannon, E., *Bell System Technol. J.,* **379,** 623 (1948)
7. Sharp, W. E. and Fan, P., 'A sorting index', *J. Geol.,* **71,** 76 (1963)
8. Sharp, W. E., 'Entropy as a parity check', *Earth Res.,* **1,** 27 (1973)
9. Full, W. E., Ehrlich, R. and Kennedy, S., 'Optimal definition of class intervals of histograms or frequency plots'. In *Morphological Analysis, Particle Characterisation in Technology* (J. K Beddow, Ed.), Vol. II, Chap. 11, CRC Press Inc., Boca Raton, FL (1984)
10. Woodcock, L., private communication (1988)
11. Figurovskii, N. A., and Sokolov, N. V., 'Thermodynamics of sedimentation', *Russian J. Phys. Chem.,* **55,** 111–112 (1981)
12. Kutepov, A. M. and Sokolov, N. V., 'Thermodynamics of separation processes', *Theor. Found. Chem. Eng.,* **76,** 287–291 (1982)

Index